Springer Tracts on Transportation and Traffic

Volume 21

About this Series

The book series "Springer Tracts on Transportation and Traffic" (STTT) publishes current and historical insights and new developments in the fields of Transportation and Traffic research. The intent is to cover all the technical contents, applications, and multidisciplinary aspects of Transportation and Traffic, as well as the methodologies behind them. The objective of the book series is to publish monographs, handbooks, selected contributions from specialized conferences and workshops, and textbooks, rapidly and informally but with a high quality. The STTT book series is intended to cover both the state-of-the-art and recent developments, hence leading to deeper insight and understanding in Transportation and Traffic Engineering. The series provides valuable references for researchers, engineering practitioners, graduate students and communicates new findings to a large interdisciplinary audience.

Indexed by SCOPUS, WTI Frankfurt eG, zbMATH, SCImago.

Petros Ioannou · Andreas A. Malikopoulos
Editors

Transportation Mobility in Smart Cities

Springer

Editors
Petros Ioannou
Department of Electrical Engineering
University of Southern California
Los Angeles, CA, USA

Andreas A. Malikopoulos
School of Civil and Environmental Engineering
Cornell University
Ithaca, NY, USA

ISSN 2194-8119 ISSN 2194-8127 (electronic)
Springer Tracts on Transportation and Traffic
ISBN 978-3-031-64768-0 ISBN 978-3-031-64769-7 (eBook)
https://doi.org/10.1007/978-3-031-64769-7

This Springer imprint is published by the registered company Springer Nature Switzerland AG
The registered company address is: Gewerbestrasse 11, 6330 Cham, Switzerland

Preface

Cities are becoming bigger and bigger and more complex due to increasing urban population. The population density in most cities has been continuously increasing adversely affecting mobility and environment and sustainability in the long term. In addition, any type of disaster may cause major disruptions to the movement of people and goods with negative consequences on the survivability of people. The use of technology to make cities smarter appears to be one of the most important viable options. Addressing these pressing challenges related to the increasing demand for mobility, we must make fundamental transformations in how communities use and access transportation. A mobility system encompasses the interactions of three heterogeneous dimensions: (1) technological developments on transportation systems and modes that lead to emerging mobility systems, e.g., autonomous vehicles, shared mobility; (2) social behavior of drivers and travelers interacting with these systems; and (3) institutional behavior of organized units such as administrators that govern the transportation systems through policies. The constellation of these dimensions constitutes a sociotechnical system that should be analyzed holistically. Yet current methods analyze, design, and optimize a mobility system without considering the social and institutional dimensions, resulting in mobility systems that might not be acceptable to travelers or might impose inequities.

This book is not just about emerging technologies but a book on important dimensions of future mobility systems for smart cities, providing indications on what innovations are needed, how to address the challenges, and on holistically exploring a research path moving forward. It is a mix of topics that include engineering, policy, and social behavior subjects that demonstrate the complexity of mobility in smart cities while addressing some crucial issues. The purpose of the book is not to solve all problems of mobility in smart cities which requires many more books but to provide the reader some of the issues involved and motivate new research directions and problems to be addressed. The book makes a strong point that technological developments alone without considering policy, social and economic issues may not lead to practical solutions. It is the interplay between these diverse issues and associated trade-offs where cost–benefit analysis, social acceptance, ethical and environmental considerations, resilience, and sustainability all play a role in reaching the best solutions.

The book consists of 19 chapters written by a diverse group of prominent researchers with different research interests addressing problems, solutions, and future research directions in mobility for smart cities. A brief description of each chapter is presented below:

Chapter 1 provides an overview of road vehicle control towers and how to guarantee safe interaction between automated driving systems and human operators. While the human driver is removed from each vehicle a human operator is introduced to monitor and manage fleets of automated vehicles for safe operations and dealing with emergencies.

Chapter 2 reviews cooperative control approaches for Connected Automated Vehicles (CAVs) at conflict areas such as merging roadways, intersections, and lane-changing maneuvers. It examines the extension of online control methods where performance and safety constraints are described using control barrier functions to mixed traffic environments where the presence of human behavior poses new challenges. The goal is to provide foundations and specific new techniques aimed at optimizing automated mobility in mixed traffic.

Chapter 3 proposes the use of vehicle platoons as actuators to address the issue of compliance, control, and optimization. The vehicle platoon receives instructions from the road network supervisor in order to affect speed and density and therefore flow for the purpose of optimizing and controlling traffic flow.

Chapter 4 provides an overview of emerging sensing technologies for traffic data with a focus on fine-grained traffic density estimation. A general streaming-data-driven probabilistic approach is designed to utilize different types of measurements obtained from (i) fixed-location sensors, (ii) connected and automated vehicles, and (iii) unmanned aerial vehicles (UAVs). The proposed methodology integrates a Gaussian process model within a Bayesian framework to effectively calculate traffic density estimates for different measurement types, even in cases where only a limited number of measurements are available within the designated time-space area being examined.

Chapter 5 introduces a route reservation architecture aimed at alleviating congestion and enhancing the overall efficiency of the road network in terms of the average time vehicles spend within the road network. This architecture integrates traffic management, steering drivers through the network, with demand management, influencing drivers' departure timings. In addition, it offers the flexibility in providing routing and departure suggestions to drivers or, alternatively, utilizing dynamic pricing mechanisms to influence their route and departure decisions. The chapter presents extensive validation results generated using detailed micro-simulations.

Chapter 6 deals with a new approach of lane-free traffic where vehicles are not bound to follow a lane and operate based on certain movement strategies with nudging generated using optimal and nonlinear feedback control and multiagent reinforcement learning. Specialized macroscopic and microscopic models are developed using theoretical and empirical techniques which are used for joint path optimization of multiple vehicles in a lane-free environment.

Chapter 7 provides an overview on electric buses, by highlighting the main modeling and control challenges, and presents some possible traffic and consumption models, as well as some traffic control algorithms for optimizing bus speeds and dwell/charging times. The use of electric buses in smart cities is an upcoming technology to offer optimized mass transit operations with zero emissions with improved quality of service.

Chapter 8 presents major results and findings obtained from a multi-year, large-scale naturalistic study that addresses the important issues of moving patterns of e-scooters in diverse road environments and locations; interaction of e-scooter riders with vehicles and other road users in different scenarios; and common scenarios for crashes or near-misses involving e-scooter riders at different risk levels. The collected data and analytical results can be used to develop behavior prediction models for e-scooter riders which will support the development of automated driving systems in challenging urban environment to improve road safety.

Chapter 9 deals with the various challenges facing connected autonomous vehicles and proposes algorithms to address such challenges and enhance vehicle safety and mobility in smart cities. Specifically, it discusses cooperative adaptive cruise control subject to temporary communication or target detection loss, connected lane-change advance warning systems for improving traffic flow, probabilistic trajectory forecasting for vulnerable road users, especially pedestrians, and, finally, approaches to collision avoidance and mitigation for autonomous vehicles.

Chapter 10 presents a synopsis of tools and methodologies that can be used for reconfigurable metasurfaces for the design of new generation of connected cooperative and automated vehicles in smart cities. The chapter considers the HyperSurface as the enabling technology for reconfigurable metasurfaces and presents recent results demonstrating the feasibility and the associated performance gains achieved from the integration of reconfigurable metasurfaces in connected automated driving applications.

Chapter 11 presents the history and background of two curb management pilot projects in Southern California and highlights the methodology and results of a series of interviews used to identify the characteristics, expectations, and limitations of these projects and to learn about the stakeholders involved and their decision-making processes before, during, and after the pilot projects. In addition to understand what outcomes each of the municipalities and stakeholders wanted, and how extensively the pilot could impact the future of curb management. This is the first step in understanding the potential of curb management to promote use of zero-emission vehicles, and more generally increase the productivity of the curb.

Chapter 12 proposes a smart active sustainable mobility model to promote accessibility, connectivity, and social inclusion, among people in urban areas. Active mobility could be one of the most important transport modes for sustainable urban development as it considers all modes of transportation including pedestrians, bicyclists, wheelchair use, and manually powered scooters in addition to different classes of vehicles.

Chapter 13 deals with a centrally coordinated freight routing system for smart cities where the routes of all trucks transporting freight are generated by a central coordinator in order to achieve system optimality by balancing all freight loads in time and space across the road network. The proposed methodology allows the mix of diesel and electric trucks where the constraints of battery life, charging times, and location of chargers can also be incorporated.

Chapter 14 provides a control framework to distribute travel demand in a given transportation network, resulting in a socially optimal mobility system that travelers would be willing to accept. A "socially optimal mobility system" implies a mobility system that is efficient (in terms of energy consumption and travel time), it mitigates rebound effects, and ensures equity in transportation.

Chapter 15 reviews the working principles and the properties of the core cost sharing mechanisms that are commonly used in transportation and logistics as well as some of the most recent advancements in this field to facilitate social optimum decisions among users. The focus of the chapter is on vehicle sharing, ride-sharing, shipping, and supply chain collaboration but it also identifies possible future research directions in this area by surveying a large number of articles published between 2016 and 2023.

Chapter 16 is based on a survey in Greece regarding the impact of self-driving vehicles in providing equity to all social groups. People with disabilities currently face unequal conditions in transportation that restrict their freedom of movement; unfavorably impact their medical, social, and economic life; and even accentuate conditions of social exclusion. Autonomous driving carries the potential to deliver accessibility to transportation and create more favorable living conditions for people with mobility impairments. The importance of the chapter is to promote relevant research and policies and influence the design of transportation systems in advancing social justice and equity.

Chapter 17 describes several intelligent transportation system solutions and their potential for improving many cities around the world with the focus on disadvantaged communities. Smart city solutions should also aim in improving conditions of disadvantaged and environmental justice communities, whose residents have suffered the most from unmitigated urban sprawl and its environmental and health impacts. The chapter focuses on the Inland Southern California region as a study case and addresses a number of topics that include improved traffic flow along major arterial roadways, smart intersections, shared mobility solutions, and improved routing of heavy-duty trucks.

Chapter 18 proposes a framework for multi-passenger transport that allows dynamic routing and behavioral dynamics of passengers and leads equitable micro-transit services for members of disadvantaged communities. The implications of disadvantaged communities and how their needs may be met by the proposed framework are discussed. A numerical study based on a community in a Food Desert in the Western United States is presented.

Chapter 19 examines the institutional environment in which new transportation technologies are being introduced within the United States. It focusses on four broad technology categories: Shared mobility, integrated mobility, priced mobility, and autonomous mobility to illustrate the potential transformations to the transportation system and the institutional obstacles to adoption at scale. It formulates a set of recommendations to steer innovations in the direction of equitably distributed benefits.

The editors would like to thank all the contributing authors of chapters for their valuable time and efforts to share their research ideas in solving important problems associated with mobility in future smart cities where technology will play a crucial role in revolutionizing the transport of people and goods. The authors would also like to thank the IEEE Intelligent Transportation Society for supporting the workshop on the "Road to Emerging Mobility Systems for Smart Cities" which took place on June 30, July 1, 2023 in Ayia Napa, Cyprus and brought most of the contributing authors together and motivated most of the ideas presented in this book.

The authors would like to thank the Springer Editor Mr. Oliver Jackson for his careful reading of the manuscript and providing guidance to improve it. His suggestions helped us make significant improvements to the manuscript and are highly appreciated.

Los Angeles, USA Petros Ioannou
Ithaca, USA Andreas A. Malikopoulos

Contents

About the Editors

Petros Ioannou received B.Sc. degree from University College, London in 1978 and the M.S. and Ph.D. degrees from the University of Illinois, Urbana, Illinois, in 1980 and 1982, respectively. In 1982, he joined the Department of Electrical Engineering, University of Southern California where he is currently the A. V. "Bal" Balakrishnan Chair Professor. He is the founder and Director of the Center of Advanced Transportation Technologies and co-founder and Associate Director for Research of METRANS, a University Transportation Center. He holds courtesy appointments with the Department of Aerospace and Mechanical Engineering and the Department of Industrial Systems Engineering. He is the founder and Director of the Master Program on Financial Engineering. He is the recipient of the 1984 IEEE Control System Society Axelby Best Paper Award, the 1985 Presidential Young Investigator Award, the 2009 IEEE Intelligent Transportation System (ITS) Society Outstanding ITS Application Award, the 2009 Institution of Engineering and Technology (IET) Heaviside Medal for Achievement in Control, the 2012 IEEE ITS Society Outstanding ITS Research Award, the 2016 IEEE Transportation Technologies Field Award, and the 2016 IEEE Control System Society Transition to Practice Award. He is a Fellow of IEEE, IFAC, AAAS and the author/co-author of nine books and over 400 research papers in the areas of controls and intelligent transportation systems. He is the VP for membership activities of the ITS Society and served as Editor in Chief of the IEEE Transactions on ITS. In 2022, he was elected to the National Academy of Engineering, the

National Academy of Inventors and Foreign member of the Academy Europaea. In 2023, he was elected Foreign Member of the European Academy of Sciences.

Andreas A. Malikopoulos is a Professor in the School of Civil and Environmental Engineering and the Director of the Information and Decision Science Lab at Cornell University. Prior to these appointments, he was the Terri Connor Kelly and John Kelly Career Development Professor in the Department of Mechanical Engineering (2017–2023) and the founding Director of the Sociotechnical Systems Center (2019–2023) at the University of Delaware (UD). Before he joined UD, he was the Alvin M. Weinberg Fellow (2010–2017) in the Energy and Transportation Science Division at Oak Ridge National Laboratory (ORNL), the Deputy Director of the Urban Dynamics Institute (2014–2017) at ORNL, and a Senior Researcher in General Motors Global Research and Development (2008–2010). He received a Diploma from the National Technical University of Athens, Greece, and his M.S. and Ph.D. degrees from the University of Michigan, Ann Arbor, in 2004 and 2008, respectively, all in Mechanical Engineering. Dr. Malikopoulos is the recipient of several prizes and awards, including the 2007 Dare to Dream Opportunity Grant from the University of Michigan Ross School of Business, the 2007 University of Michigan Teaching Fellow, the 2010 Alvin M. Weinberg Fellowship, the 2019 IEEE Intelligent Transportation Systems Young Researcher Award, and the 2020 UD's College of Engineering Outstanding Junior Faculty Award. He has been selected by the National Academy of Engineering to participate in the 2010 German-American Frontiers of Engineering (FOE) Symposium and organize a session on transportation at the 2016 European-American FOE Symposium. He has also been selected as a 2012 Kavli Frontiers of Science Scholar by the National Academy of Sciences. He has been an Associate Editor of the IEEE Transactions on Intelligent Vehicles and IEEE Transactions on Intelligent Transportation Systems from 2017 through 2020. He is an Associate Editor of Automatica and IEEE Transactions on Automatic Control, and a Senior Editor IEEE Transactions on Intelligent Transportation Systems. He is a Senior Member of the IEEE, a Fellow of the ASME, and a member of the Board of Governors of the IEEE Intelligent Transportation Systems Society.

Chapter 1
A Safe Interaction Framework for Road Vehicle Control Towers

Frank J. Jiang and Karl H. Johansson

Abstract Automated vehicles have to interact to make urban traffic safe. To handle many unplanned traffic issues, such as occlusions and system failures, automated vehicles need to cooperate with each other and the connected road infrastructure. A key component in a connected road infrastructure is the road vehicle control tower, where human operators can supervise and manage fleets of automated vehicles to ensure their safe and efficient operation. By introducing such operators, the transport system is able to mitigate a variety of complex scenarios, but at the expense of potentially introducing new human errors. In this chapter, we provide an overview of road vehicle control towers and how to guarantee safe interaction between automated driving systems and human operators. Specifically, we present our work on a behavior tree-based interaction framework that allows operators to intuitively create specifications for vehicle behavior. Using the human-defined behavior specification tree, we compute temporal logic trees, which allow us to both formally verify the feasibility of the specifications and to synthesize control sets that guarantee the specifications' completion. Finally, we overview the implementation of a shared-autonomy approach to remote driving under the interaction framework to illustrate how it can be used for guaranteeing safety in practical use-cases.

1.1 Introduction

In the last decade, one of the most rapid advancements in the transportation industry is the development of vehicle automation technologies. Automated vehicles are equipped with advanced sensors, powerful computations, and automation algorithms

F. J. Jiang (✉) · K. H. Johansson
Division of Decision and Control Systems, EECS, KTH Royal Institute of Technology, Malvinas väg 10, 10044 Stockholm, Sweden
e-mail: frankji@kth.se

K. H. Johansson
e-mail: kallej@kth.se

P. Ioannou and A. A. Malikopoulos (eds.), *Transportation Mobility in Smart Cities*, Springer Tracts on Transportation and Traffic 21,
https://doi.org/10.1007/978-3-031-64769-7_1

Fig. 1.1 Einride's control tower desk (used with permission from Einride)

that come together to create a system that has the potential to change both freight and passenger transport. However, despite the rapid progress, full automation remains largely confined to well-defined use-cases and environments, reflecting the challenges associated with ensuring safety and operational efficiency in more uncertain environments, such as urban ones. Often, in more uncertain or poorly defined use-cases, there are many exceptions, such as the unexpected effects of kangaroos on a vision system [28], that can lead to automation failure that needs to be mitigated before we can achieve a widespread deployment of automated vehicles.

There is a growing interest in the deployment of road vehicle control towers for supervising the operation of automated vehicles in confined areas, such as construction sites, mines, and logistics hubs. To mitigate automation failures, many have deployed infrastructure for human operators to remotely supervise and manage the operation of automated vehicles. We refer to this infrastructure as the road vehicle control tower. An example of such a road vehicle control tower is illustrated in Fig. 1.1. These control towers are analogous to air traffic control towers and are equipped with advanced communication and monitoring technologies that enable a human to supervise fleets of automated vehicles. Just as we see in the case of air traffic control, human operators possess inherent capabilities for adaptive problem-solving, contextual decision-making, and managing uncertainty [25]. Leveraging these capabilities, by deploying a network of road vehicle control towers, human operators can be incorporated into the operation of automated vehicles and provide fallback support in the case of automation failure. In a sense, the human operators form an important layer of resiliency that can enhance safety and operational efficiency in environments with uncertainty.

While the integration of human operators is key for resiliency, by adding human operators into the automation loop, we risk re-introducing human errors into vehicle operations. Since one of the primary goals of the deployment of automated vehicles is to remove the risk of human error, we need to be diligent about how we approach the integration of human operators. Even simple human errors due to miscalculations to decision-making under stress or fatigue might disrupt an operation and even lead to safety violations. At the same time, there is a clear need for the incorporation of human decision-making to mitigate automation failures. To achieve a system that is both reliable and resilient, these competing needs should be simultaneously designed for. In other words, we need to design a system that balances the benefits of both automated and human decision-making.

In this chapter, we will overview a recently developed framework that provides an approach for safely incorporating both automated and human decision-making in vehicle operations. The overall objective of the framework is to facilitate the interactions between automated driving systems and human operators in a way that leverages the precise sensing and fast computations of automated vehicles and the adaptive and reactive decision-making of human operators. Our key methodology for facilitating these interactions is a human-centric use of formal specifications. As the basis of our approach, formal specifications provide an interface between the automated driving systems and human operators. By leveraging formal specifications, we can implement systems that allow a human operator to specify a high-level requirement for an automated driving system, such as "Eventually reach that parking spot, but never collide with any objects," and the automated driving system is able to precisely interpret this and perform the necessary computations to fulfill that requirement. Moreover, as we will show, well-defined graphical interfaces, such as behavior trees, can be used as human-centric interfaces that human operators can use to create rich, complex specifications without explicitly knowing the details of the formal specification formulation or the computations performed on the automated driving system.

Other than providing a well-defined interface for the human to create specifications, a novel aspect of our approach is the way we are formally verifying and synthesizing control for the designed specification. Traditional approaches to these problems is through the use of automata theory, where the vehicle behavior and the task environment is discretized and modeled as automata. Then, verification and synthesis is performed on these automata. This approach is primarily designed for the efficient verification and synthesis for problems that have tasks and environments that are static, such as circuits or computer systems [14]. We take an alternative approach and use reachability analysis to directly analyze the full continuous vehicle behavior and task environment. We use a computational tool called temporal logic trees that will let us combine and update the verification and synthesis results in real time. Additionally, by using reachability analysis, we also are able to synthesize a set of feasible control policies instead of just a single control policy. After we present the core interaction framework, we will overview an implementation of a shared-autonomy teleoperation system that takes advantage of this set of policies and demonstrates the human-centric aspect of doing control set synthesis.

Before we delve into the details of the interaction framework, we will first provide more context around the design choices that were made. Once the context has been established, we will continue to present the interaction framework and the implementation of the teleoperation system. Explicitly, the rest of the chapter is organized as follows. In Sect. 1.2, we introduce the current status of road vehicle control towers and the challenges around their safe, human-centric implementation; moreover, we detail a practical, running example that exemplifies these challenges that will be used throughout the rest of the chapter. In Sect. 1.3, we overview the interaction framework and illustrate its application to the running example. In Sect. 1.4, we present the implementation of a safe teleoperation system that builds off of our interaction framework and is used on a 1/10th-scale version of the running example.

1.2 Road Vehicle Control Towers

In this section, we will discuss road vehicle control towers and the current state of related research and development. We will start with discussing the analogy and comparison between road vehicle control towers and air traffic control towers. Then, we will identify the key challenges of designing road vehicle control towers.

When designing road vehicle control towers, it's important to take inspiration from air traffic control towers. The primary inspiration from air traffic control towers comes from the fact that despite the high levels of automation for airplanes, safety regulations still require that there are two pilots onboard the airplane to perform fallback for system failures and human operators in the air traffic control tower to perform fallback for routing failures. Notably, it is not unreasonable to say that airplanes have higher levels of autonomy than road vehicles do today. Despite this, the safety engineering around airplanes is still highly dependent on pilots and air traffic operators. This is indicative of the clear role that humans have as a fallback layer for automation. Although this motivates the investigation into the integration of human operators into the design of automated road vehicles, there are some key differences between the roles the human operator has in road vehicle and air traffic control towers.

One of the striking differences is the need to perform teleoperation. As discussed thoroughly in [17] and illustrated in Fig. 1.2, a fundamental difference between fallback support for automated road vehicles and airplanes is that most automated road vehicle deployments need to be "driverless," as one of the primary business cases for automated vehicles is the removal of the onboard driver. Since fallback support for system failure, like the support provided by pilots, is a critically important form of support to ensure safe and efficient vehicle operation, this means the remote human operator will need to provide this support. While this support might take forms of supervision other than teleoperation, there may be cases where the human operator has no choice but to remotely drive a vehicle that is exhibiting automation failure. Due to this, over the years, the need for teleoperation has become well established and most major automated vehicle deployments incorporate teleoperation by human

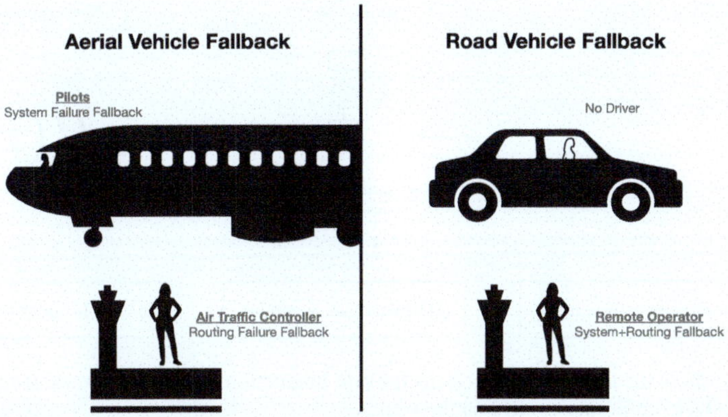

Fig. 1.2 Illustration of the key difference between fallback systems for aerial vehicles and road vehicles. Originally published in [17] (Used with permission from John Wiley & Sons)

operators in some form today [4, 8]. That said, the teleoperation systems that are built today focus on incorporating technologies for replicating the real driving experience for the remote driver (e.g., [18, 27]) or dealing with latency (e.g., [12, 20, 23]) and do not necessarily focus on the safety-critical aspect of teleoperation [17].

While teleoperation is an important mode of supervision, there are other approaches human operators might take to affect the behavior of an automated vehicle. These approaches might also vary significantly in the level of interaction. For example, instead of taking direct control over the vehicles, human operators might draw paths for the vehicle to follow or might completely change the goals of the vehicle. In this chapter, we focus on highlighting the overarching challenge of creating a framework that ensures all of these interactions are safe and human centric, as we will further unpack in the next section.

1.2.1 Road Vehicle Control Tower Interaction Challenges

Before we unpack the interaction challenges, we start by understanding how the safety case is specified between remote human operators and automated vehicles. The Society of Automotive Engineers (SAE), who are well known for their definitions of Levels of Driving Automation, specify in their 2021 specification that the safety case between remote human operators and automated vehicles is built on whether the vehicle is able to complete its task on its own or not [22]. To be able to determine whether a vehicle is able to complete its task or not, the task needs to be specified in a way that both the vehicle and human are able to make decisions about whether or not the task can be completed safely and correctly. Moreover, the operator also needs to know that the control system of the vehicle will behave in way that is consistent with the task given, mandating that there is a clear relationship between the defined

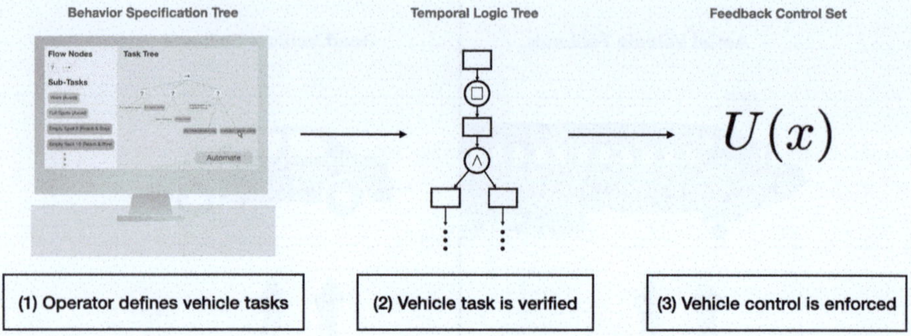

Fig. 1.3 We show the three necessary steps for safe interaction in our framework, from definition of vehicle tasks to enforcing control on the vehicle

task and the final behavior. We can break down a safe interaction into three necessary steps (illustrated in Fig. 1.3):

1. operator defines/updates vehicle task specification,
2. vehicle task specification is verified for correctness and safety,
3. vehicle control system is deployed with guarantees that the task specification will be followed.

It's important we co-design the full framework with regard to these three steps as they are necessary for facilitating safe interactions between the operator and the vehicle. For example, a useful representation for representing formal specifications is temporal logic, however, since laypersons will not have familiarity with temporal logic, it's important that we also include an intuitive interface for the operators to use. This interface must be well defined or correct-by-design, since poorly defined interface might lead to behavior that is not verifiable or different from the expectations of the operator. These interaction steps and their respective requirements impose joint challenges for the framework:

1. **Human Centered.** Vehicle driving tasks need to be easy to understand, define, and modify by a human operator.
2. **Verified Formally.** Designed vehicle driving tasks need to have a representation that can be formally verified for correctness and safety.
3. **Enforced in Real Time.** Control boundaries need to be computed for the vehicle control system to guarantee the vehicle driving tasks are satisfied.

1.2.1.1 Running Example

Before we begin developing the interaction framework, we first introduce a running example that we will revisit throughout the chapter to illustrate each step in the framework. We draw from the remote parking scenarios studied in [15, 16]. This scenario is inspired by the confined area scenarios where remote supervision by human operators is commonly required in industry [9]. In this scenario, the ego vehicle has arrived to a confined area where it should park into one of the empty parking spots, which is illustrated in Fig. 1.4. This vehicle has some automated functions, but due to the possible exceptions that can arise in the confined area, it's determined that for safety a remote operator should supervise and possibly remotely operate the vehicle if required. The presented interaction framework is used by the human operator to define or modify the parking task and facilitate the decisions made by the human operator to ensure the human's decisions about the real-time parking maneuver are implemented in a guaranteed safe manner. As we progress through the chapter, we will detail the application of the proposed interaction framework to this example and highlight approach's human-centered aspects, ability to formally verify the correctness, and the real-time enforcement of the designed parking task.

For the sake of clarity throughout this chapter, we will also determine the motion model of the ego vehicle. In our example, we track the ego vehicle's XY-position, heading, and velocity. The control inputs to the ego vehicle are steering and acceleration. Moreover, to further enhance safety, we will also consider bounded disturbances to the ego vehicle. Explicitly, let $x = [p_x,\ p_y,\ \theta,\ v]$ be the x-position, y-position, heading, and velocity, respectively. Then, let $u = [\delta,\ a]$ be the steering angle and acceleration, respectively. We denote the additive disturbance affecting the vehicle with w, which lets us include factors such as slip or localization noise into our analysis. We model the ego vehicle's motion using the following simple nonlinear bicycle model:

$$f(x, u, w) = \begin{bmatrix} \dot{p}_x \\ \dot{p}_y \\ \dot{\theta} \\ \dot{v} \end{bmatrix} = \begin{bmatrix} v \cos\theta \\ v \sin\theta \\ \frac{v \tan\delta}{L} \\ a \end{bmatrix} + w, \tag{1.1}$$

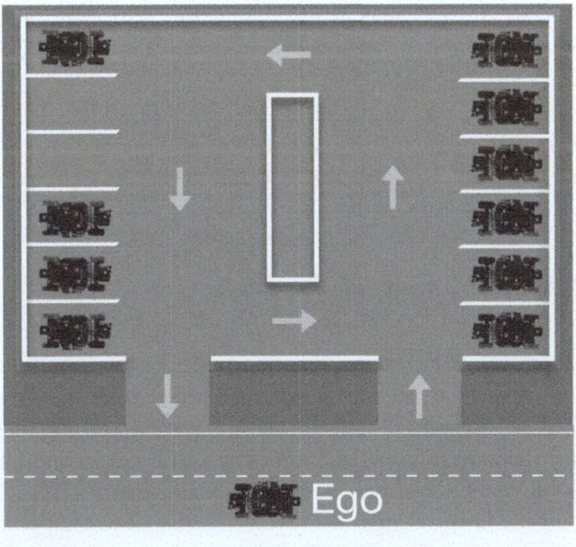

Fig. 1.4 An illustration of the parking lot for the running example used throughout the chapter

where L is the wheel-base length of the ego vehicle. For the rest of this chapter, it will be useful to formally denote several aspects of this model. As indicated by (1.1), $x \in \mathbb{R}^4$, $u \in \mathbb{U} \subset \mathbb{R}^2$, and $w \in \mathbb{W} \subset \mathbb{R}^4$. For the proposed approach, it's important to note that f is uniformly continuous, bounded, and Lipschitz continuous in x for fixed u and w. Given measurable control and disturbance functions $u(\cdot) \in U$ and $w(\cdot) \in W$, we denote the unique solution solving (1.1) with $\zeta(\cdot; x_0, t_0, u(\cdot), w(\cdot))$, which describes a trajectory starting from an initial state $\zeta(t_0; x_0, t_0, u(\cdot), w(\cdot)) = x_0$ and tracks the motion of the vehicle under $u(\cdot)$ and $w(\cdot)$. For the rest of the chapter, we will sometimes write $\zeta(\cdot)$ to denote a trajectory for notational simplicity. The basis of verification and real-time enforcement is built on the tracking and managing of possible trajectories the ego vehicle can follow. Specifically, during formal verification, we will guarantee that there exists at least one control policy that can drive the ego vehicle from it's initial condition into one of the empty parking spots while fulfilling other conditions that the human operator designed into the parking task, such as avoiding collisions with parked vehicles or the parking lot walls. Then, when enforcing the designed parking task in real time, we will constrain the control inputs the vehicle is allowed to apply to guarantee that the parking task is fulfilled. In the next section, we will develop an approach that allows us to intuitively and safely connect a human operator's designed task to the operation of the ego vehicle with real-time guarantees that the task will be completed, addressing the three challenges outlined in the previous section.

1.3 Safe Interaction Framework

In this section, we will go through the overall framework we propose in this chapter. We start with an overview where we discuss the different stages in the framework, starting from the task design and ending with the computation or synthesis of control sets to be implemented on the ego vehicle. Then, we detail and motivate the approach taken in each stage. Moreover, as we go through each stage we continuously develop the running parking example and set it up for the next section where we dig deeper into an implementation of a shared-autonomy-based teleoperation system that builds on the framework presented in this section.

1.3.1 Framework Overview

We illustrate an overview of the full framework in Fig. 1.3. As we mentioned previously, a key to our framework is the use of formal specifications. In particular, we will use temporal logic to describe formal specifications. Using temporal logic, we will be able to take different propositions corresponding to the environment of the vehicle and describe temporal requirements the vehicle needs to fulfill with regard to those propositions. However, before that can happen we need to first do a sub-task preparation step, where the operator goes through the environment and defines relevant sets for the sub-task. For example, in parking, this would involve marking one of the parking spots and giving it a particular number. The sub-task preparation step is then completed by the operator indicating the simple temporal requirements tied to each of the propositions. Then, after the sub-task preparation is complete, the operator moves on to using the prepared sub-tasks for defining the behavior specification tree that forms the specification for the full task. After the behavior specification tree is defined, the automated steps of the framework take over. The designed behavior specification tree is translated into one full formal specification that is verified using a tool called temporal logic trees. Then, if the full task is valid and feasible, the temporal logic trees are used to synthesize safe feedback control sets for guaranteeing the vehicle adheres to the designed behavior.

In the following subsection, we start delving into the framework and go through the design process behind task planning.

1.3.2 Sub-task Preparation

In this first step, we detail the process a human operator would go through to prepare sub-tasks that can later be used to assemble the full task for the vehicle. A human operator's sub-task preparation process is broken down into two steps: (1) defining and labeling regions in the vehicle's operating domain and (2) describing a plan over time for the vehicle. We start with the definition and labeling of regions in the vehicle's state space. In this step, the human operator goes through the operating domain and section regions of the operating domain off to create subsets of the domain. When sectioning off regions of the operating domain, the operator also labels the sectioned off region with a name that is useful for the task planning. For example, in Fig. 1.5, the operator is creating the blue rectangle over one of the empty spots and is labeling that empty spot "Empty Spot 10." This blue rectangle is not only defined over the XY-plane but also over the velocity and heading of the vehicle, allowing the operator to specify that a vehicle is only "in" the empty spot when the ego vehicle's XY-position is within the spot, speed is zero, *and* heading is facing left. Assuming the model (1.1), the result of Fig. 1.5 is a subset of $\mathbb{R}^4$. This result is akin to what is referred to as atomic propositions in the formal methods community.

When the operator has defined labeled regions (atomic propositions), they proceed to describe a plan for the region. This plan will create a sub-task for that region that the operator will later compose with other sub-tasks to create the full driving task. We note that an operator can describe multiple sub-tasks using the same labeled region. In Fig. 1.5, the operator is able to select three types of plans for a region: "Goal," "Avoid," "Stay." Although these are simple sub-tasks, they will allow for the composition of relatively complex tasks. For "Empty Spot 10," the operator has decided that it should be both a "Goal" and to "Stay" within the parking spot when the vehicle enters into it. After repeating this to create all of the sub-tasks, the operator will end up with a sub-task list similar to the one shown in Fig. 1.6.

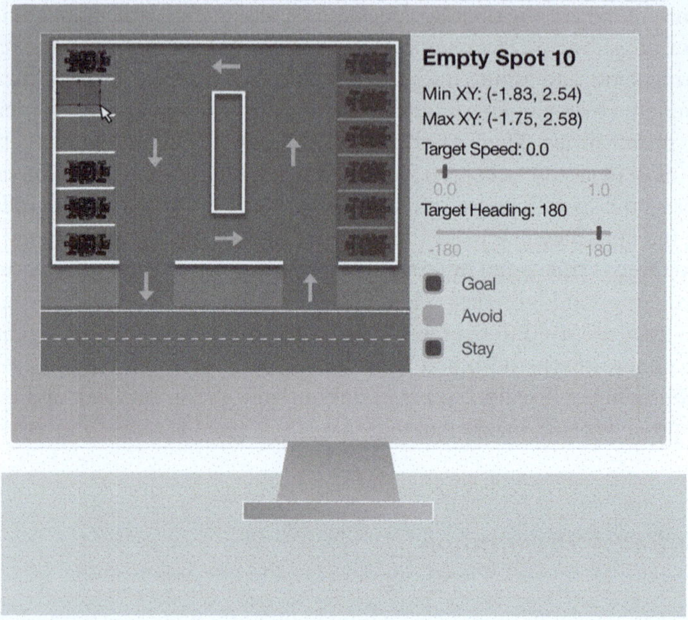

Fig. 1.5 An example of an operator's view when performing sub-task preparation

When the operator has both defined labeled regions and sub-tasks tied to the regions, we can then formalize the sub-task under-the-hood. Since the sub-task has both spatial and temporal aspects, the natural formalization is through temporal logic formulae. Temporal logic formulae allow us to make logical statements about the truthfulness of a property over time. For example, a temporal logic formula for an "Empty Spot 10" sub-task with both "Goal" and "Stay" would state that it should be eventually true that the vehicle's state is inside the region of "Empty Spot 10" and that after that point it is always true that the vehicle's state is inside of the region of "Empty Spot 10." An important outcome of describing tasks with temporal logic is that we get a formalism that allows for formal checking whether a task is feasible or not and for synthesizing controllers that will complete the task, given the model or dynamics of the vehicle. There are several types of temporal logic to choose from: some of the most common being linear temporal logic (LTL) and signal temporal logic (STL). While the verification and control methods detailed in this chapter have been applied to both LTL [11] and STL [26], for the sake of simplicity, we will use LTL in this chapter. In general, LTL is a relatively simple logic that provides a powerful way to express sub-tasks with spatial and temporal requirements, such as obstacle avoidance or stability. Despite its simplicity, LTL has been shown to be a useful formalism for setting up practical safe interaction frameworks [10, 13, 16]. For more details about LTL, we refer the readers to [1, 2, 14].

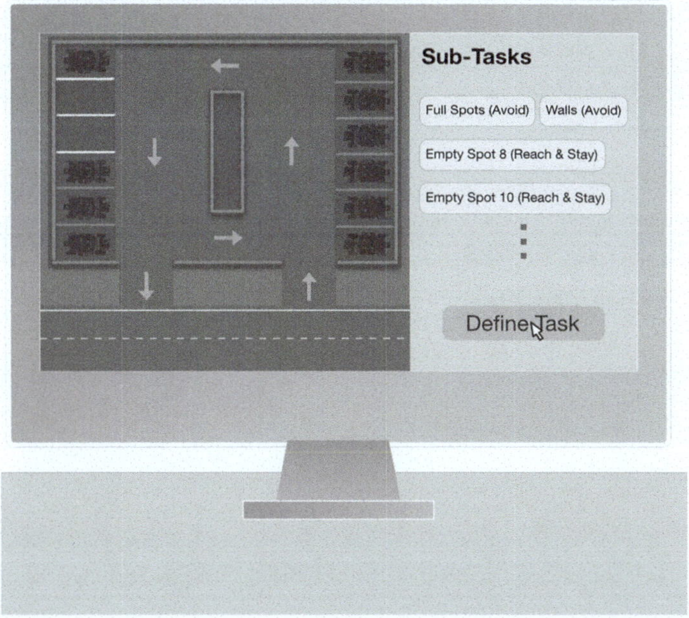

Fig. 1.6 We illustrate the labeled regions and corresponding sub-tasks in the running parking example. The shown sub-tasks are the particular ones we will formalize and work with in the chapter, however, the operator could create more sub-tasks depending on the specific operation they want the vehicle to complete

An LTL formula is defined over a finite set of atomic propositions (regions), denoted by $\mathcal{AP}$, with both logic and temporal operators. We can describe the syntax of LTL with:

$$\varphi ::= \text{true} \mid p \in \mathcal{AP} \mid \neg\varphi \mid \varphi_1 \wedge \varphi_2 \mid \varphi_1 \mathsf{U} \varphi_2,$$

where U denotes the "until" operators. By using the negation operator and the conjunction (and) operator, we can define disjunction (or), $\varphi_1 \vee \varphi_2 = \neg(\neg\varphi_1 \wedge \neg\varphi_2)$. Then, by employing the until operator, we can define (1) eventually, $\Diamond\varphi = \text{true} \cup \varphi$ and (2) always, $\Box\varphi = \neg\Diamond\neg\varphi$. Readers that are familiar with LTL might notice that we omit the next operator $\bigcirc$. In this work, we use the "LTL minus next step" fragment [19], since we will evaluate our approach using a continuous-time model. However, the presented approach is still valid when $\bigcirc$ is included and a discrete-time model is used [11].

Revisiting the parking scenario, we can use LTL to define the sub-tasks that an operator might create. On a high level, the full parking task the operator wants the ego vehicle to complete can be broken down into the following sub-tasks: parking into one of the empty spots, avoiding collisions with parked vehicles and the walls of the parking lot. To define these sub-tasks, we introduce some notation for the

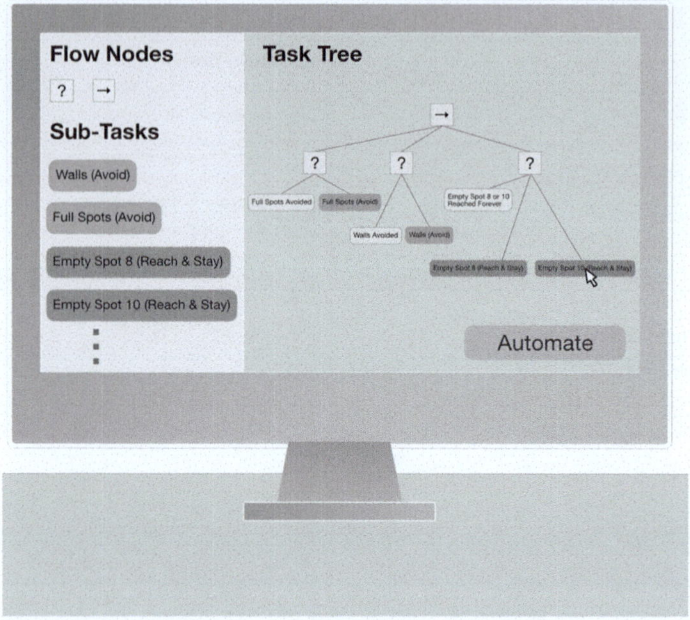

Fig. 1.7 An example of an operator's view when designing the full task tree

regions labeled in the parking sub-tasks shown in Fig. 1.6. Let the full set of states in our parking scenario be $\mathcal{D} \subset \mathbb{R}^4$. We call the set of states of the parking lot walls $\mathcal{W} \subset \mathcal{D}$, which excludes the block in the middle of the parking lot. We denote the set of states for the ith parking spots as $p_i \subset \mathcal{D}$. Denote the indices of full parking spots with $\mathcal{F}$ and the indices of empty parking spots as $\mathcal{E}$.

Now, we use LTL operators to formalize the sub-tasks shown in Fig. 1.6: "Full Spots (Avoid)," "Walls (Avoid)," "Empty Spot 8 (Reach & Stay)," "Empty Spot 10 (Reach & Stay)." In this particular parking situation, $\mathcal{F} = \{1, \ldots, 7, 9, 11, 12\}$ and $\mathcal{E} = \{8, 10\}$, assuming the spots are indexed starting from 1 in the bottom right of the parking lot. We write the formalization of the listed sub-tasks as follows:

- "Full Spots (Avoid)": $\square \left(\bigvee_{i \in \mathcal{F}} \neg p_i \right)$,
- "Walls (Avoid)": $\square \neg \mathcal{W}$,
- "Empty Spot 8 (Reach & Stay)": $\diamond \square p_8$,
- "Empty Spot 10 (Reach & Stay)": $\diamond \square p_{10}$.

Using these formalized sub-tasks, we will be able to take the behavior trees designed in the next section and formalize the entire behavior tree in a way that allows us to do formal verification and control set synthesis. We emphasize that while we will track the formal representation of the task throughout this chapter, the framework is designed so the human operators only need to work with labels such as "Full Spots (Avoid)."

1.3.2.1 Behavior Specification Tree Design

For our framework, we explore a behavior tree-based interface as the starting point of our human-centric toolchain. We are inspired by the early movement incited by the authors in [6] to use behavior trees for designing the behaviors of AI and robotic systems. They leverage a framework developed by the gaming industry, which have allowed game developers to program complex, interactive, and robust behaviors for artificial agents in games at very large scales. Interestingly, there is similarity between game development and introducing cyber-physical systems into society in terms of high-level behavior design.

Behavior trees are typically utilized as control architectures, but in our framework we will consider them as specification architectures. Traditionally, behavior trees are used to collect controllers (commonly referred to as "Actions") and conditions as leaf nodes and structures them in a tree structure using several types of internal, control flow nodes that manage the selection of which controllers to execute at each time instance. In Fig. 1.7, we have drawn an example behavior tree that might be used to park a vehicle. For brevity, we will introduce behavior trees using Fig. 1.7, assuming it was a traditional behavior tree. To learn more about behavior trees as control architectures, we refer readers to [6]. When a behavior tree is used as a control architecture, the tree is executed by sending 'ticks' down the tree (starting from the root) at each time step. When controller nodes (right-most children nodes in Fig. 1.7) are ticked, the corresponding controller executes and returns one of three values: $\{Success, Failure, Running\}$. For example, the "Full Spots (Avoid)" controller node in Fig. 1.7 will return *Success* if the vehicle has avoided a collision with a full parking spot, *Failure* if the vehicle has already crashed, and *Running* if the controller is actively avoiding a collision. When condition nodes (yellow nodes in Fig. 1.7) are ticked, they compare a value to a designer-chosen criterion and return one of two values: $\{Success, Failure\}$. For example, the "Empty Spot 8 or 10" will return *Success* if the vehicle is in parking spot 8 or 10 and *Failure* otherwise. When internal nodes are ticked, they tick their children. The two types of internal nodes we introduce are sequence ($\rightarrow$ in Fig. 1.7) and fallback (? in Fig. 1.7) nodes, as they are the most relevant for structuring specifications. Sequence nodes tick their children from left to right and will only return *Success* if all its children return *Success*; otherwise, it will return the first $\{Failure, Running\}$ it receives. Fallback nodes also tick their children left to right and will only return *Failure* if all its children return *Failure*; otherwise, it will return the first $\{Success, Running\}$ it receives. Although we will not use behavior trees control architectures, we will use the same node types when constructing behavior specification trees.

Before we start introducing our reformulation of behavior trees into behavior specification trees, it's also important to introduce the work presented in [3], where authors introduce formalisms for behavior trees that allow for the formal verification of the behavior trees using linear temporal logic. The design process prescribed by [3] goes as following: (1) designer constructs a behavior tree with pre-built controllers, (2) the designer models the success and failure conditions of the leaf nodes of the behavior tree using propositional logic, (3) the designer models the guarantees or runtime conditions of the leaf nodes using linear temporal logic, (4) the designer uses off-the-shelf verification tools to verify the model of the designed behavior tree, (5) the model is verified or the designer corrects the logical fallacies in their tree and repeat the process until they get a correctly constructed tree. The guarantees provided by this approach are "high level" since the modeling that occurs in step (2) and (3) is not explicitly based on the low-level dynamics of the controlled system. While the low-level dynamics are also a model, verifying a systems behavior against the dynamics of the system is an important approach to provide low-level safety guarantees such as collision avoidance. Thus, in order to verify a vehicle behavior against its dynamics to get low-level safety guarantees in this framework, we reformulate behavior trees into specification architectures.

In particular, the biggest change is that we consider action nodes to be sub-tasks that were created during the sub-task preparation step, instead of pre-defined controllers. Explicitly, this means that instead of having controllers in the action nodes, we have temporal logic statements. With this change, the behavior tree is effectively a switching architecture for temporal logic statements instead of controllers. By making this switch, we will be able to do a formal analysis on the tasks and provide direct guarantees for the completion and success of each sub-task. Moreover, we will be able to directly synthesize control sets that are guaranteed to complete the designed task. This change is significant, as in traditional behavior trees, the properties of the action controllers are assumed, meaning if the implemented behavior of the controllers are different in any way from the assumed behavior, then the real-time behavior of the vehicle may not exactly reflect the designed behavior. Given an accurate model of the vehicle behavior, by using the behavior tree as a specification architecture, we are able to compute controllers that are guaranteed reflect the designed behavior.

Specifically, we use the back-chained approach to building behavior trees. That is, the operator starts by setting the pre- and post-conditions of each of the sub-tasks. Once these pre- and post-conditions are set, we utilize the back-chaining algorithm presented in [7, 21]. This process consists of first creating a high-level mission plan that describes the goals that should be fulfilled and the order that they should be fulfilled in; this takes form as high-level behavior tree that orders the goals under a sequence node. Then recursively replacing the post-conditions with the corresponding atomic behavior trees that are known as postcondition–precondition–action or make-sure behavior trees. In the parking example, the high-level mission plan could consist of the following goals (listed in order):

1. Avoid full parking spots.
2. Avoid walls.
3. Park in an empty spot.

By taking the high-level mission plan and recursively inserting make-sure behavior trees for each sub-task, we get the full behavior tree shown in Fig. 1.7.

1.3.3 Formal Verification

In this section, we walk through the formal verification process for the designed back-chained behavior tree. We will start by converting the designed back-chained behavior tree to a single LTL formula that reflects the full task. Then, we will overview the construction of a temporal logic tree using reachability analysis. The successful construction of a temporal logic tree will verify that the LTL formula that represents the full task is feasible for the vehicle model. Then, in the following section, we will use the temporal logic tree to synthesize control sets that guarantee the completion of the full task.

1.3.3.1 Extracting the Formal Task from the Behavior Tree

To extract the formal task from the behavior tree, we take the back-chained behavior tree and replace the sequence ($\rightarrow$) and fallback (?) nodes with AND and OR nodes, respectively, as is done in [21]. Moreover, the action nodes are temporal logic formula. We can simply flatten the tree and get an LTL formula.

By converting the task tree in Fig. 1.7 into an AND–OR tree and flattening it into a single formula, we get the following LTL formula for our full task:

$$\varphi = \Box(\vee_{i \in \mathcal{F}} \neg p_i) \wedge \Box \neg \mathcal{W} \wedge (\Diamond \Box p_8 \vee \Diamond \Box p_{10}). \tag{1.2}$$

Upon careful inspection, one might notice that the condition nodes are not represented in the formula. This is because during verification, they are implicitly represented by the sets that will be propagated while constructing the temporal logic trees in the next step. During the real-time execution of the behavior tree and while visualizing the execution to the human operator, the condition nodes are important, however, they do not need to be explicitly represented in the verification step.

1.3.3.2 Verifying the Task Via Temporal Logic Trees

Now that we have an LTL formula, we can verify the full task the human operator designed is feasible and correct. This is the first guard we have against human error. In the case the human designs a task where all the goals cannot be completed, including safety-related goals, then the verification step will uncover this flaw. Several approaches can be taken to verify LTL formula, such as automata-based approaches [1, 2, 14]; however, in this chapter, we will use temporal logic trees. Unlike automata, temporal logic trees are computational structures that are built directly using reachability analysis. This will allow us to directly analyze and reason about our logical specification in continuous state spaces. To illustrate this point, we start by defining the notion of reachability we will use in this chapter. Denote the operating set we allow our vehicle to operate within as O. Then, denote a goal set that we want our vehicle to reach as $\mathcal{G}$. Then,

$$\mathcal{R}(\mathcal{O}, \mathcal{G}) = \{x : \exists u(\cdot) \in U, \forall w(\cdot) \in W, \\ \exists s > 0, \zeta(s; x, 0, u(\cdot), w(\cdot)) \in \mathcal{G}, x \in \mathcal{O}\}. \quad (1.3)$$

$\mathcal{R}(\mathcal{O}, \mathcal{G})$ denotes the set of states within the operating domain for which there exists a control policy that leads the vehicle to the goal, despite any disturbance within the assumed disturbance bounds. As we mentioned earlier, we are able to use this reachability operator to reason about the temporal behavior of a vehicle. For example, if we use the set $\mathcal{O} \subset \mathbb{R}^4$ to be all the drivable space in the parking scenario (e.g., excluding the grass and the walls) and p_8 be empty spot 8 we previously introduced, then $\mathcal{R}(\mathcal{O}, p_8)$ will give the set of drivable states the vehicle can be within to *eventually* reach p_8. If this set of states is not empty and the vehicle starts within the set, then this is equivalent to the satisfaction of the LTL specification $\Diamond p_8$. Then, if we want to stay in a particular set, such as p_8 after we have parked successfully, we can use the same operator and just let the operating domain to also be the goal:

$$\mathcal{RCI}(\mathcal{G}) = \mathcal{R}(\mathcal{G}, \mathcal{G}). \quad (1.4)$$

The set resulting from (1.4) is commonly known as the robust control invariant (RCI) set of the set $\mathcal{G}$. The set corresponds to the set of states within $\mathcal{G}$ that the vehicle can be within where there exists a control policy, despite worst-case disturbances, to *always* stay within $\mathcal{G}$. Explicitly, this set excludes states where, for example, the vehicle is not able to decelerate fast enough to prevent leaving the set or states where disturbances such as slip might accidentally push the vehicle out of the set it wants to stay within. If this set of states is not empty and the vehicle is within the set, then this is equivalent to the satisfaction of the LTL specification $\Box p_8$. Using the notions of reachability defined in (1.3) and (1.4), we can systematically analyze LTL formula by recursively applying the two reachability operators to build a temporal logic tree by following [11]. On a high level, the temporal logic tree facilitates reachability analysis and the combination of the reachable sets with set unions (for $\vee$) and intersections (for $\wedge$) to find the set of states that the vehicle can be within to fully satisfy the specified task.

After constructing the temporal logic tree, for the formal verification, we simply need to check whether the root set is empty or not and if the vehicle is able to start in the root set. If the root set is not empty and the vehicle is able to start in the root set, then that means there exists a control policy that can drive the vehicle from its initial position through all of the goals in the specification. With this result, we know the human operator's designed task is feasible. Then, in the next section, we compute the set of control inputs to constrain the vehicle in real-time to enforce the completion of the designed task.

1.3.4 Synthesizing Safe Feedback Control Sets

Once we have constructed the temporal logic tree and verified that the designed task is feasible, we continue to compute safe feedback control sets that can be used to enforce the completion of the designed task in real time. In general, this can be challenging to compute, as many forms of reachability analysis do not explicitly keep track of the control constraints that ensure the vehicle will stay within the computed reachable sets that satisfy the designed task. One form of reachability analysis that does implicitly keep track of the control constraints is Hamilton–Jacobi reachability analysis. Thus, for the remaining portion of the chapter, we will specify the use of Hamilton–Jacobi reachability analysis. However, we note that any form of reachability analysis can be used, as long as the control constraints can be recomputed or recovered.

Before we synthesize control sets, we first detail the computation of (1.3) and (1.4) with Hamilton–Jacobi reachability analysis. We start with the same two state sets as before: $\mathcal{O}$ and $\mathcal{G}$ as the zero superlevel sets of bounded, Lipschitz continuous functions $h_{\mathcal{O}} : \mathbb{R}^4 \to \mathbb{R}$ and $h_{\mathcal{G}} : \mathbb{R}^4 \to \mathbb{R}$. We can also write both sets as $\mathcal{O} = \{x \mid h_{\mathcal{O}}(x) \geq 0\}$ and $\mathcal{G} = \{x \mid h_{\mathcal{G}}(x) \geq 0\}$. Then, we solve for the value function $V_{\mathcal{O},\mathcal{G}}(x, \tau)$ that satisfies the following Hamilton–Jacobi–Isaacs variational inequality:

$$\min\{\frac{\partial V_{\mathcal{O},\mathcal{G}}(x,\tau)}{\partial \tau} + H(V_{\mathcal{O},\mathcal{G}}(x,\tau), f(x,u,w)),$$
$$h_{\mathcal{O}}(x) - V_{\mathcal{O},\mathcal{G}}(x,\tau)\} = 0, \tag{1.5}$$
$$V_{\mathcal{O},\mathcal{G}}(x,0) = h_{\mathcal{G}}(x), \ \tau \leq 0, \tag{1.6}$$

where the Hamiltonian is given by

$$H(V_{\mathcal{O},\mathcal{G}}(x,\tau), f(x,u,w)) =$$
$$\max_{u\in\mathbb{U}} \min_{w\in\mathbb{W}} \nabla V_{\mathcal{O},\mathcal{G}}(x,\tau) \cdot f(x,u,w). \tag{1.7}$$

Once the value function $V_{\mathcal{O},\mathcal{G}}(x,\tau)$ is computed, let $V^*_{\mathcal{O},\mathcal{G}}(x) = \lim_{\tau\to-\infty} V_{\mathcal{O},\mathcal{G}}(x,\tau)$. Then, we can compute (1.3) as

$$\mathcal{R}(\mathcal{O},\mathcal{G}) = \{x \mid V^*_{\mathcal{O},\mathcal{G}}(x) \geq 0\}. \tag{1.8}$$

Then, we can use this definition to also compute (1.4). Now, we can find the safe feedback control sets corresponding to the reachable sets. We refer to the safe feedback control set as $\hat{\mathbb{U}}(x)$. Given the state of the vehicle x at the time instant t, for a reachable set $\mathcal{R}$ of the form (1.3) or (1.4), the safe feedback control set can be computed as

$$\hat{\mathbb{U}}_{\mathcal{R}}(x) = \{u \in \mathbb{U} \mid V^*_{\mathcal{R}}(x) - \min_{w\in\mathbb{W}} \delta\nabla V^*_{\mathcal{R}}(x) \cdot f(x,u,w) \geq 0\},$$

where δ is a small sampling period. Instead of computing the control set based on the current time step, we compute it based on the next sampled time step. This ensures that the chosen control input will guarantee the specification set is not violated in consequence of that input.

Once we have the feedback control sets for each reachable set used to build the temporal logic tree, we can follow the control set synthesis algorithm in [11], which follows the following steps: (1) check if the state of the vehicle x belongs to a state set node and replace each state set node in the TLT with the corresponding control set according to the temporal operator; (2) compress the newly built control tree and backtrack starting from a single control set through a bottom-up traversal. This process will collect all the current, feasible control inputs such that the LTL specification is recursively feasible.

The set that results from this computation captures all of the control policies that fulfill the designed behavior tree. The set is particularly useful for control implementations since a controller just needs to pick a control input within this set and the vehicle will be guaranteed to satisfy the original specification. This means that regardless of what level the interaction between the operator and the vehicle is at, it's always possible to constrain the final behavior in a way that is guaranteed safe. In the next section, we will better see this in the implementation of a shared-autonomy teleoperation system for the parking example.

1.4 Shared-Autonomy Teleoperation Example

In this section, we utilize the feedback control sets and temporal logic tree that were computed in the safe interaction framework and implement a shared-autonomy teleoperation interaction between the human operator and the vehicle. Specifically, we will overview the implementation done in [16]. In the implementation, we first add an intention inference to the interaction. The intention inference will allow the automated system to predict the intention of the human operator during teleoperation, so it can compute an assistive control input to help the operator complete the task. Then, to give the operator full control over how the authority is distributed between their input and the automated input, we also add a policy blender where the human operator can tune how much assistance they receive from the automated system. Finally, we discuss the results of implementing the full system on the Small-Vehicles-for-Autonomy (SVEA) platform.

1.4.1 Intention Inference

Given the safe feedback control set $\hat{\mathbb{U}}$, the implemented inference module infers beliefs about which sub-task the operator is intending to complete. We denote the inferred beliefs as $b(t)$. There are many inference strategies that can be used for this inference in the existing literature. We refer readers to the extensive survey presented by [5].

In our parking scenario, we implement a simple intention inference that is based on time optimality. We denote the pairs of state sets for each sub-tasks with Π. Then, given that we used Hamilton–Jacobi reachability analysis to construct our temporal logic tree, the minimal time-to-reach for each sub-task is

$$\tau_i(x) = \max\{\tau | \tau < 0, V^*_{\mathcal{R}}(x, \tau) \geq 0\}, \tag{1.9}$$

where $\mathcal{R}$ is the reach set of the ith sub-task. Then, denote $\tau^*(x)$ as the minimum time-to-reach across all sub-tasks. Let the corresponding reachable set be $\mathcal{R}^*$. Then a time-optimal control policy that the automated vehicle could use based on the inferred time-optimal goal is

$$u^*(x) = \arg \max_{u \in \hat{\mathbb{U}}(x)} \min_{w \in \mathbb{W}} \nabla V_{\mathcal{R}^*}(x, \tau^*(x)) \cdot f(x, u, w). \tag{1.10}$$

Then, we define the inferred belief of the human operator's preference over each of the ith sub-tasks as

$$b_i(t) = \begin{cases} 1, & \text{if } i = \text{argmin}_i \, \tau_i(x(t)), \\ 0, & \text{otherwise.} \end{cases} \tag{1.11}$$

By defining the belief of the intention inference, we can then integrate it with a policy blender that will combine the control inputs from the human and the automated system.

1.4.2 Policy Blender ($\mathcal{PB}$)

Our implemented policy blender takes in beliefs $b(t)$ computed in the previous section, least-restrictive control sets $\hat{\mathbb{U}}$ from the core interaction framework and a decision from the remote human operator in the form of $u_h(t)$ and $\alpha(t)$, then outputs a control input $u(t)$ that will be implemented on the vehicle. The first step in the policy blending is to compute an automated driving input using (1.10). Let the control input computed by the automated driving system be $u_a(t)$. Then, to blend the two inputs together we take the weighted sum of $u_a(t)$ and the control input from the human, $u_h(t)$ using the weight $\alpha(t)$ given by the human operator, as follows:

$$\hat{u}(t) = (1 - \alpha(t)) \cdot u_h(t) + \alpha(t) \cdot u_a(t). \tag{1.12}$$

Then the operator can freely change $\alpha(t)$ to change the balance between the automated driving system and her control over the vehicle.

Fig. 1.8 Snapshots from the experiment conducted in [16] (used with permission from IFAC-PapersOnline)

Finally, since we still need to pick a control input that satisfies the designed task, we find the closest input inside of the safe feedback control set to $\hat{u}(t)$, which looks like

$$u(t) = \begin{cases} \hat{u}(t) & \text{if } \hat{u}(t) \in \hat{\mathbb{U}} \\ \underset{u \in \hat{\mathbb{U}}(x(t)), \forall i}{\operatorname{argmin}} \frac{\|u - u_h(t)\|}{b_i(t)}, & \text{otherwise.} \end{cases} \tag{1.13}$$

Using (1.13), we find control input that closest reflects the inferred belief of the human operator's objective while guaranteeing the full specification is satisfied.

1.4.3 Results on SVEA

To illustrate the implemented shared-autonomy teleoperation example, we present the results from the implementation done in [16] on a 1/10th-scale version of our running parking example. The experimental setup where a human operator is sitting at a control tower desk and remote driving a SVEA can be seen in the series of snapshots from the experiment shown in Fig. 1.8. The operator is remote driving the SVEA while their control inputs are being blended with the automated control inputs that prevent them from colliding with the obstacles in the environment. In Fig. 1.9, the times when the control inputs are constrained due to the possibility of violating the designed task are circled.

1.5 Discussion

In this chapter, we overviewed a safe interaction framework for road vehicle control towers. The designed framework enables human operators to intuitively create and update rich, complex tasks that are formally verified for correctness and enforced by synthesized safe feedback control sets. By including a behavior tree interface, the human operators can use a well-defined interface to design formal specifications that are then interpretable by formal verification algorithms. The framework provides

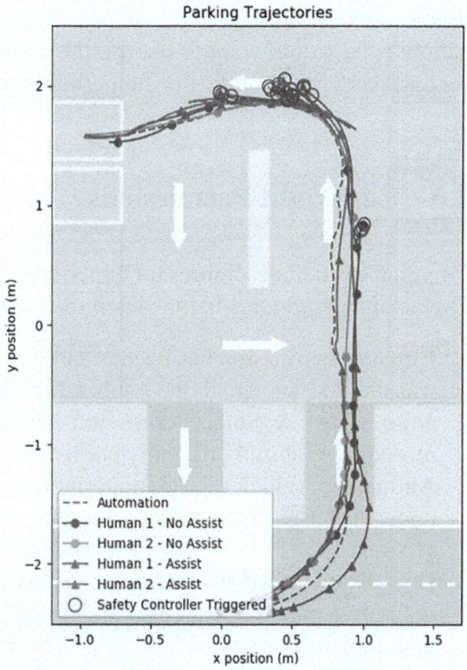

Fig. 1.9 Plotted trajectories from trial runs in experiment conducted in [16] (used with permission from IFAC-PapersOnline)

end-to-end safety guarantees from design to control. Moreover, the safe feedback control sets computed at the end can be used to design a variety of control methods and interaction modes that will be guaranteed to satisfy the designed task. There are several important future directions for extending this work. In particular, the two important future directions we have identified are (1) implementation of different levels of interaction and (2) integration of more advanced control approaches.

In this chapter, we only explore the implementation of a remote driving interaction. This is a low-level form of interaction, as the remote operator is able to directly affect the steering and speed of the controlled vehicle. In a sense, the remote operator is directly collaborating with the control system on the vehicle. However, there are important types of interaction that are higher level, such as ones where the human operator collaborates with the planning system. For example, if the current path plan of the vehicle is shown to the human operator, they could modify some of the waypoints before the vehicle begins following the path. Additionally, the human operator and the vehicle are not the only agents on the network. If the vehicle is also connected to infrastructure, such as intelligent intersections, that is making decisions, then an important consideration is how to include those decisions into interaction frameworks like the one presented in this chapter. One possibility is to develop different types of multiagent teams, as is outlined by [24].

Past the implementation of new types of interactions, there is also future work that can be on the integration of more advanced controllers. In our implementation of the shared-autonomy teleoperation, the control selection simply chooses the time-optimal control input out of the safe feedback control set. In many vehicle applications, one would want to choose the control input based on other considerations, such as passenger comfort or fuel efficiency.

1.6 Copyright Permissions

This chapter includes figures for which specific copyright permissions were obtained. The acknowledgments below detail the sources and nature of these permissions:

1. **Einride.** Permission has been granted by Einride to use Fig. 1.1. This permission was granted via email on October 11, 2023.
2. **John Wiley & Sons.** Permission has been granted by John Wiley & Sons to reproduce material originally published in "Safe Teleoperation of Connected and Automated Vehicles." The material in question, specifically Fig. 1.2, is included under the terms of this permission. The original work is cited in full in the reference section.
3. **IFAC-PapersOnLine.** Permission has been granted by IFAC-PapersOnline to reproduce material from the paper titled "Human-Centered Design for Safe Teleoperation of Connected Vehicles." This material in question, specifically Figs. 1.8 and 1.9, is reproduced with the acknowledgment: "© 2020 Frank J. Jiang, Yulong Gao, Lihua Xie, Karl H. Johansson. Reproduced with permission from 'Human-Centered Design for Safe Teleoperation of Connected Vehicles.' IFAC-PapersOnLine, 53/5 (2020), pp. 224–231." The original work is cited in full in the reference section.

References

1. Baier C, Katoen JP (2008) Principles of model checking. MIT press
2. Belta C, Yordanov B, Gol EA (2017) Formal methods for discrete-time dynamical systems. Springer
3. Biggar O, Zamani M (2020) A framework for formal verification of behavior trees with linear temporal logic. IEEE Robot Autom Lett 5(2):2341–2348
4. Bogdoll D, Orf S, Töttel L, Zöllner JM (2021) Taxonomy and survey on remote human input systems for driving automation systems. arXiv:2109.08599
5. Brown K, Driggs-Campbell K, Kochenderfer MJ (2020) A taxonomy and review of algorithms for modeling and predicting human driver behavior. arXiv:2006.08832
6. Colledanchise M, Ögren P (2018) Behavior trees in robotics and AI: an introduction. CRC Press
7. Colledanchise M, Almeida D, Ögren P (2019) Towards blended reactive planning and acting using behavior trees. In: 2019 international conference on robotics and automation (ICRA). IEEE, pp 8839–8845
8. Davies A (2017) Nissan's Path to Self-Driving Cars? Humans in Call Centers. Wired Transportation. https://www.wired.com/2017/01/nissans-self-driving-teleoperation/

9. Einride (2020) Einride will hire its first remote operator in 2020. https://www.einride.tech/insights/einride-will-hire-its-first-remote-autonomous-truck-operator-in-2020
10. Gao Y, Jiang FJ, Ren X, Xie L, Johansson KH (2020) Reachability-based human-in-the-loop control with uncertain specifications. In: Proceedings of 21st IFAC world congress
11. Gao Y, Abate A, Jiang FJ, Giacobbe M, Xie L, Johansson KH (2021) Temporal logic trees for model checking and control synthesis of uncertain discrete-time systems. IEEE Trans Autom Control 1. https://doi.org/10.1109/TAC.2021.3118335
12. Georg JM, Feiler J, Hoffmann S, Diermeyer F (2020) Sensor and actuator latency during teleoperation of automated vehicles. In: 2020 IEEE Intelligent Vehicles Symposium (IV), pp 760–766, https://doi.org/10.1109/IV47402.2020.9304802
13. Guo M, Andersson S, Dimarogonas DV (2018) Human-in-the-loop mixed-initiative control under temporal tasks. In: 2018 IEEE international conference on robotics and automation (ICRA), pp 6395–6400
14. Huth M, Ryan M (2004) Logic in computer science: modelling and reasoning about systems. Cambridge University Press
15. Jiang FJ, Gao Y, Xie L, Johansson KH (2020a) Ensuring safety for vehicle parking tasks using Hamilton-Jacobi reachability analysis. In: 2020 59th IEEE conference on decision and control (CDC), pp 1416–1421
16. Jiang FJ, Gao Y, Xie L, Johansson KH (2020b) Human-centered design for safe teleoperation of connected vehicles. In: IFAC conference on cyber-physical & human-systems, Shanghai, China
17. Jiang FJ, Mårtensson J, Johansson KH (2023) Safe teleoperation of connected and automated vehicles. In: Fundamentals and applications, cyber-physical-human systems
18. Le QS, Arogeti S, Borowsky A (2021) Remote driving testbed with force feedback based on slip angle estimation. In: 2021 European conference on mobile robots (ECMR), pp 1–8. https://doi.org/10.1109/ECMR50962.2021.9568828
19. Liu J, Ozay N, Topcu U, Murray RM (2013) Synthesis of reactive switching protocols from temporal logic specifications. IEEE Trans Autom Control 58(7):1771–1785. https://doi.org/10.1109/TAC.2013.2246095
20. Neumeier S, Walelgne EA, Bajpai V, Ott J, Facchi C (2019) Measuring the feasibility of teleoperated driving in mobile networks. In: 2019 network traffic measurement and analysis conference (TMA), pp 113–120. https://doi.org/10.23919/TMA.2019.8784466
21. Özkahraman Ö, Ögren P (2020) Combining control barrier functions and behavior trees for multi-agent underwater coverage missions. In: 2020 59th IEEE conference on decision and control (CDC). IEEE, pp 5275–5282
22. SAE International (2021) Taxonomy and definitions for terms related to driving automation systems for on-road motor vehicles. Technical Report, SAE International
23. Saeed U, Hämäläinen J, Garcia-Lozano M, David González G (2019) On the feasibility of remote driving application over dense 5G roadside networks. In: 2019 16th international symposium on wireless communication systems (ISWCS), pp 271–276. https://doi.org/10.1109/ISWCS.2019.8877213
24. Samad T (2022) Human-in-the-loop control and cyber-physical-human systems: applications and categorization. In: Annaswamy A, Khargonekar P, Lamnabhi-Lagarrigue F, Spurgeon SK (eds) Cyber-physical-human systems: fundamentals and applications. Wiley, in Press, UK
25. Wickens CD, Mavor AS, McGee JP (1997) Flight to the future human factors in air traffic control. National Academy Press
26. Yu P, Gao Y, Jiang FJ, Johansson KH, Dimarogonas DV (2023) Online control synthesis for uncertain systems under signal temporal logic specifications. arXiv:2103.09091. https://doi.org/10.48550/arXiv.2103.09091
27. Zhao L (2023) Teleoperation and the influence of driving feedback on drivers' behaviour and experience. PhD thesis, KTH Royal Institute of Technology
28. Zhou N (2017) Volvo admits its self-driving cars are confused by kangaroos. https://www.theguardian.com/technology/2017/jul/01/volvo-admits-its-self-driving-cars-are-confused-by-kangaroos

Chapter 2
Automated Mobility and Cooperation Compliance in Mixed Vehicle Traffic Environments

Christos G. Cassandras, Andres S. Chavez Armijos, Anni Li, and Ehsan Sabouni

Abstract Automated vehicles have the potential to transform transportation systems through enhanced safety, efficiency, and sustainability. This chapter first reviews cooperative control approaches for Connected Automated Vehicles (CAVs) at conflict areas such as merging roadways, intersections, and lane-changing maneuvers. Formulating and solving optimal control problems with hard safety constraints enables the derivation of cooperative trajectories. When solutions to such problems become computationally intractable, online control methods based on Control Barrier Functions (CBFs) can be used to still guarantee all constraint satisfaction at the expense of some possible performance loss. The chapter examines extending these approaches to mixed traffic environments, introducing new techniques to guarantee safety despite the presence of unpredictable Human-Driven Vehicles (HDVs). Finally, a cooperative compliance framework is proposed to incentivize HDVs to align their behavior with CAV objectives using virtual, refundable tokens, without any monetary transactions. The goal is to provide foundations and specific new techniques aimed at optimizing automated mobility in mixed traffic.

C. G. Cassandras (✉) · A. S. Chavez Armijos · A. Li · E. Sabouni
Division of Systems Engineering and Center for Information and Systems Engineering, Boston University, Brookline, MA 02446, USA
e-mail: cgc@bu.edu

A. S. Chavez Armijos
e-mail: aschavez@bu.edu

A. Li
e-mail: anlianni@bu.edu

E. Sabouni
e-mail: esabouni@bu.edu

P. Ioannou and A. A. Malikopoulos (eds.), *Transportation Mobility in Smart Cities*, Springer Tracts on Transportation and Traffic 21,
https://doi.org/10.1007/978-3-031-64769-7_2

2.1 Introduction

Automated mobility [1] has become possible as a result of Connected Automated Vehicles (CAVs) emerging as a viable mode of transportation. CAVs have the potential to improve safety, as well as reduce congestion, energy consumption, and air and noise pollution [2]. They enable intelligent traffic management through better information utilization and more precise trajectory design, especially in conflict areas, such as intersections [3], roundabouts [4, 5], and merging roadways [6]. The key to automated mobility is the *cooperation* among CAVs facilitated by connectivity and resulting in controllers designed to optimize a combination of travel time, energy consumption, and comfort metrics while ensuring that all safety constraints are enforced. In contrast, a transportation system consisting entirely of Human-Driven Vehicles (HDVs) operates based on the vehicle-centric (selfish) behavior of each driver who competes, rather than cooperating, with other drivers. This prevents a traffic network from achieving a much more efficient system-centric (socially optimal) equilibrium based on CAVs.

To date, most of the research involving the optimization of automated mobility has been based on the assumption that all vehicles in a traffic network are CAVs and can cooperate with each other. However, this assumption is unlikely to hold in the foreseeable future; therefore, developing new control and coordination algorithms for CAVs that can handle partial CAV penetration rates is essential, in view of the fact that human-driven vehicles are often unpredictable [7, 8] and their behavior can be stochastic and selfish [9], hence leading to added congestion and safety issues. A common assumption in modeling HDV behavior in such mixed traffic is that HDVs maintain constant speed in free-flow traffic and adopt one of several car-following models, e.g., [10, 11]. The safety of CAVs in mixed traffic environments is studied using the Wiedemann car-following model [12] for intersections and lane-change maneuvers in [13]. The problem of safe merging in mixed traffic is studied in [14–16], with [14] using an intelligent car-following model for HDVs, while [15] adopts a game theoretic approach without offering any safety guarantees, and [16] tackles the problem of safe merging without offering any coordination scheme.

It remains a challenge to develop both safety-guaranteed control and efficient coordination schemes for CAVs in *mixed traffic* at various conflict areas of a transportation system. There are at least three ways to proceed that we envision. First, there is a need for improved models of HDV behavior that can benefit from advances in perception and learning-based methods. In parallel, one can prioritize safety and then develop optimal CAV controllers that are robust or even independent of arbitrary HDV behavior as recently shown in [17, 18]. Finally, one can seek mechanisms to achieve *cooperation compliance* of HDVs to rules that can provably lead to socially optimal traffic operation.

The goal of this chapter is to first review in Sect. 2.2 the cooperative control of CAVs in conflict areas of a traffic network through the formulation and solution of appropriate Optimal Control Problems (OCPs) that include hard safety constraints. When the online solution of such problems becomes infeasible due to their computational complexity, we review the use of Control Barrier Functions (CBFs) as a way to replace the safety constraints with simpler ones which still guarantee their enforcement at the expense of some loss in performance. We will use the merging problem and the problem of lane-changing maneuvers to illustrate these approaches.

The next step is to consider in Sect. 2.3 these problems in mixed traffic and describe some new approaches to achieve socially optimal solutions that are robust to HDV behavior while guaranteeing safety. Finally, in Sect. 2.4, we propose a cooperative compliance framework in which HDVs are incentivized to cooperate with CAVs without any explicit monetary exchange transactions but rather using the following premise: Every driver (user) owns a digital wallet from which it commits a number of (virtual) tokens at the start of a certain driving period. These can be thought of as a toll that vehicles currently pay when accessing some transportation resources, but the key differences here are: (i) the toll is fully refundable as long as the user complies with a certain traffic "rule," and (ii) the toll is virtual in the form of "token bonds" or "deposits" that each user commits. It is important to note that the tokens are returned in their entirety to the user upon completion of the driving period as long as full compliance occurs, hence the concept of "refundable tolls."

2.2 Cooperative Control of CAVs

This section provides an overview of how the cooperative control of Connected Autonomous Vehicles (CAVs) operates within different conflict areas of a traffic network. We introduce the concept of a *Control Zone* (CZ), which designates regions where CAVs can communicate with each other and with a coordinator, such as a Road-Side Unit (RSU). The coordinator facilitates information exchange among CAVs, but does not control individual vehicles. The Mcity facility (a map of which can be seen at https://mcity.umich.edu/our-work/mcity-test-facility/) is a small-scale smart city test bed which exemplifies the various conflict areas that arise. Specifically, we identify common CZs that present opportunities for cooperation among CAVs. These CZs are critical areas where cooperative compliance among CAVs can yield substantial benefits in terms of traffic efficiency, safety, and overall travel experience. However, each specific CZ presents different challenges and benefits as summarized below:

- *Merging Roads:* Merging points, such as entrance ramps or merging lanes, facilitate cooperative interactions. CAVs cooperate to ensure smooth merging maneuvers by coordinating their speeds and positions, and optimizing traffic flow while minimizing traffic disruptions by maintaining desired vehicle speeds to the maximum possible extent.
- *Signal-Free Intersections:* Intersections without traffic signals, typically found in residential or low-traffic areas, require careful vehicle coordination. CAVs can navigate these intersections efficiently by sharing information and negotiating right-of-way, reducing the risk of accidents, and improving overall traffic efficiency.

Ideally, traffic through such intersections never stops so as to prevent travel delays and wasted energy consumption when vehicles need to accelerate from rest to desired speeds.
- *Roundabouts:* Roundabouts pose intricate traffic dynamics that require vehicle coordination to ensure safe and efficient passage. CAVs utilize cooperative strategies to synchronize their movements, negotiate merging priorities, and avoid conflicts, resulting in improved traffic flow and reduced congestion. One can view a roundabout as a combination of several tightly coupled merging areas.
- *Highway Lane Changes:* Lane change maneuvers on highways, especially during peak traffic hours, present considerable challenges. Through cooperation and information sharing, CAVs coordinate their actions, maintaining safe distances and optimizing lane usage. This leads to smoother traffic flow and minimized disruptions. In this case, a major challenge is to balance the "selfish" goal of an ego vehicle to perform a fast and safe lane change with the system-centric "social" objective of maintaining a smooth traffic flow without the disruptions that a lane-change maneuver may cause.

Using advanced communication and control mechanisms within CZs, CAVs can actively collaborate to navigate these areas, ensuring optimal performance (in terms of time, energy, and passenger comfort) while prioritizing safety. A CZ can be static, as in merging roads, intersections, and roundabouts. However, a CZ can also be dynamic, such as in highway lane changes where the CZ is defined around an ego vehicle performing the lane-changing maneuver. Additionally, there may exist multiple interconnected CZs depending on the complexity of the traffic network.

2.2.1 Optimal Control Problem Formulation

Assuming all vehicles in the network are CAVs, we can formulate a finite-horizon constrained OCP that includes all objectives and constraints. The goal is to determine trajectories that jointly minimize travel time within the CZ, energy consumption, and passenger comfort by minimizing jerk and centrifugal forces. Ensuring consistent adherence to safety constraints is essential throughout the optimal trajectory. Next, we present a general mathematical formulation that can be specialized and applied to each of the CZs described above.

Let $F(t)$ be the set of indices of all CAVs located in the CZ at time t with cardinality $N(t)$. The vehicle dynamics for each CAV $i \in F(t)$ along the lane to which it belongs in a given CZ are assumed to take the form:

$$\left[\dot{x}_i(t)\ \dot{v}_i(t)\right] = \left[v_i(t)\ u_i(t)\right], \tag{2.1}$$

where $x_i(t)$ denotes the distance traveled by CAV i from the origin of the CZ, $v_i(t)$ denotes the velocity, and $u_i(t)$ denotes the control input (acceleration). These simple dynamics can be replaced by more elaborate ones (e.g., see [19]), but we maintain this simple form as we initially attempt to determine an analytically tractable solution to the OCP.

2.2.1.1 Objectives and Constraints

There are three objectives to be minimized for each CAV $i \in F(t)$, as detailed next:

Objective 2.1 (*Minimize travel time*) Let t_i^0 and t_i^f denote the time that CAV $i \in F(t)$ arrives at its origin and leaves the CZ at its exit point, respectively. We wish to minimize the travel time $t_i^f - t_i^0$ for CAV i.

Objective 2.2 (*Minimize energy consumption*) We also wish to minimize the energy consumption for each CAV i:

$$\min_{u_i(t),t_i^f} \int_{t_i^0}^{t_i^f} \mathcal{L}(u_i(t))dt, \tag{2.2}$$

where $\mathcal{L}(\cdot)$ is a strictly increasing function of its argument.

Objective 2.3 (*Maximizing centrifugal comfort*) When a CZ includes a curved road, we want to minimize centrifugal acceleration to minimize centrifugal discomfort:

$$\min_{u_i(t)} \int_{t_i^0}^{t_i^f} \kappa(x_i(t))v_i^2(t)dt, \tag{2.3}$$

where $\kappa : \mathbb{R} \longrightarrow \mathbb{R}^{\geq 0}$ is the curvature of the road at position $x_i(t)$, determined by $\frac{1}{r(x_i(t))}$, where r is the radius of the road at $x_i(t)$.

Constraint 2.1 (*Safety constraints*) Let i_p denote the index of the CAV that physically immediately precedes i in the CZ (if one is present). We require that the distance $z_{i,i_p}(t) := x_{i_p}(t) - x_i(t)$ be constrained by

$$z_{i,i_p}(t) \geq \varphi v_i(t) + \delta, \forall t \in [t_i^0, t_i^f], \tag{2.4}$$

where φ denotes the reaction time (as a rule, $\varphi = 1.8$ s is used, e.g., [20]) and δ is a given minimum safe distance. If we define z_{i,i_p} to be the distance from the center of CAV i to the center of CAV i_p, then δ depends on the length of these two CAVs (generally dependent on i and i_p but taken to be a constant over all CAVs for simplicity).

Constraint 2.2 (*Safe merging*) Whenever a CAV crosses a *Merging Point* (MP), a lateral collision is possible and there must be adequate safe space for the CAV at this MP to avoid such collision, i.e.:

$$z_{i,i_c}(t_i^m) \geq \varphi v_i(t_i^m) + \delta, \tag{2.5}$$

where $i_c \in F(t)$ is the index of the CAV that may collide with CAV i at merging point $m = \{1, \ldots, n_i\}$ where n_i is the total number of MPs that CAV i passes in the CZ. Each CZ can have either one MP like in the single-lane merging roadways (which will be discussed in this section later) or multiple MPs like multiple-lane merging roads [21], multi-lane intersections [22], and roundabouts [23]. It is important to note that MP m can either be fixed, as in the case of merging on-ramps and intersections, or dynamic as in the case of lane-changing maneuvers on highway segments. The determination of CAV i_c depends on the policy adopted for sequencing CAVs through the CZ, such as First-In-First-Out (FIFO) based on the arrival times of CAVs, or any other desired policy [24]. It should be noted that this constraint only applies at a certain time t_i^m, which obviously depends on how the CAVs are controlled.

Constraint 2.3 (*Centrifugal acceleration constraint*) There is a constraint on the centrifugal acceleration to avoid lateral rollover for each CAV $i \in F(t)$:

$$\kappa(x_i(t))v_i^2(t) \leq \frac{w_i^h}{h_i} g, \quad \forall t \in [t_i^0, t_i^f], \tag{2.6}$$

where $w_i^h > 0$ denotes the half-width of the vehicle, $h_i > 0$ denotes the height of the center of gravity with respect to the ground, and g is the gravity constant.

Constraint 2.4 (*Vehicle limitations*) Finally, there are constraints on the speed and acceleration for each CAV $i \in F(t)$:

$$v_{\min} \leq v_i(t) \leq v_{\max}, \quad u_{\min} \leq u_i(t) \leq u_{\max}, \quad \forall t \in [t_i^0, t_i^f], \tag{2.7}$$

where $v_{\max} > 0$ and $v_{\min} \geq 0$ denote the maximum and minimum speed allowed in the CZ for CAV i, $u_{\min} < 0$ and $u_{\max} > 0$ denote the minimum and maximum control for CAV i, respectively.

The main goal of cooperative control algorithms developed for CAVs in transportation networks is to ensure safety through the satisfaction of Constraints 2.1–2.3 while optimizing the transportation network performance by minimizing Objectives 2.1–2.3 with respect to the physical limitations of every vehicle. The approaches for solving such OCPs, however, can exhibit fundamental differences despite having the same goal. In the next subsection, a brief review of the existing approaches will be provided.

2.2.1.2 Optimal Cooperative Control Problem Solution Approaches

There exist different approaches to solve the OCP formulated in each CZ, such as learning-based methods (e.g., [25–27]) and optimization-based methods (e.g., [28–30]). However, every approach can be categorized as either centralized or decentralized:

Centralized Approach: In the centralized approach, an optimization problem is formulated involving all CAVs $i \in F(t)$ in the CZ. The solution determines a centralized control law for all vehicles so as to achieve Objectives 2.1–2.3 subject to Constraints 2.1–2.4 and vehicle dynamics (2.1).

We define the controllable vectors $\mathbf{u} = \{u_i | i \in F(t)\}$ and $\boldsymbol{t}^f = \{t_i^f | i \in F(t)\}$. Choosing $\mathcal{L}(u_i(t)) = \frac{1}{2}u_i^2(t)$ and using weights $\alpha_1, \alpha_2, \alpha_3$ to create a convex combination of objectives (i.e., $\alpha_1 + \alpha_2 + \alpha_3 = 1$), we formulate the nonlinear optimization problem:

$$\min_{\boldsymbol{u},\boldsymbol{t}^f} J(\boldsymbol{u}, \boldsymbol{t}^f) = \sum_{i=1}^{N(t)} \int_{t_i^0}^{t_i^f} \Big[\alpha_1 + \alpha_2 \bar{L}_E(u_i) + \alpha_3 \bar{L}_C(x_i, v_i)\Big] dt \tag{2.8a}$$

$$\text{s.t.: } z_{i,i_p}(t) \geq \varphi v_i(t) + \delta, \quad \forall i \in F(t) \tag{2.8b}$$

$$|z_{i,i_c}(t_i^m)| \geq \varphi v_i(t_i^m) + \delta, \quad \forall i \in F(t), \tag{2.8c}$$

$$v_{\min} \leq v_i(t) \leq v_{\max}, \forall t \in [t_i^0, t_i^f], \quad \forall i \in F(t), \tag{2.8d}$$

$$u_{\min} \leq u_i(t) \leq u_{\max}, \forall t \in [t_i^0, t_i^f], \quad \forall i \in F(t), \tag{2.8e}$$

where we define

$$\bar{L}_E(u_i) = \frac{\frac{1}{2}u_i^2(t)}{\frac{1}{2}\max\{u_{\max}^2, u_{\min}^2\}}, \quad \bar{L}_C(x_i, v_i) = \frac{\kappa(x_i(t))v_i^2(t)}{\kappa_{max} v_{max}^2} \tag{2.9}$$

to be normalized versions of Objectives 2.2 and 2.3, respectively. Constraint (2.8c) is imposed to satisfy the safe merging constraint. As the centralized controller's sequencing scheme determines which CAV i_c is selected, we use the absolute value of the distance between CAV i and CAV i_c. In particular, this formulation comprises both sequencing problems (i.e., finding the sequence for CAVs to cross MPs) and motion control problems by obtaining inputs from CAVs. As a result, in real-time applications, the computational cost of solving this problem can be prohibitive due to its complexity. This complexity can be overcome by decomposing the problem into its sequencing and individual motion control components, giving rise to decentralized methods.

Decentralized Approach: In this approach, given a fixed or dynamic sequence for the MP crossing order (i.e., CAV i_c is specified), our goal is to determine a control law achieving Objectives 2.1–2.3 subject to Constraints 2.1–2.4 for each $i \in F(t)$ governed by the dynamics (2.1). Using properly normalized costs as in (2.9), we use the weights $\alpha_1, \alpha_2, \alpha_3$ to construct a convex combination as follows:

$$\min_{u_i(t),t_i^f} J_i(u_i(t), t_i^f) = \int_{t_i^0}^{t_i^f} \left[\alpha_1 + \alpha_2 \bar{L}_E(u_i) + \alpha_3 \bar{L}_C(x_i, v_i)\right] dt \tag{2.10}$$

subject to (2.1), (2.4), (2.5), (2.6), (2.7), as well as the initial and terminal positions $x_i(t_i^0) = 0,\ x_i(t_i^f) = L$ and given $t_i^0,\ v_i^0$ (i.e., the initial speed). Problem (2.10) can be solved analytically in some cases, e.g., the merging problem [31] and a signal-free intersection [32]. However, obtaining solutions for real-time applications becomes prohibitive when an optimal trajectory involves multiple constraints becoming active.

The root of the complexity imposed by the constraints (2.4), (2.5), (2.6) is that they are expressed in terms of the system state, not the actual control required to enforce them. This difficulty can be addressed through the use of Control Barrier Functions (CBFs) [33] which map the original state-based constraints into new ones which are much simpler inequalities linear in the control $u_i(t)$. These new constraints imply the satisfaction of the original constraints; the price to pay is that they may be conservative, hence affecting the optimality properties of the resulting controllers.

The OCBF Approach: This approach uses the OCP formulation in (2.10) with one key difference: CBFs replace the original state constraints (2.4), (2.5), (2.6) and the speed limitations in (2.7), hence the term OCBF. As mentioned above, this greatly simplifies the determination of controls that satisfy these constraints, at the expense of some potential conservativeness since a CBF is only a sufficient condition for the satisfaction of its associated state constraint.

Firstly, we derive the CBFs that ensure the constraints are always satisfied, subject to the vehicle dynamics (2.1) by defining $f(\boldsymbol{x}_i(t)) = [v_i(t), 0]^T$ and $g(\boldsymbol{x}_i(t)) = [0, 1]^T$. Each of these constraints can be written in the form of $b_q(\boldsymbol{x}(t)) \geq 0,\ q \in \{1, \ldots, n\}$ where n stands for the number of constraints and $\boldsymbol{x}(t) = [\boldsymbol{x}_1(t), \boldsymbol{x}_2(t), \ldots, \boldsymbol{x}_{N(t)}(t)]$. The CBF method (details provided in [33]) maps a constraint $b_q(\boldsymbol{x}(t)) \geq 0$ onto a new constraint which is *linear* in the control input and takes the general form

$$L_f b_q(\boldsymbol{x}(t)) + L_g b_q(\boldsymbol{x}(t)) u_i(t) + \gamma_q\big(b_q(\boldsymbol{x}(t))\big) \geq 0, \tag{2.11}$$

where L_f, L_g denote the Lie derivatives along f and g, respectively (it is assumed that $L_g b_q(\boldsymbol{x}) \neq 0$ when $b_q(\boldsymbol{x}) = 0$, otherwise we have to introduce higher order CBFs as detailed in [33]). Moreover, $\gamma_q(\cdot)$ is a properly selected class $\mathcal{K}$ function, i.e., it is strictly increasing and $\gamma_q(0) = 0$.

As an example, to obtain the CBF constraint for the safety requirement (2.4) we set $b_1\big(\boldsymbol{x}_i(t), \boldsymbol{x}_{i_p}(t)\big) = z_{i,i_p}(t) - \varphi v_i(t) - \delta$ and since this is differentiable, the CBF constraint for (2.4) is

$$v_{i_p}(t) - v_i(t) - \varphi u_i(t) + k_1 b_1\big(\boldsymbol{x}_i(t), \boldsymbol{x}_{i_p}(t)\big) \geq 0, \tag{2.12}$$

where the class-$\mathcal{K}$ function $\gamma_1(\boldsymbol{x}) = k_1 \cdot b_1\big(\boldsymbol{x}_i(t), \boldsymbol{x}_{i_p}(t)\big)$ is chosen here to be linear. The derivation of the CBF constraints for the rest of the OCP constraints in (2.10) is similar and is omitted, but can be found in [33].

Next, we deal with soft constraints in the OCP (2.10) by making use of Control Lyapunov Functions (CLFs). In particular, we use a CLF to track the CAV speed to a desired value $v_i^{ref}(t)$ by setting $V(\boldsymbol{x}_i(t)) = \left(v_i(t) - v_i^{ref}(t)\right)^2$ and express the CLF constraint as follows:

$$L_f V(\boldsymbol{x}_i(t)) + L_g V(\boldsymbol{x}_i(t))u_i(t) + c_3 V(\boldsymbol{x}_i(t)) \leq e_i(t), \tag{2.13}$$

where $e_i(t) > 0$ is an added decision variable to be minimized, thus making this a *soft* constraint.

Finally, we can formulate the OCBF problem as follows:

$$\min_{u_i(t),e_i(t)} J_i(u_i(t), e_i(t)) := \int_{t_i^0}^{t_i^f} \left[\frac{1}{2}(u_i(t) - u_i^{ref}(t))^2 + \lambda e_i^2(t)\right] dt \tag{2.14}$$

subject to vehicle dynamics (2.1), all CBF constraints expressed in the form (2.11), and CLF constraint (2.13). The control to be optimally tracked, $u_i^{ref}(t)$, may be selected as any desired reference signal but is typically the solution of the *unconstrained* OCP in (2.10). Thus, the OCBF approach consists of two steps: (i) We solve the unconstrained optimal control problem in (2.10) which yields $u_i^{ref}(t)$, (ii) the resulting $u_i^{ref}(t)$ is optimally tracked such that all CBF constraints (2.11) are satisfied, hence all original constraints as well.

In practice, a common way to solve the dynamic optimization problem (2.14) is to discretize $[t_i^0, t_i^f]$ into intervals $[t_i^0, t_i^0 + \Delta], \ldots, [t_i^0 + k\Delta, t_i^0 + (k+1)\Delta], \ldots$ with equal length $\Delta > 0$ and solving (2.14) over each time interval. The decision variables $u_{i,k} = u_i(t_{i,k})$ and $e_{i,k} = e_i(t_{i,k})$ are assumed to be constant on each such interval and can be easily calculated at time $t_{i,k} = t_i^0 + k\Delta$ through solving a Quadratic Program (QP) at each time step:

$$\min_{u_{i,k},e_{i,k}} \left[\frac{1}{2}(u_{i,k} - u_i^{ref}(t_{i,k}))^2 + \lambda e_{i,k}^2\right] \tag{2.15}$$

subject to the CBF and CLF constraints of the form (2.11) and (2.13), respectively, where all constraints are *linear* in the decision variables. The sequence of QPs can be readily solved, which makes it suitable for real-time applications.

Before proceeding we need to make a few observations. First, the solution is *decentralized* in the sense that CAV i requires information only from CAVs i_p and i_c required in (2.4) and (2.5). Second, the objective cost in (2.10) pertains only to travel time, energy consumption, and comfort for CAV i, therefore CAV i "selfishly" optimizes its own performance. Finally, prior to solving (2.10), as already mentioned, CAV i_c needs to be determined through the coordinator (i.e., according to a particular sequence) as their state information is required to form constraints (2.4) and (2.5), respectively.

2.2.2 *Cooperative Control in Control Zone (CZ) Examples*

Throughout the rest of the chapter, two specific CZ examples are presented to illustrate the use of decentralized optimal control formulations, in particular, OCBF. We first focus on optimizing the traffic flow of merging roadways (which can then be extended to intersections and roundabouts) and then on the challenge of optimizing lane-change maneuvers. We consider the case of traffic consisting exclusively of CAVs before turning attention in Sect. 2.3 to the mixed traffic case where only a subset of vehicles are CAVs, while the rest are HDVs.

2.2.2.1 Decentralized Cooperative Merging Control

The case of merging roadways is shown in detail in Fig. 2.1. This is a common conflict area involving traffic that arrives from two roads merging at a single MP M where a collision may occur. The roads are indexed by $r \in \{1, 2\}$, with 1 usually reserved for the "main road." The CZ here is defined to consist of a *Sequencing Zone* (SZ) in which a desired merging sequence of vehicles will be determined, and an *Awareness Zone* (AZ) in which vehicles can interact with and observe each other directly. We initially limit ourselves to the case where all vehicles are CAVs, so that we do not differentiate between the SZ and AZ; this distinction is used when mixed traffic is considered in Sect. 2.3.

Let $F_C(t)$ contain the indices associated with CAVs in the CZ, and $F_H(t)$ is the set of indices of HDVs inside the CZ at time t (i.e., $F_C(t) \cup F_H(t) = F(t)$). In this section, we set $F_H(t) = \emptyset$. Let $N(t)$ be the cardinality of $F(t)$. Thus, if a vehicle arrives at time t, it is assigned the index $N(t) + 1$. The index 0 is reserved for a vehicle that has just left the CZ (see Fig. 2.1). All vehicle indices in $F(t)$ decrease

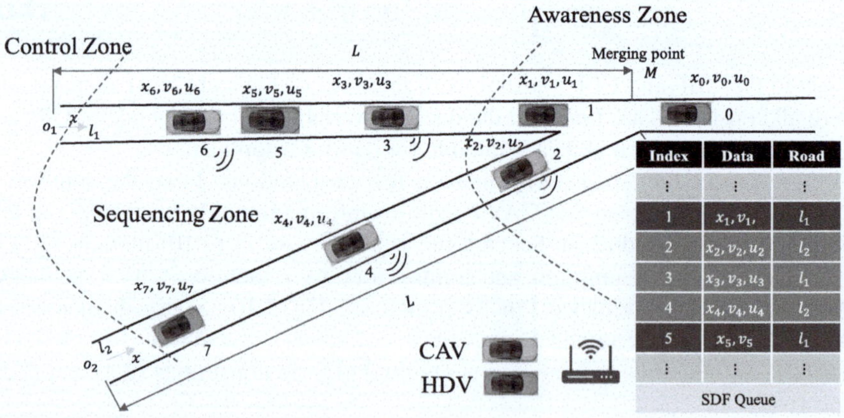

Fig. 2.1 The merging problem

by one when a vehicle passes over the MP and the vehicle whose index is -1 is dropped. The coordinator maintains a table as shown in Fig. 2.1 containing the states of all vehicles $i \in F(t)$.

Our goal is to determine a control law for all CAVs $i \in F_C(t)$ ($F(t)$ and $F_C(t)$ can be used interchangeably here, as traffic flow is limited to only CAVs) so as to achieve Objectives 2.1 and 2.2 (Objective 2.3 is excluded as the roads do not have any curvature; however, the case of merging with curved roads is analyzed in [34]). By setting $u_i^{ref}(t)$ in (2.15) as the unconstrained optimal control solution of CAV i formulated in (2.8a) (with $\alpha_3 = 0$), the QP formulated in (2.15) can be efficiently solved for the merging problem while a CAV is within the CZ. The unconstrained optimal control solution can be readily obtained using Hamiltonian analysis as shown in [31] resulting in

$$u_i^*(t) = a_i t + b_i, \quad v_i^*(t) = \frac{1}{2}a_i t^2 + b_i t + c_i, \quad x_i^*(t) = \frac{1}{6}a_i t^3 + \frac{1}{2}b_i t^2 + c_i t + d_i.$$

Then, for each $i \in F_C(t)$, we need to solve the following five nonlinear algebraic equations for a_i, b_i, c_i, d_i and t_i^m:

$$\begin{aligned}
\frac{1}{2}a_i\left(t_i^0\right)^2 + b_i t_i^0 + c_i = v_i^0, \qquad \frac{1}{6}a_i\left(t_i^0\right)^3 + \frac{1}{2}b_i\left(t_i^0\right)^2 + c_i t_i^0 + d_i = 0,\\
\frac{1}{6}a_i\left(t_i^m\right)^3 + \frac{1}{2}b_i\left(t_i^m\right)^2 + c_i t_i^m + d_i = L, \qquad a_i t_i^m + b_i = 0,\\
\beta + \frac{1}{2}a_i^2 \cdot \left(t_i^m\right)^2 + a_i b_i t_i^m + a_i c_i = 0,
\end{aligned}$$

where L is the road length. Finally, β is an adjustable weight defined as $\beta := \frac{\alpha_1 \max\{u_{max}^2, u_{min}^2\}}{2(1-\alpha_1)}$. In this case, a complete explicit *constrained* solution can be derived (see [31]) but it becomes computationally burdensome when multiple constraints become active. Thus, in practice, we can use the unconstrained solution above as the reference control $u_i^{ref}(t)$ in (2.14) and make use of the OCBF approach.

Despite the attractive features of the OCBF approach, there are several limitations to adopting this formulation, which form the basis for the remainder of this chapter. First, in both (2.14) and (2.15), the objective functions only pertain to CAV i metrics, making the solution vehicle centric (selfish) as opposed to system centric (socially optimal). In merging problems, assuming symmetric roads with equal merging priority and homogeneous CAVs, this limitation may not be severe, as opposed to the case of lane-changing maneuvers considered in the next section. Second, this formulation holds when there is cooperation among *all* vehicles, as is the case when we assume traffic consisting of 100% CAVs; this does not hold in mixed traffic where most HDVs are expected not to be cooperation compliant. For instance, according to FIFO, vehicle 4 (a CAV) in Fig. 2.1 has priority to pass the merging point earlier than vehicle 5, however since vehicle 5 is an HDV, it might not conform to the FIFO rule and possibly attempt to cross the MP before CAV 4 by simply not providing enough space for it. Motivated by these limitations, we will discuss in Sect. 2.3 how we can modify the existing decentralized framework to accommodate HDVs and CAVs in the traffic flow by introducing a *disruption metric* and new constraints that prioritize safety.

2.2.2.2 Cooperative Lane-Changing Maneuvers

Lane changing is a critical maneuver that impacts traffic flow, energy use, and travel time. To enable cooperative lane changing, a control framework allows CAVs to exchange information and optimize coordinated decision-making. By accounting for vehicle dynamics, traffic conditions, and by communicating with neighbors, CAVs can perform time- and energy-optimal maneuvers.

Different from the merging problem in Sect. 2.2.2.1, where the merging point in the control zone is fixed, the lane-changing problem has dynamic merging points for vehicles that are difficult to determine in advance and become decision variables in our OCP framework. First, we decompose the lane-changing maneuver into longitudinal and lateral phases. In the longitudinal phase, a CAV properly positions itself relative to the cooperating neighbors with enough space so that a safe lane change can be performed in the subsequent lateral phase. In the following, we focus on the longitudinal phase only and do not include the lateral component of the lane change maneuver, in which CAV C in Fig. 2.2 solves a decentralized optimal control problem seeking to jointly minimize the time and energy consumption; details on the optimal lateral maneuver are provided in [35].

As illustrated in Fig. 2.2, an overtaking example involves a CAV C cooperating with CAVs 1 and 2 to minimize maneuver time and energy when overtaking an unconnected vehicle U. Following the same setting as (2.10), the OCP for the maneuver can be formulated as

$$\min_{u_1,u_2,u_C} J(\boldsymbol{u}, t_f) = \int_{t_0}^{t_f} \left(w_t + w_{1,u}\bar{L}_E(u_1(t)) + w_{2,u}\bar{L}_E(u_2(t)) + w_{C,u}\bar{L}_E(u_C(t))\right) dt \tag{2.16a}$$

$$s.t. \;\; z_{i,i_p}(t) \geq \varphi v_i(t) + \delta, \; i = 1, 2, C, \tag{2.16b}$$

where w_t, $w_{i,u}$ denote the weights for the time and energy costs, respectively, (2.16b) is the safe merging constraint for all i in Fig. 2.2 with i_p denoting the current preceding vehicle of CAV i for $t \in [t_0, t_f]$. Finally, (2.16a) allows for a free terminal time and terminal state constraints $x_i(t_f)$, $v_i(t_f)$.

Note that this OCP combines the controls of all three cooperating CAVs. The problem can be decentralized by first solving simple nonlinear programs to compute the optimal terminal conditions (terminal time t_f^* and terminal positions $x_i(t_f^*)$) for each of the CAVs. Then, each vehicle can solve an individual OCP without needing a centralized coordinator (details are given in [35]). Thus, the decentralized longitudinal control can be determined through the following OCP for each $i = 1, 2, C$:

$$\min_{u_i(t)} \int_{t_0}^{t_f^*} \frac{1}{2} u_i^2(t) dt \tag{2.17a}$$

$$\text{s.t.} \quad (2.7), \quad x_i\left(t_f^*\right) = x_{i,f}^* \tag{2.17b}$$

$$z_{i,i_p}(t) \geq \varphi v_i(t) + \delta, \quad \forall t \in \left[t_0, t_f^*\right]. \tag{2.17c}$$

This OCP can be solved by standard Hamiltonian methods as described in [35]. It can also be shown that the optimal solution for CAVs 1, 2 has the following intuitive properties: CAV 1 never decelerates and CAV 2 never accelerates. Typically, CAV 1 accelerates and CAV 2 decelerates so that they jointly open up adequate spacing for CAV C to merge between them. However, this also highlights the *selfish* nature of this optimal maneuver executed by the ego vehicle C with the full cooperation of two given CAVs 1, 2. In particular, when CAV 2 decelerates it may affect all other vehicles behind it causing an overall slowdown of the fast lane joined by C. This motivates the need to replace such vehicle-centric optimal behavior with a system-centric approach that can achieve socially optimal behavior across a traffic network. It is worth noting that recent studies have investigated a complete elimination of lanes (hence, the challenges and limitations of lane changing), with details provided in [36].

2.3 From Selfish to Social Optimality with Cooperative Control

We have seen thus far that the advent of CAVs enables cooperative control and offers the potential for improved traffic efficiency and safety. However, as pointed out in the previous section, the *decentralized* control of a CAV within each CZ often results in optimizing "selfish" behavior rather than benefiting the entire traffic system. This outcome can occur even when all vehicles are fully cooperating CAVs as described in the case of lane-changing maneuvers in Sect. 2.2.2.2: in this case, a CAV striving to optimize its own individual maneuver often neglects the disruptive impact on surrounding traffic. Specifically, referring to Fig. 2.2, we saw that an optimal decentralized solution typically results in CAV 2 decelerating, which affects all traffic behind it and may significantly reduce the average speed in the fast lane.

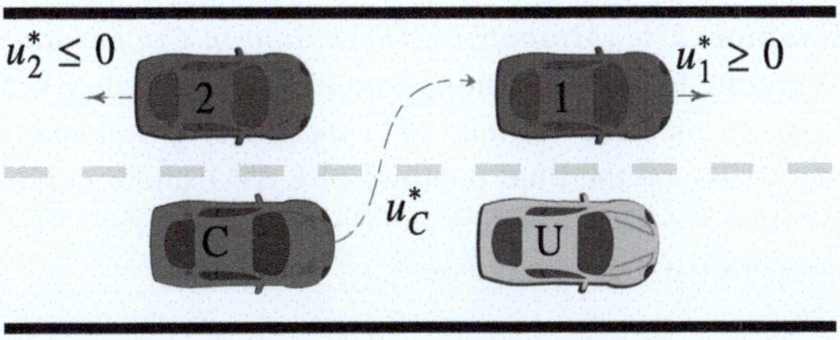

Fig. 2.2 The basic lane-changing maneuver process

While this self-centered approach may be optimal for an individual CAV, it fails to account for the overall network's well-being. True cooperation and coordination among CAVs, with consideration for the impact on total traffic flow, can unlock collective benefits like reduced travel times. When vehicles work in concert rather than focusing strictly on individual advancement, cooperative control has the potential to smooth traffic flow and improve mobility and sustainability for the system as a whole. By considering the wider network and acting in unison, CAVs can optimize movements leading to smoothing traffic flow and reducing energy use while improving mobility and sustainability.

In this section, we describe a decentralized optimal control approach that leads to socially optimal solutions by including appropriate traffic flow objectives. We begin by still assuming all vehicles are CAVs. Subsequently, we consider the problem of *mixed traffic*, where CAVs collaborate with HDVs to achieve socially optimal objectives.

2.3.1 Disruption Metric for Socially Optimal Cooperative Control

We introduce a *disruption metric* in order to quantify the adverse effects of cooperative control when driven by an ego vehicle optimizing only its own objectives. We extend the finite-horizon OCP formulation for CAVs in a CZ to minimize disruption as an additional cost component. The overall objective remains the determination of CAV trajectories that minimize travel time, energy use, and comfort while ensuring safety. In addition, however, we consider traffic disruption as an objective by defining a disruption metric $D_i(t)$ which represents the effects caused by CAV i at time t on all other CAVs within the same CZ:

$$\min_{u_i(t)} \int_{t_i^0}^{t_i^f} D_i(t)dt. \tag{2.18}$$

This depends on factors such as speed changes, lateral movements, and vehicle interactions that may cause potential cascading effects on the surrounding traffic flow. We will illustrate the effect of incorporating (2.18) into the OCP framework of (2.14) by considering next the case of cooperative lane-changing maneuvers.

Example 2.1 (*Minimizing Disruption in Cooperative Lane-Changing Maneuvers*) We return to the cooperative lane-changing problem in Sect. 2.2.2.2 and define disruption as a convex combination of *speed disruption* and *position disruption*. Let $t_f > t_0$ and $x_i(t_f)$ be the final position of CAV i under control policy $u_i(t)$ for $t \in [t_0, t_f]$. At time $t > t_0$, we define the disruption metric $D_i(t)$ for any CAV i involved in a cooperative lane change as

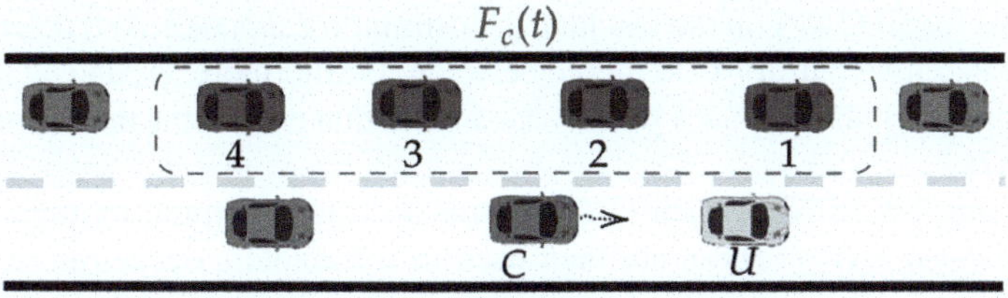

Fig. 2.3 Expanded lane-changing maneuver cooperation set diagram

$$\mathcal{D}_i(t) = \gamma_x \mathcal{D}_i^x(t) + \gamma_v \mathcal{D}_i^v(t), \tag{2.19a}$$

$$\mathcal{D}_i^x(t) = (x_i(t) - (x_i(0) + v_i(0) \cdot t))^2, \tag{2.19b}$$

$$\mathcal{D}_i^v(t) = (v_i(t) - v_d)^2. \tag{2.19c}$$

Here, $\mathcal{D}_i^x(t)$ and $\mathcal{D}_i^v(t)$ represent disruption metrics for *position* and *speed*, respectively. The position disruption $\mathcal{D}_i^x(t)$ quantifies the deviation in the positions compared to maintaining initial speed over $[0, t]$ without cooperating. The speed disruption $\mathcal{D}_i^v(t)$ measures deviation from desired flow speed v_d. The weights γ_x and γ_v normalize the terms with $\gamma_x + \gamma_v = 1$ and adjust their relative importance.

Now consider a general case with a set $F_C(t)$ of N CAVs that can cooperate with CAV C as shown in Fig. 2.3. This expands the possible cooperating vehicles beyond the original three-CAV formulation in Fig. 2.2.

To determine a controller that minimizes disruption, we formulate a mixed-integer nonlinear program (MINLP) that yields the optimal terminal time t_f^*, terminal positions $x_i^*(t_f)$, and terminal speeds $v_i^*(t_f^*)$ for each CAV i in the set $F_C(t)$ containing all cooperating CAVs. This determines the terminal positions for all N vehicles in $F_C(t)$. Binary variables B_i determine the optimal merging point for CAV C. With maximum allowable time T_{max}, the MINLP is

$$\underset{t_f, B_i, \mathbf{x}_i(t_f), v_i(t_f)}{\operatorname{argmin}} \left\{ \frac{1}{T_{\max}} t_f + \sum_{i \in F_C(t_f) \cup \{C\}} \mathcal{D}_i(t_f) \right\}$$

$$\text{s.t. } x_U\left(t_f\right) - x_C\left(t_f\right) \geq \varphi v_C\left(t_f\right) + \delta \tag{2.20a}$$

$$x_j\left(t_f\right) - x_{j+1}\left(t_f\right) \geq \varphi v_{j+1}\left(t_f\right) + \delta, \quad \forall j \in F_C(t) \backslash \{N\}, \tag{2.20b}$$

$$x_C\left(t_f\right) - x_i\left(t_f\right) + (1 - B_i) M \geq \varphi v_i\left(t_f\right) + \delta, \quad \forall i \in F_C(t) \tag{2.20c}$$

$$x_i\left(t_f\right) - x_C\left(t_f\right) + B_i M \geq \varphi v_C\left(t_f\right) + \delta, \quad \forall i \in F_C(t) \tag{2.20d}$$

$$p^{\mathrm{u}}\left(x_i, v_i, t_f\right) \leq 0 \cup p^{\mathrm{l}}\left(x_i, v_i, t_f\right) \geq 0, \quad \forall i \in F_C(t) \cup \{C\} \tag{2.20e}$$

$$0 \leq v_{\min} \leq v_i\left(t_f\right) \leq v_{\max}, \quad \forall i \in F_C(t) \cup \{C\}. \tag{2.20f}$$

Here, B_i are binary variables indicating if CAV C merges ahead of CAV i and M is a sufficiently large constant. Constraints (2.20a)–(2.20b) enforce rear-end safety at t_f, with (2.20b) for all CAVs on the fast lane. Constraints (2.20c)–(2.20d) determine the optimal cooperative pair for C to merge between. Furthermore, p^{u} and p^{l} in (2.20e) represent the reachable set for a point mass model that constrains the solution set to be finite [37].

By solving this MINLP, each CAV obtains a desired optimal terminal position $x_i(t_f^*)$ and speed $v_i(t_f^*)$. These are then used to formulate a decentralized OCP in accordance with (2.10). This guarantees that the maneuver is completed with minimal disruption. This approach was adopted in [38, 39], and the resulting controllers were implemented and shown to guarantee safety while also providing significant improvements in maneuver times and traffic throughput compared to the selfish controllers that do not account for disruption.

A similar disruption metric can be defined for the cooperative merging problem. However, unlike the lane-changing example above, where the terminal positions are optimized, in the merging case described in Sect. 2.2.2.1 the terminal position for the ego vehicle is fixed given that there exists a pre-defined merging point. Nonetheless, the speeds of the ego and surrounding vehicles at the terminal time can still be optimized to minimize disruption, using a formulation analogous to the lane-changing example.

Thus far, our cooperative controllers have been designed under the assumption that all vehicles are cooperating CAVs. In the next sections, we address the problem of controlling cooperating autonomous vehicles that need to interact with HDVs a various CZs. In particular, we examine the challenge of achieving socially optimal traffic flow objectives in such mixed traffic scenarios, where CAVs must interact with HDVs.

2.3.2 Socially Optimal Cooperative Control in Mixed Traffic

When considering the interaction of CAVs with HDVs, one often adopts a robust optimal control perspective for CAVs, in which HDVs are treated as stochastic obstacles (e.g., [40, 41]). However, this approach still focuses on satisfying individual CAV objectives and constraints rather than overall traffic network optimality. Another approach uses game theory where each HDV solves an optimal control problem with individual objectives and CAV satisfaction depends on the HDV responses (e.g., [18, 42, 43]). However, total network optimality is still not guaranteed.

In the next two subsections, we describe two decentralized approaches to the socially optimal cooperative control problem for CAVs in mixed traffic. First, we return to the optimal merging problem of Sect. 2.2.2.1 and focus on treating HDVs as non-cooperating merging options for CAVs that otherwise cooperate with each other. The problem boils down to generating a *safe merging sequence* for CAVs determined by a higher-level controller. Then, a lower-level decentralized motion controller for each CAV jointly minimizes travel time and energy over a receding horizon, subject to hard safety constraints dependent on the given safe sequence. The second approach is not only based on the cooperation among CAVs but also involves their interaction with HDVs to help the ego CAV perform an optimal lane-changing maneuver as in Sect. 2.2.2.2.

2.3.2.1 Safe Sequencing for Optimal Merging in Mixed Traffic

We return to the merging control setting of Sect. 2.2.2.1 where we now allow mixed traffic consisting of both CAVs and HDVs. As an extension to the single autonomous vehicle (AV) case in [44], our focus here is on suppressing (and possibly eliminating) the risk of safety violations due to the unpredictability of merging HDVs when they are faced by CAVs that conflict with them. To accomplish this, we introduce the concept of *Safe Sequencing* (SS) for merging CAVs which we use as the basis for a technique aiming at the elimination of those cases where a CAV merges in front of an HDV. Instead, we enforce a sequence such that each CAV merges in front of another CAV, hence the two can safely cooperate. To make effective use of this concept, we decompose the complex nonlinear mixed-integer programming problem which defines the joint coordination and motion control of CAVs by implementing a hierarchical control scheme: first, an upper level controller at a local coordinator determines the best possible *safe* sequence at each discrete time step. Next, this information is provided to each CAV which solves a lower-level decentralized optimization problem. We use a Model Predictive Controller (MPC) subject to constraints based on CBFs (overviewed in Sect. 2.2.1.2) to enforce hard safety constraints, thus exploiting their forward invariance property and the computational efficiency resulting from the linearity of the CBF-based constraints. Because CAVs now interact with HDVs, before describing the hierarchical control scheme, one more constraint needs to be added to those presented in Sect. 2.2:

Constraint 2.5 Let i^- be the vehicle that CAV i has to merge ahead of. Therefore, there has to have enough space for CAV i to safely merge ahead of i^-, i.e.,

$$z_{i^-,i}(t_i^m) - \varphi v_{i^-}(t_i^m) - \delta \geq 0. \tag{2.21}$$

This new constraint ensures that the trailing HDV (if present) has enough space to avoid any collision; otherwise, CAV i cannot merge ahead of it and must yield. Note that if $i^- \in F_C(t)$ this constraint is rendered redundant as CAV i is already a safe merging constraint for CAV i^-. Throughout the rest of this subsection, we use the notation of i^+ instead of i_c as the index of the vehicle merges ahead CAV i.

(1) Upper Level Control: Safe Sequencing: At every time step t, the CZ coordinator determines a *safe* (precisely defined below) merging sequence, denoted by s_F, in a centralized manner for all the CAVs inside the SZ (see Fig. 2.1). This is accomplished through a precise algorithm given in [17] where we use s^0 to denote the sequence determined based on Shortest Distance First (SDF) at time t. Additionally, we use $D(s, s^0)$ to measure the "disruption" caused by a safe sequence s relative to the SDF sequence s^0 as one with the least number of changes in indices compared to s^0.

Definition 2.1 (*Safe sequence*) A merging sequence of $i \in F^{SZ}(t)$ is safe if $i^- \notin \{HDV\}$ for all $i \in F_C^{SZ}(t)$.

Details on the determination of the (i^-, i^+) pair and on the disruption measure $D(s, s^0)$ can be found in [17].

(2) Lower-Level Control: Based on the sequence selected by the coordinator (i.e., the specific (i^-, i^+) assigned to each CAV i), a decentralized low-level controller onboard each CAV i adopts a control policy to follow the sequence. In particular, we use an MPC formulation to minimize Objectives 2.1 and 2.2 for the CAV's travel time in the SZ, as well as to satisfy Constraints 2.1, 2.2, 2.4, and 2.5 (since the roads do not have any curvature, Constraint 2.3 is excluded). However, instead of formulating the MPC problem with the original constraints, we replace them with new ones based on CBFs. There are two reasons for this approach: (i) exploiting the forward invariance property of CBFs [33], the CBF-based control ensures that the original constraints are satisfied over the CAV's travel time and (ii) exploiting the CBF constraint linearity in the control, the computational cost of the MPC problem subject to CBFs is significantly lower and enables its online use.

MPC-CBF Control Problem: We discretize the travel time of CAV i using time steps with length T_d. The dynamics in (2.1) are also discretized and we set $\boldsymbol{x}_{i,k} = \boldsymbol{x}_i(t_{i,k})$ where $k = 1, 2, \ldots$ is the kth time step. The decision variables $u_{i,k}$ and $e_{i,k}$ are assumed to be constant over each time step. Thus, the MPC-CBF problem formulation is as follows:

$$\min_{\boldsymbol{x}_{i,k}(\cdot), \boldsymbol{u}_{i,k}(\cdot), \boldsymbol{e}_{i,k}(\cdot)} \sum_{h=0}^{H-1} \left(u_{i,k+h} - u_i^{ref}(t_{i,k+h})\right)^2 + \beta_1 e_{i,k+h}^2 \tag{2.22}$$

$$\text{s.t. } \boldsymbol{x}_{i,k+h+1} = f(\boldsymbol{x}_{i,k+h}, \boldsymbol{u}_{i,k+h}),\ h = 0, \ldots, H-1$$
$$(2.11), (2.13), (2.7) \quad \boldsymbol{x}_{i,k} = \boldsymbol{x}_i(t_{i,k}),$$

where $\boldsymbol{u}_{i,k} = [u_{i,k}, \ldots, u_{i,k+H-1}]$, $\boldsymbol{e}_{i,k} = [e_{i,k}, \ldots, e_{i,k+H-1}]$, H is the number of prediction horizon steps, β_1 is a weight for the CLF constraint in (2.13), and $\boldsymbol{x}_i(t_{i,k})$ is given. The resulting control from (2.22) aims at tracking a given reference control trajectory $u_i^{ref}(t_{i,k})$, $k = 1, 2, \ldots$.

Lastly, due to the unpredictability of HDVs, there is a possibility that safe sequences cannot be achieved. This can be accounted for by considering the awareness zone AZ part of the CZ, which is close to the MP and where vehicles can interact directly with each other to resolve conflicts. Details on the CAV operation within the AZ can be found in [17].

Extensive simulation results for the average performance of vehicles within a CZ as a function of CAV penetration rates are provided in [17], where performance is measured for two different sequencing options: the proposed safe sequencing (SS) and the SDF sequence (note that only CAVs merge following the SDF sequence since HDVs are unpredictable). The performance of the system is evaluated in terms of average travel time, average energy consumption, and average fuel consumption.

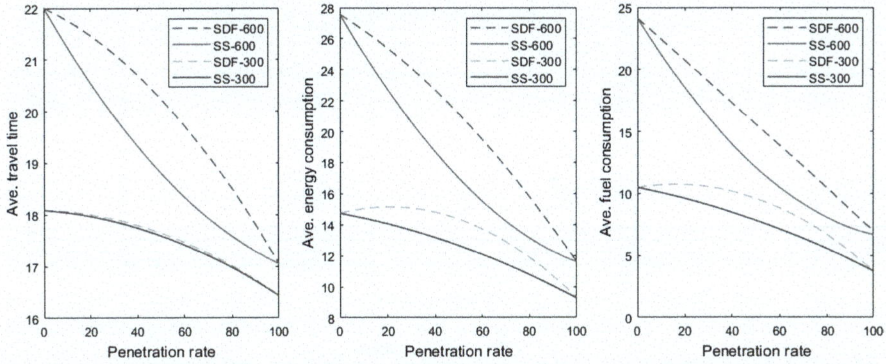

Fig. 2.4 Average travel time, energy consumption, and fuel consumption of vehicles under SDF and SS

Typical simulation results are reproduced from [17] in Fig. 2.4 for 80 vehicles under different arrival rates of 300 and 600 vehicles per hour (see also Table I in [17]).

The significance of these results is twofold: (i) increasing the CAV penetration rate under conventional sequencing policies, such as SDF, can potentially worsen the network performance until the penetration rate exceeds a certain threshold; (ii) even at low penetration rates, the proposed SS scheme can improve the performance of the entire network while favoring safety. Therefore, even at low penetration rates, the benefits of CAVs can be exploited.

It is noteworthy and somewhat surprising in Fig. 2.4 that we observe a uniform performance improvement, despite the conservative nature of SS adopted in order to guarantee safety. However, some insights explaining this behavior are provided in Fig. 2.5 where 3 CAVs and 2 HDVs are present in the SZ similar to Fig. 2.1. Based on the current relative position of the vehicles, the SDF sequence is $\boldsymbol{s}^0 = \{3, 4, 5, 6, 7\}$ in which CAVs with indices 4 and 6 may need to be re-sequenced as vehicles 5 and 7 are HDVs. If we consider the case for which the safety constraints for both CAVs 4 and 6 are violated, i.e., $4^- = 5$ and $6^- = 7$, then based on the SS controller, the new sequence will be $\boldsymbol{s}_F = \{3, 5, 6, 4, 7\}$. The speed profile of the vehicles with the safe sequence is depicted in the lower plot in Fig. 2.5. The observation here is that *the new safe sequence prevents excessive slowing down*, especially in the main road. However, with SDF, after arriving in the AZ, CAV 4 decelerates rapidly coming to a complete halt to make room for HDV 5 since it does not yield to CAV 4, as illustrated in Fig. 2.5. Similarly, since CAV 6 must cooperate with CAV 4, it also decelerates rapidly to allow CAV 4 to merge ahead of it.

2.3.2.2 Robust Lane-Changing Maneuvers in Mixed Traffic

In this section, we revisit the lane-changing maneuver in Sect. 2.2.2.2, where we now assume that vehicle 1 is a CAV while vehicle 2 is an HDV in Fig. 2.2. In

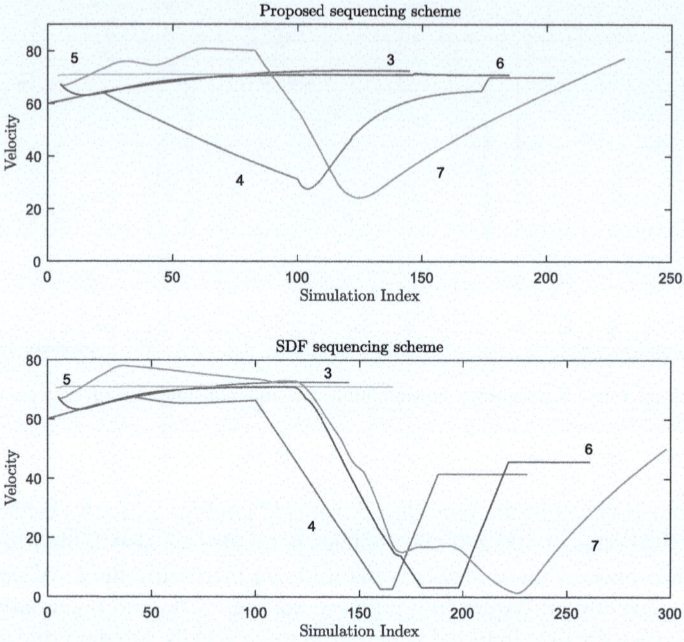

Fig. 2.5 Velocity profile of vehicles present in the SZ under SDF and SS

such a mixed traffic scenario, acquiring accurate information on the surrounding vehicles for CAVs is necessary to guarantee safety and efficiency. However, this is rendered very challenging in the presence of uncontrollable HDVs whose behavior is unpredictable: not only can an HDV be uncooperative, but it may even assume an adversarial position toward a CAV attempting to merge ahead of it. Our approach is to control an ego CAV performing a maneuver by always cooperating with surrounding CAVs and achieving a safe and optimal maneuver *regardless of HDV behavior*. In what follows, the optimal lane-change trajectories for CAV C in a mixed traffic setting are determined based on its interaction with HDVs. We will show how it is possible for C to merge ahead of either the HDV or CAV 1 to guarantee safety while also achieving optimality and minimum disruption to the traffic flow. To simplify the problem, the initial time is set at t_1 where $x_H(t_1) = x_C(t_1)$ to formulate vehicle interactions.

Case 1: CAV C Merges Ahead of HDV: Given that the HDV is uncontrollable, we estimate its trajectory by assuming that a human driver considers three factors: (i) maintaining a constant speed that minimally deviates from some desired value $v_{d,H}$; (ii) if it needs to change speeds, it does so by minimizing its acceleration/deceleration, which also saves fuel; and (iii) guaranteeing its safety (collision avoidance). To model the latter, we define a *risk function* $s(\cdot)$ as a decreasing function with respect to the vehicle distances $x_C(t) - x_H(t)$ since a closer distance between

HDV H and CAV C corresponds to a higher collision risk. We adopt the sigmoid function:

$$s(x_C(t) - x_H(t)) = \frac{1}{1 + \mu \exp\left(\mu(x_C(t) - x_H(t) - d)\right)}, \tag{2.23}$$

where μ and d are adjustable parameters to capture different unsafe regions for different drivers and define the size of the unsafe region, respectively.

The interaction between the ego CAV and the HDV is formulated by using a bilevel optimization setting with an appropriate behavioral model for an HDV. An Iterated Best Response (IBR) method is used to determine a Nash equilibrium as described next:

(1) Lower-Level Problem: The estimated trajectory of the HDV is obtained by jointly minimizing Objectives 2.1 and 2.2, the speed deviation, and the collision risk while guaranteeing the rear-end safety in Constraint 2.1. The corresponding problem formulation is as follows:

$$J_{lower} = \min_{u_H(t)} \int_{t_1}^{t_f^*} \left[\frac{\beta_u}{2} u_H^2(t) + \beta_v(v_H(t) - v_{d,H})^2 + \beta_s s(x_C^*(t) - x_H(t))\right] dt \tag{2.24}$$

$$s.t.\ (2.1),\ (2.7),\ x_1^*(t) - x_H(t) \geq d_H(v_H(t)),$$

where $\beta_{\{u,v,s\}}$ are the non-negative appropriately normalized weights that describe the characteristics of the HDV, i.e., the behavior of the driver. We immediately note that the optimal positions $x_C^*(t)$ and $x_1^*(t)$ for CAVs C and 1 are unknown to the HDV. In fact, these are determined by the upper level problem.

(2) Upper Level Problem: Based on the estimated HDV behavior, the optimal trajectories of CAVs 1 and C can be obtained by jointly minimizing Objectives 2.1–2.2 and the speed deviation relative to the traffic flow. The corresponding problem formulation is

$$\min_{u_C(t), u_1(t)} \int_{t_1}^{t_f^*} \frac{\alpha_u}{2}(u_C^2 + u_1^2)dt + \alpha_v[(v_C(t_f^*) - v_{d,C})^2 + (v_1(t_f^*) - v_{d,1})^2] \tag{2.25a}$$

$$s.t.\ (2.1),\ (2.7),$$

$$x_C(t_f^*) - x_H^*(t_f^*) \geq d_H\left(v_H^*\left(t_f^*\right)\right) \tag{2.25b}$$

$$x_1(t_f^*) - x_C^*(t_f^*) \geq d_C\left(v_C^*\left(t_f^*\right)\right) \tag{2.25c}$$

$$u_H^*(t) = \arg\min_{u_H(t)} \ J_{lower}, \tag{2.25d}$$

where $\alpha_{\{u,v\}}$ are the non-negative appropriately normalized weights; $v_{d,C}$, $v_{d,1}$ are desired speed for C and 1, respectively; and the optimal trajectories $[x_H^*(t), v_H^*(t)]$ of the HDV are obtained from the optimal control $u_H^*(t)$ in the lower-level problem (2.24). The solution to this bilevel optimization problem is complicated but it can be obtained by applying the Iterated Best Response (IBR) method outlined next.

Iterated Best Response (IBR) Process: We decentralize problem (2.25a) for CAVs 1 and C to optimize their individual objectives, respectively. Given the initial states $[x_i, v_i]$, $i = 1, C, H$, the process starts from the HDV problem (2.24) and proceeds to each CAV's optimal problem based on the results of (2.24) to complete one iteration round. A more detailed explanation of the IBR process can be found in [18].

Case 2: CAV *C* Merges Ahead of CAV 1: If CAV C is to merge ahead of the cooperating CAV 1, then the HDV is constrained to merely "follow" C. Thus, in this case the cooperation of CAVs is robust to the HDV, making *the HDV behavior irrelevant while safety can still be guaranteed.* The optimal trajectories of CAVs can be solved by solving OCPs which jointly minimize Objectives 2.1–2.2 and the speed deviation with respect to the rear-end safety Constraint 2.1.

It remains to compare the total costs of the two control policies for CAV C: "merge ahead of HDV" and "merge ahead of CAV 1." This decision is based on the initial distance $d_{H,1} := x_1(t_1) - x_H(t_1)$ between the lane-changing CAV C and the rear vehicle (HDV) in the target lane where C merges. Simulation results obtained in [18] are shown in Fig. 2.6 where we can see how this increasing initial distance affects the optimal decision for C to merge ahead of HDV or CAV 1. When the initial gap increases, the optimal policy switches from "merge ahead of CAV 1" to "merge ahead of HDV" if C aims to complete a minimal-cost maneuver. In fact, due to the monotonicity of the cost of the "merge ahead of CAV 1" policy, there exists an optimal threshold determining when CAV C should switch from one policy to another based only on an observation of the distance $d_{H,1}$. Moreover, by taking these two options into consideration, the robustness of CAV C to perform a feasible lane-changing maneuver can be guaranteed by either cooperating with another CAV to get ahead of it or choosing to get ahead of the HDV when the distance $d_{H,1}$ is sufficiently large to ensure a safe lane-changing maneuver regardless of the HDV's behavior. This framework can also be extended to other CZs, such as merging, intersections, and roundabouts.

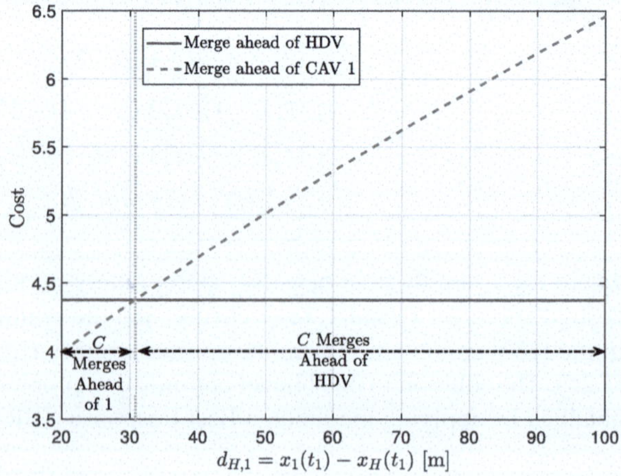

Fig. 2.6 Optimal policy determination for CAV C

2.4 Incentive Mechanisms for Cooperation Compliance in Mixed Traffic

In this section, we address the problem of HDV cooperation compliance by encouraging them to cooperate with each other or with CAVs through some incentive mechanism. This facilitates the coordination between CAVs and HDVs, improving overall traffic flow. A key variable in this framework is the probability that human drivers comply with social contracts or recommendations issued by a social planner; this "compliance rate" greatly influences the performance of the traffic network. We proceed by considering two examples of CAV–HDV cooperation in a mixed traffic environment. The first one illustrates the importance of vehicle cooperation by quantifying how the overall traffic performance varies with respect to different HDV compliance rates. The second example provides a feedback control mechanism through which potentially uncooperative vehicles improve their overall compliance rates.

Example 2.2 (*Effect of Compliance Rates in Adaptive Cruise Control*) Consider an Adaptive Cruise Control (ACC) problem, where five vehicles form a platoon. The leader of the platoon is an HDV and the four following vehicles are CAVs. There is a social planner that provides the optimal policy for the platoon to travel through a road segment with length L. The compliance rate for the HDV is $p_H \in [0, 1]$, i.e., the HDV will follow the social planner's guidance with probability p_H. On the other hand, the CAVs comply with all requests from the social planner. This example aims to investigate the effect of the HDV's compliance rate on system performance measured through Objectives 2.1 and 2.2, i.e., minimizing travel time and energy while guaranteeing safety.

The HDV's dynamics are defined as in (2.1) with $x, v \in \mathbb{R}$ denoting the vehicle's position and speed, respectively. Letting the state vector be $\mathbf{x} = [x, v]^T$, the error $\mathbf{e}_H$ between the real HDV trajectory $\mathbf{x}_H^{real}$ and the optimal trajectory $\mathbf{x}_H^{ref}$ satisfies the following relationship:

$$\mathbf{x}_H^{real}(t) = \mathbf{x}_H^{ref}(t) + (1 - p_H)\mathbf{e}_H(t), \tag{2.26}$$

where the state error is bounded by $||\mathbf{e}_H(t)|| \leq \mathbf{e}_{\max}$. The CAVs are assumed to follow the HDV based on the Intelligent Driver Model (IDM) [45] with dynamics given by

$$\dot{x}_i = v_i, \quad \dot{v}_i = u_{\max}\left[1 - \left(\frac{v_i}{v_d}\right)^{\epsilon} - \left(\frac{s_i^*}{s_i}\right)\right], \tag{2.27}$$

where v_d is the desired speed of i, ϵ is the acceleration exponent constant (typically set $\epsilon = 4$). s_i^*, s_i denote the optimal and current distance between i and its preceding vehicle $i - 1$, respectively, and s_i^* is given as

$$s_i^* = s_0 + \max\left(0, \varphi v_i + \frac{v_i \Delta v_i}{2\sqrt{u_{\max} b}}\right),$$

where s_0 is the minimum safety distance for vehicle i to maintain safety and avoid rear-end collision, φ is the headway time, Δv_i is the speed difference between vehicles i and $i-1$, and $b > 0$ is a comfortable deceleration constant. Considering the stochastic nature of the HDV trajectory with different driver compliance rates, in order to guarantee rear-end safety constraint (2.4), we define the minimum safety distance s_0 as

$$s_0(t) = \varphi v_i(t) + \delta + (1 - p_H)e_{\max}. \tag{2.28}$$

Based on the above setting, we assume that $L = 3\,\text{Km}$ and the optimal policy for the platoon is to travel with constant speed $v_d = 30\,\text{m/s}$. The initial speed for all vehicles is set as 20 m/s, and the initial distance between each two vehicles is 50 m. While CAVs comply with the social planner's guidance, we vary the compliance probability p_H for the HDV and the effect in the total system cost (travel time and energy) is shown in Fig. 2.7. We can see that the cost is drastically decreasing when increasing p_H from 0.1 to 0.8, hence a higher compliance rate corresponds to a lower overall cost for the platoon. Therefore, the HDV's cooperation compliance expressed through its compliance probability in this traffic scenario is critical to the entire system's performance.

To achieve the goal of improving the compliance rates of HDVs, we propose an incentive mechanism similar to the one in [46] (see also [47–49]) as illustrated in the following example.

Example 2.3 (*Cooperation Compliance Control in Lane Changing*) Let us return to the lane-changing maneuver setting by considering n uncontrollable HDVs on a road segment. When one of them (the ego vehicle) wishes to change lanes, it is

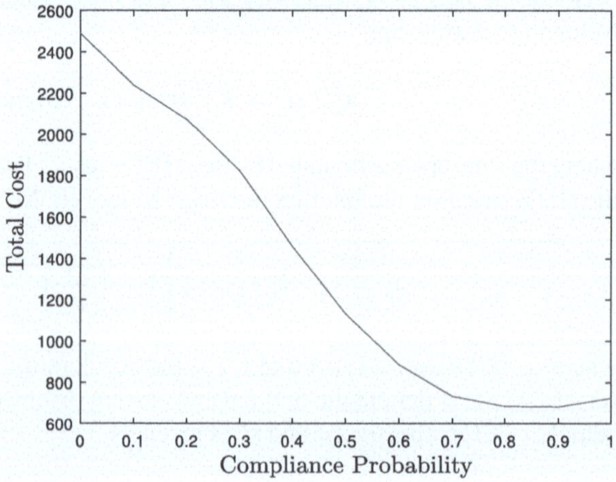

Fig. 2.7 Total cost of system under different fixed compliance probability of HDV

often the case that such a maneuver requires cooperation from its neighbors, as we saw in Sect. 2.2.2.2, i.e., at least one vehicle must create adequate space for the ego vehicle to merge ahead of it when a request for such a maneuver is made (e.g., by the use of a blinker or through some simple form of V2V communication). To describe the cooperation compliance of such vehicles, we define binary random variables $\{M_i(k)\}_{i=1}^n$ over discrete time instant k where $M_i(k) = 1$ means that vehicle i complies with the request at time k. Our goal is to improve the compliance probability $Q(k) := P(M_i(k) = 1)$ through a *refundable toll* concept that operates as follows: (i) Every user (driver) owns a digital wallet consisting of virtual tokens that may only be used in the context of the cooperation compliance process. (ii) A user commits a number of tokens at the start of a given period, i.e., pays a *toll* in advance. (iii) The toll is fully refundable as long as the user complies with every cooperation request received. Otherwise, it is subject to a feedback control mechanism that updates the number of tokens as detailed next.

We define a global-level toll $C(k)$ that every user contributes toward global (social) optimality and an additional local-level (class i) toll $c_i(k)$ that every member of a given user class i contributes toward his/her individual (selfish) optimality. These two toll components are updated as follows:

$$C(k+1) = C(k) + \alpha(Q^* - \frac{1}{n}\sum_{i=1}^{n} M_i(k)), \tag{2.29}$$

$$c_i(k+1) = c_i(k) + \beta(Q_i^* - \bar{M}_i(k)), \quad s.t. \ \frac{1}{n}\sum_{i=1}^{n} Q_i^* = Q^*, \tag{2.30}$$

where $\bar{M}_i(k) = (1-\gamma)\sum_{j=1}^{k} \gamma^{k-j} M_i(j)$ denotes the windowed time average of the compliance probability of vehicle i and $\gamma \in (0, 1)$ adjusts the length of the window used for this average. Q^*, Q_i^* denote the desired compliance probability for the entire traffic system and for vehicle i, respectively. The controllable parameters α, β are selected appropriately for the variables $C(k)$, $c_i(k)$ to converge to the desired compliance probabilities.

The compliance probability function $Q(k) := P(M_i(k) = 1)$ is assumed to have the form $P(M_i(k+1) = 1) = p(a_1 q_i + a_2 C(k) + a_3 c_i(k))$, where $p : \mathbb{R} \to [0, 1]$ is a monotone increasing function, and a_1, a_2, a_3 are weighted parameters to balance the initial compliance proclivity of a user and the relative importance of the global and local tolls.

The results of applying this refundable toll scheme are shown in Figs. 2.8 and 2.9, where we have set $Q^* = Q_i^* = 0.9$, $\alpha = 0.025$, $\beta = 0.01$, $n = 1000$, and $a_1 = 0.2$, $a_2 = a_3 = 0.4$, while the probability function is selected to be of the form $p(x) = \text{mid}\{0, x, 1\}$. As seen in the figures, the two-component feedback mechanism increases the toll values and ultimately leads to the convergence of the compliance rates to their desired values of 0.9. Specifically, with different class toll $c_i(k)$, the converged compliance probability for each class is also different while the overall compliance converges to Q^*. It is worth stressing that ignoring the local or global

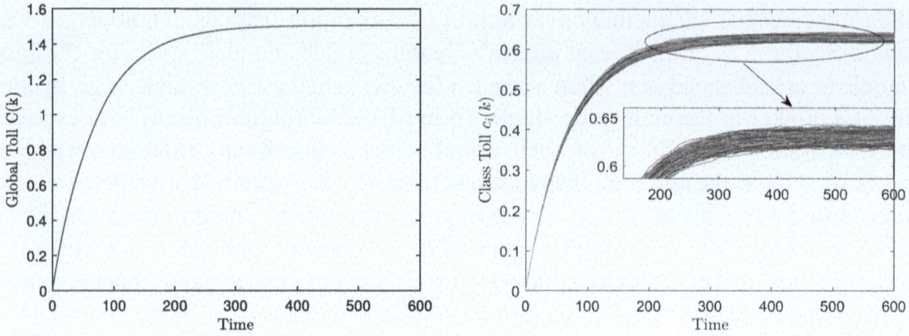

Fig. 2.8 Global toll $C(k)$ (left) and class tolls $c_i(k)$ (right)

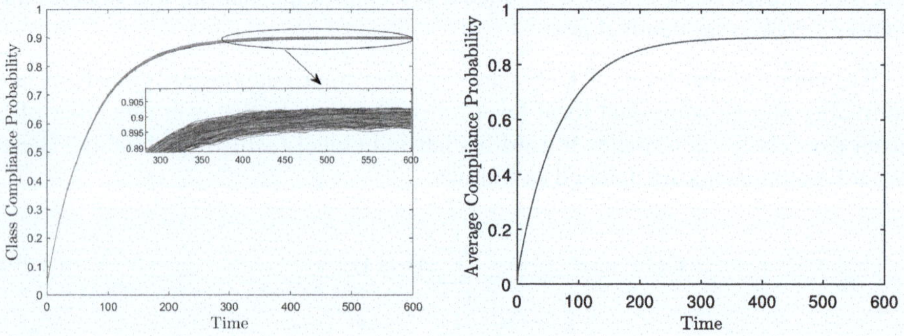

Fig. 2.9 Class compliance probabilities (left) and overall average compliance rate for $n = 1000$ vehicles (right)

toll component may not lead to such convergence as it fails to balance the trade-off between the social and selfish effects of a user's compliance rate considerations.

With the above framework, drivers are incentivized to cooperate with CAVs or each other or with a social planner setting the desired compliance rates for any given operation time period. As already mentioned, the token-based tolls are fully refundable as long as a vehicle remains in compliance, otherwise, the resulting token loss reduces its priority over access to the traffic system resources.

2.5 Conclusion

This chapter has provided an overview of cooperative control for Connected Autonomous Vehicles (CAVs) at conflict zones like intersections and highway merges. Optimal Control Problems (OCPs) with hard safety constraints have been formulated and solved to generate cooperative trajectories. However, the computational complexity of such problems motivates the use of Control Barrier Functions

(CBFs), which guarantee safety at the possible expense of performance. Building on these methods for traffic networks of only CAVs, the challenges of mixed traffic have been addressed, where CAVs interact with unpredictable Human-Driven Vehicles (HDVs). To handle such environments, new approaches have been introduced to achieve social optimality while robustly maintaining safety. Furthermore, a cooperative compliance framework has been proposed to incentivize HDVs to align behavior with CAV objectives. This utilizes refundable digital tokens without monetary transactions.

Looking ahead, research will continue to focus on enabling socially optimal automated mobility through new methods that enhance safety, accessibility, and equity in transportation systems.

References

1. Papageorgiou M, Diakaki C, Dinopoulou V, Kotsialos A, Wang Y (2003) Review of road traffic control strategies. Proc IEEE 91(12):2043–2067
2. Wen D, Li L, Yao D (2013) A survey of traffic control with vehicular communications. IEEE Trans Intell Trans Syst 15(1):425–432
3. Lee J, Park B (2012) Development and evaluation of a cooperative vehicle intersection control algorithm under the connected vehicles environment. IEEE Trans Intell Transp Syst 13(1):81–90
4. Xu H, Zhang K, Zhang D (2016) Multi-level traffic control at large four-leg roundabouts. J Adv Transp 50:10
5. Zhao L, Malikopoulos A, Rios-Torres J (2018) Optimal control of connected and automated vehicles at roundabouts: an investigation in a mixed-traffic environment. IFAC-PapersOnLine 51(9):73–78. 15th IFAC Symposium on Control in Transportation Systems CTS 2018
6. Rios-Torres J, Malikopoulos AA (2017) A survey on the coordination of connected and automated vehicles at intersections and merging at highway on-ramps. IEEE Trans Intell Transp Syst 18(5):1066–1077
7. Zheng F, Liu C, Liu X, Jabari SE, Lu L (2020) Analyzing the impact of automated vehicles on uncertainty and stability of the mixed traffic flow. Trans Res Part C: Emerg Technol 112:203–219
8. Giammarino V, Baldi S, Frasca P, Monache M (2021) Traffic flow on a ring with a single autonomous vehicle: an interconnected stability perspective. IEEE Trans Intell Transp Syst 22(8):4998–5008
9. Delle P, Pricing S (2021) of connected and autonomous vehicles in mixed-traffic networks. Trans Res Record 2675(5):178–192
10. Olstam J, Tapani A (2004) Comparison of Car-following models, vol 960. Swedish National Road and Transport Research Institute Linköping, Sweden
11. Munigety CR, Mathew TV (2016) Towards behavioral modeling of drivers in mixed traffic conditions. Transp Devel Econ 2:1–20
12. Yao H, Li X (2020) Decentralized control of connected automated vehicle trajectories in mixed traffic at an isolated signalized intersection. Trans Res Part C: Emerg Technol 121:102846
13. Li Q, Li X, Huang Z, Halkias J, McHale G, James R (2022) Simulation of mixed traffic with cooperative lane changes. Comput-Aid Civil Infrastruct Eng 37(15):1978–1996
14. Zhao Z, Wang Z, Wu G, Ye F, Barth M (2019) The state-of-the-art of coordinated ramp control with mixed traffic conditions, pp 1741–1748
15. Liao X, Zhao X, Wang Z, Han K, Tiwari P, Barth MJ, Wu G (2022) Game theory-based ramp merging for mixed traffic with unity-sumo co-simulation. IEEE Trans Syst Man Cybern: Syst 52(9):5746–5757

16. Liu H, Zhuang W, Yin G, Li Z, Cao D (2023) Safety-critical and flexible cooperative on-ramp merging control of connected and automated vehicles in mixed traffic. IEEE Trans Intell Transp Syst 24(3):2920–2934
17. Sabouni E, Ahmad HM, Cassandras CG, Li W (2023) Merging control in mixed traffic with safety guarantees: a safe sequencing policy with optimal motion control. In: 26th IEEE conference on intelligent transportation systems ITSC. (To appear). arXiv:2305.16725
18. Li A, Armijos ASC, Cassandras CG (2023) Cooperative lane changing in mixed traffic can be robust to human driver behavior. In: 62nd IEEE conference on decision and control. (to appear). arXiv:2303.16948
19. Xiao W, Cassandras CG, Belta C (2021) Bridging the gap between optimal trajectory planning and safety-critical control with applications to autonomous vehicles. Automatica 129:109592
20. Vogel K (2003) A comparison of headway and time to collision as safety indicators. Accid Anal & Prev 35(3):427–433
21. Xiao W, Cassandras CG, Belta C (2020) Decentralized optimal control in multi-lane merging for connected and automated vehicles. In: 2020 IEEE 23rd international conference on intelligent transportation systems (ITSC), pp 1–6
22. Xu H, Xiao W, Cassandras CG, Yi Z, Li L (2022) A general framework for decentralized safe optimal control of connected and automated vehicles in multi-lane signal-free intersections. IEEE Trans Intell Transp Syst 23(10):17382–17396
23. Xu K, Cassandras CG, Xiao W (2021) Decentralized time and energy-optimal control of connected and automated vehicles in a roundabout. In: 2021 IEEE international intelligent transportation systems conference (ITSC), pp 681–686
24. Xu H, Cassandras CG, Li L, Zhang Y (2022) Comparison of cooperative driving strategies for cavs at signal-free intersections. IEEE Trans Intell Transp Syst 23(7):7614–7627
25. Zhuang H, Lei C, Chen Y, Tan X (2023) Cooperative decision-making for mixed traffic at an unsignalized intersection based on multi-agent reinforcement learning. Appl Sci 13(8):5018
26. Jiang L, Xie Y, Evans NG, Wen X, Li T, Chen D (2022) Reinforcement learning based cooperative longitudinal control for reducing traffic oscillations and improving platoon stability. Transp Res Part C: Emerg Technol 141:103744
27. Wu C, Kreidieh A, Vinitsky E, Bayen AM (2017) Emergent behaviors in mixed-autonomy traffic. In: Conference on robot learning. PMLR, pp 398–407
28. Rios-Torres J, Malikopoulos AA (2016) Automated and cooperative vehicle merging at highway on-ramps. IEEE Trans Intell Transp Syst 18(4):780–789
29. Malikopoulos AA, Cassandras CG, Zhang YJ (2018) A decentralized energy-optimal control framework for connected automated vehicles at signal-free intersections. Automatica 93:244–256
30. Xu H, Zhang Y, Li L, Li W (2019) Cooperative driving at unsignalized intersections using tree search. IEEE Trans Intell Transp Syst 21(11):4563–4571
31. Xiao W, Cassandras CG (2021) Decentralized optimal merging control for connected and automated vehicles with safety constraint guarantees. Automatica 123:109333
32. Zhang Y, Cassandras CG (2019) Decentralized optimal control of connected automated vehicles at signal-free intersections including comfort-constrained turns and safety guarantees. Automatica 109:108563
33. Xiao W, Cassandras CG, Belta C (2023) Safe autonomy with control barrier functions: theory and applications. Synthesis lectures on computer science. Springer International Publishing
34. Xiao W, Cassandras CG (2021) Decentralized optimal merging control for connected and automated vehicles on curved roads. In: 2021 60th IEEE conference on decision and control (CDC), pp 2677–2682
35. Chen R, Cassandras CG, Tahmasbi-Sarvestani A, Saigusa S, Mahjoub HN, Al-Nadawi YK (2020) Cooperative time and energy-optimal lane change maneuvers for connected automated vehicles. IEEE Trans Intell Transp Syst 23(4):3445–3460
36. Yanumula V, Typaldos P, Troullinos D, Malekzadeh M, Papamichail I, Papageorgiou M (2023) Optimal trajectory planning for connected and automated vehicles in lane-free traffic with vehicle nudging. IEEE Trans Intell Veh 8(3):2385–2399

37. Haddad S, Halder A (2022) Boundary and taxonomy of integrator reach sets. In: Proceedings of the American control conference. IEEE, pp 4133–4138
38. Chalaki B, Tadiparthi V, Mahjoub HN, D'sa J, Moradi-Pari E, Armijos ASC, Li A, Cassandras CG (2023) Minimally disruptive cooperative lane-change maneuvers. IEEE Control Syst Lett
39. Armijos ASC, Li A, Cassandras CG (2023) Maximizing safety and efficiency for cooperative lane-changing: a minimally disruptive approach. In: 26th IEEE conference on intelligent transportation systems ITSC. (To appear) arXiv:2305.17883
40. Nakamura K, Bansal S (2023) Online update of safety assurances using confidence-based predictions. In: 2023 IEEE international conference on robotics and automation (ICRA). IEEE, pp 12765–12771
41. Nyberg T, Pek C, Dal Col L, Norén C, Tumova J (2021) Risk-aware motion planning for autonomous vehicles with safety specifications. In: 2021 IEEE intelligent vehicles symposium (iv). IEEE, pp 1016–1023
42. Le V, Malikopoulos AA (2023) Optimal weight adaptation of model predictive control for connected and automated vehicles in mixed traffic with Bayesian optimization. In: 2023 American control conference (ACC). IEEE, pp 1183–1188
43. Buckman N, Pierson A, Schwarting W, Karaman S, Rus D (2019) Sharing is caring: socially-compliant autonomous intersection negotiation. In: 2019 IEEE/RSJ international conference on intelligent robots and systems (IROS). IEEE, pp 6136–6143
44. Sabouni E, Cassandras CG (2023) Optimal merging control of an autonomous vehicle in mixed traffic: an optimal index policy. In: 22nd IFAC world congress, Yokohama, Japan, July 9–14
45. Treiber M, Hennecke A, Helbing D (2000) Congested traffic states in empirical observations and microscopic simulations. Phys Rev E 62(2):1805
46. Ferraro P, Zhao L, King C, Shorten R (2023) Personalised feedback control, social contracts, and compliance strategies for ensembles. IEEE Internet Things J
47. Fioravanti AR, Mareček J, Shorten RN, Souza M, Wirth FR (2019) On the ergodic control of ensembles. Automatica 108:108483
48. Elokda E, Cenedese C, Zhang K, Lygeros J, Dörfler F (2022) Carma: fair and efficient bottleneck congestion management with karma. arXiv:2208.07113
49. Shakarami M, Cherukuri A, Monshizadeh N (2022) Steering the aggregative behavior of non-cooperative agents: a nudge framework. Automatica 136:110003

Chapter 3
Control of Traffic Networks Exploiting Vehicle Platooning: State of the Art, Opportunities, and Challenges

Cecilia Pasquale, Silvia Siri, and Simona Sacone

Abstract Connected and automated vehicles represent a technological breakthrough that can change the concept of mobility, paving the way for a new paradigm of traffic control in which the vehicles themselves can be used to implement traffic control strategies. Indeed, early field experiments have shown that even a small percentage of these vehicles traveling together in the form of *platoons* acts by limiting the occurrence of traffic instabilities that lead to congestion. In this context, the development of *prediction models* able to represent the mutual interactions between platoons and vehicular traffic is a key factor. It is then possible to develop model-based control schemes in which the speed of platoons is the adopted control action. The architectures of these control schemes can be of different types, with the control actions defined by a single controller with complete knowledge of the system state, giving rise to *centralized control schemes*, or with the control problem divided into subproblems, in *decentralized control schemes*, which are slightly less performant but more computationally efficient. In turn, these control schemes may involve the adoption of a *hierarchical architecture* in which different levels of control are defined, making the overall control framework more suitable for practical applications.

3.1 Introduction

In recent years, we are witnessing a new wave of innovation that is leading to the development and production of vehicles with an increasingly high level of automation and communication technology. Vehicles of the future can be imagined as

C. Pasquale · S. Siri · S. Sacone (✉)
Department of Informatics, Bioengineering, Robotics and Systems Engineering (DIBRIS), University of Genova, Genoa, Italy
e-mail: simona.sacone@unige.it

C. Pasquale
e-mail: cecilia.pasquale@unige.it

S. Siri
e-mail: silvia.siri@unige.it

P. Ioannou and A. A. Malikopoulos (eds.), *Transportation Mobility in Smart Cities*, Springer Tracts on Transportation and Traffic 21,
https://doi.org/10.1007/978-3-031-64769-7_3

increasingly connected, both with each other and with infrastructure, and with some autonomous driving capabilities.

These innovations represent a technological breakthrough that is not limited to the automotive sector alone, but will also change the concept of mobility as we know it. Indeed, these vehicles, if properly designed and controlled, can bring significant benefits to society both in terms of efficiency, implementing the concept of mobility as a service, and in terms of the sustainability of road transport, exploiting the potential these vehicles have in reducing accidents and energy consumption. These aspects have been thoroughly analyzed in [1, 2]; however, the effective impacts due to the massive introduction of these vehicles in a real-world context are not yet completely clear, and thus when traditional vehicles and CAVs coexist in mixed traffic, they will require new traffic management strategies.

One of the first real-world applications of these intelligent vehicles involves the formation of platoons. Platooning consists of forming a convoy of vehicles traveling in close formation to achieve some common goal. The idea of platooning originated in the 1950s by studying the behavior of vehicles following a leading vehicle by means of simple models [3], while early practical implementations have involved the creation of convoys of trucks traveling less than a second apart thanks to the adoption of wireless communication systems, radar equipment, and intelligent communication system technologies [1].

Freight transport has been one of the first sectors to seize the opportunities offered by vehicle platooning, as this travel technique substantially reduces aerodynamic drag, allowing significant fuel savings for trucks traveling within the platoon [4, 5]. Due to these advantages, many efforts have been made to prove the practical applicability of truck platooning, which has involved the execution of test cases, such as the one conducted in the German freeway network [6], and the development of research projects (see, for example, [7–9]). Traditionally, scientific research in this area has focused on two lines of research dealing with truck platoon planning and control, respectively.

On the planning side, the goal is to define the cooperation policies and the plans to implement them so that all platoon members can benefit directly or indirectly from the collaboration. Many works have been conducted in this direction, mainly with the aim of reducing fuel consumption (see, for example, [10]) or maximizing the number of vehicles in the platoon. For a complete survey on truck platoon planning see [11].

Truck platoon control strategies, in turn, can be distinguished into several research directions. The first approaches have concerned, on the one hand, the definition of control actions aimed at achieving the objectives established at the planning level [12, 13] and, on the other hand, the definition of control actions, concerning both lateral and longitudinal control [14], that allows tracking the trajectory of the vehicle in front, while maintaining a predetermined intra-vehicular distance [15] and ensuring string stability properties [16].

This chapter focuses on a third line of research, which involves the use of platoons as mobile actuators of freeway control strategies. As it is well known, freeway traffic management has been typically addressed through the definition and implementation of *road-based traffic control* schemes. In these schemes, control actions are defined based on traffic conditions that are detected through measurement systems located along the freeways and implemented through fixed actuators. The most popular road-based freeway traffic control methodologies are those involving ramp metering control, mainstream control, and route guidance (for more details, the interested reader may refer to the survey in [17]). All these methodologies have proven to be very effective in limiting the occurrence of congestion when due to recurring factors, while they perform less well when congestion is due to nonrecurring causes.

The idea behind *vehicle-based traffic control*, in this case actuated by vehicle platoons, is to provide more dynamic and adaptive solutions to deal with congestion that can appear anywhere in the freeway network. Vehicle-based control strategies, in particular, are mainly inspired by mainstream control strategies implemented with Variable Speed Limits (VSLs). The purpose of VSLs control is to suggest a speed, which is displayed through variable message signs, that is lower than the speed that would be implemented by drivers in the absence of control, thus allowing the inflow into a possible bottleneck to be controlled so as not to exceed the capacity of the bottleneck itself, and thus preventing the onset of congestion. In practical implementation, the effectiveness of VSLs is limited by the willingness of drivers to comply with the suggested speed and the location where the variable message signs are placed.

In control schemes where platoons are actuators, the speed and/or lanes to be occupied are the control variables used to avoid the activation of bottlenecks. Similar to control with VSLs, the control implemented by platoons can be defined based on prediction traffic models. Note that these models must include the presence of the platoons themselves in order to evaluate their impact on vehicular traffic.

The idea of platoon-actuated control is very promising because it appears to be more flexible, as it can handle both ordinary and non-ordinary congestion, and potentially more effective, as drivers are forced to follow the speed considered optimal to minimize congestion, whereas in traditional road-based speed control, they generally only receive not compulsory speed indications.

The purpose of this chapter is to provide an overview of traffic control strategies that exploit vehicle platooning, paying particular attention to the new opportunities that this type of control can offer, but also to the challenges that must be faced before these control methods become common practice. To this end, in Sect. 3.2, a brief survey on traffic models with platoons is provided. In particular, two macroscopic discrete time-space traffic models including platoons are presented in detail. Section 3.3 proposes a review of control schemes in which platoons are used as actuators, while Sect. 3.4 provides some insights about the opportunities and challenges related to traffic control with platoons.

3.2 A Brief Overview of Traffic Models with Platoons

Connected and Automated Vehicles (CAVs) and even self-driving vehicles represent a unique chance to make road transport safer, more efficient, and greener. Indeed, studies and tests realized in this direction show that even a limited percentage of CAVs in vehicular traffic may act by regulating traffic movements and limiting the circumstances that lead to congestion (see, for instance, [18]). These preliminary experiments suggest the idea that controlling clusters of CAVs arranged in platoon formations can result in benefits for the entire traffic flow. Yet, since the real-world applications of these intelligent vehicles are almost limited to testing in protected environments with prototype vehicles, appropriate models are needed to evaluate the interactions of these vehicles with the rest of traffic in large-scale networks.

This section presents a review of the main modeling approaches involving elements comparable to platoons of CAVs. Since the primary aim of this chapter is to investigate the use of platoons to control traffic as a whole, the main approaches belonging to the category of macroscopic models will be analyzed below, while microscopic [19, 20] and mesoscopic [21] models with platoons will not be the subject of this chapter.

Early modeling approaches of this type originated from the idea of modeling the impact of one or more slow-moving vehicles in traffic by representing them as moving bottlenecks. One of the first approaches in this direction dates back to the late 1990s, [22], and modifies the well-known first-order continuous macroscopic Lighthill–Whitham–Richards model [23, 24] to include the presence of buses that temporarily reduce the capacity of the roadway while they are passing. Starting from this work, other studies have been carried on in this direction, extending both first-order [25–27] and second-order [28] continuous traffic models. The main peculiarity of the approaches [22–28] is that the slow vehicles are represented as pointlike moving bottlenecks and modeled as flow constraints. More recent works extend these previous works by explicitly considering the platoon length, as done, for instance, in [29]. The work in [29] proposes a novel PDE-ODE model in which the equations of the Lighthill–Whitham–Richards model are coupled with two ordinary differential equations describing the trajectory of platoons. Further extensions of the PDE-ODE model of [29] are presented in [30], to consider the presence of platoons of CAVs, and in [31], to represent platoons of electric CAVs.

Other traffic models oriented to the control of platoons of vehicles have been developed starting from discrete time-space models. In [33], an extension of the well-known second-order macroscopic traffic model METANET [32] is proposed to include platoons of trucks. Specifically, [33] presents a hybrid model, named Micro-Macro-METANET model, which combines aspects related to the platoon of connected and automated trucks and other macroscopic aspects related to the traffic flow as a whole. Some other approaches extend the discrete time-space first-order Cell Transmission Model (CTM) [34] to tackle the presence of platoons. For example, in works [35–37] the dynamic equations of the CTM are enriched to consider the presence of multiple platoons, while in [38, 39] the CTM is combined with tandem queuing models [40] to represent the stochastic arrivals of platoons in a freeway stretch.

In the following sections, two discrete time-space traffic models with platoons, of the first and second order, are given as examples.

3.2.1 *A Discrete Time-Space First-Order Traffic Model with Platoons*

The following version of the CTM is the one proposed by [35] to consider the presence of multiple platoons. More specifically, in this model, the platoons are represented as moving bottlenecks traveling in a freeway stretch between entering and exiting ramps. As in the original CTM model, the stretch is discretized both in time and in space. Regarding time discretization, the time horizon is divided into K steps with T the sampling time [h]. The space discretization has been done by dividing the stretch length into N portions, called cells, whose length, in [km], is denoted with L. Unlike the original model, this length is defined so that the presence of a platoon in a cell can affect a set of neighboring vehicles that are in the cell, while not violating the condition $T v_i^{\text{free}} \leq L$, in which v_i^{free} [km/h] is the free-flow speed of cell i, with $i = 1, \ldots, N$.

The main variables of the model are those that describe the overall traffic flow behavior in the stretch; a sketch depicting the stretch subdivision and the main variables of the CTM with platoons is shown in Fig. 3.1. Hence, for each cell $i = 1, \ldots, N$, and for each time step $k = 0, \ldots, K$, the quantities adopted to describe the dynamics of the traffic flows are: the traffic density $\rho_i(k)$ of cell i at time kT [veh/km], the traffic flow $\phi_i(k)$ entering cell i from cell $i-1$ during time interval $[kT, (k+1)T)$ [veh/h], the demand $D_i(k)$ of cell i, that is, the traffic flow from cell i that wants to enter cell $i+1$ during time interval $[kT, (k+1)T)$ [veh/h], and the supply $S_i(k)$ of cell i, that is, the traffic flow that can be received by cell i from cell $i-1$ during time interval $[kT, (k+1)T)$ [veh/h].

In addition to the free-flow speed introduced earlier, other parameters are included in the model. Thus, for each cell the main parameters are: the congestion wave speed w_i [km/h], the capacity $q_i^{\max}$ [veh/h], and the maximum density $\rho_i^{\max}$ [veh/km].

The presence of platoons is modeled by denoting with Z the total number of platoons and by z a generic platoon, i.e., $z = 1, \ldots, Z$. Each platoon is associated

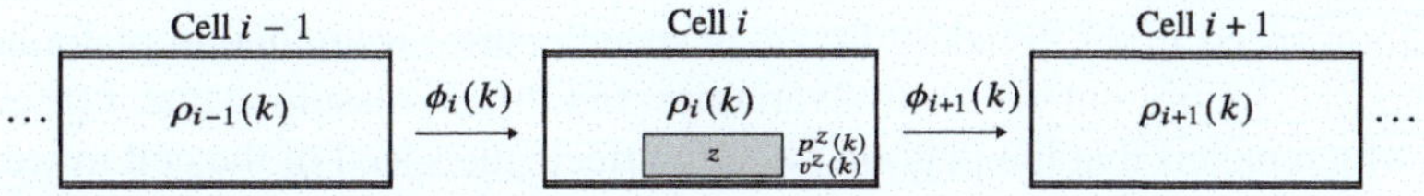

Fig. 3.1 Stretch discretization and main variables of the CTM with platoons

with the parameter o^z, which denotes its occupancy [veh], and with the binary variable $\delta_i^z(k)$ that is used to model the entry (or exit) of the platoon z into a cell i. Specifically, this variable is defined as follows:

$$\delta_i^z(k) = \begin{cases} 1 & \text{if the platoon } z \text{ enters cell } i \text{ at time step } k \\ 0 & \text{otherwise.} \end{cases} \tag{3.1}$$

Note that in this model it is assumed that only one platoon can be present in a cell i for each time step k. Therefore, the condition $\sum_{z=1}^{Z} \delta_i^z(k) \leq 1$ must be satisfied.

According to this, the dynamic evolution of the traffic density, for each cell $i = 1, \ldots, N$, and for each time step $k = 0, \ldots, K-1$ is given by

$$\rho_i(k+1) = \rho_i(k) + \frac{T}{L}\left[\phi_i(k) - \phi_{i+1}(k)\right] + \sum_{z=1}^{Z}\left[\delta_i^z(k)\frac{o^z}{L} - \delta_{i+1}^z(k)\frac{o^z}{L}\right]. \tag{3.2}$$

Furthermore, considering a generic cell i, the traffic flow $\phi_i(k)$ that enters the cell i is the minimum between the demand of the preceding cell $i-1$ and the supply of cell i, where the demand $D_{i-1}(k)$ is the flow from cell $i-1$ that wants to get into cell i in the interval $[kT, (k+1)T)$, while the supply $S_i(k)$ is the flow that cell i can receive in the same time period. These quantities are then calculated as

$$\phi_i(k) = \min\{D_{i-1}(k), S_i(k)\} \tag{3.3}$$

$$D_{i-1}(k) = \min\left\{v_{i-1}(k)\rho_{i-1}(k), q_{i-1}^{\max}\right\} \tag{3.4}$$

$$S_i(k) = \min\left\{w_i(\rho_i^{\max} - \rho_i(k)), q_i^{\max}\right\}. \tag{3.5}$$

Unlike the original CTM formulation, Eq. (3.4) has been modified to consider the presence of a possible platoon presence in the cell. Therefore, the speed $v_i(k)$ in (3.4) is given by the following relationship:

$$v_i(k) = \begin{cases} f(\bar{v}^z(k)) & \text{if the platoon } z \text{ is in cell } i \text{ at time } k \\ v_i^{\text{free}} & \text{otherwise,} \end{cases} \tag{3.6}$$

where $\bar{v}^z(k)$ denotes the speed of platoon z at time step k. Note that, according to (3.6), the traffic speed $v_i(k)$ of a cell i in which a platoon z is present is a function of the platoon speed $\bar{v}^z(k)$. In addition, when speed control is applied to platoons, the variables $\bar{v}^z(k)$ represent the control variables. In [35], it is assumed that $f(\bar{v}^z(k)) = \bar{v}^z(k)$, which means that the platoon directly affects the speed of the cell in which it travels.

The trajectory of each platoon in the freeway stretch can be reconstructed from the dynamic equation describing the position of platoon $p^z(k)$, that is:

$$p^z(k+1) = p^z(k) + \bar{v}^z(k) \cdot T. \tag{3.7}$$

In the case that the platoon position is close to crossing the boundary between two consecutive cells, its trajectory is updated according to the amount of flow that the downstream cell can receive. Therefore, two cases are distinguished in the model which correspond to the *Free-Flow Case* and the *Congested Case*.

Free-Flow Case: In this case, the demand of cell $i-1$ can be completely hosted by cell i, which implies that the condition $D_{i-1}(k) \leq S_i(k)$ is satisfied. Therefore the platoon z present in cell $i-1$ can enter cell i in its full length continuing its travel with speed $\bar{v}^z(k)$ and updating its position $p^z(k)$ by Eq. (3.7). Furthermore, according to Eq. (3.1), the binary variable $\delta_i^z(k)$ is set equal to 1.

Congested Case: If the condition $D_{i-1}(k) > S_i(k)$ is verified, it implies that cell i is unable to receive the entire demand coming from cell $i-1$. In this case, it is necessary to assess whether platoon z is involved in the portion of the flow that can access the downstream cell or not. In this revised version of the CTM, $\bar{p}^z(k)$ indicates the position the platoon would get to if it could enter cell i, which is calculated as

$$\bar{p}^z(k+1) = L \cdot (i-1) + (T - \Delta T^z) \cdot \frac{\phi_i(k)}{\rho_i(k)} \tag{3.8}$$

with

$$\Delta T^z = \frac{L \cdot (i-1) - p^z(k)}{\bar{v}^z(k)}. \tag{3.9}$$

Then, denoting by l^z the length of the zth platoon, two further subcases of the congested case can be considered.
Case 1—If the condition $(T - \Delta T^z) \cdot \frac{\phi_i(k)}{\rho_i(k)} \geq l^z$ is satisfied (i.e., the platoon can enter the cell completely), platoon z enters cell i with probability $\frac{S_i(k)}{D_{i-1}(k)}$. The probability term is introduced to model the uncertainties in the position of platoons due to the use of a discrete macroscopic model. With this term, the higher the supply of the receiving cell the higher the probability that the platoon enters the cell. When this scenario occurs, the position at time step $k+1$ is updated as follows:

- If the platoon can access cell i, the corresponding binary variable $\delta_i^z(k)$ is set equal to 1 and the covered distance becomes

$$p^z(k+1) = \bar{p}^z(k+1). \tag{3.10}$$

This leads to the fact that when platoon z enters the congested cell i, its speed is no longer controlled and becomes

$$\bar{v}^z(k) = \frac{\phi_i(k)}{\rho_i(k)}. \tag{3.11}$$

- If the zth platoon cannot enter cell i, the binary variable $\delta_i^z(k)$ remains equal to 0 and the platoon is stopped at the end of cell $i-1$. Then, the updated platoon position is given by

$$p^z(k+1) = L \cdot (i-1) \tag{3.12}$$

and, accordingly, the platoon speed is

$$\bar{v}^z(k) = \frac{L \cdot (i-1)}{T}. \tag{3.13}$$

In the latter case, it is supposed that if at time step $k+1$ the receiver cell is able to accommodate the platoon, i.e., $S_i(k+1) \geq S^{\min}$ (with $S^{\min}$ the minimum supply of the receiving cell that has properly defined), the platoon that was in cell $i-1$ at time step k can access cell i. Therefore, the binary variable $\delta_i^z(k+1)$ is placed equal to 1 and the position of the platoon is updated following (3.7), where the speed $\bar{v}^z$, at time step $k+1$, must be defined by the controller if cell i is under free-flowing conditions, otherwise it is determined by (3.11). If $S_i(k+1) < S^{\min}$, the platoon still remains in cell $i-1$ and the same condition is verified at $k+2$.

Case 2—If $(T - \Delta T^z) \cdot \frac{\phi_i(k)}{\rho_i(k)} < l^z$, the platoon cannot enter without splitting into two parts, and therefore it stops at cell $i-1$. This implies that the position of the platoon at time step $k+1$ is given by (3.12), while its speed is updated with (3.13) and the corresponding binary variable $\delta_i^z(k)$ is posed equal to 0.

Again, if at time step $k+1$, the condition $S_i(k+1) \geq S^{\min}$ is satisfied, the platoon enters cell i, updating its position according to (3.7), and setting $\delta_i^z(k+1)$ equal to 1. Also in this case, the platoon speed at time step $k+1$ must be defined according to the state of the following cell. Therefore, if $S_i(k+1) < S^{\min}$, the platoon still remains in cell $i-1$ and the same condition is verified at $k+2$.

It should be noted that the main weakness of CTM-based models is related to the fact that these models are unable to capture the capacity drop, and therefore for some applications it is preferable to use second-order models, such as the one given in Sect. 3.2.2, which, besides providing the traffic speed dynamics, capture this important phenomenon. However, in some recent approaches present in the literature, the CTM model has been modified to consider this detriment of capacity in presence of congestion (see the survey paper [41] and references therein).

Specifically, the work proposed in [36] extends the model reported in this section by considering the capacity drop phenomenon. In more detail, in [36] $q_i^{\max}(k)$ denotes the reduced capacity of a cell i at time step k when a bottleneck is active in cell i, while the parameter c_i denotes the free-flow capacity of cell i, i.e., the full capacity of the cell in the absence of congestion. Therefore, if a bottleneck is active in a cell, the reduced capacity $q_i^{\max}(k)$ is less than the free-flow capacity c_i; otherwise, in the absence of congestion, the reduced capacity is equal to the free-flow capacity. Therefore, the reduced capacity $q_i^{\max}(k)$ is computed as follows:

$$q_i^{\max}(k) = \min \left\{ c_i, c_i + (\alpha - 1)\, c_i \frac{\rho_i(k) - \rho_i^{\text{cr}}}{\rho_i^{\max} - \rho_i^{\text{cr}}} \right\}, \tag{3.14}$$

where ρ_i^{cr} is the critical density of cell i, α is a parameter with values in $]0, 1[$ that indicates the magnitude of the reduction in outflow from the cell where the active bottleneck is present.

Then, in [36] the demand and supply of a cell are defined by the following relations:

$$D_i(k) = \min \left\{ v_i(k)\rho_i(k), q_i^{\max}(k) \right\} \tag{3.15}$$

$$S_i(k) = w_i(\rho_i^{\max} - \rho_i(k)). \tag{3.16}$$

3.2.2 A Discrete Time-Space Second-Order Traffic Model with Platoons

In this section, the model *Micro-Macro METANET* (also known as M^3-net) first introduced in [33] is shown. In particular, the version for freeway stretches adopted in [42] is reported here.

Similar to the model described in the previous section, discretization in time and space is performed by partitioning the time horizon into K time instants of width T, in [h], and subdividing the stretch into N portions of length L_i, which in this model are also characterized by the presence of λ_i lanes. The presence of Z platoons is also considered in this model, and each platoon z is characterized by the occupancy o^z, (expressed in [veh/km/lane]), and by the number of vehicles n^z composing it.

Since this is a hybrid model, microscopic and macroscopic variables are used to describe the overall traffic behavior in presence of platoons, see Fig. 3.2. Specifically, the microscopic variables are those introduced to track the trajectories of the platoons in the freeway and are the speed $v^z(k)$ that is referred to each platoon z at time instant kT in [km/h] and the position $p^z(k)$, that is, the distance covered by platoon z at time instant kT, in [km].

Macroscopic variables describe traffic behavior as a whole, and in turn can be distinguished into variables referring to flow in the mainstream, variables referring to entering and exiting ramps, and variables referring to vehicle flows entering from the mainstream.

More in details, the macroscopic variables related to the mainstream are the traffic density $\rho_i(k)$, that is, referred to section i at time instant kT, in [veh/km/lane], the traffic density including platoons $\bar{\rho}_i(k)$, which is specified for each section i and for each time instant kT, in [veh/km/lane], the mean traffic speed $v_i(k)$, for each section

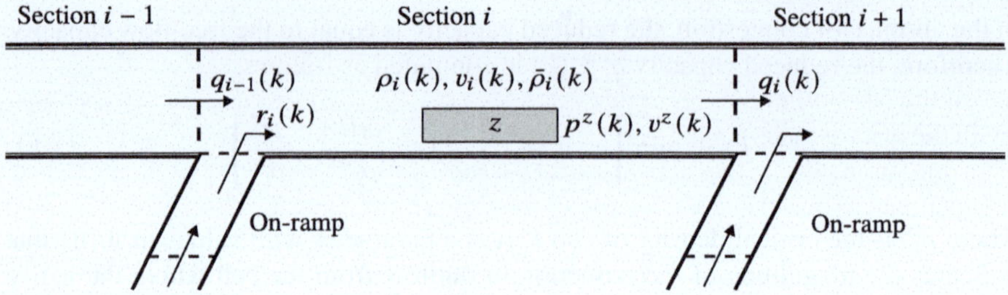

Fig. 3.2 Stretch discretization and main variables of the M^3-net model

i and for each time instant kT, in [km/h] and the traffic volume $q_i(k)$ that is the flow which leaves section i during time interval $[kT, (k+1)T)$, in [veh/h].

With respect to the macroscopic variables describing the behavior in on-ramps, the main ones are the queue length $l_i(k)$ that represents the number of vehicles waiting on the on-ramp of section i at time instant kT, in [veh], the queue length of platoons $b_i(k)$, which is the number of platoons waiting on the on-ramp of section i at time instant kT, in [veh], the traffic demand $d_i(k)$, i.e., the traffic volume that would want to enter section i from the on-ramp during time interval $[kT, (k+1)T)$, in [veh/h], the on-ramp traffic volume $r_i(k)$ that is associated with the on-ramp in section i during time interval $[kT, (k+1)T)$, in [veh/h], and the off-ramp traffic volume $s_i(k)$ exiting at the section i during time interval $[kT, (k+1)T)$, in [veh/h].

Finally, some macroscopic variables are needed to model inflows from the mainstream: the queue length $l_0(k)$, that is, the number of vehicles waiting in the mainstream to enter the first section at time kT, in [veh]; the queue length $b_0(k)$, that is, the number of platoons waiting in the mainstream to enter the first section at time kT, in [veh]; the demand $d_0(k)$, that is, the traffic volume requiring to access the first section from the mainstream during time interval $[kT, (k+1)T)$, in [veh/h]; and the flow $q_0(k)$, that is, the traffic volume entering the first section from the mainstream during time interval $[kT, (k+1)T)$, in [veh/h].

Starting from the description of the microscopic aspects, the position of each platoon z for each time step k is given by

$$p^z(k+1) = p^z(k) + v^z(k)T. \tag{3.17}$$

Even in this model, if platoon speed control is applied, the variables $v^z(k)$ are the control variables, whereas in the absence of control, the platoon speed $v^z(k)$ is assumed to lie between the maximum speed that the legislators impose for platoons, i.e., $v^{z,\max}$, and the average traffic speed in the portion of the stretch in which the platoon is traveling. To impose this condition, the index $i^z(k)$ is introduced to denote the section that corresponds to the position of platoon z at time step k. Hereafter, $v_{i^z(k)}(k)$ denotes the average traffic speed that platoon z finds at time step k. When no control is applied, the speed of each platoon is updated as follows:

$$v^z(k) = \min\left\{v^{z,\max}, v_{i^z(k)}(k)\right\}. \tag{3.18}$$

In the M^3-net model, the dynamic equations referring to platoons are combined with the macroscopic equations of the traditional METANET model, as shown below. Starting from the traffic density, the equation of conservation of vehicles in the mainstream in the absence of platoons, for $i = 1, \ldots, N, k = 0, \ldots, K - 1$, is given by

$$\rho_i(k+1) = \rho_i(k) + \frac{T}{L_i \lambda_i}\Big[q_{i-1}(k) - q_i(k) + r_i(k) - s_i(k)\Big], \tag{3.19}$$

while the traffic density including platoons is defined as

$$\bar{\rho}_i(k) = \rho_i(k) + \sum_{z=1}^{Z} \gamma_i^z(k) o^z, \tag{3.20}$$

where $\gamma_i^z(k)$ is a binary variable adopted to indicate the presence of platoon z, at time step k, in section i, i.e.:

$$\gamma_i^z(k) = \begin{cases} 1 & \text{if } i^z(k) = i \\ 0 & \text{otherwise.} \end{cases} \tag{3.21}$$

Similarly, the speed dynamics is modified to account for the presence of platoons, so for each section i and for each time step k, the average speed is calculated by considering the density $\bar{\rho}_i(k)$ as follows:

$$v_i(k+1) = v_i(k) + \frac{T}{\tau}\Big[V_i(k) - v_i(k)\Big] + \frac{T}{L_i} v_i(k)\Big[v_{i-1}(k) - v_i(k)\Big] - \frac{\nu T\big[\bar{\rho}_{i+1}(k) - \bar{\rho}_i(k)\big]}{\tau L_i\big[\bar{\rho}_i(k) + \chi\big]} - \Delta T \frac{v_i(k) r_i(k)}{L_i\big[\bar{\rho}_i(k) + \chi\big]}, \tag{3.22}$$

where τ, ν, χ, and Δ are model parameters and where the steady-state speed–density relation in (3.22) is given by

$$V\left(\bar{\rho}_i(k)\right) = v_i^{\mathrm{f}} \cdot \left[1 - \left(\frac{\bar{\rho}_i(k)}{\rho_i^{\max}}\right)^l\right]^m \tag{3.23}$$

in which $\rho_i^{\max}$ is the maximum density [veh/km/lane], v_i^{f} is the free-flow speed [km/h], while l, m are other model parameters. Note that the traffic flow to be used in (3.19) is computed as $q_i(k) = \rho_i(k) v_i(k) \lambda_i$.

Finally, it is important to note that all parameters appearing in Eqs. (3.22), (3.23) must be properly calibrated in order to correctly represent the presence of platoons in the freeway stretch. In absence of real field data, simulation campaigns obtained by means of traffic simulations, such as those realized in [43], can be used.

Also the flow dynamics entering from the on-ramps and the mainstream must be appropriately modified to consider the presence of platoons. In [42], it is assumed that platoons can enter the mainstream through the on-ramps or through the first section of the stretch. In this model, the arrival of platoons at the on-ramps is assumed to be known and is modeled through a binary constant $a_i^z(k)$ that is defined as follows:

$$a_i^z(k) = \begin{cases} 1 & \text{if platoon } z \text{ arrives at the on-ramp } i \\ 0 & \text{otherwise} \end{cases} \tag{3.24}$$

similarly the entrance from the first section of the mainstream is modeled with the binary constant $a_0^z(k)$, therefore:

$$a_0^z(k) = \begin{cases} 1 & \text{if platoon } z \text{ arrives at section 1} \\ 0 & \text{otherwise.} \end{cases} \tag{3.25}$$

However, the effective entrance of platoons into the freeway depends not only on the arrival time of the platoons, but also on the traffic conditions in the sections downstream the on-ramps or in the first section of the mainstream. The conditions (3.28)–(3.31) are introduced in order to model the entry or waiting of platoons according to the traffic conditions in the mainstream.

Therefore, the entry from an on-ramp of a generic platoon z at time step k is given by the binary variable $y_i^z(k)$, defined as

$$y_i^z(k) = \begin{cases} 1 & \text{if platoon } z \text{ enters from on-ramp } i \\ 0 & \text{otherwise.} \end{cases} \tag{3.26}$$

Similarly, the variable $y_0^z(k)$ models the entry from the mainstream of platoon z at time step k:

$$y_0^z(k) = \begin{cases} 1 & \text{if platoon } z \text{ enters from section 1} \\ 0 & \text{otherwise.} \end{cases} \tag{3.27}$$

Then, given the arrival schedule of platoons, for the on-ramp i, platoon z enters the freeway, i.e., $y_i^z(k)$ is set equal to 1, if and only if the following conditions are met:

$$r_i(k) + \frac{1}{T}\left[a_i^z(k)n^z + b_i^z(k)\right] \le r_i^{\max} \tag{3.28}$$

$$r_i(k) + \frac{1}{T}\left[a_i^z(k)n^z + b_i^z(k)\right] \le r_i^{\max} \cdot \frac{\rho_i^{\max} - \bar{\rho}_i(k)}{\rho_i^{\max} - \rho_i^{\mathrm{cr}}}. \tag{3.29}$$

In the same way, platoon z enters from the mainstream, i.e., $y_0^z(k)$ is set equal to 1, if the following conditions are both fulfilled:

$$q_0(k) + \frac{1}{T}\left[a_0^z(k)n^z + b_0^z(k)\right] \le q_0^{\max} \tag{3.30}$$

$$q_0(k) + \frac{1}{T}\left[a_0^z(k)n^z + b_0^z(k)\right] \le q_0^{\max} \cdot \frac{\rho_1^{\max} - \bar{\rho}_1(k)}{\rho_1^{\max} - \rho_1^{\mathrm{cr}}}, \tag{3.31}$$

where ρ_i^{cr} is the critical density of the section i, $r_i^{\max}$ is the maximum flow which can enter the on-ramp i, and $q_0^{\max}$ is the maximum flow that can enter the mainstream.

The waiting of platoons that cannot enter from the on-ramps or the mainstream is represented by buffers, named $b_i^z(k)$ and $b_0^z(k)$, which are defined for each platoon z and which can have lengths equal to 0 or equal to the number of vehicles n^z composing the platoon. This implies that, in this model, the platoon cannot be split to enter the freeway, and therefore if it cannot enter for its full length, it must wait. In accordance with this, the dynamic evolution of buffers is given by

$$b_i^z(k+1) = b_i^z(k) + a_i^z(k)n^z - y_i^z(k)n^z \tag{3.32}$$

$$b_0^z(k+1) = b_0^z(k) + a_0^z(k)n^z - y_0^z(k)n^z. \tag{3.33}$$

Note that if conditions (3.28)–(3.29) and (3.30)–(3.31) are not satisfied, the corresponding binary variables $y_i^z(k)$ and $y_0^z(k)$ are set equal to 0. This implies that the succeeding platoon $z+1$ cannot be processed until conditions (3.28) and (3.29) are satisfied, if platoon z cannot enter from the on-ramps, or until conditions (3.30) and (3.31) are satisfied, if platoon z cannot enter from the mainstream. In case the latter scenarios occur, the buffer lengths are updated as follows:

$$b_i(k) = \sum_{z=1}^{Z} b_i^z(k), \quad b_0(k) = \sum_{z=1}^{Z} b_0^z(k). \tag{3.34}$$

In conclusion, the queue lengths at the on-ramps and at the mainstream are computed as in the METANET model, and therefore

$$l_i(k+1) = l_i(k) + T\left[d_i(k) - r_i(k)\right] \tag{3.35}$$

$$l_0(k+1) = l_0(k) + T\left[d_0(k) - q_0(k)\right], \tag{3.36}$$

where the on-ramp entering flow is obtained as

$$r_i(k) = \min\left\{d_i(k) + \frac{l_i(k)}{T}, r_i^{\max}, r_i^{\max} \cdot \frac{\rho_i^{\max} - \bar{\rho}_i(k)}{\rho_i^{\max} - \rho_i^{\mathrm{cr}}}\right\} \tag{3.37}$$

while the traffic flow entering the first section from the mainstream is given by

$$q_0(k) = \min\left\{d_0(k) + \frac{l_0(k)}{T}, q_0^{\max}, q_0^{\max} \cdot \frac{\rho_1^{\max} - \bar{\rho}_1(k)}{\rho_1^{\max} - \rho_1^{\mathrm{cr}}}\right\}. \tag{3.38}$$

3.3 Traffic Regulation Exploiting Vehicle Platoons

The possibility of regulating traffic by exploiting the presence of vehicle platoons in a portion of the freeway network is the subject of this section. The general idea is that of considering platoons as actuators of a specific control action which has shown to be effective in traffic regulation, that is, the adoption of vehicle speed limitations (see [44, 45]). The definition of a control scheme deploying this idea has recently been addressed also in [33, 35, 42, 47]. All these works are based on a predictive approach, thus requiring the use of specific models describing the presence of platoons in the traffic flow and, specifically, representing the influence of the behavior of platoons on the rest of traffic. Possible modeling approaches have been described in the previous section. In this section, assuming that an effective modeling framework exists, we concentrate on the design of possible model-based control architectures regulating traffic on a freeway stretch.

Considering predictive control, several control architectures can be designed and adopted to regulate traffic in a freeway network. The most straightforward and conceptually most effective structure is the so-called *centralized control* scheme, in which the overall system state is accessible and known, and the control actions on the whole network are jointly determined on the basis of the present system state and on predictions of its future behavior. In Model Predictive Control (MPC), this is realized by solving a finite-horizon optimization problem in which the control requirements are expressed as an objective function to be optimized and the system model providing predictions of the system behavior is included as a set of constraints in the problem itself. The optimization problem is run at any time step on the basis of updated measurements of the system state, only the first values of the sequence of the optimal control actions are applied to the process, and, then, time and state are updated and the process is iteratively repeated.

The adoption of MPC in centralized control schemes makes it possible to couple the optimization of the process performance with the guarantee of the system stability properties. The drawback is the required huge computational complexity, especially for large-scale systems as freeway traffic networks definitely are. As effectively described in [48], in large-scale systems composed of several local units with weak interactions, it may be convenient to design local regulators (always with a predictive nature) active on single units, in such a way to break the structure of the overall optimization problem related to the centralized scheme into a set of smaller and less cumbersome problems. If the local regulators are defined independently, this gives rise to the so-called *decentralized control* architecture.

A variation with respect to decentralized control is a scheme in which each local controller determines the control actions to be applied in a single subsystem also using information about the state of other subsystems. A *distributed control* architecture is, in this case, adopted, for which a further distinction is made between the so-called *fully connected* structures in which each regulator has knowledge about all the subsystems and the *partially connected* schemes in which, instead, each local controller uses information about a subset of the remaining subsystems.

Moreover, when information about other subsystems is received by the local regulators only once within each sampling time, the control algorithm is defined as a *noniterative algorithm* which becomes an *iterative algorithm* when the information is received many times in the same sampling time. Finally, a distributed scheme is defined as *independent* when each local regulator adopts a local performance index and as *cooperating* when each local regulator optimizes a global cost function.

Whenever a completely centralized architecture in which the control algorithm determines all the control actions to be applied in the system, results too cumbersome for the considered process, as is often the case in large-scale processes, a possible alternative with respect to fully decentralized control schemes or distributed architectures is a two-level *hierarchical control* structure. In this case, at a higher level a control algorithm coordinates the actions of local regulators placed at a lower level. The number of supervisors/coordination algorithms acting at a higher level again is a design choice. The presence of a single supervisor, knowing the whole system state and coordinating several local regulators, gives rise to the so-called centralized hierarchical control scheme. If, instead, several supervisors coexist, distributed or decentralized hierarchical schemes are adopted.

In this section, two control schemes aimed at regulating the movements of platoons in the considered freeway network are presented (analogous structures are in [46]). For both schemes, a hierarchical architecture is adopted in order to make use of two control levels: the lower level acts on each single platoon and the higher level determines some crucial parameters to be adopted at a lower level. More specifically, at the low level, a feedback control algorithm, defined for each platoon, determines the platoon speed on the basis of traffic measurements in the sections downstream the platoon position and using parameters that are provided to the platoon by the high control level.

The difference between the two schemes lies in the fact that the first scheme is indicated as centralized, since a single *network supervisor* constitutes the high control level. This supervisor determines the controller parameters for each platoon traveling in the network by using a full knowledge of the system state (including the state of platoons) and by making predictions on the system evolution.

The second control scheme presents a decentralized architecture since the overall network is divided into portions, named clusters, and one supervisor is dedicated to each cluster of the network and determines the parameters of the controller of platoons possibly present in the cluster itself. Then, the *cluster supervisor* provides the platoons traveling in the cluster with the necessary control parameters, which are determined by only using local information about the state in the cluster (including the state of platoons inside the link) and predictions again only related to the conditions of traffic in the cluster.

In the rest of this section, the proposed control schemes are described, starting from the definition of the low-level controllers that are common to the two schemes. Then, the control structures and algorithms used in the two control schemes are discussed.

3.3.1 The Low-Level Control Algorithm

The control algorithm acting at the low level, named as the *platoon controller*, is presented in the following. It is important to note that such an algorithm does not deal with the control of single vehicles inside the platoon. Several approaches exist for vehicles' control in platoons [13, 14], also addressing crucial properties as the string stability [15, 16, 49, 50].

The platoon controller here presented determines the platoon speed, as the control variable acting on platoons. The way in which the platoon controller is actuated is through the technology mounted onboard the platoon leader. Then, in this case, the platoon technology is exploited not only to regulate the vehicle behavior inside the platoon, but also to communicate with the infrastructure, and specifically with the adopted high-level controllers.

It is assumed that, at any time step k, the platoon controller receives measurements of the traffic densities in a subset of sections downstream its position. It provides its position, indicated with $i^z(k)$, to the high control level, which computes, as it will be described in the next sections, the parameters needed to completely compute the low-level control variable (the platoon speed).

The low-level controller can be realized with different control methodologies. A simple but effective algorithm is proposed here, computing the control action with a PI-type feedback law. More precisely, the control law determines the platoon speed at time step k with the following equation:

$$v^z(k) = v^z(k-1) + K_P \left[e^z(k) - e^z(k-1) \right] + K_I \cdot e^z(k), \tag{3.39}$$

where K_P and K_I are the gain parameters of the controller, and $e^z(k)$ is the control error, computed as

$$e^z(k) = \hat{\rho}^z(k) - \varrho^z(k), \tag{3.40}$$

where $\hat{\rho}^z(k)$ and $\varrho^z(k)$ are the density set point and the average measured density for platoon z. Then, the control error is the difference between a density set point received from the high control level and an average density value $\varrho^z(k)$ computed by using the downstream measurements. Such downstream information is related to a set $I^z(k)$ of sections downstream the position of platoon z. Then, it holds

$$\hat{\rho}^z(k) = \sum_{i \in I^z(k)} \alpha_i(k)\hat{\rho}_i(k) \tag{3.41}$$

$$\varrho^z(k) = \sum_{i \in I^z(k)} \alpha_i(k)\bar{\rho}_i(k), \tag{3.42}$$

where $\hat{\rho}_i(k)$, $i \in I^z(k)$ are the density set points computed by the high control level and communicated to the platoon together with the weighing parameters $\alpha_i(k) \in [0, 1]$ (as detailed in the next subsections), for which it must hold

$$\sum_{i \in I^z(k)} \alpha_i(k) = 1. \tag{3.43}$$

The controller parameters (the weights in (3.41) and (3.42) and the density set points) are not indexed also with platoon z since it is assumed that only one platoon leader can be present in a section at each time step. This assumption appears to be definitely realistic, anyway it would not be critical to remove it.

Finally, the speed obtained with the feedback control law (3.39) has to be bounded between the minimum values of speed allowed for platoon z and the current speed of the section in which the platoon is located at time step k or the maximum platoon speed, i.e.:

$$v^{z,\min} \leq v^z(k) \leq \min\left\{v^{z,\max}, v_{i^z(k)}(k)\right\}. \tag{3.44}$$

As already mentioned, different control methods can be used for the platoon controller. A MPC scheme would also be effective, but it would make the controller computations much more complex, thus requiring a suitable technology onboard the leader vehicle. The proposed algorithm couples a simple structure and simple computations with a suitable use of lookahead measurements and an effective parameterization realized by the higher control level.

3.3.2 *The Hierarchical Centralized Control Scheme*

The hierarchical centralized control scheme includes two levels: at the low level, each platoon is controlled by the feedback speed regulator described in Sect. 3.3.1, whereas a *network supervisor* acts at the higher level optimally determining, at each time step, the parameters of all the platoon controllers on the basis of its knowledge of the whole system state.

Figure 3.3 represents the hierarchical centralized scheme in which the supervisor receives measurements of the system state of the whole network at each time step k. The traffic densities $\bar{\rho}_i(k)$ and the mean traffic speeds $v_i(k)$ in all the freeway sections, i.e., $\forall i$, constitute the state measurements, together with the queue lengths $l_i(k)$ in the on-ramps.

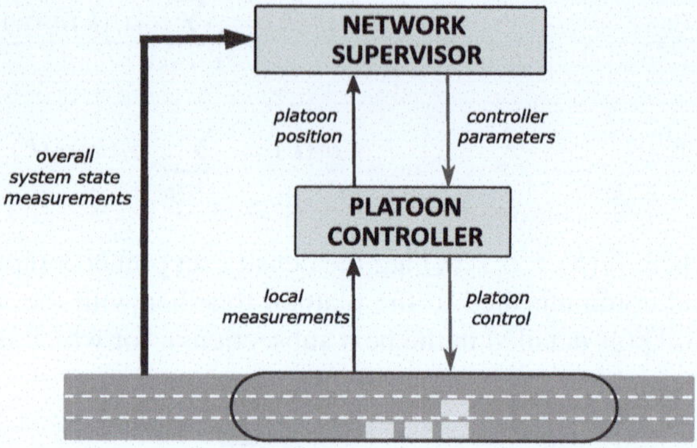

Fig. 3.3 Structure of the hierarchical centralized control scheme

The information available to the supervisor also includes the schedule of platoons in terms of time instants at which they reach a specific on-ramp or the mainstream, then, the supervisor knows the values of $a_i^z(k)$, $\forall z$, $\forall i$, $\forall k$, and it also receives information related to the platoon size n^z and occupancy o^z, $\forall z$. Finally, at each time step k, each platoon communicates to the supervisor its position in terms of section $i^z(k)$.

The high control level runs an optimization procedure consisting in the solution of a finite-horizon optimization problem whose decision variables are the density set point and the measurement weights to be indicated to each platoon moving in the network, that is, the parameters of all the low-level controllers. The optimal values of such parameters are, then, determined depending on the platoon position, on the current traffic conditions, on the whole present system state, and also by making predictions about the system behavior.

The optimization problem is solved at each time step at which a platoon enters the network and, also, periodically, i.e., at each time step $\bar{k} = nP$, where $n = 0, 1, 2, \ldots$ and P is an integer representing the number of time steps between one computation and the subsequent one. The time horizon considered in the optimization procedure is denoted with H.

The objective function optimized inside the problem is the total time spent by vehicles in the network, denoted with TTS [veh·h] which, with reference to an initial time step h, is computed as

$$\mathcal{J}(h) = \sum_{k=h}^{h+H-1} T\left[\sum_{i=1}^{N} (L_i \bar{\rho}_i(k) + l_i(k))\right]. \tag{3.45}$$

The statement of the finite-horizon optimization problem solved at time step $\bar{k}$, with $\bar{k} = nP$ or $\bar{k} : \exists i = 1, \ldots, N, \exists z = 1, \ldots, Z : a_i^z(\bar{k}) = 1$ is the following.

Problem 3.1 Given the system state $\bar{\rho}_i(\bar{k})$, $v_i(\bar{k})$, $l_i(\bar{k})$, $i = 1, \ldots, N$, the estimated demands $d_i(k)$, $i = 1, \ldots, N$, $k = \bar{k}, \ldots, \bar{k} + H - 1$, the position of platoons $i^z(\bar{k})$, $z = 1, \ldots, Z$, find

$$\min_{\substack{\hat{\rho}_i(k),\alpha_i(k),i=1,\ldots,N \\ k=\bar{k},\ldots,\bar{k}+H-1}} \mathcal{J}(\bar{k}) \tag{3.46}$$

subject to (3.17), (3.19)–(3.38), (3.39)–(3.44).

The solution of Problem 3.1 at any time step $\bar{k}$ provides the configuration of the controllers of all the platoons traveling in the network at that time step. Specifically, the configuration consists in the definition of the controller parameters $\hat{\rho}_i(k)$, $\alpha_i(k)$, $i \in I^z(k)$, $k = \bar{k}, \ldots, \bar{k} + H - 1$ found as solution of Problem 3.1 and communicated to any platoon z.

The solution of Problem 3.1, which has the structure of a mixed-integer nonlinear optimization problem, can be determined by derivative-free algorithms as the ones adopted in [51] for problems with an analogous complexity. The problem is anyway a complex problem, whose solution requires a high computational load, also due to the large-scale nature of the considered networks.

3.3.3 The Hierarchical Decentralized Control Scheme

As already discussed, when dealing with traffic networks, it is reasonable to define control structures in which the overall system is subdivided into subsystems separately controlled, in order to reduce the computational complexity of the centralized optimization problem, thus improving the practical applicability of the control scheme. This is what is proposed in this section in which it is assumed that the traffic network is composed of several portions named *clusters* that have dedicated control architectures. This yields to the definition of a decentralized control scheme in which we anyhow preserve the hierarchical two-level structure already adopted in the centralized case. It can be noted that the definition of clusters in a traffic network was already proposed in [52], where a discussion of several (centralized, distributed, and decentralized) control schemes was included, and in [53] where hierarchical architectures were addressed.

The hierarchical decentralized control scheme is again composed of two levels: the low control level is completely analogous to the one described in Sect. 3.3.1, whereas the high level includes one supervisor for each cluster of the considered freeway network. This scheme actually replicates the same concept used in the centralized scheme, but each cluster is equipped with one supervisor acting locally. Of course, the information available for each cluster supervisor is only referred to the state in the cluster, the cluster supervisor, then, has a limited knowledge with respect to the network one and it also has a less effective predictive capability. On the other hand, the optimization problems to be solved by cluster supervisors have much smaller dimensions and, then, also lower computational requirements.

It is important to note that in decentralized schemes, not only is the computational load distributed on different local supervisors, but also the overall communication effort is decreased. This is due to the fact that data transmissions are limited in the

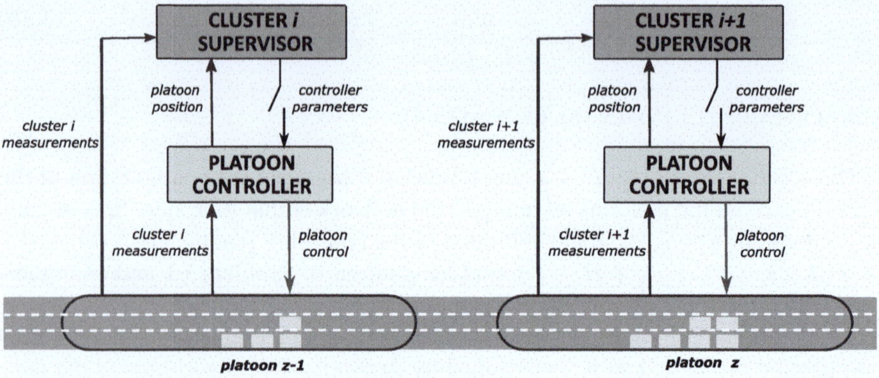

Fig. 3.4 Structure of the decentralized control scheme

cluster, while in the centralized scheme the network supervisor is located in a control center possibly far away from clusters, thus requiring a higher communication effort.

A sketched picture of a portion of the hierarchical decentralized control scheme is provided in Fig. 3.4 in which two subsequent clusters with two platoons traveling in such clusters are included.

The cluster supervisor determines the controller parameters to be communicated to platoons reaching the cluster, depending on the platoons position, on the current traffic conditions in the cluster and also by making predictions about the traffic behavior in the cluster itself.

The platoon controller parameters are determined by solving a finite-horizon optimization problem at any time step in which at least one platoon is traveling in the cluster. The considered time horizon considered is denoted with H_c which just indicates the maximum time necessary for a platoon to travel along the cluster (it is computed on the basis of the platoon minimum speed).

In the optimization problem formulation, the time spent by vehicles in the cluster constitutes the objective function to be minimized, computed, at time step h, as

$$\mathcal{J}_c(h) = \sum_{k=h}^{h+H_c-1} T\left[\sum_{i=1}^{N_c} L_i \bar{\rho}_i(k)\right], \tag{3.47}$$

where N_c is the number of freeway sections included in cluster c. Moreover, let us denote with $\mathcal{Z}_c(k)$ the set of indices of platoons present in cluster c at time step k. The statement of the local optimization problem solved in cluster c at time step $\bar{k}$, with $\bar{k} : \mathcal{Z}_c(\bar{k}) \neq \emptyset$, is the following.

Problem 3.2 Given the system state $\bar{\rho}_i(\bar{k})$, $v_i(\bar{k})$, $i = 1, \ldots, N_c$, the position $i^z(\bar{k})$ of each platoon z, $z \in \mathcal{Z}_c(\bar{k})$, find

$$\min_{\substack{\hat{\rho}_i(k),\alpha_i(k),\\ i=1,\ldots,N_c,\\ k=\bar{k},\ldots,\bar{k}+H_c-1}} \mathcal{J}_c(\bar{k}) \tag{3.48}$$

subject to (3.17), (3.19)–(3.38), referred to cluster c, and (3.39)–(3.44), $\forall z$, $z \in \mathcal{Z}_c(\bar{k})$.

The cluster supervisor determines, by solving Problem 3.2, the controller parameters for each platoon traveling inside the cluster, thus updating the platoon controllers any time the problem is solved.

A control scheme similar to the one here presented has been reported in [46] where some simulation results are discussed, showing that a decentralized control architecture can constitute a good compromise between a very cumbersome centralized scheme and the use of fixed parameters platoon controllers.

3.4 Opportunities and Challenges

Some modeling approaches and control architectures designed to leverage the presence of platoons of CAVs in freeway networks have been proposed in this chapter. The idea is that *CAVs can organize their movements forming vehicle platoons which can be beneficial for the vehicles themselves to reduce fuel consumption and for the whole traffic if vehicle platoons are exploited as speed control actuators*.

In this sense, platoons moving with a controlled speed can be assimilated to variable speed control actuators. As already mentioned, speed control in freeways is generally actuated by variable speed indications shown on panels that are present at specific positions along the road. The adoption of vehicle platoons inducing a reduction of the speed of the overall traffic flow with their controlled behavior would yield two main advantages: the first one is the fact that the position of platoons varies, thus actuating speed control along the road in a more frequent and continuous way. The second advantage is that the influence that the platoon speed induces on the speed of the rest of traffic can be much stronger than the one obtained with speed indications displayed on variable message panels. The presence of platoons along the road can provide advantages both from an individual point of view (the one of single vehicles reducing fuel and energy consumptions) and at a system level (that is, for the overall traffic which can better counteract congestion phenomena). Some simulation results about the implementation of this opportunity have been reported in the referenced works provided in this chapter, showing that it is a promising direction.

In order to determine the most effective regulation of traffic, from a methodological point of view, there exist several control techniques and schemes that can be used, such as the ones proposed in this chapter. The controller structure can vary from centralized to distributed and decentralized approaches. The control law can be determined with simple feedback laws or it can be the solution of suitable optimal control problems, also adopting predictive capabilities. Then, depending on the network topology, on the kind of traffic conditions characterizing the road network and on the objectives of the users and of the operators managing the road network, it is possible to evaluate and design the most proper control approach. The challenge, in the Authors' opinion, is not so much related to the control method or the architecture used, which can be ad hoc designed and customized for the considered process, but rather it is important to define effective **prediction methods**, to evaluate the system behavior in different conditions and with different regulators and to effectively design and configure the traffic controllers both for single platoons and for the overall traffic flow. This implies the **need for modeling approaches** able to replicate the dynamic evolution of traffic in a flow of vehicles including CAVs and, specifically, platoons.

Actually, the definition of models that represent mixed traffic, in which traditional human-driven vehicles and CAVs coexist, is an open problem. Modeling the interactions between vehicles of different nature still needs the researchers' attention. As discussed in Sect. 3.2 there exist some approaches, but still suitable **calibration and validation procedures** for the existing modeling methods cannot be fully realized due to the lack of extensive real case studies and, consequently, of available datasets. Also, the newly techniques related to Machine Learning methods, designed to determine surrogate models on the basis of large sets of data, again cannot be exploited yet. This means that the effectiveness of the models for mixed traffic is not yet established, which makes it difficult also to deeply understand the behavior of mixed traffic.

Besides being able to model and replicate the behavior of traffic including CAVs and vehicle platoons, which is matter of ongoing research, the following aspects still need to be analyzed and represented in order to be able to adopt the speed actuation capabilities of vehicle platoons:

- The **influence of the platoon speed on the speed of the rest of traffic**. In the models here presented, this influence is considered by basing the speed–density relationship on the evaluation of the overall density considering the different vehicle classes. Present work is devoted to a variation of the speed–density relationship making the presence of a platoon traveling at a certain speed (possibly controlled) equivalent to the presence of speed limitations (as it is done in variable speed control). Several relations exist for the case with variable speed limitations [44, 45, 54] that can be adapted to the speed actuation realized by means of platoons.
- The **influence of the platoon occupancy on the speed of the rest of traffic**. It can be argued that the number of lanes occupied by a platoon can vary (it can be matter of decisions as in [47]) and that the speed of the overall flow moving in a section where a platoon is present, adapts to the platoon speed depending on the number of lanes it occupies. The greater the number of occupied lanes compared to the total number of lanes, the more closely the traffic speed matches that of the platoon. In the extreme case when the platoon occupies all lanes, the upstream traffic moves at the same speed as the platoon.

The above aspects must first of all be effectively translated in the relations included in the modeling frameworks to be defined.

Finally, the effectiveness of using CAVs and vehicle platoons for traffic control is strictly dependent on the penetration rate of such specific classes of vehicles in the overall traffic flow. The low presence of CAVs and, consequently, the difficulty in forming platoons (besides truck platooning that already is a reality in several countries) make it impossible to exploit their speed actuation capabilities whenever it is needed, that is, in presence of bottlenecks or when congestion is forming. As soon as a high percentage of CAVs will be available in freeway networks, it will instead be possible to have platoons where needed or even to plan the formation of ad hoc defined platoons to mitigate or prevent congestion situations.

As a whole, it can be concluded that specific vehicle formations as platoons can be very useful in freeway networks to impose speed limitations when needed, thus freeing the speed control from the need to use indications on fixed panels and, instead, forcing speed reductions when and where it can be useful. From the scientific viewpoint, modeling approaches must still be completely defined, calibrated, and validated and, on the basis of such models, the best control approaches can be devised. From an application point of view, instead, vehicle technology is very advanced, but the penetration of this technology is not yet very high, while its widespread use will be the key to exploiting all the consequent individual and systemic opportunities.

3.5 Conclusions

A new paradigm of traffic control can be considered in which the vehicles themselves, equipped with advanced automation and connection technology, can be used to implement freeway traffic control strategies. Specifically, the presence of formation of vehicles traveling together in the so-called vehicle platoons is considered in this chapter which presents some modeling approaches and control schemes used to deploy freeway traffic control by platooning, while also providing a critical analysis of the future prospects of this control methodology.

In the presented models a discretization both in space and in time is operated and the presence of platoons is represented by a microscopic layer included in classical macroscopic modeling frameworks (both of first order and of second order).

As regards control, two control architectures are presented, both with a hierarchical structure in which, at the low level, the speed of platoons is determined by a feedback control law, whose parameters are computed by the high control level. In the first scheme, which is a centralized one, the high control level is constituted by a network supervisor acting on the whole system. In the second scheme, which is, instead, a decentralized one, the network is subdivided into portions named clusters and a supervisor is dedicated to each cluster.

The preliminary works existing in the literature already show promising results, thus encouraging researchers to keep working on the definition of modeling and control schemes exploiting CAVs and, specifically, platoons to enhance the effectiveness of existing traffic control schemes on freeways.

References

1. Diakaki C, Papageorgiou M, Papamichail I, Nikolos I (2015) Overview and analysis of vehicle automation and communication systems from a motorway traffic management perspective. Transp. Res. Part A 75:147–165
2. Alonso Raposo M et al (2019) The future of road transport - implications of automated, connected, low-carbon and shared mobility. EUR 29748 EN. Publications Office of the European Union, Luxembourg
3. Pipes LA (1953) An operational analysis of traffic dynamics. J Appl Phys 24:274–281
4. Alam A, Besselink B, Turri V, Mårtensson J, Johansson KH (2015) Heavy-duty vehicle platooning for sustainable freight transportation: a cooperative method to enhance safety and efficiency. IEEE Control Syst 35:34–56
5. Janssen GR, Zwijnenberg J, Blankers IJ, de Kruijff JS (2015) Truck platooning: driving the future of transportation. TNO Technical Report
6. Ramakers R, Henning K, Gies S, Abel D, Max HM (2009) Electronically coupled truck platoons on German highways. In: Proceeding of the IEEE international conference on systems, man and cybernetics, pp 2409–2414
7. Fritz H, Gern A, Schiemenz H, Bonnet C (2004) CHAUFFEUR assistant: a driver assistance system for commercial vehicles based on fusion of advanced ACC and lane keeping. In: Proceeding of the IEEE intelligent vehicles symposium, pp 495–500
8. Kunze R, Ramakers R, Henning K, Jeschke S (2011) Organization and operation of electronically coupled truck platoons on German motorways. Autom Commun Cybern Sci Eng. Springer, Berlin Heidelberg, pp 427–439
9. Tsugawa S, Jeschke S, Shladover SE (2016) A review of truck platooning projects for energy savings. IEEE Trans Intell Veh 1:68–77
10. Liang K-Y, Mårtensson J, Johansson KH (2013) When is it fuel efficient for a heavy duty vehicle to catch up with a platoon? In IFAC Proc Vol 46:738–743
11. Bhoopalam AK, Agatz N, Zuidwijk R (2018) Planning of truck platoons: a literature review and directions for future research. Transp Res Part B 107:212–228
12. Liang K-Y, van de Hoef S, Terelius H, Turri V, Besselink B, Mårtensson J, Johansson KH (2016) Networked control challenges in collaborative road freight transport. Europ J Control 30:2–14
13. van de Hoef S, Johansson KH, Dimarogonas DV (2018) Fuel-efficient en route formation of truck platoons. IEEE Trans Intell Transp Syst 19:102–112
14. Bayuwindra A, Ploeg J, Lefeber E, Nijmeijer H (2019) Combined longitudinal and lateral control of car-like vehicle platooning with extended look-ahead. IEEE Trans Control Syst Technol 28:790–803
15. Besselink B, Johansson KH (2017) String stability and a delay-based spacing policy for vehicle platoons subject to disturbances. IEEE Trans Autom Control 62:4376–4391
16. Feng S, Zhang Y, Li SE, Cao Z, Liu HX, Li L (2019) String stability for vehicular platoon control: definitions and analysis methods. Ann Rev Control 47:81–97
17. Siri S, Pasquale C, Sacone S, Ferrara A (2021) Freeway traffic control: a survey. Automatica 130:109655
18. Stern RE, Cui S, Delle Monache ML, Bhadani R, Bunting M, Churchill M et al (2018) Dissipation of stop-and-go waves via control of autonomous vehicles: field experiments. Transp Res Part C: Emer Technol 89:205–221
19. Jia D, Ngoduy D, Vu HL (2018) A multiclass microscopic model for heterogeneous platoon with vehicle-to-vehicle communication. Transp Dyn Transp B 311–335
20. Seraj M, Li J, Qiu Z (2018) Modeling microscopic car-following strategy of mixed traffic to identify optimal platoon configurations for multiobjective decision-making. J Adv Transp 7835010
21. Ngoduy D (2013) Platoon-based macroscopic model for intelligent traffic flow. Transp B: Transp Dyn 1:153–169

22. Lebacque JP, Lesort JB, Giorgi F (1998) Introducing buses into first-order macroscopic traffic flow models. Transp Res Record 1644:70–79
23. Lighthill MJ, Whitham JB (1955) On kinematic waves II: a theory of traffic flow in long crowded roads. Proc R Soc 229:317–345
24. Richards PI (1956) Shockwaves on the highway. Oper Res 4:42–51
25. Lattanzio C, Maurizi A, Piccoli B (2011) Moving bottlenecks in car traffic flow: a PDE-ODE coupled model. SIAM J Math Anal 43:50–67
26. Delle Monache ML, Goatin P (2014) Scalar conservation laws with moving constraints arising in traffic flow modeling: an existence result. J Diff Equ 257:4015–4029
27. Simoni MD, Claudel CG (2017) A fast simulation algorithm for multiple moving bottlenecks and applications in urban freight traffic management. Transp Res Part B: Methodol 104:238–255
28. Dymski NS, Goatin P, Rosini MD (2018) Existence of BV solutions for a non-conservative constrained Aw-Rascle-Zhang model for vehicular traffic. J Math Anal Appl 467:45–66
29. Piacentini G, Goatin P, Ferrara A (2020) A macroscopic model for platooning in highway traffic. SIAM J Appl Math 80:639–656
30. Piacentini G, Goatin P, Ferrara A (2021) Traffic control via platoons of intelligent vehicles for saving fuel consumption in freeway systems. IEEE Control Syst Lett 5:593–598
31. Ferrara A, Incremona GP, Birliba E, Goatin P (2022) Multi-scale model-based hierarchical control of freeway traffic via platoons of connected and automated vehicles. IEEE Open J Intell Transp Syst 3:799–812
32. Kotsialos A, Papageorgiou M, Diakaki C, Pavlis Y, Middelham F (2002) Traffic flow modeling of large-scale motorway networks using the macroscopic modeling tool METANET. IEEE Trans Intell Transp Syst 3:282–292
33. Pasquale C, Sacone S, Siri S, Ferrara A (2018) A new micro-macro METANET model for platoon control in freeway traffic networks. In: Proceedings of the 21st IEEE international conference on intelligent transportation systems, pp 1481–1486
34. Daganzo CF (1995) The cell transmission model: network traffic. Transp Res Part B 29:79–93
35. Piacentini G, Pasquale C, Sacone S, Siri S, Ferrara A (2019) Multiple moving bottlenecks for traffic control in freeway systems. In: Proceedings of the 18th IEEE European control conference, pp 3662–3667
36. Piacentini G, Ferrara A, Papamichail I, Papageorgiou M (2019) Highway traffic control with moving bottlenecks of connected and automated vehicles for travel time reduction. In: Proceedings of the 58th IEEE conference on decision and control, pp 3140–3145
37. Chaanine T, Ferrando A, Pasquale C, Sadeghi Garjan M, Pastore VP, Siri S (2023) A control-oriented highway traffic model with multiple clusters of CAVs. In: Proceedings of the 26th IEEE international conference on intelligent transportation systems
38. Jin L, Čičić M, Amin S, Johansson KH (2018) Modeling the impact of vehicle platooning on highway congestion: a fluid queuing approach. In: Proceedings of the 21st international conference on hybrid systems: computation and control, pp 237–246
39. Čičić M, Xiong X, Jin L, Johansson KH (2021) Coordinating vehicle platoons for highway bottleneck decongestion and throughput improvement. IEEE Trans Intell Transp Syst 23:8959–8971
40. Jin W-L (2015) Point queue models: a unified approach. Transp Res Part B: Methodol 77:1–16
41. Kontorinaki M, Spiliopoulou A, Roncoli C, Papageorgiou M (2017) First-order traffic flow models incorporating capacity drop: overview and real-data validation. Transp Res Part B: Methodol 106:52–75
42. Bozzi A, Graffione S, Pasquale C, Sacile R, Sacone S, Siri S (2022) A hierarchical control scheme to improve the travel performance of truck platoons in freeways. In: Proceeding of the 25th IEEE international conference on intelligent transportation systems, pp 2063–2068
43. Moridpour S, Mazloumi E, Mesbah M (2015) Impact of heavy vehicles on surrounding traffic characteristics. J Adv Transp 49:535–552
44. Hegyi A, De Schutter B, Hellendoorn H (2005) Optimal coordination of variable speed limits to suppress shock waves. IEEE Trans Intell Transp Syst 6:102–112

45. Carlson RC, Papamichail I, Papageorgiou M (2011) Local feedback-based mainstream traffic flow control on motorways using variable speed limits. IEEE Trans Intell Transp Syst 12:1261–1276
46. Sacone S, Pasquale C, Siri S, Ferrara A (2021) Centralized and decentralized schemes for platoon control in freeway traffic systems. In: Proceedings of the IEEE conference on decision and control, pp 2665–2670
47. Čičić M, Pasquale C, Siri S, Sacone S, Johansson KH (2022) Platoon-actuated variable area mainstream traffic control for bottleneck decongestion. Eur J Control 68
48. Scattolini R (2009) Architectures for distributed and hierarchical model predictive control: a review. J Process Control 19:723–731
49. Ploeg J, van de Wouw N, Nijmeijer Henk (2014) $\mathcal{L}_p$ string stability of cascaded systems: application to vehicle platooning. IEEE Trans Control Syst Technol 22:786–793
50. Monteil J, Russo G, Shorten R (2019) On $\mathcal{L}_\infty$ string stability of nonlinear bidirectional asymmetric heterogeneous platoon systems. Automatica 105:198–205
51. Pasquale C, Anghinolfi D, Sacone S, Siri S, Papageorgiou M (2016) A comparative analysis of solution algorithms for nonlinear freeway traffic control problems. In: Proceedings of the 19th IEEE intelligent transportation systems conference, pp 1773–1778
52. Ferrara A, Nai Oleari A, Sacone S, Siri S (2015) Freeways as systems of systems: a distributed model predictive control scheme. IEEE Syst J 9:312–323
53. Pasquale C, Sacone S, Siri S, Ferrara A (2021) Hierarchical centralized/decentralized event-triggered control of multi-class traffic networks. IEEE Trans Control Syst Technol 29:1549–1564
54. Frejo JRD, Papamichail I, Papageorgiou M, De Schutter B (2019) Macroscopic modeling of variable speed limits on freeways. Transp Res Part C 100:15–33

Chapter 4
Streaming-Data-Driven Traffic Density Estimation Using Gaussian Processes

Yiolanda Englezou, Christos G. Panayiotou, and Stelios Timotheou

Abstract Inefficient mobility management leads to congestion with several socioeconomic and environmental adverse effects. Despite numerous research efforts and deployed technological solutions to address congestion, the problem still persists. One of the main reasons is attributed to the sparsity and low quality of traffic data. Recent advances in information and communication technologies have led to the emergence of new sensing devices with numerous capabilities for the collection of traffic data, that introduce new opportunities to traffic monitoring. This work provides an overview of such emerging sensing technologies and focuses on fine-grained traffic density estimation. A general streaming-data-driven probabilistic approach is designed to utilize different types of measurements. In this work, we examine how to utilize measurements from (i) fixed-location sensors, (ii) Connected and Automated Vehicles (CAVs), and (iii) Unmanned Aerial Vehicles (UAVs). The proposed methodology integrates the Gaussian process model within a Bayesian framework to effectively calculate traffic density estimates for different measurement types, even in cases where only a limited number of measurements are available within the designated time–space area being examined. This is achieved by using the available information of the time–space vehicle trajectory diagram that yields traffic density measurements which are extrapolated in time and space where no information has been obtained. A simulation study is performed to evaluate the estimation methodology utilizing measurements from the three different sensing technologies. Estimation results show that each sensing technology can be effective under different settings with their main difference being that CAVs and UAVs offer information with higher spatiotemporal resolutions compared to fixed-location sensors, while UAV-based sensing is effective even under low penetration rates of UAVs flying above the network and low percentages of network coverage.

Y. Englezou (✉) · C. G. Panayiotou · S. Timotheou
KIOS Research and Innovation Center of Excellence, and the Department of Electrical and Computer Engineering, University of Cyprus, Nicosia, Cyprus
e-mail: englezou.yiolanda@ucy.ac.cy

C. G. Panayiotou
e-mail: christosp@ucy.ac.cy

S. Timotheou
e-mail: stimo@ucy.ac.cy

P. Ioannou and A. A. Malikopoulos (eds.), *Transportation Mobility in Smart Cities*, Springer Tracts on Transportation and Traffic 21,
https://doi.org/10.1007/978-3-031-64769-7_4

4.1 Introduction

Cities are facing a continuous surge in urbanization and motorization, resulting in an excessive number of vehicles that lead to undesirable everyday phenomena, such as traffic congestion and accidents. The lack of effective traffic management has multiple negative consequences, including prolonged travel times, higher travel expenses, reduced quality of life for residents, substantial greenhouse gas emissions, and excessive fossil fuel consumption. Despite various research efforts and technological interventions implemented to address congestion, the issue still persists. One of the primary reasons behind this ongoing problem is the inadequate quality of traffic data [1] caused by insufficient sensing within the traffic network.

4.1.1 Sensing Technologies: An Overview

Traffic management tasks require traffic data collected using either fixed-location (or stationary) sensors, termed *fixed-location sensor data* [2, 3], or probe sensors (e.g., GPS, smartphones, connected vehicles) mounted on vehicles, termed *Floating Car Data (FCD)* [4, 5]. Such data differ both in the quality they provide, as well as the level of information detail. Further, data can be categorized as disaggregated and aggregated data [6]. Following, an overview of traditional and emerging sensing technologies is presented.

Fixed-Location Sensors: Fixed-location or stationary sensors are generally grouped into two categories: (i) intrusive and (ii) non-intrusive methods. Intrusive methods consist of a sensor placed on, or in, the road, e.g., inductive loop detectors, embedded magnetometers, while non-intrusive methods are based on remote observations, e.g., surveys, interviews, and cameras. Typically, fixed-location sensors collect traffic data such as vehicle speeds, flow, and occupancy at their installed locations, while other important traffic variables such as the density, can then be estimated by making some assumptions, e.g., knowledge of the average vehicle length. These data are disaggregated data that contain information of each individual vehicle and can be visualized through time–space diagrams. However, it is often necessary to aggregate this information due to storage and transmission limitations such that an average traffic measure is calculated at a specific temporal resolution, e.g., from a few minutes up to days. Road administrators have extensively employed fixed-location sensors on highways since the early 90s [7] as they are commonly available both on highways and urban networks. Nevertheless, fixed-location sensors have inherent limitations stemming from their sparse deployment [2, 3] and high installation costs [8]. Moreover, a major drawback is that they can only measure traffic parameters at their location [9]. Furthermore, the information they provide can be unreliable due to their susceptibility to disruptions caused by system errors [10].

Recent advances in Information Communication Technologies (ICT), e.g., navigation systems, have enabled collection of mobile data, i.e., data collected from mobile sensors (e.g., global navigation satellite system or GNSS, speedometer) mounted on floating (probe) vehicles. Such data, termed *Floating Car Data (FCD)* [4], offer several advantages compared to fixed-location sensors, for example, the information collected represents a wider spatiotemporal domain as probe vehicles continuously collect information while driving [11], either offline or online. The most common on-vehicle mobile sensor is the GNSS, such as GPS, GLONASS, and Galileo, which typically provides a vehicle's position and speed, yielding the vehicle's trajectory (disaggregated data), while aggregation of the raw data includes the average speed for a specific spatiotemporal resolution. The main drawbacks of FCD are that they are sparse, bound to the uncontrollable route of the collecting vehicle, hard to associate with traffic volumes [1], and suffer from privacy issues [12]. The resulting low quality of traffic data limits the estimation and prediction performance of traffic monitoring, ultimately leading to ineffective traffic control strategies.

Connected and Automated Vehicles: Advanced on-vehicle sensors, such as LiDARs and radars, are used for vehicle automation purposes, such as advanced driver assistance system and autonomous vehicles. Connected and Automated Vehicles (CAVs) present innovative capabilities for significantly improved traffic sensing. These vehicles enable the collection of extensive data concerning their surroundings, encompassing important information such as the space and time headway from neighboring vehicles, and the surrounding traffic density. Such data are termed *extended FCD (xFCD)*. The potential and efficiency of CAVs to estimate important traffic variables and disseminate traffic information to the surrounding vehicles were explored by [13]. On the contrary, as all probe sensors, CAVs suffer from privacy issues, because the position of a given vehicle is continuously recorded [12], and can only provide information of the traffic at specific subsets of the network, i.e., they provide traffic data only within their route of travel in the specific time period under study [14].

Unmanned Aerial Vehicles: Unmanned Aerial Vehicles (UAVs) or drones, have gained significant attention as an alternative to overcome the aforementioned limitations of the currently used sensing technologies [15–19]. Their effectiveness in monitoring vehicular and pedestrian movements [20] and recent advances in communication technology and computer vision [21, 22] place them on the top of the list for future intelligent transportation systems [16, 23], despite their need for recharging [24]. UAVs have been proven useful for large infrastructure monitoring and environment surveillance [25]. They are equipped with a diverse range of imaging devices, such as day and night video cameras, infrared cameras, laser scanners, and road weather systems and they serve multiple transportation-related functions, including estimation of the Origin–Destination (OD) matrices, monitoring freeway conditions, promptly responding to incidents, and efficiently tracking vehicle movements at intersections [15] at a lower cost and faster than other technologies. They offer several key advantages in monitoring traffic networks, due to their ability to fly between areas at high altitudes and collect high-quality spatiotemporal traffic-related data [23, 26–30]. First, they are not bounded to traverse solely within the network, enabling them to deliberately move across various locations above the network to collect data [15, 32].

Second, UAVs have *multi-sensing* capabilities able to simultaneously measure multiple parameters at distinct aggregation levels such as vehicle trajectories and spacing at *vehicle-level* [33, 34], traffic density, turning ratios at intersections, and queue lengths at road *link level* [16, 28], and average speeds and densities at *regional level* [28, 30]. In addition, UAVs can even provide data associated with the pedestrian movement [35, 36]. Third, UAVs is a non-invasive technology offering economic viability, fast deployment, and reliability. Research concerning traffic monitoring using UAVs has primarily centered around remote capturing of traffic information over the network [16, 23]. The collected data serves various purposes, including surveillance, traffic congestion and road safety management, traffic signal optimization, highway infrastructure management, and the derivation of traffic parameters [26–28, 37]. While most studies have traditionally focused on a single UAV following a fixed trajectory, some research has explored solutions involving multiple UAVs with adaptive trajectories, e.g., [17]. Despite these advantages, their applicability has so far been limited to occasional surveillance of road networks through the provision of live video feeds to traffic operators, or the recording of videos for offline processing [28, 37] and extraction of historical traffic data [30, 31]. A recent review paper presents the developments in relation to the application of UAVs in transportation and identifies the challenges of the deployment of the UAV technology [38].

4.1.2 Traffic Monitoring and Novel Sensing Technologies

Traffic monitoring has been one of the major tools used for transportation operations and planning as it helps road operators and management systems in decision-making, enabling the implementation of various control strategies [39]. The most important monitoring tasks include Traffic State Estimation (TSE) [6, 40–47], origin–destination (OD) matrix estimation [48–51, 54], path prediction [55, 56], travel-time estimation [57–60], queue estimation [61, 62], and traffic state and demand prediction [63–65]. In this work, we focus on TSE utilizing measurements from different sensing technologies.

Traditional traffic sensing and estimation have heavily relied on data obtained from fixed-location sensors due to its widespread availability [2]. However, fixed-location methods come with inherent limitations, including sparse coverage, high installation costs [66], and the restriction of traffic state measurements to specific positions [9]. Moreover, fixed-location sensors tend to provide unreliable and low-quality information due to susceptibility to system errors [10]. On the other hand, FCD-based TSE methods offer significant advantages. These methods allow continuous observation and analysis of traffic conditions and consequently they provide higher spatiotemporal resolution [11]. FCD-based approaches leverage traffic flow models designed to accept vehicle trajectory data. A novel model-based approach that utilizes the Lighthill–Whitham–Richards model and integrates GNSS mobile data into highway traffic models was developed by [67], while an extended Kalman filtering approach was later introduced for the solution of this problem [68]. Several research works have also explored TSE methods that leverage both data from fixed-location sensors and FCD. Data-driven approaches that combine historical data from fixed-location sensors with mobile data to extract traffic information, such as the fundamental diagram, were proposed to estimate current traffic states [69, 70]. In addition, Kalman filtering model-based TSE approaches were explored, utilizing average speed measurements from connected vehicles on single-lane and multi-lane highways and flow measurements from fixed-location sensors [5, 71, 72].

Recently, novel traffic sensing and estimation methods utilizing xFCD have been introduced. These methods involve the collection of data from CAVs equipped with various sensors to measure their net space headway, or gap from the leading vehicle, along with their position and speed. One such approach employs a streaming-data-driven method to estimate traffic density on highways, based on Edie's definitions, using only xFCD [73]. Although this methodology exhibits moderate estimation performance, it has been successfully applied for TSE in an urban area of Barcelona [74]. Streaming-data-driven methods typically rely on streaming data, and no historical data are needed, while traffic modeling assumptions are not required. To tackle microscopic vehicle behaviour fluctuations, an extended approach was developed, leveraging xFCD to estimate traffic states and fundamental diagram parameters. The traffic states are then further refined using a model-based data assimilation method [75]. In another recent study, two methods based on Bayesian inference and deep learning were proposed for estimating traffic flow rate under free-flow conditions using headway data acquired by CAVs [76]. These methods assume error-free measurements and rely on a large amount of historical data (spanning from multiple days to years) to establish an initial probability density function of traffic flow, which is updated as new measurements are acquired. Last, a fully Bayesian data-driven approach for TSE in multi-lane highways was proposed based only on xFCD, that achieves high-quality estimation results even at a low penetration rate of CAVs [47].

Practical reasons generated by inherent disadvantages, such as the quality and quantity of measurements utilized for monitoring tasks, the assumed model's inadequacy to adapt to real situations and the associated computational cost, privacy, and security concerns, and the requirement of a vast amount of historical data, lead to the necessity to consider more suitable approaches, such as combining models in a common framework, or obtaining a more detailed variety of measurements. In this work, we focus on traffic density estimations using streaming-data-driven measurements from three different sensing technologies to evaluate the quality of information obtained from the time–space vehicle trajectory diagram formed from each technology. A general probabilistic approach is developed that does not consider any underlying traffic model or historical data, while the traffic density is given as a probability density function along with probability bounds to quantify the uncertainty of the estimations.

Structure: The remainder of this chapter is organized as follows. In the next section, we formulate the traffic density estimation problem for the three different types of measurements. In Sect. 4.3, we introduce the Gaussian process model tailored to the estimation of traffic density and develop a Bayesian streaming-data-driven framework for the solution of the traffic density estimation problem. In Sect. 4.4, we present a simulation study that is performed to evaluate the proposed estimation methodology using measurements from three different sensing technologies. Last, Sect. 4.5 summarizes and outlines the main outcomes of this work.

4.2 Problem Formulation

Assume a traffic network modeled as a directed graph $\mathcal{G} = (\mathcal{V}, \mathcal{E})$, where the set of vertices, $\mathcal{V}$, represents road junctions and the set of edges, $\mathcal{E}$, represents road segments, where $N^E = |\mathcal{E}|$ is the total number of segments. Each segment consists of N^L distinct road lanes. We monitor the traffic network for a fixed time period T and denote as $\mathcal{M}^\tau \subset \mathcal{E}$ the set of monitored segments at time-point τ^{sense}, sense $=$ {fixed, CAV, UAV},[1] while the total number of monitored segments is $N^\tau = |\mathcal{M}^\tau|$ such that $N^\tau \leq N^E$.

The objective of this work is to derive fine-grained traffic density estimations for all road segments of a given network, over a fixed time period T. Traffic density estimations correspond to fixed time-window durations T^W, such that $N^W = T/T^W$ different density values are obtained in T, with $\mathcal{N}^W = \{1, \ldots, N^W\}$.

A time–space region $\mathcal{A}_{ij}$ is defined through a space region $\mathcal{R}_i = \{x | x_{i-1} \leq x \leq x_i\}$, $i \in \mathcal{N}^E$, and a time-window $\mathcal{T}_j = \{\tau | \tau_{j-1} \leq \tau \leq \tau_j\}$, $j \in \mathcal{N}^W$. The true traffic density, ρ^{true}, in a specific time–space region $\mathcal{A}_{ij}$ is computed according to Edie's definition [77] as

$$\rho^{true} = \frac{t(\mathcal{A}_{ij})}{|\mathcal{A}_{ij}|}, \tag{4.1}$$

where $t(\mathcal{A}_{ij})$ is the total time spent of all vehicles in $\mathcal{A}_{ij}$ and $|\mathcal{A}_{ij}|$ is the time–space area of region $\mathcal{A}_{ij}$. Following, we formulate the traffic density estimation problem with respect to the available measurements in the network under study, and define the time–space region with respect to three considered sensing technologies: fixed-location sensors, CAVs, and UAVs.

Because the analysis is uniform across all time–space regions, we drop indices i, j in the remainder of this work.

4.2.1 Fixed-Location Sensor Measurements

Assume that fixed-location sensors are deployed on specific road segments, i.e., monitored segments, in a traffic network such that they measure the traffic density in different time-windows. Traditionally, fixed-location sensors observe the traffic continuously at their installed location and they report (i) vehicle counts, $c(t)$, i.e., the number of vehicles that crossed the sensor during the last period of T_s seconds and (ii) occupancy, $o(t)$, i.e., the percentage of time in the last T_s seconds where a

[1] The time-points may differ for each sensing technology used to monitor specific segments of the traffic network under study.

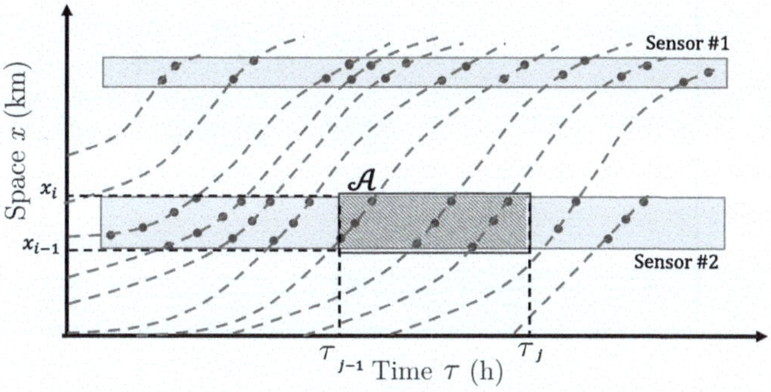

Fig. 4.1 Fixed-location sensor traffic data on time–space diagram

vehicle was driving above the detector, for each lane of the road segment. Using occupancy measurements, estimates of the traffic density for each lane can be obtained, subject to some uncertainty, through $\rho_i(t) = o_i(t)/\bar{l}_v$, $\forall t \in \mathcal{T}$ [78], where $\bar{l}_v$ is the average effective vehicle length at the sensor location, that can vary with time, but is sometimes estimated precisely and available as public information [79]. In this work, we utilize Edie's definition of traffic density (4.1) and obtain density measurements through the vehicle trajectory time–space diagram as shown in Fig. 4.1. In the figure, the blue dashed lines represent vehicle trajectories, the green shaded areas represent the monitored regions over time and space from different fixed sensors, and the green bullets indicate measurements obtained in a specific time–space region. The space region corresponds to a road segment of the network. The time-window duration T^W is the aggregation time unit of the measurements such that we seek to find a new traffic density estimation every T^W, yielding the following data:

- The *traffic density* of road segment i, given as $\rho_i^{fixed} = [\rho_{i,1}^{fixed}, \ldots, \rho_{i,N^W}^{fixed}]^{\mathrm{T}}$ [veh/km].

Above, N^W is the total number of measurements obtained in time, equal to the number of all time-windows $\mathcal{T}_j$, $j \in \mathcal{N}^W$. Without loss of generality, we assume that the measured density at the fixed point of the road segment is constant across the whole segment. Note that the set of monitored segments remains fixed over time and no information is provided for road segments that are not equipped with a sensor.

4.2.2 CAV-Based Measurements

In the traffic network under study, we assume that there are two types of vehicles: CAVs with the ability to measure the distance to their leading vehicle, and Human-Driven Vehicles (HDVs) without this measuring capability. We denote as $\mathcal{N}^V$ the set of all vehicles and $\mathcal{N}^C$ the set of CAVs in a given time–space region $\mathcal{A}$, with $\mathcal{N}^C \subseteq \mathcal{N}^V$, and $N^V = |\mathcal{N}^V|$, $N^C = |\mathcal{N}^C|$. Each CAV $c \in \mathcal{N}^C$ in $\mathcal{A}$ records the following measurements every $T^S \ll T^W$ time units:

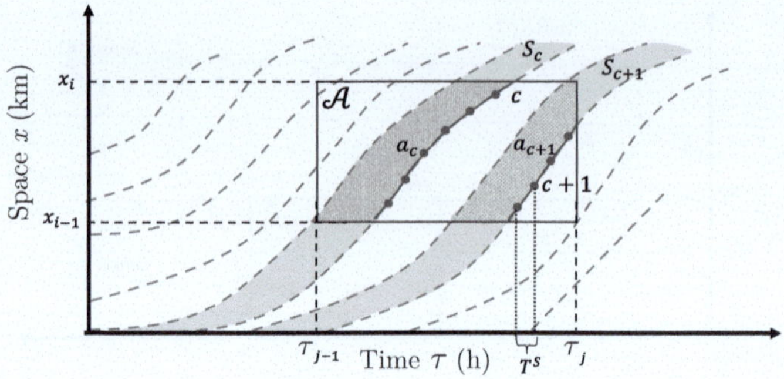

Fig. 4.2 CAV-based data in time–space diagram. Red bullets indicate the position recorded from a CAV. Red solid lines and blue dashed lines indicate the derived trajectories of CAVs and HDVs, respectively. The pink shaded area is the time–space region between CAV c and its leading vehicle, $\mathcal{S}_c$, and the hatched area is the time–space region between CAV c and its leading vehicle in $\mathcal{A}$, a_c

- The *speed* $\mathbf{u}_c = [u_{c,1}, \ldots, u_{c,M_c}]^{\mathrm{T}}$ [km/h].
- The *time spent* of CAV $c \in \mathcal{N}^C$ in the specific region $\mathcal{A}$, approximately given by $t_c = \hat{\tau}_{c,M_c} - \hat{\tau}_{c,1}$ [h], where $\hat{\tau}_{c,M_c}$ and $\hat{\tau}_{c,1}$ are the time recordings of the last and first measurements of the CAV, respectively.
- The *area* of the time–space region between CAV $c \in \mathcal{N}^C$ and its leading vehicle in region $\mathcal{A}$, denoted as $a_c = \mathcal{A} \cap \mathcal{S}_c$ [km×h]. By using the position and the distance between the front-bumper of the ego vehicle and the rear-bumper of its leading vehicle, we derive $\mathcal{S}_c$, the time–space region between CAV c and its leading vehicle. Finally, we calculate a_c, which is the value of the intersecting area between $\mathcal{A}$ and $\mathcal{S}_c$, as shown in Fig. 4.2.

For CAV-based measurements, the true traffic density, ρ^{true}, in a specific time–space region can be re-written as

$$\rho^{true} = \frac{\sum_{n \in \mathcal{N}^V} t_n}{\sum_{n \in \mathcal{N}^V} a_n}, \tag{4.2}$$

which indicates that, if all vehicles in the specific road segment are CAVs, then the traffic density is known. However only a subset of vehicles in the network are CAVs hence Equation (4.2) becomes

$$\rho^{CAV} = \frac{\sum_{c \in \mathcal{N}^C} t_c}{\sum_{c \in \mathcal{N}^C} a_c}, \tag{4.3}$$

indicating the traffic density in a specific time–space region.

Without loss of generality, we assume that each road segment in the network is a distinct time–space region. Given that a penetration rate of CAVs is deployed in the network under study, the measurements provided by CAVs, $\mathbf{u}_c$, t_c and a_c, $\forall c \in \mathcal{N}^C$, yield the following data:

- The *traffic density* of road segment i calculated using Equation (4.3), given as $\boldsymbol{\rho}_i^{CAV} = [\rho_{i,1}^{CAV}, \ldots, \rho_{i,L_i}^{CAV}]^{\mathrm{T}}$ [veh/km].
- The time-window indices associated with the traffic density measurements, given as $\boldsymbol{\tau}_i^{CAV} = [\tau_{i,1}^{CAV}, \ldots, \tau_{i,L_i}^{CAV}]$.

Above, $\rho_{i,l}^{CAV}$ is the lth density measurement obtained for road segment i; $\tau_{i,l}^{CAV}$ denotes the index of the time-window of the lth density measurement such that $\tau_{i,l}^{CAV} = j'$ indicates that the lth measurement corresponds to time-window $\mathcal{T}_{j'}$. In addition, L_i gives the total number of measurements for the specific road segment. The absence of index $\hat{j}$ in $\boldsymbol{\tau}_i^{CAV}$ indicates that no CAV has observed any part of road segment i over time-window $\mathcal{T}_{\hat{j}}$.

4.2.3 UAV-Based Measurements

We consider a collection of N^U UAVs hovering over a network under investigation, with $\mathcal{N}^U$ the set of all UAVs such that $N^U = |\mathcal{N}^U|$. We adopt the assumption that all UAVs share identical flying and sensing capabilities. The network being studied is discretized into P UAV sampling points, represented by $\mathcal{S}^U = \{s_1, \ldots, s_P\}$. Each UAV achieves complete coverage of the road infrastructure by visiting all sampling points within a fixed sampling cycle. In addition, UAVs hover at an altitude of h^U meters above the network. Without loss of generality assume that the field of view of the UAV cameras is 90 degrees, resulting in a circular line-of-sight with a radius of R meters, where $R = h^U$. Accounting for the impact of obstacles within the UAVs' line-of-sight [29], or the UAVs' battery life [21] is beyond the scope of this work.

Every UAV moves between sampling points at a constant velocity equal to v^{UAV} [m/s] such that it requires $T_{ss'}^U = d_{ss'}/v^{UAV}$ [s] to travel from sampling point s to s', with $s, s' \in \mathcal{S}^U$. Here, $d_{ss'}$ represents the distance between these two sampling points. Figure 4.3 illustrates the time–space diagram outlining the UAV-based mea-

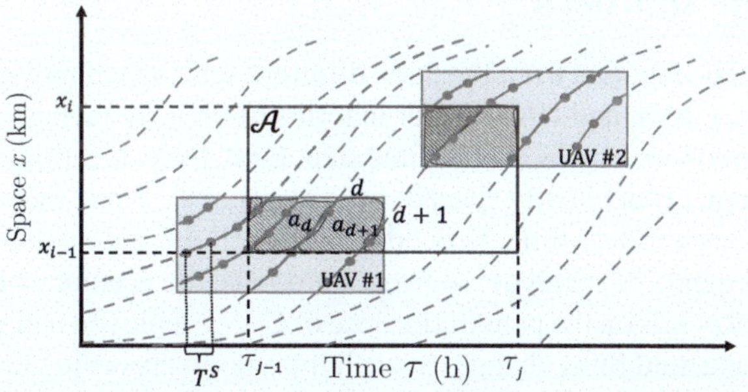

Fig. 4.3 UAV-based traffic data on time–space diagram. The orange-shaded area is the time–space region between a vehicle d and its leading vehicle within the line-of-sight of a UAV, and the hatched area is the time–space region between a vehicle d and its leading vehicle within the line-of-sight of a UAV, a_d

surements, where each rectangle corresponds to the time–space region observed by an individual UAV.

UAVs are equipped with cameras and have computer vision capabilities for measuring vehicle speed, position, acceleration, etc., as well as road link density and flow. We define as $\mathcal{N}^D$ the set of all vehicles that are within the line-of-sight of any UAV in time–space region $\mathcal{A}$. Hence the traffic density of $\mathcal{A}$ is given as

$$\rho^{UAV} = \frac{\sum_{d \in \mathcal{N}^D} t_d}{\sum_{d \in \mathcal{N}^D} a_d}, \tag{4.4}$$

where a_d, $d \in \mathcal{N}^D$ is the area of the time–space region between a vehicle and its leading vehicle, and t_d is the time spent of vehicle d in region $\mathcal{A}$, within the line-of-sight of any UAV, as shown in Fig. 4.3.

At different sampling points, UAVs can observe specific subsets of the network, such that multiple UAVs observe potentially overlapping subsets of time–space region $\mathcal{A}$. Consolidating these time–space subsets from different UAVs yields the following measurements:

- The *density* of road segment i calculated using Equation (4.4), given as $\boldsymbol{\rho}_i^{UAV} = [\rho_{i,1}^{UAV}, \ldots, \rho_{i,L_i'}^{UAV}]^{\mathrm{T}}$ [veh/km].
- The time-window indices associated with traffic density measurements, given as $\boldsymbol{\tau}_i^{UAV} = [\tau_{i,1}^{UAV}, \ldots, \tau_{i,L_i'}^{UAV}]^{\mathrm{T}}$ [s].

As before, $\rho_{i,l}^{UAV}$ is the lth measurement obtained for road segment i, $\tau_{i,l}^{UAV} = j'$ indicates that the lth measurement corresponds to time-window $\mathcal{T}_{j'}$ and L_i' is the total number of measurements in the specific road segment. The absence of index $\hat{j}$ in $\boldsymbol{\tau}_i^{UAV}$ indicates that no UAV has observed any part of road segment i over time-window $\mathcal{T}_{\hat{j}}$.

4.3 Solution Approach

The measurements obtained from the three different sensing technologies, i.e., fixed-location sensors, CAVs, and UAVs differ in terms of their frequency obtained, while the set of measured links varies at each time step. First, fixed-location sensors provide density measurements at equally spaced time-points in $\mathcal{T}$, while the set of measured links remains constant over time. Second, CAV-based measurements provide traffic density measurements of segment i at irregular points, the accuracy of traffic density measurements depends on the penetration rate of CAVs deployed in the network [73], and the set of measured links changes at each time-point according to the position of the CAVs. Last, similar to the CAV-based measurements, UAV-based measurements are not continuous with respect to different time-points and the set of measured links changes at each time-point according to the sampling position of the set of UAVs.

The objective of this study is to utilize probabilistic interpolation methods, specifically a Gaussian process (GP) interpolator to derive fine-grained density estimations within a specific time interval duration. The GP model is a valuable tool for statistical and machine learning applications due to its flexibility in modeling complex relationships without assuming a functional form for the available data; it has been applied to address transportation-related tasks in the literature. Further, the covariance matrix of the GP posterior distribution serves as an indicator of uncertainty for particular time steps and locations where neither CAVs nor UAVs have gathered measurements, while they perform well even for relatively small datasets. A recent study introduced a data-driven approach to estimate traffic states at signalized intersections, employing a Bayesian filtering method and leveraged GPs to represent the traffic dynamics' state space model, while the model relied on a substantial volume of historical data [52]. That research investigated a single four-legged signalized intersection, utilizing data from either loop detectors or connected vehicles, and was expanded to incorporate a traffic model into the data-driven framework [53].

This work aims to generate fine-grained *virtual measurements*, i.e., measurements that have not been directly observed, along with their quantified uncertainty, a useful aspect for efficient decision-making in real-world applications. In contrast with other works in the literature, we aim to derive traffic density estimations using only streaming-data-driven measurements obtained from (i) fixed-location sensors, (ii) CAVs, and (iii) UAVs, without reliance on historical data or a specific traffic model. Virtual measurements can then be employed in traffic monitoring and control applications for improved results [54].

4.3.1 Gaussian Process Model

The Gaussian process (GP) model extends the concept of the Gaussian probability distribution. While a probability distribution characterizes random variables in the form of scalars or vectors, a GP is employed to define a distribution encompassing functions, as detailed in [80]. GPs were used to model the behavior of an unknown mathematical function and GP modeling is characterized as the Bayesian[2] alternative for classical nonparametric methods such as splines. In general, a GP represents prior information about an unknown function, updated using data.

Given data in the format of $(\boldsymbol{\rho}_i, \boldsymbol{\tau}_i)$, where $\boldsymbol{\tau}_i \in \mathbb{Z}_+^{Li}$ and $\boldsymbol{\rho}_i$ represents the vector containing the traffic density for segment i, obtained at irregularly spaced sampling time-points denoted by $\boldsymbol{\tau}_i$, we make the assumption that

$$\boldsymbol{\rho}_i = f(\boldsymbol{\tau}_i) + \varepsilon_i. \tag{4.5}$$

In this equation, function $f(\cdot)$ is unknown, and each ε_i signifies the measurement noise, independently drawn from a normal distribution $\varepsilon_i \sim N(\mathbf{0}, \sigma_{i,\varepsilon}^2 \mathbf{I})$. It is important to note that each road segment i is treated independently, and a distinct Gaussian

[2] Bayesian inference is a natural way to update prior beliefs for unknown parameters through the posterior distribution and obtain marginal distributions of interest.

Process (GP) model is considered for each road segment. Consequently, for simplicity, we will subsequently omit the subscript i.

A GP model is fully defined by its mean and variance functions. In our approach, we assume the GP prior as follows:

$$f(\boldsymbol{\tau}) \sim \mathrm{GP}\left[\mathbf{h}^{\mathrm{T}}(\boldsymbol{\tau})\mathbf{w}, \sigma^2 k(\boldsymbol{\tau}, \boldsymbol{\tau}'; \gamma)\right]. \tag{4.6}$$

In this expression, $\mathbf{h}(\boldsymbol{\tau}) = [h_0(\boldsymbol{\tau}), h_1(\boldsymbol{\tau}), \ldots, h_{q-1}(\boldsymbol{\tau})]^{\mathrm{T}}$ represents a q-dimensional vector of regression functions which we assume are known.[3] The vector $\mathbf{w} = (w_0, w_1, \ldots, w_{q-1})^{\mathrm{T}} \in \mathcal{W}$ comprises unknown regression coefficients, commonly referred to as trend parameters. Moreover, $0 \leq k(\boldsymbol{\tau}, \boldsymbol{\tau}'; \gamma) \leq 1$ is the known correlation function, $\gamma \in \Gamma = (0, \infty)$ is the unknown correlation parameter, and $\sigma^2 > 0$ is the unknown constant variance. We have opted for constant regression functions $\mathbf{h}(\boldsymbol{\tau})$, with their corresponding coefficients $\mathbf{w}$ to be determined from the data. Given the aforementioned GP prior (4.6), it follows that any finite set of function evaluations $\mathbf{f} = [f(\tau_1), \ldots, f(\tau_{Li})]^{\mathrm{T}}$ conforms to a multivariate normal distribution, expressed as

$$\mathbf{f} \sim \mathrm{N}\left[\mathbf{H}\mathbf{w}, \sigma^2\mathbf{C}(\gamma)\right]. \tag{4.7}$$

Here, $\mathbf{H} = [\mathbf{h}(\tau_1); \mathbf{h}(\tau_2) \ldots \mathbf{h}(\tau_{Li})]^{\mathrm{T}}$ denotes the $Li \times q$ model matrix, and $\mathbf{C}(\gamma)$ is the correlation matrix, with the (r, l)th element $\mathbf{C}(\gamma)_{rl} = k(\tau_r, \tau_l, \gamma)$, where $\tau_r, \tau_l \in \mathbb{Z}_+$. We choose the squared exponential correlation function, a commonly used kernel, given by

$$k(\tau, \tau'; \gamma) = \exp\left[-\gamma \mid\mid \tau - \tau' \mid\mid^2\right], \tag{4.8}$$

where $\gamma > 0$ is the unknown smoothness (correlation) parameter and $||\cdot||$ is the Euclidean norm.

Combining (4.5) and (4.7), we get

$$\boldsymbol{\rho} \mid \mathbf{w}, \sigma^2, \gamma, \sigma_\varepsilon^2 \sim \mathrm{N}(\mathbf{H}\mathbf{w}, \sigma^2\mathbf{C}(\gamma) + \sigma_\varepsilon^2\mathbf{I}_{Li}),$$

where $\boldsymbol{\rho} = [\rho_1, \ldots, \rho_{Li}]^{\mathrm{T}}$ is the Li-vector of measurements and $\mathbf{I}_{Li}$ is the $Li \times Li$ identity matrix. For computational purposes, the state of the art is to reparameterize the above equation to

$$\boldsymbol{\rho} \mid \mathbf{w}, \sigma^2, \gamma, \nu^2 \sim \mathrm{N}\left(\mathbf{H}\mathbf{w}, \sigma^2[\mathbf{C}(\gamma) + \nu^2\mathbf{I}_{Li}]\right),$$

where ν^2 is defined as the ratio of the noise variance σ_ε^2 to the signal variance σ^2, representing the noise-to-signal ratio. Consequently, moving forward, we will work with the updated covariance matrix denoted as $\sigma^2\boldsymbol{\Sigma}$, where

$$\boldsymbol{\Sigma} = [\mathbf{C}(\gamma) + \nu^2\mathbf{I}_{Li}].$$

[3] Typically the mean function of the GP prior is assumed to be a constant value set to zero [80].

With this adjustment, the likelihood can be expressed as

$$\pi(\boldsymbol{\rho}|\mathbf{w},\sigma^2,\gamma,\nu^2)=\frac{\exp\left\{\frac{1}{2\sigma^2}[(\boldsymbol{\rho}-\mathbf{H}\mathbf{w})^{\mathrm{T}}\boldsymbol{\Sigma}^{-1}(\boldsymbol{\rho}-\mathbf{H}\mathbf{w})]\right\}}{(2\pi\sigma^2)^{Li}|\boldsymbol{\Sigma}|^{\frac{1}{2}}}. \tag{4.9}$$

4.3.2 Gaussian Process Inference

An important aspect of the GP model involves estimating the traffic density, denoted as $\tilde{\rho}$, at an unobserved time-point $\tilde{\tau}$. To achieve this, we formulate the predictive distribution of $\tilde{\rho}$ using the statistical model (4.9). The joint prior distribution of $\boldsymbol{\rho}$ and $\tilde{\rho}$, given all unknown parameters $\mathbf{w}$, σ^2, γ, ν^2, can be expressed as follows:

$$\begin{pmatrix}\tilde{\rho}\\ \boldsymbol{\rho}\end{pmatrix}\Bigg|\,\mathbf{w},\sigma^2,\gamma,\nu^2\sim \mathrm{N}\left(\begin{pmatrix}\mathbf{h}^{\mathrm{T}}(\tilde{\tau})\mathbf{w}\\ \mathbf{H}\mathbf{w}\end{pmatrix},\ \sigma^2\begin{pmatrix}(1+\nu^2) & \mathbf{c}(\tilde{\tau})^{\mathrm{T}}\\ \mathbf{c}(\tilde{\tau}) & \boldsymbol{\Sigma}\end{pmatrix}\right). \tag{4.10}$$

Here, $\mathbf{c}(\tilde{\tau})=[k(\tilde{\tau},\tau_1;\gamma),\ldots,k(\tilde{\tau},\tau_{Li};\gamma)]^{\mathrm{T}}$ represents the Li-dimensional vector of correlations between the traffic density at each sampling time-point $\boldsymbol{\tau}$ and the traffic density at the new time-point $\tilde{\tau}$ [81]. By employing standard results, we obtain the subsequent conditional posterior distribution:

$$\tilde{\rho}\mid\boldsymbol{\rho},\mathbf{w},\sigma^2,\gamma,\nu^2\sim\mathrm{N}\left(\mu(\tilde{\tau}),s^2(\tilde{\tau})\right), \tag{4.11}$$

$$\mu(\tilde{\tau})=\mathbb{E}(\tilde{\rho}\mid\boldsymbol{\rho},\mathbf{w},\sigma^2,\gamma,\nu^2)=\mathbf{h}(\tilde{\tau})^{\mathrm{T}}\mathbf{w}+\mathbf{c}(\tilde{\tau})^{\mathrm{T}}\boldsymbol{\Sigma}^{-1}[\boldsymbol{\rho}-\mathbf{H}\mathbf{w}], \tag{4.12}$$

$$s^2(\tilde{\tau})=\mathrm{var}(\tilde{\rho}\mid\boldsymbol{\rho},\mathbf{w},\sigma^2,\gamma,\nu^2)=\sigma^2[(1+\nu^2)-\mathbf{c}(\tilde{\tau})^{\mathrm{T}}\boldsymbol{\Sigma}^{-1}\mathbf{c}(\tilde{\tau})]. \tag{4.13}$$

Equations (4.12)–(4.13) reveal that if all parameters $\mathbf{w}$, σ^2, γ, and ν^2 were known, the traffic density at $\tilde{\tau}$ would follow a normal distribution with a predetermined mean $\mu(\tilde{\tau})$ and variance $s^2(\tilde{\tau})$. However, since these parameters are not known and require integration, the unconditional posterior predictive distribution of the traffic density $\pi(\tilde{\rho}|\boldsymbol{\rho})$ needs to be derived. In the context of representing uncertainty regarding unknown parameters, the Bayesian approach offers a natural framework. The core principle of Bayesian inference revolves around continually refining prior beliefs about events as new measurements are acquired. In the subsequent section, we describe the Bayesian predictive inference of the traffic density. For a more comprehensive understanding of Bayesian inference, further insights can be found in [82].

4.3.3 Bayesian Inference

Prior distributions need to be specified for all the parameters whose values are unknown. A common approach is to assume that trend parameters and the GP variance are independent of the correlation parameter and the noise-to-signal ratio, which are also independent from each other. Hence the joint prior distribution assumed in this work is $\pi(\mathbf{w}, \sigma^2, \gamma, \nu^2) = \pi(\mathbf{w}|\sigma^2)\pi(\sigma^2)\pi(\gamma)\pi(\nu^2)$ [80].

For the parameters $\mathbf{w}$ and σ^2, we opt for conjugate[4] prior distributions. Specifically, we assign a normal distribution to $\mathbf{w}|\sigma^2$ and an inverse-gamma distribution to σ^2, $\mathbf{w}|\sigma^2 \sim N(\mathbf{w}_0, \sigma^2\mathbf{S})$ and $\sigma^2 \sim \mathrm{IG}(a, b)$. In these expressions, $\mathbf{w}_0$ represents the q-dimensional vector denoting known prior means, $\mathbf{S}$ is a known symmetric, positive definite $q \times q$ matrix, and $a, b > 0$ are known hyperparameters. Their combined prior density forms a normal-inverse-gamma distribution:

$$(\mathbf{w}, \sigma^2) \sim \mathrm{NIG}\,(\mathbf{w}_0, \mathbf{S}, a, b)\,. \tag{4.14}$$

By applying Bayes' rule along with Eqs. (4.9) and (4.14), we can compute the posterior density of $\mathbf{w}$ and σ^2, denoted as $\pi(\mathbf{w}, \sigma^2|\boldsymbol{\rho}, \gamma, \nu^2)$. This posterior distribution conforms to a normal-inverse-gamma distribution:

$$\pi(\mathbf{w}, \sigma^2|\boldsymbol{\rho}, \gamma, \nu^2) \sim \mathrm{NIG}(\mathbf{w}_*, \boldsymbol{\Sigma}_*, a_*, b_*) \tag{4.15}$$

$$\mathbf{w}_* = \left(\mathbf{H}^{\mathrm{T}}\boldsymbol{\Sigma}^{-1}\mathbf{H} + \mathbf{S}^{-1}\right)^{-1}\left(\mathbf{H}^{\mathrm{T}}\boldsymbol{\Sigma}^{-1}\boldsymbol{\rho} + \mathbf{S}^{-1}\mathbf{w}_0\right) \tag{4.16}$$

$$\boldsymbol{\Sigma}_* = \left(\mathbf{H}^{\mathrm{T}}\boldsymbol{\Sigma}^{-1}\mathbf{H} + \mathbf{S}^{-1}\right)^{-1} \tag{4.17}$$

$$a_* = a + \frac{Li}{2} \tag{4.18}$$

$$b_* = b + \frac{1}{2}\left[(\boldsymbol{\rho} - \mathbf{H}\mathbf{w}_0)^{\mathrm{T}}\left(\boldsymbol{\Sigma} + \mathbf{HSH}^{\mathrm{T}}\right)^{-1}(\boldsymbol{\rho} - \mathbf{H}\mathbf{w}_0)\right]. \tag{4.19}$$

To derive the posterior predictive distribution for $\tilde{\rho}$ at a new time-point $\tilde{\tau}$, it is necessary to integrate out all the unknown parameters by considering their marginal posterior distributions. Beginning with the integration of the trend parameters $\mathbf{w}$, we utilize the conditional posterior predictive distribution (4.11). This distribution's mean and variance are defined by (4.12)–(4.13), respectively. In addition, we factor in the marginal posterior distribution $\mathbf{w}|\boldsymbol{\rho}, \sigma^2, \gamma, \nu^2 \sim \mathrm{N}(\mathbf{w}_*, \sigma^2\boldsymbol{\Sigma}_*)$, which can be easily deduced from (4.15). The reformulated form of Eq. (4.12) can be expressed as

$$\mu(\tilde{\tau}) = \mathbf{a}_{\tilde{\rho}} + \mathbf{b}_{\tilde{\rho}}^{\mathrm{T}}\mathbf{w},$$

where $\mathbf{a}_{\tilde{\rho}} = \mathbf{c}(\tilde{\tau})^{\mathrm{T}}\boldsymbol{\Sigma}^{-1}\boldsymbol{\rho}$, $\mathbf{b}_{\tilde{\rho}}^{\mathrm{T}} = \mathbf{f}(\tilde{\tau})^{\mathrm{T}} - \mathbf{c}(\tilde{\tau})^{\mathrm{T}}\boldsymbol{\Sigma}^{-1}\mathbf{H}$ and

$$\mu(\tilde{\tau}) \mid \boldsymbol{\rho}, \sigma^2, \gamma, \nu^2 \sim N\left[\mathbf{a}_{\tilde{\rho}} + \mathbf{b}_{\tilde{\rho}}^{\mathrm{T}}\mathbf{w}_*,\ \mathbf{b}_{\tilde{\rho}}^{\mathrm{T}}(\sigma^2\boldsymbol{\Sigma}_*)\mathbf{b}_{\tilde{\rho}}\right].$$

Also, from (4.11), we have that

$$\tilde{\rho} - \mu(\tilde{\tau}) \mid \boldsymbol{\rho}, \mathbf{w}, \sigma^2, \gamma, \nu^2 \sim N\left[0, s^2(\tilde{\tau})\right], \tag{4.20}$$

[4] A conjugate prior is a distribution that produces a posterior distribution of the same family as the prior.

where $s^2(\tilde{\tau})$ is given by (4.13). It is easily observed from (4.20) that $\tilde{\rho} - \mu(\tilde{\tau})$ is statistically independent of $\mu(\tilde{\tau})$ and hence independent from $\mathbf{w}$, and consequently $\tilde{\rho} - \mu(\tilde{\tau})$ is independent of $\mu(\tilde{\tau})$ conditional on $\boldsymbol{\rho}$, σ^2, γ, and ν^2. Thus, conditional on $\boldsymbol{\rho}$, σ^2, γ, and ν^2, $\tilde{\rho}$ can be re-written as the sum of two independent random variables as

$$\tilde{\rho} = \left[\tilde{\rho} - \mu(\tilde{\tau})\right] + \mu(\tilde{\tau}).$$

Hence,

$$\tilde{\rho} \mid \boldsymbol{\rho}, \sigma^2, \gamma, \nu^2 \sim \mathrm{N}(\tilde{\mu}, \tilde{\sigma}^2),$$

$$\begin{aligned}\tilde{\mu} &= \mathbf{a}_{\tilde{\rho}} + \mathbf{b}_{\tilde{\rho}}^{\mathrm{T}} \mathbf{w}_* \\ &= (\mathbf{f}^{\mathrm{T}}(\tilde{\boldsymbol{\tau}}) - \mathbf{c}(\tilde{\tau})^{\mathrm{T}} \boldsymbol{\Sigma}^{-1}\mathbf{H})\mathbf{Q}^{-1}\mathbf{S}^{-1}\mathbf{w}_0 \\ &\quad + [\mathbf{c}(\tilde{\tau})^{\mathrm{T}} \boldsymbol{\Sigma}^{-1} + (\mathbf{f}^{\mathrm{T}}(\tilde{\tau}) - \mathbf{c}(\tilde{\tau})^{\mathrm{T}} \boldsymbol{\Sigma}^{-1}\mathbf{H})\mathbf{Q}^{-1}\mathbf{H}^{\mathrm{T}} \boldsymbol{\Sigma}^{-1}]\rho, \end{aligned} \tag{4.21}$$

$$\begin{aligned}\tilde{\sigma}^2 &= s^2(\tilde{\tau}) + \mathbf{b}_{\tilde{\rho}}^{\mathrm{T}}(\sigma^2 \boldsymbol{\Sigma}_*)\mathbf{b}_{\tilde{\rho}} \\ &= \sigma^2 \left\{(1+\nu^2) - \mathbf{c}(\tilde{\tau})^{\mathrm{T}} \boldsymbol{\Sigma}^{-1}\mathbf{c}(\tilde{\tau}) \right. \\ &\quad \left. + [\mathbf{f}(\tilde{\boldsymbol{\tau}})^{\mathrm{T}} - \mathbf{c}(\tilde{\tau})^{\mathrm{T}} \boldsymbol{\Sigma}^{-1}\mathbf{H}]\mathbf{Q}^{-1}[\mathbf{f}^{\mathrm{T}}(\tilde{\tau}) - \mathbf{c}(\tilde{\tau})^{\mathrm{T}} \boldsymbol{\Sigma}^{-1}\mathbf{H}]^{\mathrm{T}}\right\} \\ &= \sigma^2 \tilde{\delta}_*, \end{aligned} \tag{4.22}$$

where $\mathbf{Q} = \mathbf{H}^{\mathrm{T}} \boldsymbol{\Sigma}^{-1}\mathbf{H} + \mathbf{S}^{-1}$.

Subsequently, we proceed with integrating out the unknown GP variance, σ^2, utilizing its marginal posterior distribution $\pi(\sigma^2|\boldsymbol{\rho}, \gamma, \nu^2)$. This distribution is recognized as an inverse gamma density, $\pi(\sigma^2|\boldsymbol{\rho}, \gamma, \nu^2) \sim \mathrm{IG}(a_*, b_*)$, a consequence of $\pi(\mathbf{w}, \sigma^2|\boldsymbol{\rho}, \gamma, \nu^2) \sim \mathrm{NIG}(\mathbf{w}_*, \boldsymbol{\Sigma}_*, a_*, b_*)$ and $\mathbf{w}|\boldsymbol{\rho}, \sigma^2, \gamma, \nu^2 \sim \mathrm{N}(\mathbf{w}_*, \sigma^2\boldsymbol{\Sigma}_*)$. This leads us to derive the posterior predictive distribution, conditioned solely on γ and ν^2, which can be formulated as

$$\begin{aligned}\pi(\tilde{\rho} \mid \boldsymbol{\rho}, \gamma, \nu^2) &= \int_0^\infty \pi(\tilde{\rho} \mid \boldsymbol{\rho}, \sigma^2, \gamma, \nu^2)\pi(\sigma^2 \mid \boldsymbol{\rho}, \gamma, \nu^2)d\sigma^2 \\ &\propto \left[1 + \frac{(\tilde{\rho} - \tilde{\mu})^2}{2b_*\tilde{\delta}_*}\right]^{-(a_*+0.5)}. \end{aligned} \tag{4.23}$$

In this equation, $\tilde{\mu}$ and $\tilde{\delta}$ are defined in (4.21) and (4.22), respectively. The expression in (4.23), which signifies the posterior predictive density of the traffic density $\tilde{\rho}$ at a new time-point $\tilde{\tau}$ conditioned on γ and ν^2, conforms to a t-distribution. This distribution exhibits mean and variance $\tilde{\mu}$ and $b_*\delta_*/(a_* - 1)$, and is characterized by $2a_*$ degrees of freedom.

The subsequent stage involves integrating out γ and ν^2 to obtain the unconditional posterior predictive distribution, which is expressed as

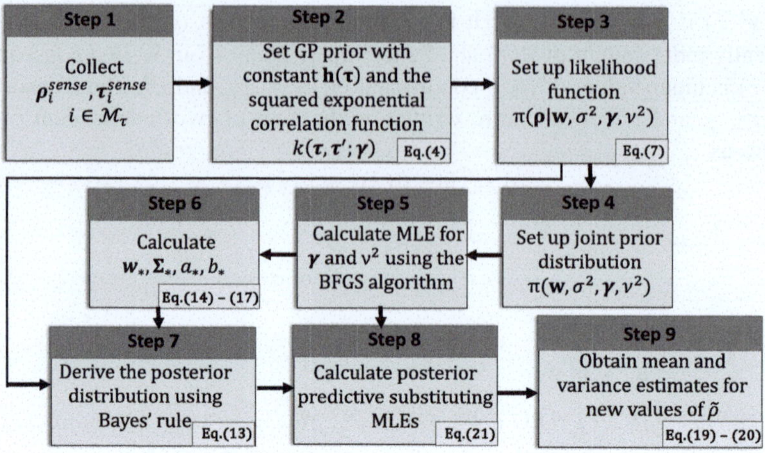

Fig. 4.4 Illustration of the proposed solution approach

$$\pi(\tilde{\rho} \mid \rho) = \int_0^\infty \int_0^\infty \pi(\tilde{\rho} \mid \rho, \gamma, \nu^2)\pi(\gamma, \nu^2 \mid \rho) d\gamma \, d\nu^2.$$

However, this integral lacks an analytical solution and necessitates numerical evaluation, often facilitated by sampling techniques based on Markov chain Monte Carlo methods. Alternatively, one can use point estimates for γ and ν^2, such as the posterior mode or the Maximum Likelihood Estimate (MLE) [83].

For the purposes of this study, we adopt a GP prior model to characterize the unknown function $f(\cdot)$ linking traffic density to time steps. Additionally, we employ prior distributions for the unknown trend parameters and GP variance, while using MLE plug-in values for γ and ν^2. Deriving the MLE for γ and ν^2 involves maximizing the logarithm of the likelihood (4.9) through numerical optimization, specifically employing the Broyden–Fletcher–Goldfarb–Shanno (BFGS) algorithm [85].

The proposed solution approach is summarized in Fig. 4.4.

4.4 Simulation Results

To assess the efficacy of information acquired from different sensing technologies, we conduct traffic simulations using the SUMO microscopic simulator [87] on a multi-lane highway scenario depicted in Fig. 4.5. This simulation scenario represents a segment of the M25 motorway in London, England, specifically the stretch between junctions J_{13} and J_{14}. The section is approximately 8 km long and is divided into 16 road segments, each spanning 0.5 km. In addition, the configuration includes two off-ramps (segments 18 and 19) and two on-ramps (segments 17 and 20). Without loss of generality, we assume unidirectional traffic progressing from segment 1 to segment 16. Traffic flow enters the network from the upstream

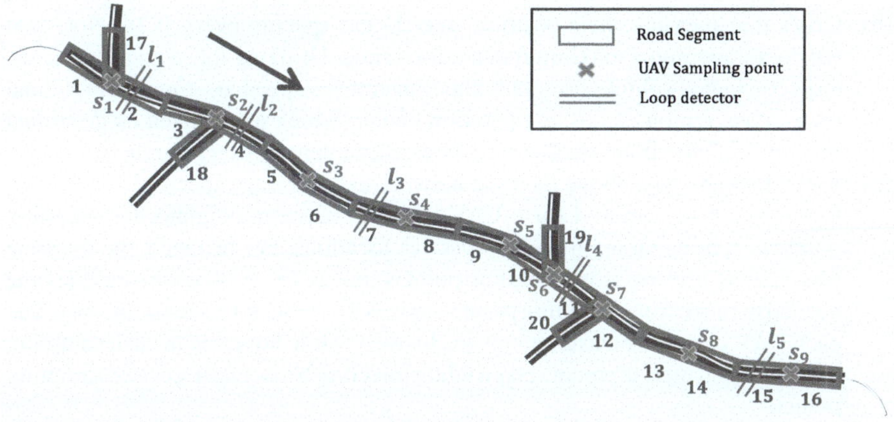

Fig. 4.5 Topological representation of the M25 Highway stretch in London, England

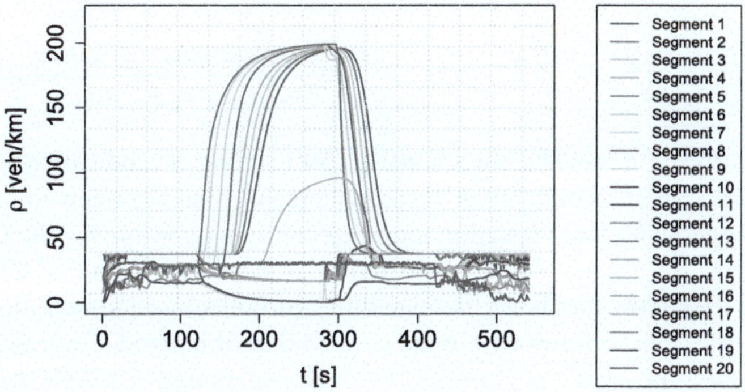

Fig. 4.6 True density of all road segments shown in Fig. 4.5

boundary of segments 1, 17, and 20, and exits from the downstream boundary of segments 16, 18, and 19. Moreover, the number of lanes varies across segments, ranging from 4 to 6 lanes.

The considered scenario involves a 1.5-hour traffic simulation that initially depicts free-flow traffic conditions. Over the course of this 1.5-hour period, an accident occurs in road segment 9, prompting the closure of two lanes within this segment for 30 min. Consequently, road segments 6, 7, 8, and 9 become congested during the 30-minute lane closure, causing vehicles to be unable to exit segment 9. The true traffic density of all road segments in the network under study is depicted in Fig. 4.6, obtained from SUMO. In this simulation setup, we consider three distinct sensing technologies:

(a) Fixed-location loop detectors (depicted as the green lines in Fig. 4.5). Fixed-location sensor measurements include a traffic density measurement every 3 min.

(b) CAVs that measure their position, speed, and spacing every 0.2 s. Note that we do not assume a maximum detection range. Utilizing these measurements, each connected vehicle computes and transmits its speed, $\mathbf{u}_c$, time spent, t_c, and time–space region, a_c, to an operation center, which are utilized as described in Sect. 4.2.2 to obtain measurements of traffic density every 3 min.
(c) UAVs that navigate above the road network at an average speed of $v^{UAV} = 40$ km/h. The time required for a UAV to travel between neighboring sampling points is contingent upon their respective locations. For instance, the duration for a UAV to traverse from sampling point s_1 to s_2 is $T^U = 90$ s, whereas the time needed to move from sampling point s_5 to s_6 is $T^U = 45$ s. Our assumption is that each UAV captures measurements exclusively while hovering at each sampling point, with no measurements taken while traveling between successive sampling points.

Each sensing technology is assessed under different percentages of coverage of the network under study. We define the percentage of coverage of a network as

$$\text{Percentage of Network Coverage} = \frac{\text{Time-space area monitored}}{\text{Total time-space area}} 100\%.$$

For the fixed-location sensors, we assume (i) 50%, (ii) 25%, and (iii) 10% coverage of the network, which is defined as the percentage of road segments in the network that are equipped with fixed-location loop detectors. In addition, for the CAV-based measurements, we assume 50%, 25%, and 10% penetration rate of CAVs in the network, i.e., vehicles that have the capability of collecting measurements. These penetration rates can be translated to the percentages of network coverage by taking the ratio of the time–space are covered by each percentage of CAVs over the total time–space region of the whole network and time duration under study. Hence, for a fair comparison among the three different sensing technologies, we have that 50% penetration rate of CAVs corresponds to 47% of network coverage, 25% penetration rate of CAVs corresponds to 23% of network coverage, and 10% penetration rate of CAVs corresponds to 9% of network coverage. Last, for the UAVs, we consider different deployment scenarios such as the percentage of network coverage per sampling time-point is (i) 50%, (ii) 25%, and (iii) 10%. The UAV deployment scenarios assumed are as follows: (i) $N^U = 3$ with fixed trajectories with UAV 1, i.e., 50% penetration rate: $s_1 \rightarrow s_2 \rightarrow s_3 \rightarrow s_2 \rightarrow s_1$, UAV 2: $s_4 \rightarrow s_5 \rightarrow s_6 \rightarrow s_5 \rightarrow s_4$, UAV 3: $s_7 \rightarrow s_8 \rightarrow s_9 \rightarrow s_8 \rightarrow s_7$, (ii) $N^U = 2$ with fixed trajectories with UAV 1, i.e., 25% penetration rate: $s_1 \rightarrow s_2 \rightarrow s_3 \rightarrow s_4 \rightarrow s_5 \rightarrow s_4 \rightarrow s_3 \rightarrow s_2 \rightarrow s_1$, UAV 2: $s_6 \rightarrow s_7 \rightarrow s_8 \rightarrow s_9 \rightarrow s_8 \rightarrow s_7 \rightarrow s_6$, and (iii) $N^U = 1$ with fixed trajectory with UAV 1, i.e., 10% penetration rate: $s_1 \rightarrow s_2 \rightarrow s_3 \rightarrow s_4 \rightarrow s_5 \rightarrow s_6 \rightarrow s_7 \rightarrow s_8 \rightarrow s_9 \rightarrow s_8 \rightarrow s_7 \rightarrow s_6 \rightarrow s_5 \rightarrow s_4 \rightarrow s_3 \rightarrow s_2 \rightarrow s_1$.

Following the steps in Fig. 4.4, we estimate the traffic density for each road segment of the network using the different sensing technologies and settings for each technology. We use the Mean Absolute Percentage Error (MAPE) between the sth estimated traffic density, $\hat{\rho}_s^{\text{sense}}$, sense $\in$ {fixed, CAV, UAV}, and the sth true traffic density ρ_s^{true} to compare the performance of each technology given by

$$\text{MAPE} = \frac{1}{S}\sum_{s=1}^{S}\frac{|\rho_s^{true} - \hat{\rho}_s^{\text{sense}}|}{\rho_s^{true}}100\%, \tag{4.24}$$

where S is the total number of estimations. The mean value of the derived probability distribution, $\pi(\tilde{\rho}|\rho)$, is calculated by drawing an independent identically distributed sample from the posterior predictive distribution using Eqs. (4.21)–(4.22). The MAPE is averaged across density estimations of all road segments for each distinct sensing technology and settings of each technologies. In addition to the MAPE, we calculate the Standard Deviation (SD) of the calculated MAPE to show the consistency of estimation across different road segments.

Initially, we obtain aggregated results of the three sensing technologies by averaging the MAPE over all road segments, for different coverage percentages. To eliminate any biases we run the estimation procedure 50 times for each sensing technology, each time assuming different settings, e.g., different locations for the fixed-location sensors, choosing different vehicles to behave as CAVs during the simulation setup and different starting sampling locations of UAVs. Note that both the CAVs and UAVs are able to obtain sparse density measurements for all road segments within the total period under study, while fixed-location sensors by definition can only obtain measurements at their fixed locations. Hence, for a fair comparison, we extrapolate density estimations using data from the fixed locations to neighboring locations that are not equipped with such sensors. For this task, we use kriging, a well-known spatial interpolation technique.[5]

As shown in Table 4.1 for around 50% coverage of the network, i.e., ten fixed-location sensors, or three UAVs hovering above the network or 50% penetration of CAVs moving within the network, all sensing technologies yield MAPE less than 20%, with fixed-location sensors and UAVs resulting in 5% less error than CAVs. As the percentage of network coverage decreases, the MAPE increases for all technologies, with fixed-location sensors resulting in MAPE higher than 50% for 10% network coverage. Both UAVs and CAVs yield lower values of MAPE with UAVs having the best performance for around 10% network coverage, while for around 25% network coverage CAVs yield slightly better estimation results.

Note that the estimation window of CAVs plays an important role to the estimation results. A recent paper has presented a detailed analysis of the impact of the duration of time-window, concluding that larger time-window durations yield more accurate results [47]. Further, the hovering time of a UAV above a sampling location affects the estimation accuracy of the traffic density, with small hovering duration yielding

---

[5] Kriging represents a spatial interpolation technique enabling the sampling of a limited dataset and deducing the remaining values through interpolation. The fundamental concept underlying kriging involves utilizing a constrained set of data points to make predictions for other proximate points within a specified geographical region. For more details, see [88].

Table 4.1 Average Mean Absolute Percentage Error (MAPE) and MAPE Standard Deviation (SD) of all road segments for the different sensing technologies and different percentages of network coverage

MAPE (%) and MAPE SD (%)

Sensing technology	~50% coverage		~25% coverage		~10% coverage	
	MAPE	SD	MAPE	SD	MAPE	SD
Fixed-location sensors	14.9	11.7	31.3	36.7	52.3	31.2
UAVs	14.5	18.6	21.6	18.4	26.8	18.2
CAVs	18.2	16.2	20.9	12.5	34.0	13.9

more accurate results as illustrated in [89], while both the trajectory of the UAVs and the selection of UAV sampling points significantly influence the estimation process.

Following, we present estimation results per road segment for each sensing technology and the different percentages of network coverage. Figures 4.7, 4.8, and 4.9 depict the per road segment estimated density for fixed-location sensors, CAVs, and UAVs, respectively, for 50% network coverage. Similarly, Figs. 4.10, 4.11, 4.12 and 4.13, 4.14, 4.15 are the corresponding figures for 25% and 10% network coverage.

In Figs. 4.7, 4.8, 4.9, 4.10, 4.11, 4.12, 4.13, 4.14, and 4.15, the estimated density values are the posterior medians of the GP model (blue dashed lines), along with pointwise 95% probability bounds around the estimated values (black dotted lines). In addition, black bullets correspond to the density measurements obtained/derived from each sensing technology, while the solid lines correspond to the true traffic density of the particular segment, as also shown in Fig. 4.6. Subfigures (a)–(t) depict the results for road segments 1–20 as shown in Fig. 4.5. Note that for the fixed-location sensors and segments that are not equipped with sensors we only obtain an estimation of the traffic density through extrapolation of neighboring spatial measurements using kriging.

As shown, data obtained from fixed-location sensors yield highly accurate results for road segments that are equipped with a sensor, e.g., Figs. 4.7b, 4.10k, however extrapolation yields large errors, especially as the percentage of network coverage becomes smaller, e.g., Fig. 4.10h. CAV-based measurements yield accurate results for large penetration rates, e.g., Fig. 4.8q, while estimation error becomes larger as the deployed penetration rate of CAVs becomes smaller and more specifically for large values of traffic density, e.g., Figs. 4.11c, 4.14a. This increase in the estimation error might result from the fact that lower penetration rates of CAVs yield biased results of traffic density [73]. Similar to CAV-based measurements, UAV-based measurements yield accurate results for large network coverage, i.e., three UAVs flying above the network under study, e.g., Figs. 4.9c, 4.9j, while estimation error becomes larger as the number of deployed UAVs becomes smaller, i.e., the network coverage percentage

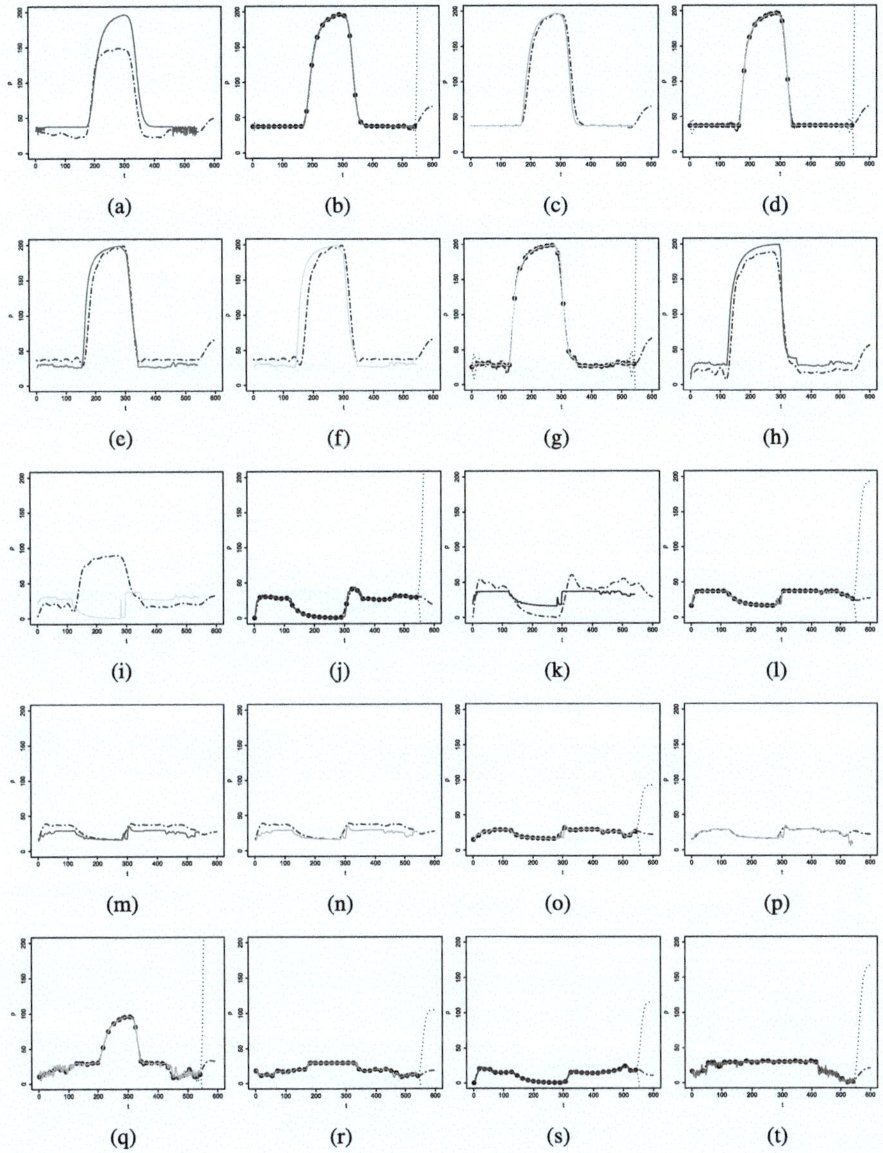

Fig. 4.7 Fixed-location sensors: 50% network coverage

decreases, e.g., Fig. 4.15c–i, while the effect of increased error for larger values of traffic density is not as obvious as with CAV-based measurements.

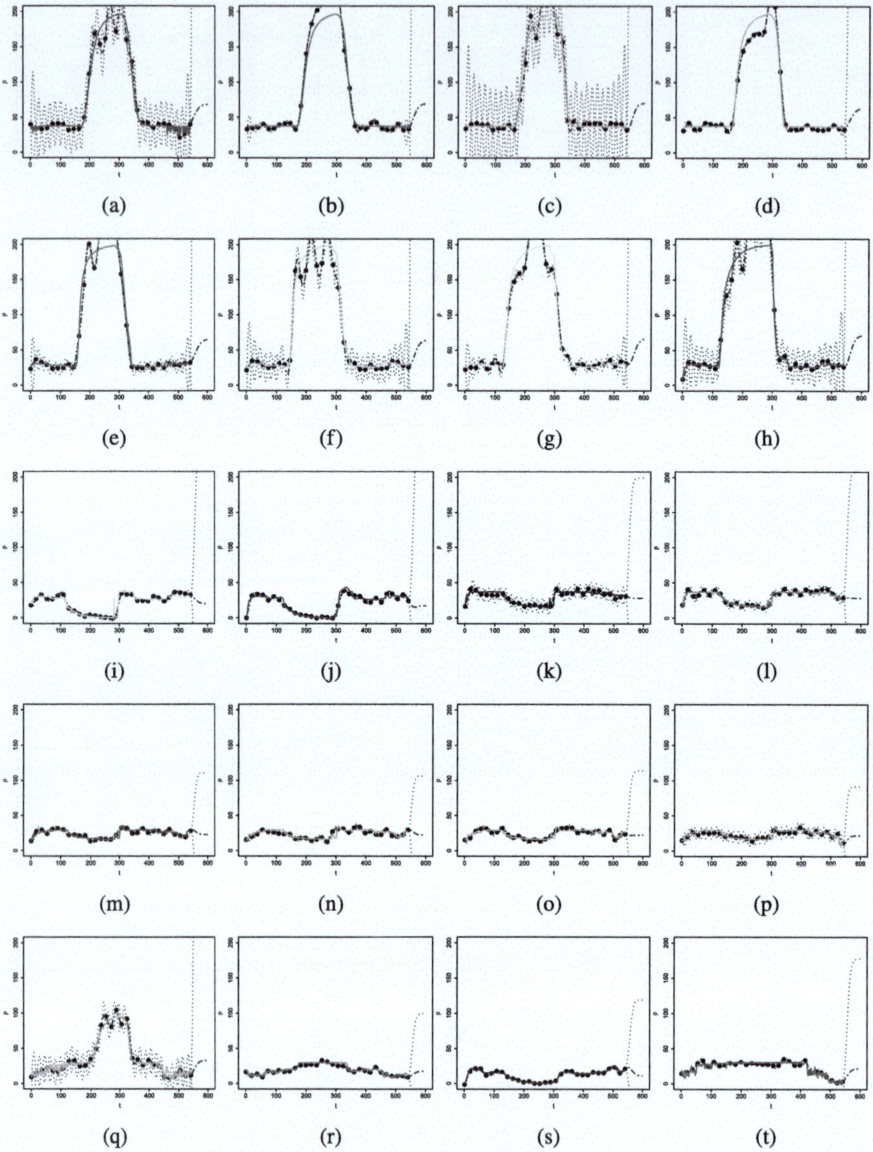

Fig. 4.8 CAVs: 50% penetration rate (47% network coverage)

4.5 Summary

In this work, we have presented novel sensing technologies and how these have been employed for traffic monitoring tasks and investigated traffic density estimation that utilizes data from three different sensing technologies: fixed-location sensors, CAVs, and UAVs. A streaming-data-driven probabilistic methodology has been developed,

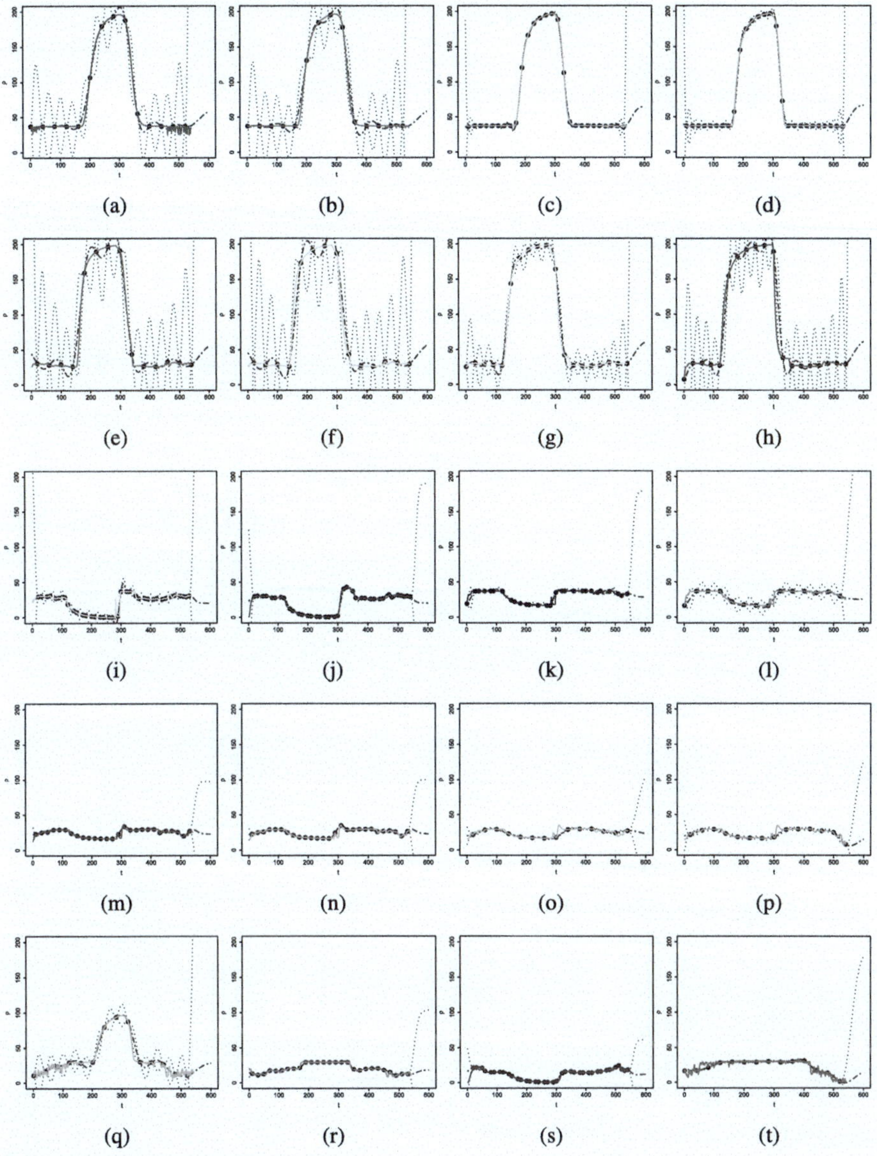

Fig. 4.9 UAVs: 50% network coverage

in which the Gaussian Process (GP) model is integrated within the Bayesian framework to effectively calculate traffic density estimates, even in cases where only a limited number of measurements are available within the designated time interval being examined. Information from each sensing technology is extracted through the time–space vehicle trajectory diagram. The estimation accuracy using measurements

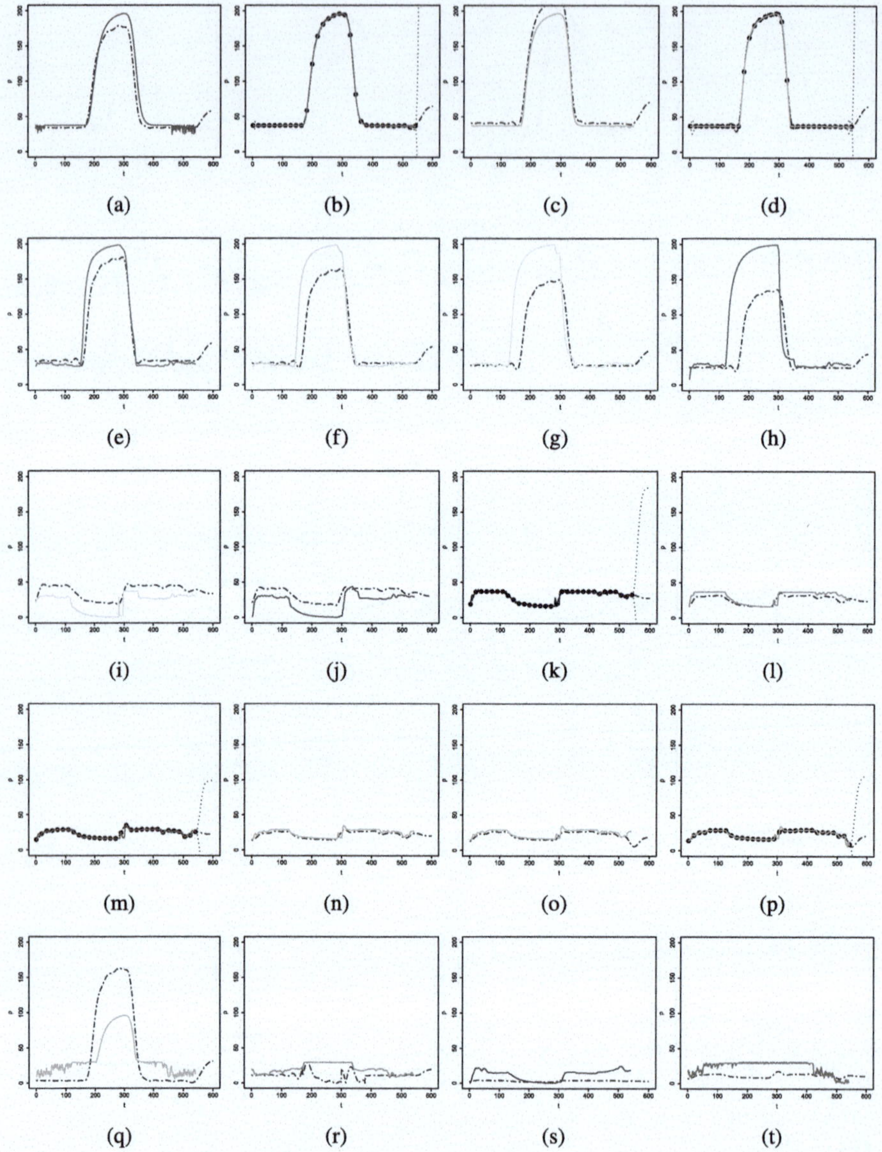

Fig. 4.10 Fixed-location sensors: 25% network coverage

from the three different sensing technologies has been investigated using a simulated multi-lane highway.

The main conclusions of this work are as follows: (i) fixed-location sensors deployed in the network yield highly accurate results for segments equipped with such sensors, however, there exists a strong dependence between the number and

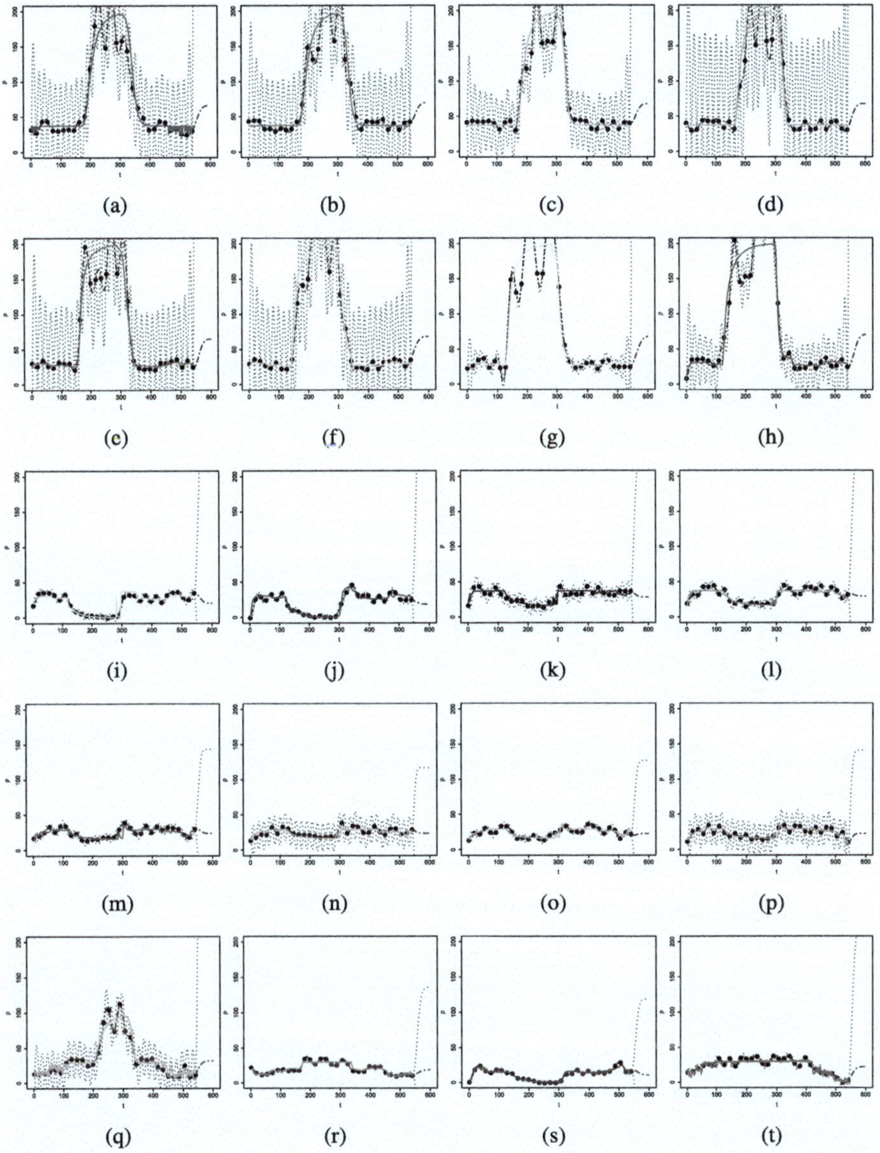

Fig. 4.11 CAVs: 25% penetration rate (23% network coverage)

location of fixed-location sensors and estimation accuracy; (ii) there exists a trade-off between the penetration rate of CAVs and estimation accuracy, especially for higher values of traffic density despite the fact that CAVs offer measurements from a higher spatiotemporal domain; (iii) UAVs yield the most accurate results among the three sensing technologies, cover high spatiotemporal resolutions of the network

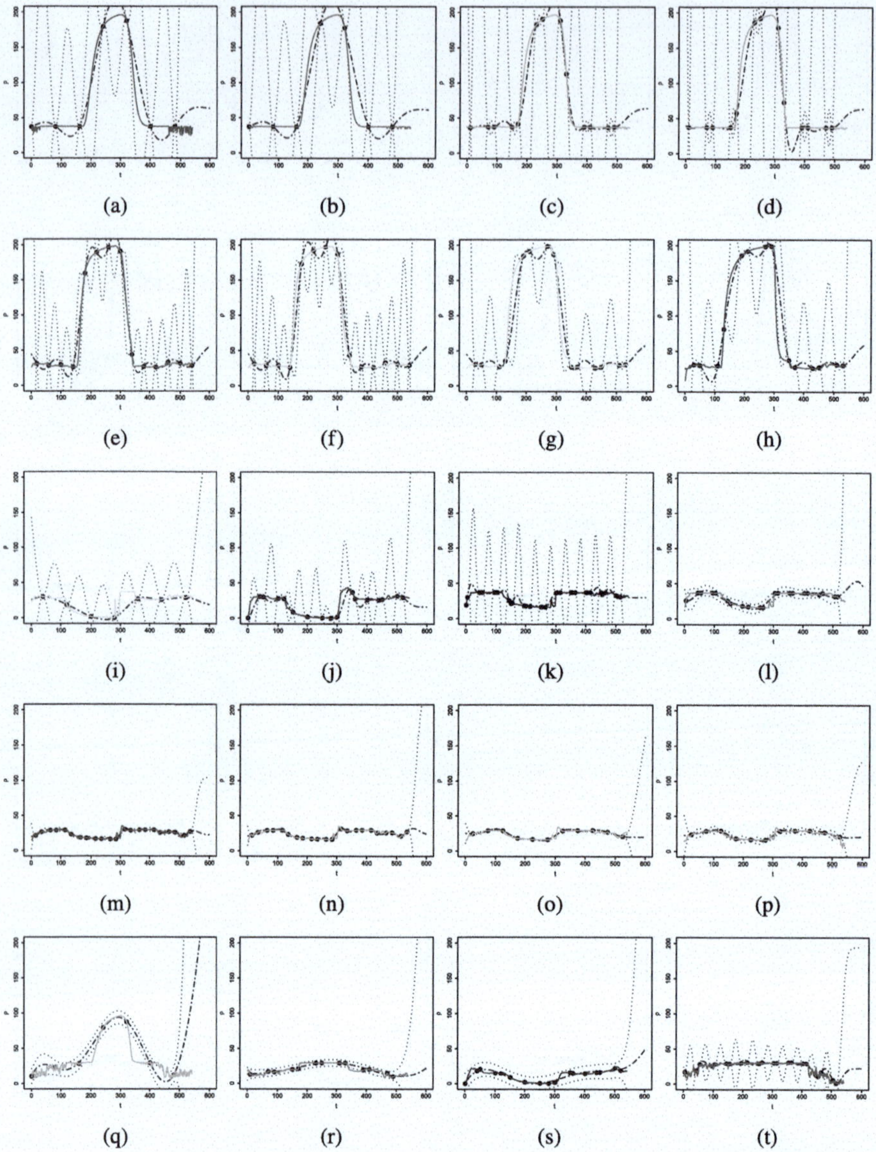

Fig. 4.12 UAVs: 25% network coverage

under study, while, as the percentage of network coverage decreases, the estimation error does not change significantly. Fusing techniques for utilizing measurements from different sensing technologies is an important task with great advantages and future work aims to investigate such techniques.

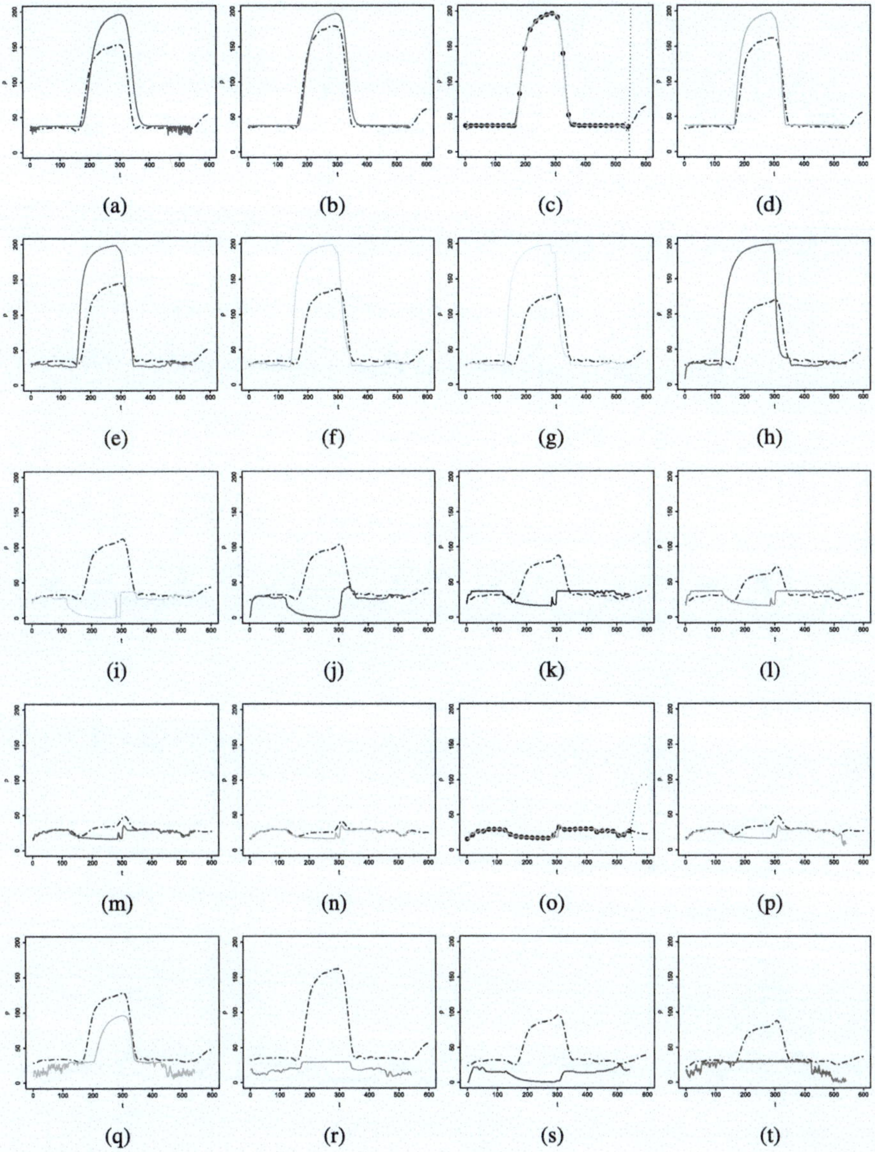

Fig. 4.13 Fixed-location sensors: 10% network coverage

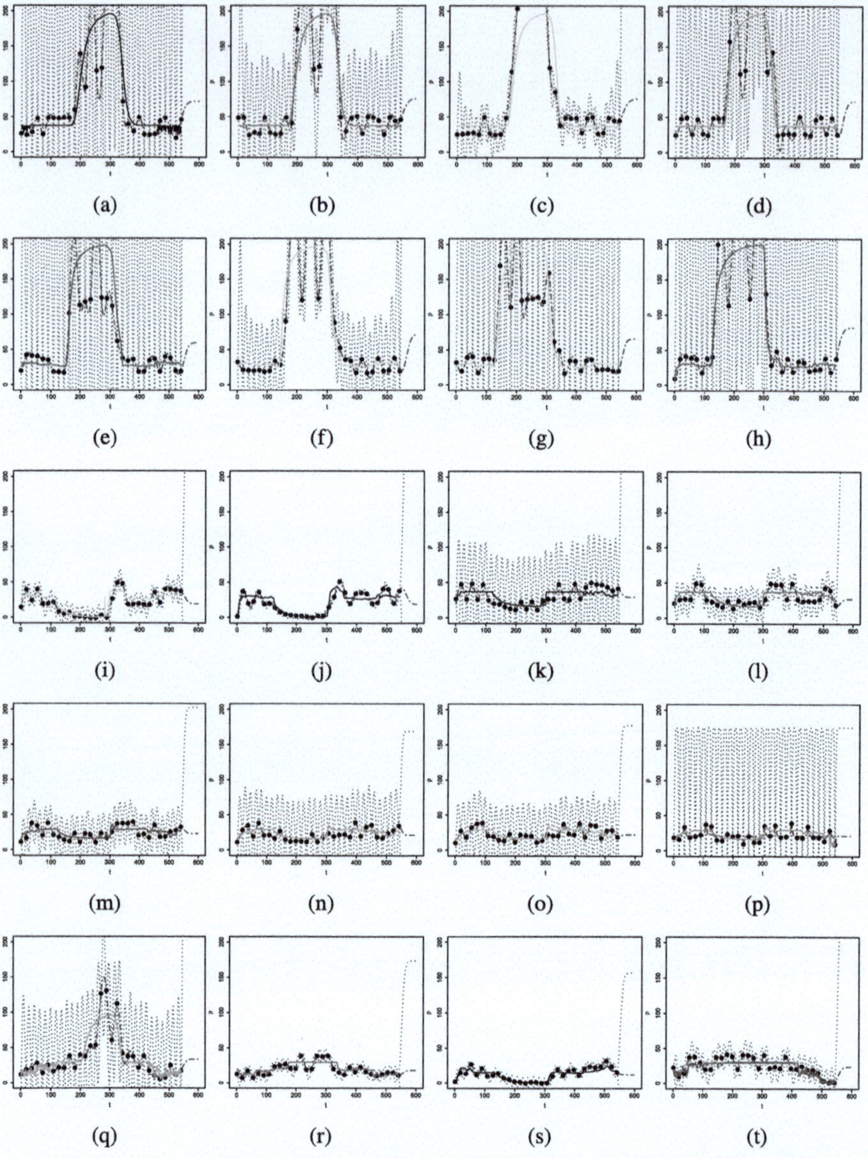

Fig. 4.14 CAVs: 10% penetration rate (9% network coverage)

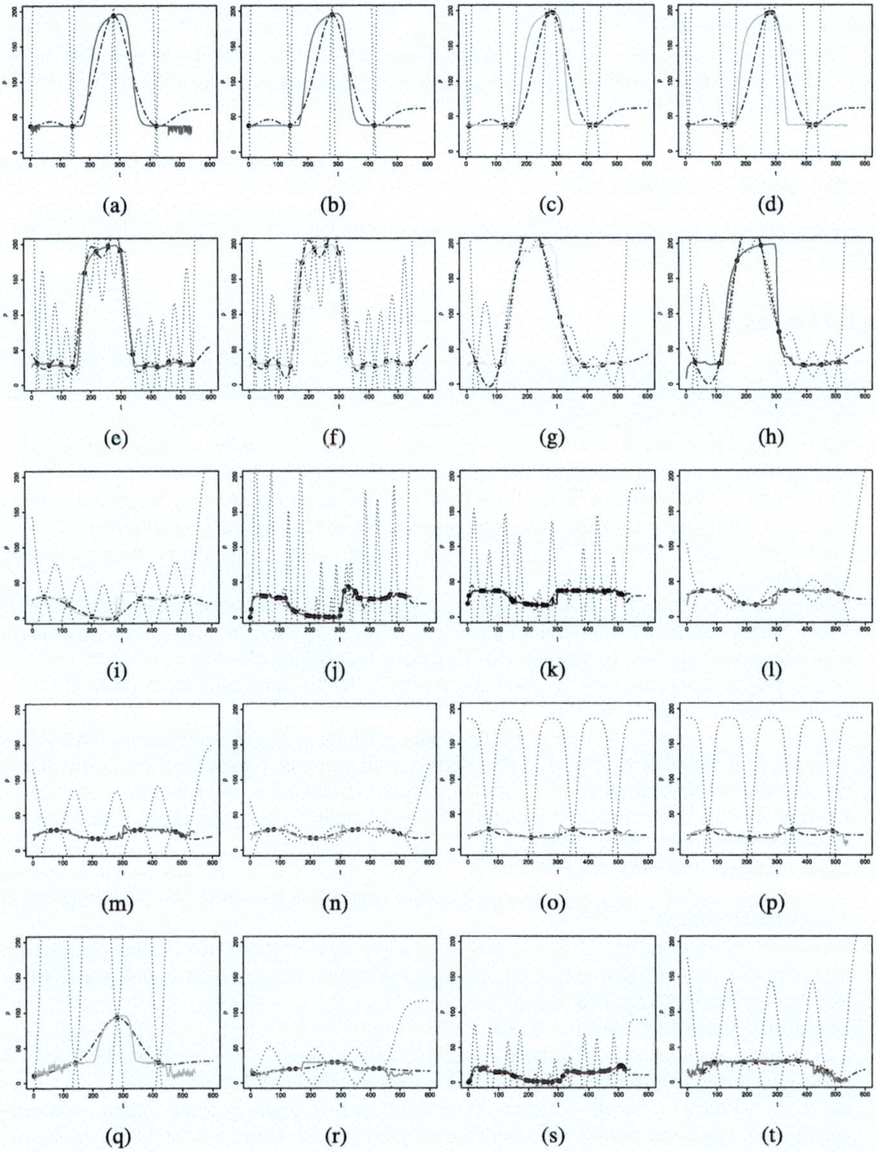

Fig. 4.15 UAVs: 10% network coverage

Acknowledgements This work is supported by the European Union (i. ERC, URANUS, No. 101088124, ii. Horizon 2020 Teaming, KIOS CoE, No. 739551 and iii. MSCA BITS, No 101003435), and the Government of the Republic of Cyprus through the Deputy Ministry of Research, Innovation, and Digital Strategy. Views and opinions expressed are however those of the author(s) only and do not necessarily reflect those of the European Union or the European Research Council Executive Agency. Neither the European Union nor the granting authority can be held responsible for them.

References

1. Kurzhanskiy AA, Varaiya P (2015) Traffic management: an outlook. Econ Transp 4:35–146. https://doi.org/10.1016/j.ecotra.2015.03.002
2. Wang Y, Papageorgiou M (2005) Real-time freeway traffic state estimation based on extended Kalman filter: a general approach. Transp Res Part B: Methodol 39:141–167
3. Van Lint JWC, Hoogendoorn SP (2010) A robust and efficient method for fusing heterogeneous data from traffic sensors on freeways. Comput-Aided Civil Infrastruct Eng 25:596–612
4. Treiber M, Kesting A (2013) Trajectory and floating-car data. In: Traffic flow dynamics. Springer
5. Papadopoulou S, Roncoli S, Bekiaris-Liberis N, Papamichail I, Papageorgiou M (2018) Microscopic simulation-based validation of a per-lane traffic state estimation scheme for highways with connected vehicles. Transp Res Part C: Emerg Technol 86:441–452
6. Seo T, Bayen AM, Kusakabe T, Asakura Y (2017) Traffic state estimation on highway: a comprehensive survey. Annu Rev Control 43:128–151
7. Greenshields BD (1935) A study of traffic capacity. Highway Res Board Proc 14:448–477
8. Darwish T, Abu Bakar K (2015) Traffic density estimation in vehicular ad hoc networks: a review. Ad Hoc Netw 24:337–351. https://doi.org/10.1016/j.adhoc.2014.09.007
9. Coifman B (2014) Revisiting the empirical fundamental relationship. Transp Res Part B: Methodol 68:173–184. https://doi.org/10.1016/j.trb.2014.06.005
10. Chen C, Kwon J, Rice J, Skabardonis A, Varaiya P (2003) Detecting errors and imputing missing data for single-loop surveillance systems. Transp Res Rec 1855:160–167. https://doi.org/10.3141/1855-20
11. Herrera JC, Work DB, Herring R, Ban XJ, Jacobson Q, Bayen AM (2010) Evaluation of traffic data obtained via GPS-enabled mobile phones: the mobile century field experiment. Transp Res Part C: Emerg Technol 18:568–583
12. Llorca DF, Satelo MA, Sánchez S, Ocaña M, Rodríguez-Ascariz JM, Garcìa-Garrido MA (2010) Traffic data collection for floating car data enhancement in V2I networks. EURASIP J Adv Signal Process 2010:425–464. https://doi.org/10.1155/2010/719294
13. Ma Y, Chowdhury M, Sadek A, Jeihani M (2009) Real-time highway traffic condition assessment framework using vehicle-infrastructure integration (VII) with artificial intelligence (AI). IEEE Trans Intell Transp Syst 10:615–627
14. Paipuri M, Leclercq L, Krug J (2019) Validation of macroscopic fundamental diagrams-based models with microscopic simulations on real networks: importance of production hysteresis and trip lengths estimation. Transp Res Rec 2673:478–492. https://doi.org/10.1177/0361198119839340
15. Coifman B, McCord M, Mishalani R, Redmill K (2004) Surface transportation surveillance from unmanned aerial vehicles. In: 83rd annual meeting of the transportation research board
16. Barmpounakis E, Vlahogianni EN, Golias JC (2016) Unmanned aerial aircraft systems for transportation engineering: current practice and future challenges. Int J Transp Sci Technol 5:111–122. https://doi.org/10.1016/j.ijtst.2017.02.001

17. Rasmussen S, Kalyanam K, Kingston D (2016) Field experiment of a fully autonomous multiple UAV/UGS intruder detection and monitoring system. In: International conference on unmanned aircraft systems, pp 1293–1302
18. Barmpounakis E, Vlahogianni EN, Golias JC (2018) Identifying predictable patterns in the unconventional overtaking decisions of PTW for cooperative ITS. IEEE Trans Intell Veh 3:102–111
19. Brahimi M, Karatzas S, Theuriot J, Christoforou Z (2020) drones for traffic flow analysis of urban roundabouts. J Traffic Transp Eng 62–71. https://doi.org/10.5923/j.ijtte.20200903.02
20. Zhou H, Kong H, Wei L, Creighton D, Nahavandi S (2015) Efficient road detection and tracking for unmanned aerial vehicle. IEEE Trans Intell Transp Syst 16:297–309. https://doi.org/10.1109/TITS.2014.2331353
21. Galkin B, Kibilda J, DaSilva LA (2019) UAVs as mobile infrastructure: addressing battery lifetime. IEEE Commun Mag 57:132–137. https://doi.org/10.1109/MCOM.2019.1800545
22. Van Nguyen C, Quyen T, Le A, Truong L, Nguyen M (2020) Advanced hybrid energy harvesting systems for unmanned aerial vehicles (UAVs). Adv Sci Technol Eng Syst J 5:34–39. https://doi.org/10.25046/aj050105
23. Kanistras K, Martins G, Rutherford MJ, Valavanis KP (2013) A survey of unmanned aerial vehicles (UAVs) for traffic monitoring. In: 2013 international conference on unmanned aircraft systems (ICUAS), pp 221–234. https://doi.org/10.1109/ICUAS.2013.6564694
24. Tsouros DC, Bibi S, Sarigiannidis PG (2019) A review on UAV-based applications for precision agriculture. In: Information. https://doi.org/10.3390/info10110349
25. Kyrkou C, Timotheou S, Kolios P, Theocharides T, Panayiotou C (2019) Drones: augmenting our quality of life. IEEE Potentials 38:30–36. https://doi.org/10.1109/MPOT.2018.2850386
26. Menouar I, Guvenc H, Akkaya K, Uluagac AS, Kadri A, Tuncer A (2017) UAV-enabled intelligent transportation systems for the smart city: applications and challenges. IEEE Commun Mag 55:22–28
27. Pham HQ, Camey M, Pham KD, Pham KV, Rirett LR (2020) GIGOS 2019, Innovation for sustainable infrastructure. Springer, 2020. Chap. Review of Unmanned Aerial Vehicles (UAVs) Operation and Data collection for driving behaviour analysis, pp 1111–1116
28. Barmpounakis E, Geroliminis N (2020) On the new era of urban traffic monitoring with massive drone data: the pneuma large-scale field experiment. Transp Res Part C: Emerg Technol 111:50–71. https://doi.org/10.1016/j.trc.2019.11.023
29. Kyrkou C, Timotheou S, Kolios P, Theocharides T, Panayiotou CG (2018) Optimized vision-directed deployment of UAVs for rapid traffic monitoring. In: IEEE international conference on consumer electronics (ICCE), pp 1–6. https://doi.org/10.1109/ICCE.2018.8326145
30. Makrigiorgis R, Kolios P, Timotheou S, Theocharides T, Panayiotou C (2020) Extracting the fundamental diagram from aerial footage. In: 2020 IEEE 91st vehicular technology conference (VTC2020-Spring), pp 1–5. https://doi.org/10.1109/VTC2020-Spring48590.2020.9128534
31. Krajewski R, Bock J, Kloeker L, Eckstein L (2018) The highD dataset: a drone dataset of naturalistic vehicle trajectories on german highways for validation of highly automated driving systems. In: 2018 21st international conference on intelligent transportation systems, pp 2118–2125
32. Salvo G, Caruso L, Scordo A, Guido G, Vitale A (2017) Traffic data acquirement by unmanned aerial vehicle. Eur J Remote Sens 50:343–351
33. Kim EJ, Park HC, Ham SW, Kho SY, Kim DK (2019) Extracting vehicle trajectories using unmanned aerial vehicles in congested traffic conditions. J Adv Transp. ISSN: 20423195. https://doi.org/10.1155/2019/9060797
34. Makrigiorgis R, Hadjittoouli N, Kyrkou C, Theocharides T (2022) AirCamRTM: enhancing vehicle detection for efficient aerial camera-based road traffic monitoring. In: 2022 IEEE/CVF winter conference on applications of computer vision (WACV), pp 3431–3440. https://doi.org/10.1109/WACV51458.2022.00349
35. Kim D (2020) Pedestrian and bicycle volume data collection using drone technology. J Urban Technol 27:5–60. ISSN: 1063-0732. https://doi.org/10.1080/10630732.2020.1715158

36. Kyrkou C (2020) YOLOpeds: efficient real-time single-shot pedestrian detection for smart camera applications. IET Comput Vision 14:417–425
37. Gu X, Abdel-Aty M, X Q, Cai Q, Yuan J (2019) Utilizing UAV video data for in depth analysis of drivers' crash risk at interchange merging areas. Accid Anal Prev 123:159–169
38. Outay F, Mengash HA, Adnan M (2020) Applications of unmanned aerial vehicle (UAV) in road safety, traffic and highway infrastructure management: recent advances and challenges. Transp Res Part A 141:116–129
39. Timotheou S, Panayiotou C, Polycarpou M (2015) Transportation systems: monitoring, control, and security. Stud Comput Intell 565:125–166. https://doi.org/10.1007/978-3-662-44160-2_5
40. Timotheou S, Panayiotou CG, Polycarpou MM (2015) Moving horizon fault-tolerant traffic state estimation for the Cell Transmission Model. In: 54th IEEE conference on decision and control (CDC), pp 3451–3456. https://doi.org/10.1109/CDC.2015.7402753
41. Kim J, Wang G (2016) Diagnosis and prediction of traffic congestion on urban road networks using Bayesian networks. Transp Res Rec: J Transp Res Rec 2595:108–118
42. Zheng F, Jabari S, Liu H, Lin D (2018) Traffic state estimation using stochastic Lagrangian dynamics. Transp Res Part B: Methodol 115:143–165
43. Timotheou S, Panayiotou C, Polycarpou M (2018) Optimization-based highway traffic state estimation with bound guarantees. In: Proceedings of the transportation research board annual meeting
44. Laña I, Lobo JL, Capecci E, Del Ser J, Kasabov N (2019) Adaptive long-term traffic state estimation with evolving spiking neural networks. Transp Res Part C: Emerg Technol 101:126–144
45. Rostami-Shahrbabaki M, Safavi AA, Papageorgiou M, Setoodeh P, Papamichail I (2020) State estimation in urban traffic networks: a two-layer approach. Transp Res Part C: Emerg Technol 115:102616
46. Shi R, Mo Z, Di X (2021) Physics informed deep learning for traffic state estimation: a hybrid paradigm informed by second-order traffic models. In: Proceedings of the 35th AAAI conference on artificial intelligence, vol 31, pp 540–547
47. Kyriacou V, Englezou Y, Panayiotou C, Timotheou S (2022) Bayesian traffic state estimation using extended floating car data. IEEE Trans Intell Transp Syst 24:1518–1532. https://doi.org/10.1109/TITS.2022.3225057
48. Bera S, Rao KVK (2011) Estimation of origin-destination matrix from traffic counts: the state of the art. Eur Transp 49:3–23
49. Englezou Y, Timotheou S, Panayiotou CG (2019) Bayesian estimation of the origin-destination matrix using traffic flow dynamics. In: 2019 IEEE intelligent transportation systems conference (ITSC), pp 2545–2550. https://doi.org/10.1109/ITSC.2019.8917143
50. Krishnakumari P, Van Lint H, Djukic T, Cats O (2020) A data driven method for OD matrix estimation. Transp Res Part C 113:38–56
51. Ma W, Pi X, Qian S (2020) Estimating multi-class dynamic origin-destination demand through a forward-backward algorithm on computational graphs. Transp Res Part C: Emerg Technol 119:102747
52. Jin J, Ma X (2019) A non-parametric Bayesian framework for traffic state estimation at signalized intersections. Inf Sci 498:21–40
53. Sederlin M, Ma X, Jin J (2021) A hybrid modelling approach for traffic state estimation at signalized intersections. In 2021 IEEE intelligent transportation systems conference (ITSC), pp 3604–3609
54. Englezou Y, Timotheou S, Panayiotou C (2021) Estimating the origin-destination matrix using link count observations from unmanned aerial vehicles. In: IEEE international intelligent transportation systems conference (ITSC), pp 3539–3544. https://doi.org/10.1109/ITSC48978.2021.9564959
55. Chootinan P, Chen A, Recker W (2005) Improved path flow estimator for origin-destination trip tables. Transp Res Rec 1923:9–17. https://doi.org/10.1177/0361198105192300102
56. Rahmani M, Koutsopoulos H (2013) Path inference from sparse floating car data for urban networks. Transp Res Part C: Emerg Technol 30:41–54. https://doi.org/10.1016/j.trc.2013.02.002

57. Rahmani M, Jenelius E, Koutsopoulos H (2015) Non-Parametric Estimation of Route Travel Time Distributions from Low-Frequency Floating Car Data. Transp Res Part C: Emerg Technol 58:343–362. https://doi.org/10.1016/j.trc.2015.01.015
58. Ma Z, Koutsopoulos H, Ferreira L, Mesbah M (2017) Estimation of trip travel time distribution using a generalized Markov chain approach. Transp Res Part C: Emerg Technol 74:1–21. https://doi.org/10.1016/j.trc.2016.11.008
59. Zhang Z, Wang Y, Chen P, He Z, Yu G (2017) Probe data-driven travel time forecasting for urban expressways by matching similar spatiotemporal traffic patterns. Transp Res Part C: Emerg Technol 85:476-493. ISSN: 0968-090X. https://doi.org/10.1016/j.trc.2017.10.010
60. Carrese S, Cipriani E, Crisalli U, Gemma A, Mannini L (2021) Bluetooth traffic data for urban travel time forecast. Transp Res Proc 52:236–243. https://doi.org/10.1016/j.trpro.2021.01.027
61. Ramezani M, Geroliminis N (2014) Queue profile estimation in congested urban networks with probe data. Comput-Aided Civil Infrastruct Eng 30:1–19. https://doi.org/10.1111/mice.12095
62. Gu Y, Zhen QS, Zhang G (2017) Traffic state estimation for urban road networks using a link queue model. Transp Res Rec 2623:29–39. https://doi.org/10.3141/2623-04
63. Zhou X, Mahmassani HS (2007) A structural state space model for real-time traffic origin-destination demand estimation and prediction in a day-to-day learning framework. Transp Res Part B: Methodol 41:823–840. https://doi.org/10.1016/j.trb.2007.02.004
64. Pan TL, Sumalee A, Zhong RX, Indra-payoong N (2013) Short-term traffic state prediction based on temporal-spatial correlation. IEEE Trans Intell Transp Syst 14:1242–1254. https://doi.org/10.1109/TITS.2013.2258916
65. Li W, Wang J, Fan R, Zhang Y, Guo Q, Siddique C, Ban ZJ (2020) Short-term traffic state prediction from latent structures: Accuracy vs. efficiency. Transp Res Part C: Emerg Technol 111:72–90 https://doi.org/10.1016/j.trc.2019.12.007
66. Darwish T, Bakar KA (2015) Traffic density estimation in vehicular ad hoc networks: a review. In Ad Hoc Netw 24:337–351
67. Herrera JC, Bayen AM (2010) Incorporation of Lagrangian measurements in freeway traffic state estimation. Transp Res Part B: Methodol 44:460–481
68. Yuan Y, van Lint JWC, Wilson RE, van Wageningen-Kessels F, Hoogendoorn SP (2012) Real-time Lagrangian traffic state estimator for freeways. IEEE Trans Intell Transp Syst 13:59–70
69. Neumann T, Bohnke PL, Touko Tcheumadjeu LC (2013) Dynamic representation of the fundamental diagram via Bayesian networks for estimating traffic flows from probe vehicle data. In: Proceedings of the 16th IEEE international conference on intelligent transportation systems (ITSC), pp 1870–1875
70. Wilby MR, Diaz JJV, Gonzalez ABR, Sotelo MA (2014) Lightweight occupancy estimation on freeways using extended floating car data. J Intell Transp Syst 18:149–163
71. Bekiaris-Liberis N, Roncoli C, Papageorgiou M (2016) Highway traffic state estimation with mixed connected and conventional vehicles. IEEE Trans Intell Transp Syst 17:3484–3497
72. Bekiaris-Liberis N, Roncoli C, Papageorgiou M (2017) Highway traffic state estimation per lane in the presence of connected vehicles. Transp Res Part B: Methodol 106:1–28
73. Seo T, Kusakabe T, Asakura Y (2015) Estimation of flow and density using probe vehicles with spacing measurement equipment. Transp Res Part C: Emerg Technol 53:134–150
74. Montero L, Pacheco M, Barceló J, Homoceanu S, Casanovas J (2016) A case study on cooperative car data for traffic state estimation in an urban network. Transp Res Rec: J Transp Res Board, pp 1–17
75. Seo T, Kusakabe T, Asakura Y (2015) Traffic state estimation with the advanced probe vehicles using data assimilation. In: Proceedings of the 18th IEEE international conference on intelligent transportation systems (ITSC), pp 824–830
76. Han Y, Ahn S (2021) Estimation of traffic flow rate with data from connected-automated vehicles using Bayesian inference and deep learning. Front Future Transp, vol 2
77. Edie L (1963) Discussion of traffic stream measurements and definitions. In: Proceedings of the 2nd international symposium on the theory of traffic flow, pp 139–154
78. Knoop V (2017) Introduction to traffic flow theory: an introduction with exercises, 2nd edn. TU Delft Open

79. Jia Z, Chen C, Coifman B, Varaiya P (2001) The pems algorithms for accurate, real-time estimates of g-factors and speeds from single-loop detectors. In: Proceedings of the 4th IEEE conference on intelligent transportation systems, pp 536–541
80. Rasmussen CE, Williams CKI (2006) Gaussian processes for machine learning. MIT Press, Cambridge
81. Banerjee S, Carlin B, Gelfand A (2004). Hierarchical modeling and analysis of spatial data. https://doi.org/10.1201/9780203487808
82. O'Hagan A, Forster J (2004) Bayesian inference, 2nd edn. Arnold, Oxford University Press
83. Bayarri MJ, Berger JO, Paulo R, Sacks J, Cafeo JA, Cavendish J, Lin CH, Tu J (2007) A framework for validation of computer models. Technometrics 49:138–154
84. OpenStreetMap contributors (2017) Planet dump retrieved from https://planet.osm.orghttps://www.openstreetmap.org
85. Bonnas JF, Gilbert JC, Lemarechal C, Sagastizabal CA (2006) Numerical optimisation: theoretical and practical aspects. Springer
86. Hazelton ML (2010) Bayesian inference for network-based models with a linear inverse structure. Transp ResPart B 44:674–685
87. Lopez PA, Behrisch M, Bieker-Walz L, Erdmann J, Flötteröd Y-P, Hilbrich R, Lücken L, Rummel J, Wagner P, Wiebner E (2018) Microscopic traffic simulation using SUMO. In: Proceedings of the 21st IEEE international conference on intelligent transportation systems (ITSC), pp 2575–2582
88. Chilés JP, Desassis L (2018). Fifty years of Kriging. In: Daya Sagar B, Cheng Q, Agterberg F (eds) Handbook of mathematical geosciences. Springer
89. Englezou Y, Timotheou S, Panayiotou CG (2022) Probabilistic traffic density estimation using measurements from unmanned aerial vehicles. In IEEE International conference on unmanned aircraft systems (ICUAS 2022), Dubrovnik, Croatia

Chapter 5
Traffic and Demand Management in the Era of Connected Vehicles

C. Menelaou, S. Timotheou, P. Kolios, C. G. Panayiotou, and M. P. Polycarpou

Abstract Traffic congestion in modern cities inflicts substantial economic and societal burdens, manifesting in issues like reduced productivity, heightened driver stress, and fuel wastage. Traffic congestion emerges when road demand surpasses available road capacity. Despite numerous proposed traffic management solutions, congestion remains an unsolved problem. A notable limitation of existing solutions is their primary focus on redistributing traffic spatially across the network. While this might momentarily relieve specific congested points, it often merely redistributes or delays the emergence of congestion. This chapter introduces an innovative route reservation architecture aimed not only at alleviating congestion but also at enhancing the overall efficiency of the road network in terms of the average time vehicles spend within it. This architecture seamlessly integrates traffic management, steering drivers through the network, with demand management, influencing drivers' departure timings. The architecture offers flexibility in application; it can provide direct routing and departure suggestions to drivers or, alternatively, utilize dynamic pricing mechanisms to influence their route and departure decisions. To ensure the robustness of the methodologies and algorithms derived from this architecture, this chapter provides extensive validation via detailed micro-simulations.

C. Menelaou · S. Timotheou · P. Kolios · C. G. Panayiotou (✉) · M. P. Polycarpou
KIOS Research and Innovation Center of Excellence, and the Department of Electrical and Computer Engineering, University of Cyprus, Nicosia, Cyprus
e-mail: christosp@ucy.ac.cy

C. Menelaou
e-mail: cmenel02@ucy.ac.cy

S. Timotheou
e-mail: timotheou.stelios@ucy.ac.cy

P. Kolios
e-mail: pkolios@ucy.ac.cy

M. P. Polycarpou
e-mail: mpolycar@ucy.ac.cy

P. Ioannou and A. A. Malikopoulos (eds.), *Transportation Mobility in Smart Cities*, Springer Tracts on Transportation and Traffic 21,
https://doi.org/10.1007/978-3-031-64769-7_5

5.1 Introduction

Traffic congestion has emerged as a significant challenge in contemporary urban settings, leading to several detrimental consequences including environmental pollution, unpredictable travel times, and undesired delays [1]. The socio-economic implications of time wasted in congestion are vast, costing approximately 1% of the EU's GDP, which translates to around 100 billion € annually [2]. On average, drivers face around 26 h of delay annually due to congestion [3]. Congestion arises when the traffic demand exceeds the available capacity of the road infrastructure [4].

This situation is predominantly observed during peak hours (high demand) and manifests as extended vehicle queues; reduced speeds; and, consequently, prolonged travel times. Presently, countermeasures for traffic congestion can be categorized into three main groups, depicted in Fig. 5.1:

1. **Modification and Expansion of Road Infrastructure**. This approach involves constructing new roads and enhancing existing road junctions. Examples include the development of grade-separated intersections [5].
2. **Traffic Management**. The primary objective here is to optimize vehicles' travel times and enhance the overall performance of the traffic network. The recent advances in information and communication technologies (ICT) and the connectivity of vehicles have empowered the efficient management control of vehicular movements across large-scale areas. Techniques such as ramp metering, gating, perimeter control, and route guidance are prime examples [6].
3. **Demand Management**. This approach aims to mitigate traffic congestion by implementing various restriction policies. Examples of these policies are economic instruments like road and congestion pricing, regulatory measures, and physical constraints such as road closures and parking restrictions [7]. Addition-

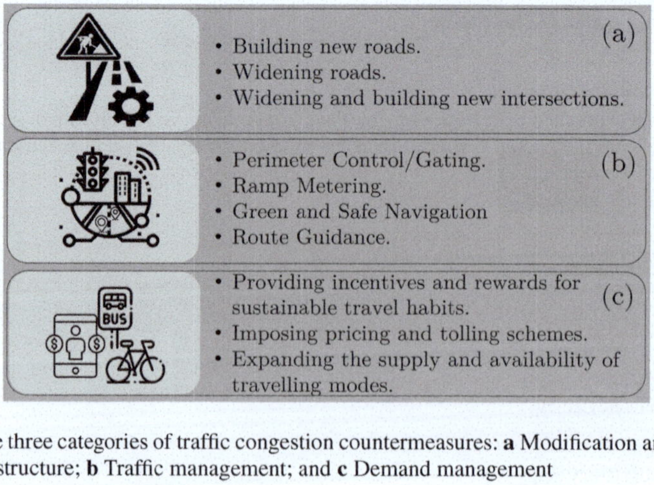

Fig. 5.1 The three categories of traffic congestion countermeasures: **a** Modification and expansion of road infrastructure; **b** Traffic management; and **c** Demand management

ally, promoting the use of alternative transportation modes, such as buses and trains, is also a strategy within demand management.

Interestingly, traffic congestion is not solely a result of inadequate network capacity. Consequently, merely expanding the road infrastructure may prove inefficient [8]. Moreover, the substantial investment and operational costs of such expansions can be prohibitive, leading recent studies to concentrate on online demand and traffic management techniques. Leveraging the recent advancements in ICT and vehicle connectivity offers promising avenues to reduce the levels of congestion through a range of intelligent traffic management strategies [6]. For instance, the existing route guidance systems can provide drivers with real-time travel updates and recommend congestion-avoiding alternative routes. The goal is to evenly distribute traffic across the network and, in turn, reducing the overall experienced travel times [9, 10]. However, there's an unforeseen side effect. When drivers are informed with real-time traffic information, many, attempting to reduce their transit times, might choose less congested routes over shorter distance ones. This can lead to traffic fluctuations, resulting in certain parts of the network becoming over-utilized, while others remain under-utilized, and this pattern can oscillate over time [4].

Traffic congestion remains a persistent issue, primarily because most existing route guidance methods target the **User Optimum** (also known as the Wardrop Equilibrium [11]). Such "selfish" routing often merely relocates congestion to alternative road segments, causing the network state to oscillate with congestion levels continuously shifting between different areas of the road network [8, 12]. On the contrary, other traffic management solutions aimed at achieving the **Social Optimum** (also known as the System Equilibrium [11]) have the potential to significantly decrease travel times by marginally reducing traffic demand[1] [4, 13]. Achieving the social optimum, however, necessitates that some vehicles be willing to change their departure time or even opt for longer routes if these decisions benefit the System Equilibrium. Interestingly, the number of vehicles making such sacrifices for delayed departures (i.e., demand management) can be vastly outnumbered by those who benefit from the improved flow [14].

Figure 5.2 depicts a typical traffic demand pattern for an urban area, illustrated by the gray line. The two peaks correspond to the morning (07:30–09:30) and evening (16:30–19:00) rush hours, during which the traffic demand surpasses the road infrastructure's capacity. Under these conditions, congestion is inevitable. In such scenarios, demand management schemes aim to influence drivers' travel patterns by employing operational strategies to spatially and temporally reduce or redistribute the demand, as represented by the blue line in Fig. 5.2. Moreover, integrating traffic management with demand management has the potential to completely eliminate congestion. This synergistic approach can redirect traffic away from congestion-prone zones, ensuring smoother flow across the entire network [3].

The aim of this chapter is to introduce the **Route Reservation Architecture**, conceived as a comprehensive solution to tackle the traffic congestion problem by

[1] Throughout this chapter, the term "demand" consistently refers to the number of vehicles intending to access the road network.

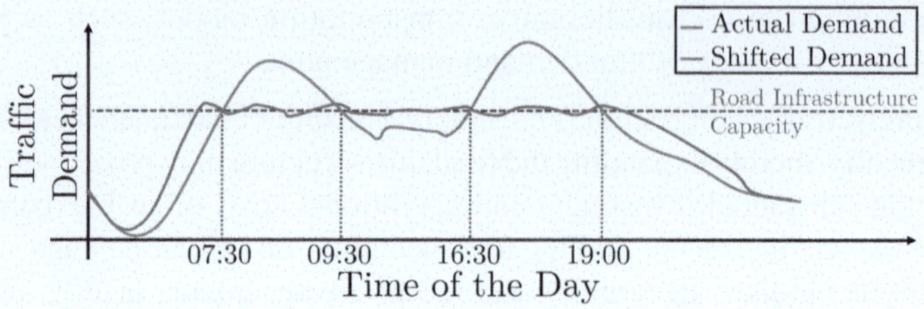

Fig. 5.2 In urban areas, a typical traffic demand pattern, represented by the gray line, shows peaks during the morning (07:30–09:30) and evening hours (16:30–19:00). Demand management schemes aim to adjust this demand, both spatially and temporally, as denoted by the blue line

jointly integrating demand with traffic management. The route reservation architecture guides drivers in choosing routes (traffic management) while concurrently adjusting departure times (demand management[2]). The presented architecture operates under the assumption that each road segment within the controlled network has a threshold that defines the critical density of each segment. By ensuring that the density of each road segment stays below a critical threshold, the overall density of the region remains manageable, resulting in a congestion-free network operation. Therefore, the core idea behind the route reservation architecture is to prevent vehicle density of each road segment from exceeding its threshold, ensuring that each vehicle is routed only through *congestion-free*[3] road segments, while at the same time avoiding state oscillations across the network. The architecture's operation is simple. Before or as a vehicle prepares to commence its journey, communicates a route reservation request to a centralized Road-Side Unit (RSU), detailing its **origin and destination** as well as its **desired departure time**. The RSU assumes responsibility for all route reservations, tracking the count of vehicles reserving different time frames on specific road segments. Road segments temporarily become unavailable once reservations reach their threshold; reconsideration occurs only after the allocated time slots have passed. Thus, these route reservations offer predictions regarding the forthcoming state of each road segment. Consequently, if the reserved density of a segment surpasses its threshold during specific time intervals, the RSU classifies it as *non-admissible*. Whenever the RSU receives a route request, it calculates drivers' paths and departure times to ensure their Earliest Destination Arrival Time (EDAT). In this chapter, we will explore how the presented architecture can completely eliminate congestion and at the same time can keep the network utilization at the highest level. Building upon the advancements of reservations, this chapter also introduces a new dynamic pricing scheme. This can be seamlessly integrated with reservations to provide vehicles with added flexibility in choosing among various departure times and route alternatives.

---

[2] For the purpose of this chapter, "demand management" pertains to regulating traffic inflow within the given network.

[3] These segments, also termed *admissible road segments*, are those whose density remains below the threshold defined for each segment.

The remainder of this chapter is structured as follows. Section 5.2 provides a brief overview of the relevant literature and highlights the benefits of route reservations in comparison to state-of-the-art traffic management strategies. Section 5.3 presents the route reservation architecture, which aims to achieve congestion-free routing within the context of intelligent transportation systems. Within this section, the EDAT problem is mathematically formulated where through rigorous complexity analysis, it is demonstrated that the EDAT problem is NP-complete in the worst-case scenarios. Further expanding on the route reservations, Sect. 5.4 introduces a dynamic pricing strategy that integrates along with the route reservation architecture. All proposed approaches undergo thorough validation via extensive microscopic simulations, demonstrating notable enhancements achievable through route reservations within realistic traffic conditions. Finally, Sect. 5.5 summarizes the principal contributions of this chapter and concludes by outlining avenues for future research.

5.2 Background

5.2.1 Admission Strategies: Gating, Perimeter Control, and Ramp Metering

Currently, the *gating* and *perimeter control* methods stand as state-of-the-art solutions for mitigating traffic congestion [15, 16]. Gating control focuses on regulating traffic within a homogeneous region [17]. It achieves this by permitting external traffic entry only if the critical density of the region has not been exceeded [18, 19]. To avoid the complex link-level dynamics, most recent attempts in the literature are relying on macroscopic traffic dynamics based on the concept of Macroscopic Fundamental Diagram (MFD) [13, 20]. Strategies include street closures and controlling traffic light phases.

Ramp metering, like perimeter and gating control, represents an admission control strategy designed to enhance freeway efficiency by managing the inflow from on-ramps to the main freeway stream [21, 22]. It operates responsively to traffic conditions, utilizing real-time measurements to coordinate ramp metering actions. These actions can focus on either a single on-ramp [22] (referred to as the uncoordinated approach) or multiple consecutive on-ramps (known as the coordinated approach) [23].

Nonetheless, admission strategies share a significant limitation: inadequate queuing space on gated links can result in queues obstructing destinations upstream within the network. As a consequence, these strategies move the problem from the controlled area to its periphery, potentially diminishing their overall benefits [24, 25]. In contrast, the introduced route reservation architecture implements a more precise control over all vehicles, preventing the buildup of lengthy queues and the occurrence of excessive delays. In addition, the mentioned approaches lack a reliable mechanism for predicting the future network state. This predictive ability, crucial for effective congestion management, is a notable feature of the proposed route reservation architecture.

5.2.2 *Route Guidance*

Route guidance strategies aim at minimizing travel times by directing drivers along the shortest travel time routes [26]. These strategies try to enhance either the user equilibrium or the system optimum, especially when faced with varying demand conditions [9, 10]. While the aforementioned approaches hold significant relevance for ITS applications, many of them fail to account for the unpredictability often exhibited in driver behavior, particularly in highly congested scenarios [4]. This arises as their scheduling decisions frequently rely on paths with the shortest travel time. These paths are derived from state estimates which may overlook the unpredictable nature of congested conditions [27]. Consequently, these algorithms fail to capture traffic state fluctuations during scheduling decisions, rendering the traffic state susceptible to congestion without any guaranteed prevention. An additional limitation of such methods that rely on traffic state estimation is the need for estimates across all road segments, which is often impractical due to the limited availability of data, usually restricted to main road segments.

Clearly, the unreliability of state estimations can significantly impact other state-dependent traffic management strategies. As a result, a considerable body of research has been dedicated to achieving accuracy in travel time predictions using both analytical and statistical methods [28]. These methods can be applied to predict travel times in either deterministic or stochastic manners [29, 30], drawing on data collected from various traffic surveillance sensors, such as loop detectors, mobile detectors, radars, and cameras [31]. Practical aspects of stochastic routing have also garnered substantial interest. These aspects revolve around identifying paths with the least expected travel time or the highest reliability. In this context, each road segment's travel time is treated as a random variable with an associated probability distribution [32, 33]. However, the primary challenge with these approaches lies in the fact that routing decisions determine traversal paths again without adequately accounting for the detrimental impacts of congestion and the variations in traffic state stemming from unreliable travel time estimates [27].

The presented route reservation architecture offers a solution to address the aforementioned shortcomings. In contrast to current state-of-the-art time-dependent routing strategies, route reservations yield accurate travel time estimates. This accuracy is achieved by adjusting the density of each road segment below the critical threshold, thereby ensuring an operation that remains free from congestion. In such congestion-free conditions, vehicles can traverse at maximum possible speed ensuring accurate travel time estimations.

5.2.3 *Infrastructure Reservations*

The concept of reservations is not new to the literature. It has been employed to address the ground-holding problem within Air Traffic Management and Control Systems (ATM/ATC) for efficient airport utilization, where runway capacity remains constant as airport traffic increases [34]. In this context, airport ground control assigns specific time slots for each airplane's take-off or landing requests. These time slots are shared among both arrival and departure flights, and planes adhere to their schedules without deviations or delays. This disciplined approach leads to a marked enhancement in airport efficiency [35].

The concept of reservations has found its way into road transportation networks as well. Initial attempts in this direction were made in [36], where the idea of trip reservations was put forth to address congestion on rural motorways by proposing a reservation system similar with seat reservations on trains. The quantitative evaluation of the impact of flexible departure times on trip reservations was conducted in [36] through a stated-preference survey. Building on this notion, [37] introduced a conceptual highway booking system that collaborates with other driver information systems. Further advancement in the realm of trip-booking methods is evident in [38], where an infrastructure manager employs a slot allocation algorithm to manage demand (departure time allocation) for pre-defined routes. Recent literature also delves into the concept of trip reservations on highways, as evidenced by works like [39, 40]. These studies establish that the effectiveness of a reservation system can outpace existing traffic management approaches.

5.2.4 Demand Management and Pricing

As traffic demand surges within road networks, demand management strategies come into play to effectively and economically handle the increased traffic [41]. Initial efforts in travel demand management focused on persuading single-occupant vehicle drivers to adopt alternative transportation modes. Additionally, incentive reward programs have been proposed to motivate travelers to consider alternative transportation options, such as ride-sharing or other modes like public transport and soft mobility [3]. Likewise, recent studies have delved into the idea of shared vehicles, allowing users to access vehicles on-demand with charges based on travel time or distance [42]. Notably, demand management strategies go beyond influencing travel modes; they also encompass promoting alternative routes and even suggesting delayed departure times to spatially and temporally distribute the demand [3].

Congestion pricing is an economic policy often utilized to mitigate congestion [43]. This strategy entails the imposition of charges to regulate the influx of vehicles into controlled areas [41]. In essence, congestion pricing strategies prove effective as techniques for managing travel demand, prompting drivers to explore alternative transportation modes [44, 45]. Moreover, these strategies hold the potential to sway drivers toward selecting different trip departure times or routes [3]. An additional approach involves the application of charges based on road infrastructure usage, with the dual aim of curbing traffic congestion and lessening environmental impact [41]. Furthermore, the integration of pricing strategies with reservation-based systems merits attention. For instance, the study in [46] introduces a reservation system tailored for users who are willing to pay for access to high-priority lanes, thereby ensuring enhanced quality of service and predictability in travel times.

Despite the acknowledged effectiveness of pricing schemes and demand management strategies, their widespread adoption still poses significant challenges as their acceptance remains a substantial concern [47]. In this chapter, we also present a novel dynamic pricing strategy designed to enhance route reservations. This strategy offers drivers the flexibility to choose their departure times and routes, with each route option coming at a specific price. Thereby this pricing strategy enables drivers to tailor their choices based on their preferences. Such features elevate the route reservation architecture into a comprehensive traffic management strategy that synergistically merges demand with traffic management. In this way, it does not only preemptively tackle the issue of urban traffic congestion but also ensures that the network operates at peak utilization.

5.3 Route Reservation Architecture

The distinctive feature of the **Route Reservation Architecture** (RRA) is its ability to decompose the traffic network across spatial and temporal dimensions. In spatial dimension, each road segment is individually assessed. On the temporal dimension, these segments correspond to distinct time slots, spanning from the immediate present into an anticipated future. The primary focus of the RRA is to provide drivers with precise navigation guidance, on both their route selection and departure timing. This guidance is specifically formulated to sidestep road segments anticipated to reach their threshold during the expected journey time slots. As a consequence, when reservations for a particular road segment approach its threshold, then that segment is rendered temporarily non-admissible, meaning that no more vehicles are allowed to enter it. It becomes admissible only once the reservations recede below the threshold. By adopting this approach, the density of each road segment is maintained below the threshold, ensuring the overall region's density stays near the critical density levels. In this way,the route reservation architecture ensures that vehicles travel at free-flow speeds without facing congestion, while also maintaining the network's utilization at its peak level [4].

Figure 5.3 provides an illustration of the RRA, emphasizing the central role of the Road-Side Unit (RSU). Not only does the RSU provide navigation instruction to vehicles, but it also monitors each road segment's utilization, and manages route reservations for each Origin–Destination (OD) pair. In the operational workflow of the RRA, a vehicle, before initiating its journey, forwards a reservation request to the RSU. This request specifies its preferred departure time and the associated OD pair. Taking into account both the current and future reservation state of each road segment, the RSU derives an optimal route coupled with a recommended departure time. The obtained route and departure time strategically bypass "congested" road segments[4] and also ensures the vehicle's earliest possible arrival at its destination. The suggested route, as showcased with red line in Fig. 5.3, is then relayed to the vehicle, paired with the suggested departure time. Subsequently, the RSU updates its reservation database, integrating the path designated to the newly scheduled vehicle.

[4] In the RRA framework, "congested" segments denote those on the verge of reaching their threshold density.

Fig. 5.3 The route reservation architecture over the Nicosia's historic city center

A key advantage of the RRA lies in the utilization of the reservation states. Through the monitoring of reservation states, the RSU is empowered to preemptively identify road segments anticipated to hit their threshold, subsequently deeming them as non-admissible. This capability is complemented by another significant feature of the RRA—its ability to implicitly manage demand by proposing either early or late departures. Hence, during peak periods where congestion is unavoidable RRA spreads the demand in time dimension to effectively curb the onset of in-network queues and potential congestion hotspots. Hence, RRA ensures that the density of each segment remains below its threshold. With such measures in place, the RSU can reliably project vehicular traffic on each road segment for upcoming time frames, operating under the assumption of vehicles maintaining free-flow speeds [48]. While the RRA operates on the assumption of constant vehicles' speed, real-world scenarios introduce inevitable variations. Nonetheless, simulation results indicate that the RRA is robust, even when confronted with these inherent inconsistencies. In the subsequent section, we present a formulation of the Earliest Destination Arrival Time (EDAT) problem within the context of RRA and analyze its complexity.

5.3.1 Problem Formulation

An urban network is represented as a directed graph $\mathcal{G} = (\mathcal{V}, \mathcal{E})$. Here, the vertices $\mathcal{V}$ with cardinality $N_V = |\mathcal{V}|$ signify the road junctions, while the edges $\mathcal{E}$ with cardinality $N_E = |\mathcal{E}|$ correspond to the road segments. Specifically, for each edge $(i, j) \in \mathcal{E}$, the connected vertices are $\{v_i, v_j\} \in \mathcal{V}$. Each road segment is characterized by the parameters λ_{ij} and l_{ij} representing the number of lanes and segment's length, respectively. The network under consideration is associated with a well-defined MFD [49], characterized by parameters ρ^C, ρ^J, u_c, u_f, $u_r(k)$ and $\rho_r(k)$

which represent the critical density corresponding to maximum flow, jam density, speed-at-capacity, free-flow speed, and the instantaneous speed and density at time slot k, respectively.

For each road segment, $(i, j) \in \mathcal{E}$, with associated junctions $\{v_i, v_j\} \in \mathcal{V}$, we define the macroscopic parameters: ρ_{ij}^J, ρ_{ij}^C, and $\rho_{ij}(k)$ to represent jam density, threshold, and instantaneous density at time slot k, respectively. The threshold ρ_{ij}^C represents the maximum density at which a segment can operate at speed u^f, implying vehicles traverse at u^f only when $\rho_{ij}(k) \leq \rho_{ij}^C$. Here, ρ_{ij}^C is formulated in terms of the segment's total available density, ρ_{ij}^J, given by $\rho_{ij}^C = \zeta \rho_{ij}^J$, where $\zeta \in [0, 1]$ indicates the proportion of the segment's jam density that results in its threshold. Consequently, if the density of each segment stays below its critical value, i.e., $\rho_{ij}(k) \leq \rho_{ij}^C$, the entire network's density remains beneath its critical level, ensuring vehicles move between the free-flow speed and the speed-at-capacity. *Congestion-free* segments, labeled as **admissible**, are those satisfying $\rho_{ij}(k) \leq \rho_{ij}^C$. On such segments, vehicles are anticipated to travel at the speed-at-capacity u_c, which accounts for potential delays due to driver behaviors and intersections. To ensure this, the RSU, based on the admissibility state of each segment, formulates and solves an optimization problem to ascertain the optimal departure time and path for each vehicle that ensures the segment's **admissibility condition** for each segment is met. Key variables used in this formulation are k_0 (indicating each vehicle's preferred departure time) and d_{v_i} (the time a vehicle reaches junction v_i).

Assuming all vehicles comply with the RSU's suggestions regarding both the chosen route and the departure time, the RRA can estimate the cumulative count of vehicle arrivals and departures for each road segment $(i, j) \in \mathcal{E}$ during time slot k. These metrics are represented by the variables $\alpha_{i,j}(k)$ for arrivals and $\beta_{i,j}(k)$ for departures. Assuming that the traversal speed for each road segments is u_c, then the number of time slots required to traverse the road segment $(i, j) \in \mathcal{E}$ is given by

$$\bar{\tau}_{ij} = \left\lceil \frac{l_{ij}}{u_c T} \right\rceil, \tag{5.1}$$

where T denotes the duration of each time slot, and the expression $\lceil z \rceil$ signifies rounding z to the ceiling. Given the traversal time of each road segment and considering that the path is known, the RRA can estimate the cumulative counts of vehicle arrivals and departures for each road segment. Using these cumulative counts, RRA can further estimate the cumulative number of vehicle reservations for each segment (i, j) at time slot k, defined as

$$n_{ij}(k) = \alpha_{i,j}(k) - \beta_{i,j}(k). \tag{5.2}$$

As a result, the RRA can utilize reservations to estimate the instantaneous density of a road segment $(i, j) \in \mathcal{E}$ at any given time slot k, denoted by $\hat{\rho}_{ij}(k)$, which is mathematically expressed as

$$\hat{\rho}_{ij}(k) = n_{ij}(k)/(\lambda_{ij} l_{ij}). \tag{5.3}$$

Let $x_{ij}(k)$ represent the **admissibility state** of a road segment $(i, j) \in \mathcal{E}$ at time slot t. A segment is deemed *admissible* if a vehicle departing from the road junction v_i at time slot k can traverse the segment (i, j) without causing the density to exceed its critical value during the entirety of the traversal. In this case, $x_{ij}(k) = 1$. Otherwise, the segment is considered inadmissible, leading to $x_{ij}(k) = 0$. Mathematically, $x_{ij}(k)$ can be expressed as

$$x_{ij}(k) = \begin{cases} 1, \text{ if } \hat{\rho}_{ij}(k) \leq \rho_{ij}^C, \ \forall \tau = k, \ldots, k + \bar{\tau}_{ij} \\ 0, \text{ otherwise.} \end{cases} \tag{5.4}$$

Given the aforementioned notation, the travel time cost of a road segment, $c_{ij}(k)$, at a specific discrete time slot k can be mathematically articulated as

$$c_{ij}(k) = \begin{cases} \bar{\tau}_{ij}, \text{ if } x_{ij}(k) = 1, i \neq O \\ \bar{\tau}_{ij} + w, \text{ if } x_{ij}(k) = 0, i = O \\ \infty, \text{ if } x_{ij}(k) = 0, i \neq O, \end{cases} \tag{5.5}$$

where w represents the number of time slots during which a vehicle may wait at the origin until all segments in the selected route become admissible.

5.3.1.1 Congestion-Free Routing Under Admissibility State

When a vehicle's shortest path includes a non-admissible segment, two alternatives may be considered, the first involves vehicles to wait at their origin until all segments in the path become admissible. The second is to identify an alternative route, ensuring every segment is admissible. Occasionally, a hybrid approach, such as waiting slightly at the origin O followed by an alternate path, might offer better results. This chapter presents the *Earliest Destination Arrival Time (EDAT)* problem, addressing route reservations with consideration for both admissible and non-admissible road segments, and allowing for potential waiting at the origin.

5.3.2 *The Earliest Destination Arrival Time (EDAT) Problem*

This section mathematically formulates the EDAT problem, which aims to identify the optimal path allowing vehicles to reach their destination in the earliest possible time, while bypassing non-admissible road segments. It's noteworthy to mention that the EDAT problem is conceptualized and solved across both discrete- and continuous-time domains.

Given an Origin–Destination $(O - D)$ pair, with a time stamp t_0 indicating the preferred departure time and the admissibility states $x_{ij}(k)$, where $(i, j) \in \mathcal{E}$, for each time slot $\forall k \geq t_0$, the objective of the EDAT problem is to determine the earliest possible arrival time at the destination, transitioning from O to D. Let p_h represent the h-th distinct path from origin (O) to destination (D), expressed as $p_h = (v_0^h, v_1^h), (v_1^h, v_2^h), \ldots, (v_{L_h-1}^h, v_{L_h}^h)$. In this expression, $v_j^h \in \mathcal{V}$ denotes the j-th vertex on the h-th path. It's important to note that $v_0^h = O$, $v_{L_h}^h = D$, and L_h indicates the length of the path p_h, quantified by the number of hops. Furthermore, let w stand for the waiting duration at the origin, and $d_{v_j}^h$ for the earliest arrival time at junction v_j, accounting for a delay of w at the origin. Then, the earliest arrival time to each vertex of the path can be expressed as

$$\begin{aligned} d_{v_0^h}^h &= t_0, \quad w \geq 0 \\ d_{v_1^h}^h &= d_{v_0^h}^h + c_{v_0^h, v_1^h}(d_{v_0^h}^h), \\ &\vdots \\ d_{v_{L_h}^h}^h &= d_{v_{L_h-1}^h}^h + c_{v_{L_h-1}^h, v_{L_h}^h}(d_{v_{L_h}^h}^h), \end{aligned} \tag{5.6}$$

where $c_{ij}(k)$ is given by Eq. (5.5). Consequently, the EDAT problem can be succinctly formulated as

$$(\Pi_1)\ d_D^* = \min_{w \geq 0,\ p_h} d_D^h \tag{5.7}$$

$$\text{s.t. Constraints } (1) - (6).$$

5.3.2.1 EDAT Problem Complexity Analysis

This section delves into the main findings of the complexity analysis of the EDAT problem in its discrete-time formulation, denoted as Problem (Π_1), designed to facilitate congestion navigation under the proposed route reservations. Intriguingly, in certain scenarios, Problem (Π_1) can be classified as NP-complete. While at first glance the EDAT may appear similar to the well-researched time-dependent route problem [50], there are prominent differences to note. Specifically, the EDAT approach incorporates road segments with infinite costs (representing non-admissible segments) and allows for possible waiting intervals, but exclusively at the origin junction. This nuanced differentiation further underscores the NP-complete nature of the EDAT problem. A comprehensive proof validating the NP-completeness of Problem (Π_1) is presented in [51], leading us to the subsequent theorem.

Theorem 5.1 *The problem (Π_1) is an NP-complete problem when more than one road segments become non-admissible during certain time slots.* □

Note that if the shortest path has up to one non-admissible road segments the problem can be solved within a polynomial time. However, if more than one of the road segments becomes more admissible, then Π_1 can be reduced to the well-known number partitioning problem which is NP-complete.

5.3.3 Solution Approaches for EDAT Problem

Several alternative algorithms are proposed by the authors for solving the EDAT problem (either in the discrete or continuous time) [51–53]. This section briefly describes the three pivotal algorithmic approaches designed for the discrete-time formulation of the EDAT problem: the *Time Expansion Approach (TEA)*, the *Iterative Dijkstra Algorithm (IDA)*, and the *Directed Acyclic Graph (DAG) approach* strategies.

5.3.3.1 Time Expansion Approach (TEA)

The *TEA* approach capitalizes the concept that time-dependent networks can be seamlessly transformed into static networks via "time expansion," a notion explored in depth in [54]. This transformation facilitates the time-dependent EDAT problem to be addressed solely within the spatial dimension. Consequently, the solution unfolds in two distinct phases. In the initial phase, the graph undergoes time expansion into future time steps, taking into account incremental waiting periods at the origin junction. During this expansion, any road segments anticipated to encounter non-admissible time slots are pruned from the time-expanded graph. Subsequently, a static shortest path algorithm, for instance, Dijkstra's algorithm [55], is employed to extract a solution. Unfortunately, while *TEA* serves as an effective heuristic, it might not return the optimal solution, due to the pruning operation of the time-expanded graph.

5.3.3.2 Iterative Dijkstra Algorithm (IDA)

Another heuristic solution is provided by IDA. Central to IDA's is the application of Dijkstra's algorithm, typically associated with static networks. The effectiveness of Dijkstra's algorithm is rooted in two fundamental properties. Firstly, it operates as a label-setting algorithm [56]. During its execution, in each iteration, a label (denoted as d_{v_j}) is considered as part of the shortest path from the origin up to the junction v_j. This process continues until all nodes receive a permanent label, with the optimal path nodes stored in a predecessor array [56]. The second property is the relaxation technique.[5] Herein, each iteration recalculates the cost for all non-labeled nodes as $d_{v_i} = \min(d_{v_i}, d_{v_j} + c_{ij}(d_{v_j}))$. With these properties, Dijkstra's algorithm determines the earliest arrival time from the origin to every other road junction v_i. IDA

[5] The term "relaxation" references the iterative refinement of an upper bound solution, as elucidated in [56].

leverages these principles, yielding a feasible EDAT solution while accommodating potential wait times at the origin.

The execution of IDA is structured in two main stages: the inner loop and the outer loop. The inner loop determines the earliest destination arrival time path from O to D, accommodating vehicle waiting at any junction until a road segment transitions from non-admissible to admissible. As demonstrated in [50], when waiting intervals span all intermediate road junctions, a polynomial time-optimal solution emerges, especially under the RRA where the FIFO constraint holds true. The outer loop, in contrast, assesses if the inner loop's solution necessitates any wait periods at intermediate junctions. If no such waits are required, the algorithm converges. Conversely, if waiting does occur at one or multiple intermediate nodes, the outer loop relocates the minimum waiting interval from these nodes to the origin, updating the vehicle's departure time as $t_0 = t_0 + w_{\min}$, where $w_{\min} = \min(w_{ij}(d^*_{v_i}))$, and $w_{i,j} > 0$ denotes the minimal wait at an intermediate node in the relaxed solution. With the revised departure time, the inner loop is executed again. This iterative process continues until a path is identified that does not involve waiting at any intermediate nodes.

5.3.3.3 Directed Acyclic Graph (DAG) Approach

The optimal solution to the EDAT problem can be realized using the principles of dynamic programming [56]. For this purpose, a time-space Directed Acyclic Graph (DAG) is constructed to encapsulate all potential control actions required to solve the EDAT problem. In this graph, the space dimension signifies the indices of the road junctions, while the time dimension represents consecutive time slots, ranging from the current time slot to the future. Each node in the DAG serves as a replica of its corresponding junction created over time, reflecting the junction's state across future time slots. A solution emerges through an edge insertion procedure on the DAG. These edges represent road segments, and their lengths encapsulate the time needed to travel between adjacent junctions, under the assumption that vehicles consistently traveling at u_c. Starting from the current time slot, an edge (i, j) finds its place in the DAG if and only if two criteria reachability and admissibility are fulfilled:

- **Reachability:** This criterion ascertains whether the node v_i at the k-th time step maintains a physical connection (path) with the origin. In doing this, it employs the variable $d_{v_i,k}$ for assessment, where reachability is confirmed if $d_{v_i,k} < \infty$. Note that the index k represents the time replica of v_i at the k-th time step in the DAG.
- **Admissibility:** This indicates the admissibility of edge (i, j) during the time slots in question. It utilizes the variable $x_{ij}(k)$, where an edge is considered as admissible when $x_{ij}(k) = 1$.

When both criteria are met, the edge (i, j) is integrated into the DAG. This is an iterative procedure that continues until the first time step in the DAG where D becomes reachable, denoted by the condition $d_{D,k} < \infty$ for any given edge (v_i, D). Finally,

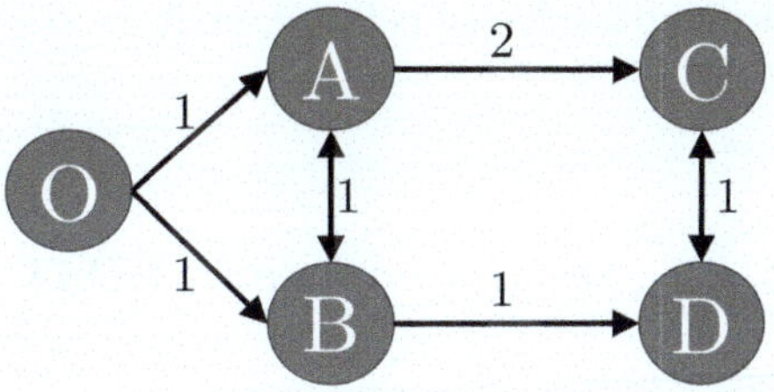

Fig. 5.4 Example network $G(V, E)$

once D becomes reachable, the algorithm halts, tracing back through the nodes in the space-time DAG to obtain the resultant route.

To better understand the DAG approach, consider the example of the graph presented in Fig. 5.4. In this graph, the lengths of the edges denote the traversal times across specific road segments. For this example, let assume $t_0 = 0$, and the admissibility along various road segments is defined by $x_{OB}(2) = x_{AC}(1) = x_{AC}(2) = x_{BD}(1) = x_{BD}(2) = 0$. Figure 5.5 provides a visual representation of the graph constructed using the DAG approach. Within this graph, the spatial dimension of each node denotes the junction index, whereas the temporal dimension indicates the replica of the junction created over time. As depicted, the initial column contains only the edges (represented by black solid lines) originating from the source O. This is attributed to the fact that, up to the time $k = 0$, other junctions remain not reachable. Similarly, the second column contains edges emerging from junctions O, A, and B since at $k = 1$ these nodes have been reached from origin O. Moreover, the edges depicted by the red dashed lines symbolize road segments that may be selected as alternatives, without impacting the overall solution cost. The algorithm prioritizes the road segments identified initially and disregards any later insertions at the specific node. Applying this procedure iteratively, by the fifth column, the destination is reached and the algorithm terminates. Within the graph, light-blue-shaded nodes

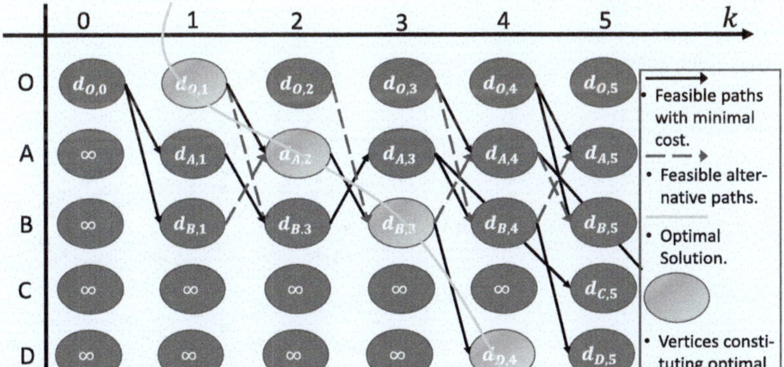

Fig. 5.5 Example of the DAG approach

depict the selected path ($O \rightarrow A \rightarrow B \rightarrow D$ with $w = 1$, highlighted by the solid green line).

5.3.4 *Simulation Results EDAT Problem*

The simulation results demonstrate the benefit of the RRA when compared against the Uncontrolled Scenario (US), and the state-of-the-art Decreasing Order of Time (DOT) algorithm [57]. US determines the cases where no control actions are taken while DOT determines the time-dependent shortest travel time path based on a user-selected departure time window.

5.3.4.1 Simulation Setup

The study focuses on an unsignalized, homogeneous region of Downtown San Francisco spanning 2.5 km^2, as depicted in Fig. 5.6. The spatial uniformity and compactness of this region have been previously explored in [58, 59]. The chosen region encompasses 143 road junctions and 319 single-lane road segments, with total network length being 9.7 km. To simulate traffic dynamics within this region, the SUMO micro-simulator [60] is employed, incorporating Krauss' car-following model [61]. Simulations adopted the following car-following parameters: a vehicle length of 5 m, a maximum speed of 15 m/s, acceleration and deceleration rates set at 2.5 m/s^2 and 4.5 m/s^2, respectively, driver imperfection of 5%, driver reaction time of 0.5 s, and a minimum gap distance of 2.5 m. In SUMO, the simulation time step was configured to 0.1 s, while the algorithm's time step was set to $T = 1$ s. Finally, in the SUMO environment, it's crucial to note that even under free-

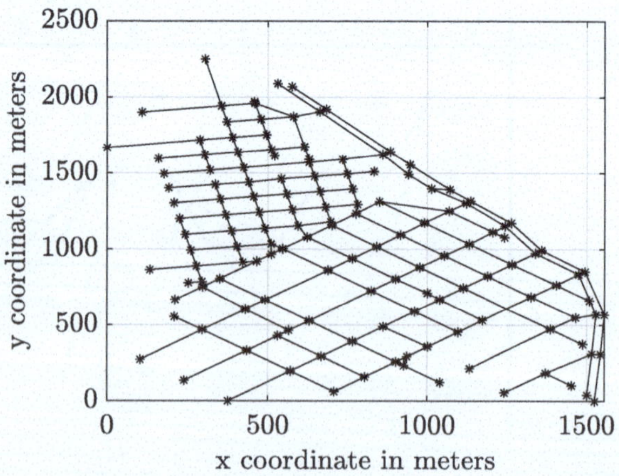

Fig. 5.6 The road network being examined spans an unsignalized, homogeneous area of 2.5 km^2 in Downtown San Francisco

flow conditions, vehicle speeds can fluctuate considerably. This variability stems from random events, including acceleration and deceleration dynamics, as well as queuing delays at intersections when yielding to higher priority vehicles. Consequently, the actual travel times may deviate from those predicted by the proposed algorithms, given the inherent uncertainties like driver imperfections and junction priorities.

5.3.4.2 MFD Extraction

As a first step, the region's MFD is extracted from the considered region in order to identify the parameters to be used by the solution approaches, including u_c and ρ_{ij}^C. To do so, a 5 h scenario was simulated within which for the first 1 h the input flow was set to 2000 veh/h and incrementally increased by 2000 veh/h every hour, for the next 3 hours assuming only exogenous flows entering the network. Subsequently, the input flow was set to 2000 veh/h for the last hour in order to discharge the network. For the results presented hereafter, 10 Monte Carlo simulations were conducted within which the $O - D$ pairs and inter-arrival times were randomly generated.

Figure 5.7a presents the Macroscopic Fundamental Diagram (MFD) for the uncontrolled scenario (US), where vehicle routes are determined solely by the shortest path. Each point in this figure represents a 1-minute measurement interval. The calibrated model, represented by the solid yellow line, utilizes the automated calibration technique introduced by [62] for the single-regime *Van Aerde* model [63]. Following the method outlined in [62], an initial set of parameters—free-flow-speed (u_f), speed-at-capacity (u_c), capacity, and jam density (ρ^J)—are iteratively adjusted to ascertain the best fitting values that minimize the sum of squared orthogonal errors. Analysis of the diagram yields the model parameters as: $u_f = 46$ km/h, $u_c = 38.5$ km/h, and $\rho^J = 140$ veh/km. Note that the average density of vehicles in the network is equivalent with the number of vehicles divided by the total length of the network. Subsequently, the RRA was configured to apply $\rho_{ij}^C = 35$ veh/km (approximating to 30% of the total regional density) and to compute travel times using $u_c = 38.5$ km/h. Figure 5.7b and c displays the MFD outcomes for the RRA employing both the IDA and DAG algorithms, respectively. Notably, neither exhibits a congested regime, a result of regulating the concurrent vehicle count on the network. Comparing the IDA and DAG algorithms in network utilization, it's evident that differences are minimal; both maintain network operations at peak capacity. Thus, subsequent results will focus solely on the IDA for in-depth performance analysis.

5.3.4.3 Performance Analysis

To assess the efficacy of IDA, this section presents the average volume metrics of flow, density, and speed, derived from each Monte Carlo realization. In particular, Fig. 5.8a depicts the mean volume of total network flow for both US and IDA, mapped against the simulation's progression. In a same manner, Fig. 5.8b presents the total number

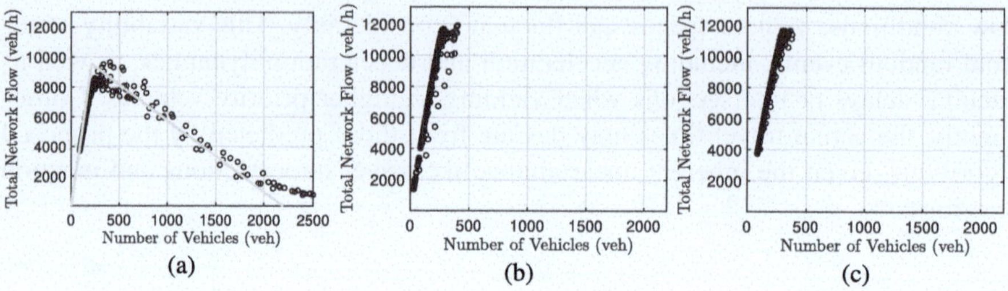

Fig. 5.7 Region MFD of: **a** US; **b** RRA:IDA; and **c** RRA:DAG

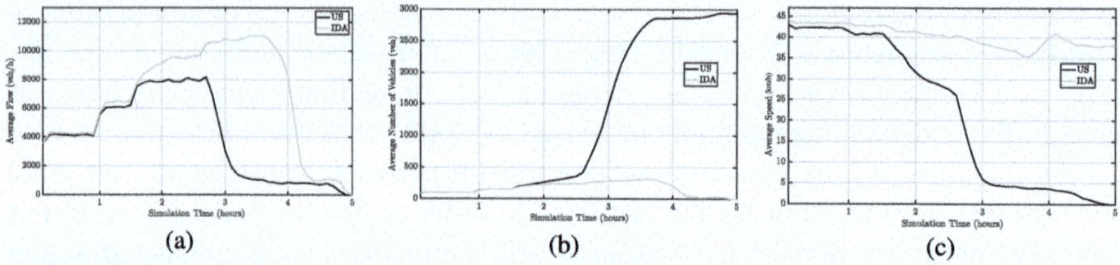

Fig. 5.8 Comparative analysis between US and IDA across various metrics: **a** Average network flow over time; **b** Average network number of vehicles over time; and **c** Average network speed over time

Fig. 5.9 Average vehicle travel time

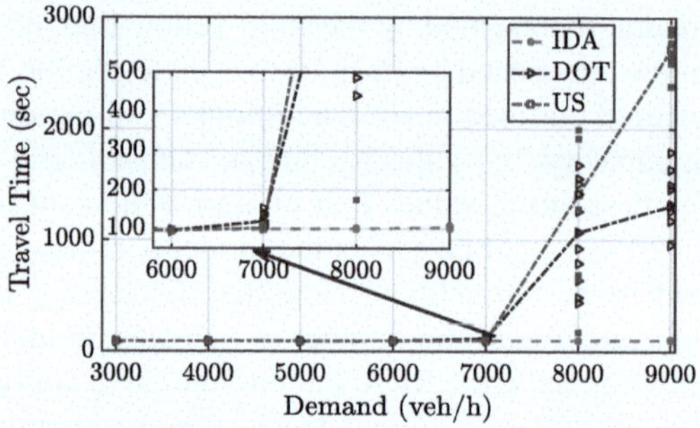

of vehicles traversing the network and Fig. 5.8c shows the mean network speeds for both approaches across the simulations. Drawing from these visualizations, IDA's deployment is seen to significantly reduce total number of vehicles (around 330 veh for IDA in contrast to over 2900 veh for US as evident in Fig. 5.8b). However, travel speeds persist at the highest rate, ensuring that flow is sustained at the same levels with what is observed in US. Figure 5.8b confirms that IDA consistently keeps traffic beneath the critical density, ρ^C (approximately 35 veh/km), even during peak demand periods (notably between 200 and 240 min). Concurrently, IDA effectively upholds vehicle speeds proximate to the speed-at-capacity throughout, as highlighted in Fig. 5.8c.

The subsequent results explore scenarios with varying flow rates, spanning from 3000 to 9000 veh/h over a period of 2 h. Figures 5.9 and 5.10 display the average

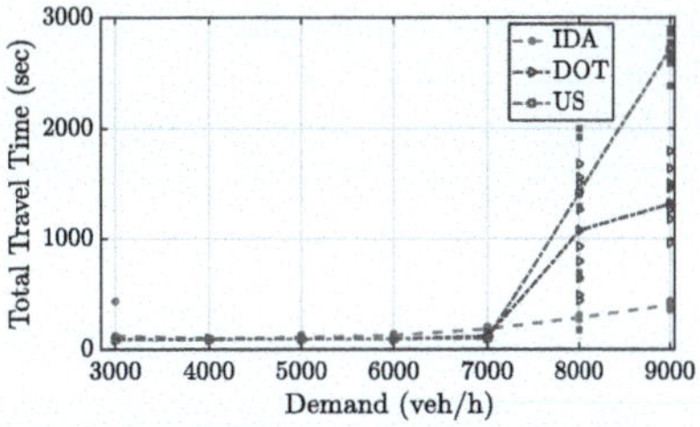

Fig. 5.10 Average total travel time

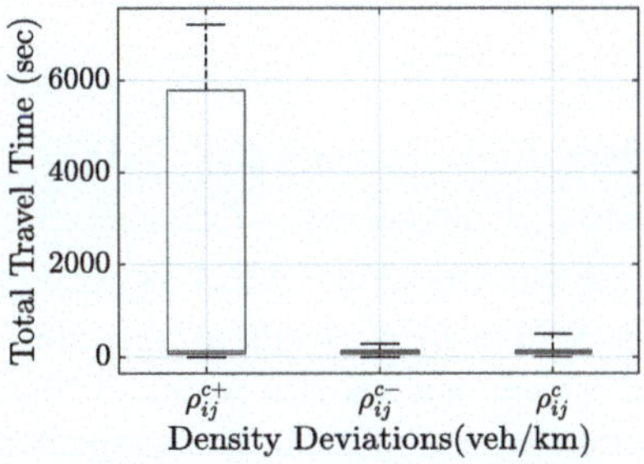

Fig. 5.11 Total travel time for q_{ij}^{c+}, q_{ij}^{c-}, and q_{ij}^{c} (9000 veh/h)

vehicle travel time and the total travel time, encompassing both the waiting duration at the origin and the travel time, for these flow rates. In these figures, scatter plots represent the mean travel and total travel time of each individual simulation while the dashed lines capture the aggregated mean across all realizations of each flow rate. A closer examination of Figs. 5.9 and 5.10 reveals that at lower flow rates (from 3000 veh/h to 6000 veh/h), congestion is minimal. Consequently, both the IDA and DOT algorithms exhibit performance metrics akin to those of the US. However, as flow rates increase, congestion becomes pronounced. In such scenarios, the IDA outperforms DOT, maintaining shorter travel times. Figure 5.10 further highlights that both IDA and DOT algorithms, when operating under lower flow rates, encounter negligible waiting periods at the origin. Yet, as congestion intensifies, the average waiting duration at the origin for IDA has an upward trend. In contrast, the average waiting time of DOT remains almost at the same levels with a slight increase of about 50 seconds. Despite this, an overall comparison in Fig. 5.10 affirms IDA's superior performance over both US and DOT. This superiority becomes especially evident when one considers that the waiting times at the origin, for IDA, form a mere fraction of the waiting duration resulting from congestion.

The performance of IDA, when subjected to varying values of ρ_{ij}^{C}, is further investigated. Figures 5.11 and 5.12 present the total travel time and the initial waiting-time vehicles endure before commencing their journey considering the highest flow rate of 9000 veh/h. These statistics are depicted using box plots.[6] Both figures consider

[6] In these plots, the bottom and top of each box represent the first and third quartiles (25% and 75%) of a ranked data set, respectively. The internal horizontal line signals the median value (second quartile). The whiskers, or horizontal lines outside the box, represent data points that remain within 1.5 times the inter-quartile range from the lower and upper quartiles.

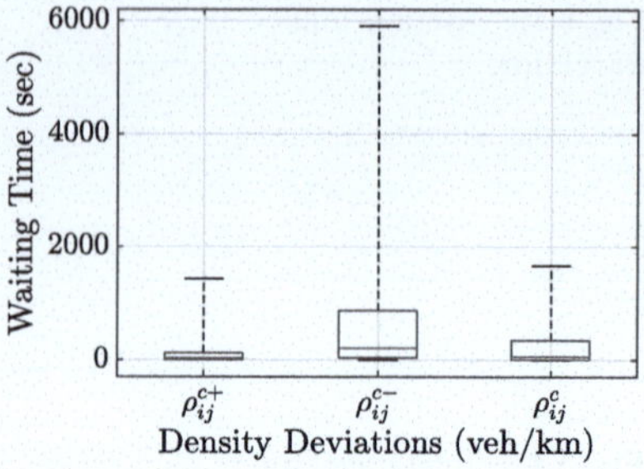

Fig. 5.12 Waiting intervals for q_{ij}^{c+}, q_{ij}^{c-}, and q_{ij}^{c} (9000 veh/h)

three distinct scenarios: $\rho_{ij}^{C} = 0.33\rho_{ij}^{J}$, $\rho_{ij}^{C-} = 0.2\rho_{ij}^{J}$, and $\rho_{ij}^{C+} = 0.45\rho_{ij}^{J}$. Here, ρ_{ij}^{C-} and ρ_{ij}^{C+} deviate by -25% and 25%, respectively, from the originally selected critical capacity value, which is ρ_{ij}^{C}. Observations from the figures suggest that a 25% escalation in ρ_{ij}^{C} leads to a decline in algorithmic efficiency. This is attributed to the simultaneous rise in travel times and reduction in waiting times, diluting the effectiveness of demand management. Furthermore, when the algorithm operates at lower capacities, its performance closely resembles that of the base scenario (ρ_{ij}^{C}). In such cases, there's no pronounced congestion, and the travel times remain relatively consistent as segment densities never surpass their critical thresholds. However, the extended waiting times imply a more conservative strategy, which inadvertently prolongs the overall travel time.

5.4 Pricing of Route Reservations

The route reservation architecture discussed up to this point offers a fixed control for all drivers, ignoring the variations in their immediate needs and priorities. For instance, some drivers might urgently need to attend a crucial meeting, making it impossible for them to wait for a congestion-free period. Conversely, some individuals might possess greater flexibility in their schedules, accommodating extended wait times. The initially proposed RRA does not accommodate such contrasting requirements.

The aim of this section is to introduce a dynamic pricing strategy that allows drivers to express their distinct preferences. Depending on the urgency to utilize the road network and the accompanying pricing strategy, each driver will be attributed a utility reflecting their specific preferences [64]. For instance, a driver aiming to reach a destination by a specific time might choose to leave earlier, even if it entails a higher cost, to guarantee timely arrival. They might perceive a greater utility in being timely rather than saving money. In contrast, a driver with a more flexible schedule might opt to leave later, following the scheduler's advice, to benefit from a reduced price and maximize their utility. It's worth noting that this strategy can either be expressed in conventional monetary units, such as euros or dollars, or in more abstract terms like karma units [65] or utils [66].

5.4.1 Price-Dependent Route Reservation Architecture

This dynamic pricing proposition aims to evolve the RRA into a versatile navigation service. This enhancement empowers drivers to select both their route and departure time from an assortment of options. Unlike the earlier approach that dictated a single path, the proposed system promotes user autonomy. Under this flexible paradigm, the RSU proposes a set of alternate routes, each associated with a distinct departure time and toll fees. Drivers then select from these options based on their immediate requirements and associated utility function. After finalizing their decision, they communicate it back to the RSU. The RSU then updates the reservation status of the road segments, taking into account the time frame during which the driver is expected to traverse them.

This section delves into three route alternatives that the RSU can offer to drivers: (i) Shortest Path (SP), (ii) Earliest Destination Arrival Time (EDAT) path, and (iii) Least Congestion (LC) path. The SP path strives to guide drivers via the route with the shortest travel time. However, this approach might inadvertently lead to congestion since it prioritizes individual travel times [4]. Conversely, the EDAT and LC paths represent more societally considerate alternatives as they factor in the departure times and route choices of other drivers. As a result, both might suggest a departure time slightly later than what the driver originally desired. Such a delay is proposed if it aids in either reducing the overall travel time for most drivers or in choosing roads with a higher available capacity.[7] Thus, by adhering to the EDAT and LC suggested times, drivers can be assured of a congestion-free journey.

5.4.2 Problem Formulation

The three alternative paths computed by the RSU are grounded in the model dynamics delineated in Sect. 5.3.1. However, the Shortest Path (SP) diverges from the others as it does not consider the admissibility state. This section offers a concise overview of the challenges associated with each path alternative.

---

[7] Aligned with the route reservation framework, reservations ensure each road segment's density remains within its capacity. The phrase "higher remaining availability" indicates a path with road segments anticipated to have the fewest reservations.

5.4.2.1 Earliest Destination Arrival Time (EDAT) Path Problem

Given a driver's $O - D$ pair, the desired departure time from the origin, t_d, and the admissibility states $x_{ij}(k)$ for all $(i, j) \in \mathcal{E}$ and $k \in \mathcal{T}$, the EDAT problem aims to identify the optimal starting time $k_E^* = t_d + w_E^*$ and the path p_E^* that ensures the earliest arrival at the destination. This is achieved while avoiding segments that have reached or exceeded their critical threshold. Here, w_E^* represents the delay in departure for the driver. A comprehensive formulation of the problem is presented in Eq. (5.7).

5.4.2.2 Least Congested Path Problem

The LC problem utilizes the same input information as the EDAT problem. It aims to identify the optimal starting time $k_L^* = t_d + w_L^*$ and the path p_L^* that favors road segments with the fewest reservations. This is achieved while ensuring links that have reached their threshold are avoided. The term w_L^* represents the departure delay that optimizes the LC problem.

For the sake of clarity, we'll omit the L index henceforth. Let p_h represent the h-th path from O to D and the variable

$$\mathrm{P}_{ij}(k) = \sum_{k}^{k+\bar{c}_{ij}(k)} \left(\hat{\rho}_{ij}(k) + \frac{\bar{c}_{ij}(k)}{l_{ij}\lambda_{ij}} \right) \tag{5.8}$$

define the total reserved density of vehicles on the road segment (i, j) when the driver departs from junction i at time k. It's essential to note that the variable $\mathrm{P}_{ij}(k)$ includes also the reservation that the driver has to make thus traverse the network via the h-th path, specifically $\frac{\bar{c}_{ij}(k)}{l_{ij}\lambda_{ij}}$.

Additionally, let $T_H = \alpha d_{D_E}^*$ be the time horizon for the LC problem, where $d_{D_E}^*$ signifies the earliest destination arrival time obtained from the EDAT problem, and $\alpha \geq 1$ is a constant. Furthermore, let variable ρ_j^h denote the total reserved density of vehicles up to the j-th junction of the h-th path. Then, the total expected density of vehicles at each junction of the path is expressed as

$$\begin{aligned}
\rho_{0^h}^h &= 0, \\
\rho_{1^h}^h &= \rho_{v_0^h}^h + \mathrm{P}_{v_0^h, v_1^h}(d_{v_0^h}^h) \\
&\vdots \\
\rho_{L_h^h}^h &= \rho_{L_h^h - 1}^h + \mathrm{P}_{L_h^h - 1, L_h^h}(d_{L_h^h}^h)
\end{aligned} \tag{5.9}$$

with $d_{L_h^h}^h \leq T_H$. Given the driver's desired departure time (i.e., t_d) and the admissible states $x_{ij}(k)$, $\forall\,(i, j) \in \mathcal{E}$, $k \in \mathcal{T}$, then the LC problem can be expressed as

$$\rho^*_{D_L} = \min_{w,\ p_h} \rho^h_D \tag{5.10}$$

$$\text{s.t. Model Dynamics } (1) - (5) \text{ and } (8) - (9),$$

$$d^h_{v^h_{L_h}} \leq T_H.$$

5.4.2.3 Shortest Path Problem

Given a driver's origin–destination pair, $O - D$, where $O,\ D \in \mathcal{V}$, and their desired departure time from the origin, t_d, the SP problem aims to identify the path p^*_{SP} that minimizes the driver's travel time, assuming departure from the origin at $k^*_{SP} = t_d$. For brevity, we will omit the SP index in subsequent discussions. Given the desired departure time, t_d, and the number of time slots to traverse a road segment, $\bar{\tau}_{ij}(k), \forall(i, j) \in \mathcal{E}, k \in \mathcal{T}$ according Eq. (5.1), the SP problem is expressed as

$$d^*_{D_{SP}} = \min_{p_h} d^h_D \tag{5.11}$$

$$\text{s.t. Model Dynamics } (1).$$

5.4.3 *Route and Departure Time Choices Under Pricing*

This section adopts the random utility discrete-choice modeling framework from [67] to ascertain the route and departure time preferences of drivers. In the context of the reservation architecture, the RSU presents drivers with three route alternatives: the SP, EDAT, and LC. Each of these is paired with a suggested departure time, k^*_s, where $s \in \mathcal{S}$ and $\mathcal{S} = \{SP, E, L\}$. Each proposed route is also associated with a toll function, designed to penalize departures that deviate (either early or late) from the RSU's recommended time. Consequently, drivers who have pressing commitments and need to depart earlier or later than the suggested time incur a toll, $u_s(t_c)$, contingent on their chosen route, i.e., s and departure time, i.e., t_c. Drivers then select the option and departure time that best align with their utility function, $U(s, t_c)$, where $s \in \mathcal{S}$ and $t_c \in \mathcal{K} = \{t_d, \ldots, k^*_s, \ldots, |\mathcal{T}|\}$. It is noteworthy that $\mathcal{K}$ enumerates all feasible departure times a driver might consider.

5.4.4 *Pricing Framework*

Let variables $k^*_{SP} = t_d$, $k^*_E = t_d + w^*_E$ and $k^*_L = t_d + w^*_L$ denote the RSU's recommended departure times for the SP, EDAT, and LC paths, respectively. Figure 5.13a, b, and c showcases representative toll functions for the SP, EDAT, and LC path

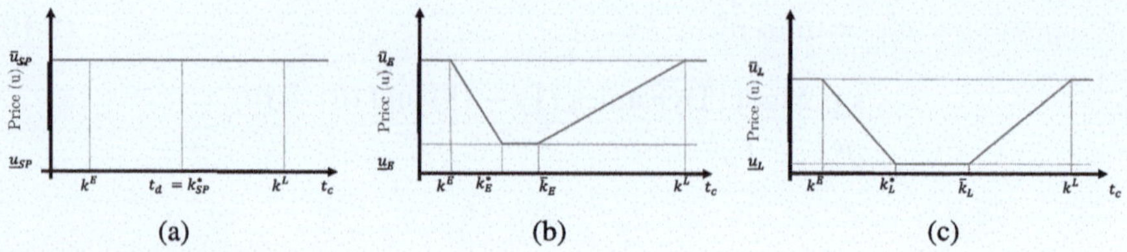

Fig. 5.13 Toll function representations for various path choices: **a** Shortest path, **b** Earliest destination arrival time, and **c** Least congested alternative

choices, symbolized as $u_{SP}(t_c)$, $u_E(t_c)$, and $u_L(t_c)$ correspondingly, where t_c is the departure time choice of the driver. The three alternatives suggest a path to follow associate with a recommended departure time, where on each case if a drive uses the recommended time then pays the minimum toll. The parameters $\underline{u}_s$ and $\bar{u}_s$ represent the minimum and maximum toll values for each alternative s in S. Otherwise, if the drives' choice differs from the recommend departure times then pay a price as given by functions depicted in Fig. 5.13a, b, and c. It is pertinent to note that, for the sake of simplicity, these tolls are represented in utility units (u), thus bypassing any need to deal with monetary trades.

5.4.5 *Route and Departure Time Choice Model*

This chapter adapts the random utility modeling framework in a manner similar to [68]. All drivers spend their day engaging in various activities and commuting between them. The theoretical model under investigation is grounded on several assumptions, devised to align the discrete-choice model with the route reservation architecture:

- Every driver possesses a distinct utility function tied to a predetermined activity. Specifically, a driver must select from the available alternatives to commute from home to work. It's essential to note that drivers can arrive at work at varying times, with some having the flexibility to arrive later than others.
- The utility associated with the work activity is influenced by the trip's travel time, denoted as t_s, for the s-th alternative. This time is equivalent to the time needed to navigate the path of the given alternative under free-flow conditions, represented as $t_s = \sum_{\forall (i,j) \in p^*} l_{ij}/v_c$.
- Each activity has a preferred start time, denoted as t_{d_a}. Regardless of a driver's chosen departure time, their ideal trip departure aligns with this activity's desired start time, which can be computed as $t_d = t_{d_a} - t$.
- Drivers experience disutility from departing either too early or too late, which subsequently impacts the start time of the activity.
- Drivers also face disutility due to the toll charges they incur.

Within this framework, drivers are confronted with the decision of selecting a route from a set of alternatives (e.g., SP, EDAT, and LC) and determining their actual departure time. A driver will opt for the s-th alternative that maximizes their utility, denoted as $U(s, t_s^c, t_d)$. This utility is associated not only with the trip but also with the planned activity the driver intends to engage in, as outlined in [69].

$$U(s, t_c, t_d) = \max\big(V_s(t_c, t_d)\big), \forall s \in \mathcal{S}, \tag{5.12}$$

where $V_s(t_c, t_d)$ denotes the utility of the s-th alternative of a driver which is defined as the sum of the trip and activity utilities expressed as

$$V_s(t_c, t_d) = V_s^T(t_c) + V_s^A(t_c, t_d), \tag{5.13}$$

where $V_s^T(t_c)$ and $V_s^A(t_c, t_d)$ denote the trip and activity utilities of a driver, respectively.

The utility associated with a trip, denoted as U_{trip}, comprises the sum of the utilities arising from the actual trip taken, D_s, the trip travel time, t_s, and the utility related to the travel cost, $V_s^C(t_c)$. Furthermore, this travel cost utility is intrinsically linked to the toll price, expressed as $u_s(t_c)$:

$$V_s^T(t_c) = D_s + \xi t_s + \zeta V_s^C(t_c), \tag{5.14}$$

where ξ and ζ are constant parameters representing a driver's sensitivity to travel time and trip cost, respectively. Consequently, $V_s^T(t_c)$ is influenced by the departure time choice for the s-th alternative. The trip cost utility, $V_s^C(t_c)$, is defined as the negation of the pricing functions given by Fig. 5.13a, b, and c, i.e., $V_s^C(t_c) = -u_s(t_c) + \bar{u}_s$.

As previously mentioned, drivers request a route from their homes to their workplaces. While some drivers are required to reach work at a precise time, others have the flexibility to arrive later. Consequently, the utility associated with an activity is influenced by its early and late schedules, which in turn are linked to the trip departure time. This relationship is mathematically expressed as follows:

$$V_s^A(t_c, t_d) = H + \alpha D + \beta V_s^S(t_c, t_d), \tag{5.15}$$

where the constants H and D represent the specific time of the day and the duration of the activity, respectively. Meanwhile, the function $V_s^S(t_c, t_d)$ denotes the utility stemming from the schedule delay, which is influenced by the desired start time of the activity, denoted by t_d. Specifically, the utility associated with these schedule delays is defined as

$$V_s^S(t_c, t_d) = \gamma_1 SDE(t_c, t_d) + \gamma_2 SDL(t_c, t_d), \tag{5.16}$$

where $SDE(t_c, t_d)$ and $SDL(t_c, t_d)$ denote the early and late schedule delays, respectively. Furthermore, the parameters γ_1 and γ_2 are negative constants, reflecting the disutility associated with excessively early or late work arrivals. As a result, both the early and late schedule delays are contingent upon the trip departure times. For every t_c in the set $\mathcal{K}^c$, these are defined as

$$SDE(t_c, t_d) = \max(0, (t_d - t_c)) \tag{5.17}$$

$$SDL(t_c, t_d) = \max(0, (t_c - t_d)). \tag{5.18}$$

Combining all above, for each alternative $s \in \mathcal{S}$, drivers will opt to depart at the time slot that maximizes their utility. This is determined by solving the following optimization problem:

$$\begin{aligned} t_c^* &= \max_{t_c \in \mathcal{K}^c} V_s(t_c, t_d) \\ &= \max_{t_c \in \mathcal{K}^c} \left(\delta_i V_s^T(t_c) + V_s^A(t_c, t_d)\right) \\ \text{s.t.}\ & \text{ Equations (5.14)} - \text{(5.18)}. \end{aligned} \tag{5.19}$$

where δ_i is constant such that $0 \leq \delta_i \leq 1$ for every $i = 1, \ldots, n$, representing n distinct driver groups. As previously noted, given the provided alternatives, a driver will select the path and departure time yielding the maximum utility, as determined by Eq. (5.12).

5.4.6 *Solving the SP, EDAT, and LC Problems*

For every request the controller is responsible to determine the solutions of SP, EDAT, and LC associate with their recommended departure times, so the drivers can make a choice.

A solution to the SP problem can be found by employing the primary procedure of a shortest path algorithm, like Dijkstra's algorithm [55], under the assumption of free-flow speed conditions. Furthermore, the solution procedure for the EDAT path problem is detailed in Sect. 5.3.3.2.

Similar to the EDAT problem, the LC problem can be solved by constructing a space-time directed acyclic graph, denoted as $\mathcal{G}_{ST}(\mathcal{V}_{ST}, \mathcal{E}_{ST})$. In this construction, the space dimension of the graph mirrors the node indices of the original graph $\mathcal{G}(\mathcal{V}, \mathcal{E})$. Conversely, the time dimension represents successive time slots, arranged in ascending order from the desired departure time, t_d, up to T_H. Consequently, a node $i_k \in \mathcal{V}_{ST}$ in the space-time graph signifies a driver's arrival at road junction $i \in \mathcal{V}$ during time slot k.

To construct the space-time DAG, one must insert all admissible segments in similar manner as in Sect. 5.3.3.3. For a given road segment $(i, j) \in \mathcal{E}$, the edge $(i_{k-\bar{c}_{ij}}, j_k)$ is incorporated into the DAG at a cost of $\mathrm{P}_{ij}(k - \bar{c}_{ij})$ only if $x_{ij}(k - \bar{c}_{ij}) = 1$. The process also necessitates the addition of two auxiliary nodes. The node $\tilde{O}$ connects to every time instance of the starting node, O_k, forming zero-cost edges

$(\tilde{O}, O_k)$. In a similar vein, node $\tilde{D}$ links to all temporal instances of the endpoint node, D_k, creating edges $(D_k, \tilde{D})$ without any associated cost. For concurrent construction of DAG and convergence of the LC problem, it's essential to define the state of node v_{i_k}, denoted as ρ_{i_k}, for all $v_{i_k} \in \mathcal{V}_{ST}$. This state represents the cumulative number of reservations along the path from the starting point O to the road junction v_{i_k}, as alluded to in Eq. (5.9). An edge's insertion during this construction phase implies the presence of a valid path from O to i_k, where the state of $v_{i_k} \in \mathcal{V}_{ST}$ is delineated by ρ_{i_k}.

$$\rho_{i_k} = \min(\rho_{i_k}, \rho_{j_{k-\bar{c}_{ji}}} + \mathrm{P}_{ji}(k - \bar{c}_{ji})), \tag{5.20}$$

where the edge $(j_{k-\bar{c_{ji}}}, i_k)$ is considered viable for the solution only if it results in the minimal value for the state ρ_{i_k}. In essence, the construction yields the optimal path from O_k to v_{i_k} via path p^*. This indicates that every node within p^* can be accessed with the lowest cost, adhering to the *optimal substructure property* outlined in Eq. (5.20) [56].

5.4.7 *Simulation Results*

We evaluate the performance of the proposed schemes using the same scenario and network described in Sect. 5.3.4. As previously discussed, drivers request departures from their origins based on Poisson arrival process. The desired start time for each user's activity is modeled as follows:

$$t_{d_a} = t_d + t + X, \tag{5.21}$$

where X is a random variable adhering to a uniform distribution $U(-t, t)$. For the evaluations, we consider three distinct toll prices: 0 u, 2 u, and 8 u, serving as the maximum toll value for each alternative while the minimum toll value is considered as 0 u. These evaluations are based on 10 Monte Carlo simulations using random $O - D$ pairs. The simulations span a 2-hour time horizon, with demand ranging from 3000 veh/h to 8000 veh/h.

Figure 5.14a illustrates the average travel times over the course of the simulation. The scatter plots depict the mean travel time for each trial at every toll price. Meanwhile, the dashed lines convey the aggregate mean travel time across all trials for each specified toll price. As evident from the figure, increasing the toll doesn't significantly elevate the travel time, ensuring no congestion arises. However, when there's no toll, travel times increase substantially. It's also apparent that higher tolls yield only a marginal boost in network efficiency: a toll of 8 u barely outpaces 2 u in travel time performance.

Figure 5.14c presents a box plot detailing the departure delays, which represent the difference between actual and desired departure times, for toll prices of 0 u, 2 u, and 8 u with a demand set to 8000 veh/h. As shown, a greater toll price corresponds to longer departure delays. This trend suggests that as the toll price rises, more drivers tend to adhere to scheduler recommendations.

5.5 Conclusions and Future Work

This chapter presents the *route reservation architecture*, tailored for the efficient management of large-scale urban traffic networks. By cohesively integrating demand and travel management, the primary objective of this architecture is to increase the efficiency of urban transport systems while concurrently mitigating the traffic congestion problem. On one hand, demand is proficiently managed by determining vehicle departure schedules, and, on the other, traffic is managed through the reservation of distinct routes for each vehicle. Hence, a key feature of the route reservation architecture is its ability to keep track for both spatial and temporal changes in vehicles densities in regions thus to promote system/social optimum and at the same time prevent traffic state oscillations. Furthermore, a distinctive feature of route reservation architecture is its inherent adaptability, allowing for seamless integration with other traffic mitigation techniques, such as pricing strategies, which employ tolls for reservation allocations. The provided simulation results indicate that this socially conscious approach can ensure a network-wide congestion-free operation.

Initial outcomes from the integration of route reservations with pricing indicate its promising potential to refine the landscape of intelligent transportation systems. Consequently, upcoming research avenues entail examining this approach under a more sophisticated dynamic pricing scheme. This model would leverage reservation predictions to calibrate toll rates in alignment with prevailing congestion levels within the targeted network. Such pricing could be employed on a per-link or per-region basis, aiming to mitigate concentrated congestion in known hotspot areas.

Furthermore, the *route reservation architecture* presents numerous unresolved research questions and challenges. A pressing challenge among these lies in identify-

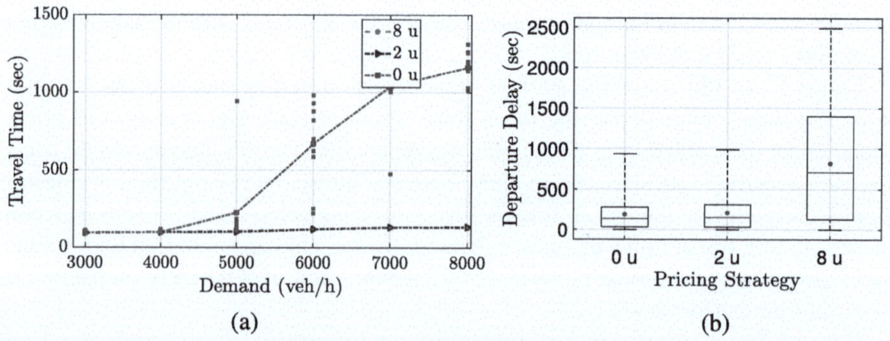

Fig. 5.14 **a** The average travel time associated with each toll price, and **b** the departure time delays for toll prices set at 0 u, 2 u, and 8 u when the demand is 8000 veh/h

ing potential fairness issues within the route reservation architecture and formulating novel approaches to tackle them.

Acknowledgements This work is supported by the European Union ((i). ERC, URANUS, No. 101088124 and (ii). Horizon 2020 Teaming, KIOS CoE, No. 739551) and the Government of the Republic of Cyprus through the Deputy Ministry of Research, Innovation, and Digital Strategy. Views and opinions expressed are however those of the author(s) only and do not necessarily reflect those of the European Union or the European Research Council Executive Agency. Neither the European Union nor the granting authority can be held responsible for them.

References

1. Arnott R, Small K (1994) The economics of traffic congestion. Am Sci 82(5):446–455
2. Commission E (2011) Roadmap to a single European transport area: towards a competitive and resource efficient transport system: white paper
3. Luten K, Katherine B, Deborah D, Tanisha H, Eric S (2004) Mitigating traffic congestion: the role of demand-side strategies
4. Çolak S, Lima A, González MC (2016) Understanding congested travel in urban areas. Nat Commun 7(10793):03
5. Geurs KT, van Wee B (2004) Accessibility evaluation of land-use and transport strategies: review and research directions. J Transp Geogr 12(2):127–140
6. Papageorgiou M, Diakaki C, Dinopoulou V, Kotsialos A, Wang Y (2003) Review of road traffic control strategies. Proc IEEE 91(12):2043–2067
7. Seik FT (1997) An effective demand management instrument in urban transport: the area licensing scheme in Singapore. Cities 14(3):155–164
8. Macfarlane J (2019) When apps rule the road: the proliferation of navigation apps is causing traffic chaos. it's time to restore order. IEEE Spectrum 56(10):22–27
9. Papageorgiou M (1990) Dynamic modeling, assignment, and route guidance in traffic networks. Transp Res Part B Methodol 24(6):471–495
10. Mahmassani H, Peeta S, Hu T-Y, Ziliaskopoulos A (1993) Dynamic traffic assignment with multiple user classes for real-time ATIS/ATMS applications. In: Large urban systems. Proceedings of the advanced traffic management conference federal highway administration
11. Wardrop JG (1952) Some theoretical aspects of road traffic research. In: Inst Civil Engineers Proc London/UK/
12. Correa JR, Schulz AS, Stier-Moses NE (2004) Selfish routing in capacitated networks. Math Oper Res 29(4):961–976
13. Daganzo CF (2007) Urban gridlock: macroscopic modeling and mitigation approaches. Transp Res Part B Methodol 41(1):49–62
14. Ziliaskopoulos AK (2000) A linear programming model for the single destination system optimum dynamic traffic assignment problem. Transp Sci 34(1):37–49
15. Keyvan-Ekbatani M, Yildirimoglu M, Geroliminis N, Papageorgiou M (2015) Multiple concentric gating traffic control in large-scale urban networks. IEEE Trans Intell Transp Syst 16(4):2141–2154
16. Aboudolas K, Geroliminis N (2013) Perimeter and boundary flow control in multi-reservoir heterogeneous networks. Transp Res Part B Methodol 55:265–281
17. Mazloumian A, Geroliminis N, Helbing D (2010) The spatial variability of vehicle densities as determinant of urban network capacity. Philos Trans R Soc Lond Math Phys Eng Sci 368(1928):4627–4647
18. Keyvan-Ekbatani M, Kouvelas A, Papamichail I, Papageorgiou M (2012) Exploiting the fundamental diagram of urban networks for feedback-based gating. Transp Res Part B Methodol 46(10):1393–1403

19. Haddad J, Geroliminis N (2012) On the stability of traffic perimeter control in two-region urban cities. Transp Res Part B Methodol 46(9):1159–1176
20. Geroliminis N, Daganzo CF (2007) Macroscopic modeling of traffic in cities. In: TRB 86th annual meeting, no. 07-0413
21. Carlson RC, Papamichail I, Papageorgiou M, Messmer A (2010) Optimal motorway traffic flow control involving variable speed limits and ramp metering. Transp Sci 44(2)
22. Papageorgiou M, Kotsialos A (2002) Freeway ramp metering: an overview. IEEE Trans Intell Transp Syst 3(4):271–281
23. Papamichail I, Papageorgiou M (2008) Traffic-responsive linked ramp-metering control. IEEE Trans Intell Transp Syst 9(1):111–121
24. Keyvan-Ekbatani M, Carlson RC, Knoop VL, Hoogendoorn SP, Papageorgiou M (2016) Queuing under perimeter control: analysis and control strategy. In: 2016 IEEE 19th international conference on intelligent transportation systems (ITSC), pp 1502–1507
25. Kouvelas A, Saeedmanesh M, Geroliminis N (2017) Enhancing model-based feedback perimeter control with data-driven online adaptive optimization. Transp Res Part B Methodol 96:26–45
26. Xiao L, Lo HK (2014) Adaptive vehicle navigation with en route stochastic traffic information. IEEE Trans Intell Transp Syst 15(5):1900–1912
27. Du L, Chen S, Han L (2015) Coordinated online in-vehicle navigation guidance based on routing game theory. In: Transportation research board 94th annual meeting, no. 15-3613
28. Lin H-E, Zito R, Taylor M et al (2005) A review of travel-time prediction in transport and logistics. In: Proceedings of the Eastern Asia Society for transportation studies, pp 1433–1448
29. Foschini L, Hershberger J, Suri S (2014) On the complexity of time-dependent shortest paths. Algorithmica 68(4):1075–1097
30. Kaysi I, Ben-Akiva M, Koutsopoulos H (1993) An integrated approach to vehicle routing and congestion prediction for real-time driver guidance. PhD dissertation, Doctoral dissertation, Massachusetts Institute of Technology Cambridge, Mass
31. Jiang Y, Li X (2013) Travel time prediction based on historical trajectory data. Ann GIS 19(1):27–35
32. Nie YM, Wu X (2009) Shortest path problem considering on-time arrival probability. Transp Res Part B Methodol 43(6):597–613
33. Miller-Hooks ED, Mahmassani HS (2000) Least expected time paths in stochastic, time-varying transportation networks. Transp Sci 34(2):198–215
34. De Neufville R, Odoni A (2004) Airport systems. Planning, design and management, vol 12. Eurocontrol
35. Panayiotou CG, Cassandras CG (2001) A sample path approach for solving the ground-holding policy problem in air traffic control. IEEE Trans Control Syst Technol 9(3):510–523
36. Akahane H, Kuwahara M (1996) A basic study on trip reservation systems for recreational trips on motorways. In: Proceeding of the 3rd world congress on intelligent transportation systems, pp 1–7
37. Wong J-T (1997) Basic concepts for a system for advance booking for highway use. Transport policy
38. De Feijter R, Evers JJ, Lodewijks G (2003) Improving travel time reliability by the use of trip booking. in Intelligent Transportation Systems. In: Proceedings. 2003 IEEE, vol. 1. IEEE 2003:205–210
39. Edara P, Teodorović D (2008) Model of an advance-booking system for highway trips. Transp Res Part C Emerg Technol 16(1):36–53
40. Su P, Park BB (2015) Auction-based highway reservation system an agent-based simulation study. Transp Res Part C Emerg Technol 60:211–226
41. Mirshahi M, Obenberger J, Fuhs CA, Howard CE, Krammes RA, Kuhn BT, Mayhew RM, Moore MA, Sahebjam K, Stone CJ et al (2007) Active traffic management: the next step in congestion management. Technical Report. United States. Federal Highway Administration
42. Fagnant DJ, Kockelman KM (2018) Dynamic ride-sharing and fleet sizing for a system of shared autonomous vehicles in Austin, Texas. Transportation 45(1):143–158
43. Pigou AC (1920) Wealth and welfare. MacMillan, UIK

44. Zheng N, Rérat G, Geroliminis N (2016) Time-dependent area-based pricing for multimodal systems with heterogeneous users in an agent-based environment. Transp Res Part C Emerg Technol 62:133–148
45. Simoni M, Pel A, Waraich R, Hoogendoorn S (2015) Marginal cost congestion pricing based on the network fundamental diagram. Transp Res Part C Emerg Technol 56:221–238
46. Liu W, Yang H, Yin Y (2015) Efficiency of a highway use reservation system for morning commute. Transp Res Part C Emerg Technol 56:293–308
47. Schade J, Schlag B (2003) Acceptability of urban transport pricing strategies. Transp Res Part F Traffic Psychol Behav 6(1):45–61
48. Immers L, Logghe S (2003) Traffic flow theory. Department of Civil engineering Section Traffic and Infrastructure, Belgium, 05, course H 111
49. Geroliminis N, Sun J (2011) Properties of a well-defined macroscopic fundamental diagram for urban traffic. Transp Res Part B Methodol 45(3):605–617
50. Orda A, Rom R (1990) Shortest-path and minimum-delay algorithms in networks with time-dependent edge-length. J ACM (JACM) 37(3):607–625
51. Menelaou C, Timotheou S, Kolios P, Panayiotou CG, Polycarpou MP (2017) Controlling road congestion via a low-complexity route reservation approach. Transp Res Part C Emerg Technol 81:118–136
52. Menelaou C, Timotheou S, Kolios P, Panayiotou CG (2017) Improved road usage through congestion-free route reservations. J Transp Res Board 2621:71–80
53. Menelaou C, Timotheou S, Kolios P, Panayiotou CG, Polycarpou MM (2019) Minimizing traffic congestion through continuous-time route reservations with travel time predictions. IEEE Trans Intell Veh 4(1):141–153
54. Bast H, Delling D, Goldberg A, Müller-Hannemann M, Pajor T, Sanders P, Wagner D, Werneck R (2015) Route planning in transportation networks. Technical Report MSR-TR-2014-4. Microsoft Research Silicon Valley
55. Dijkstra EW (1959) A note on two problems in connexion with graphs. Numer Math 1(1):269–271
56. Cormen TH, Leiserson CE, Rivest RL, Stein C et al (2001) Introduction to algorithms, vol 2. MIT press Cambridge
57. Chabini I (1998) Discrete dynamic shortest path problems in transportation applications: complexity and algorithms with optimal run time. Transp Res Rec 1645:170–175
58. Ji Y, Geroliminis N (2012) On the spatial partitioning of urban transportation networks. Transp Res Part B Methodol 46(10):1639–1656
59. Saeedmanesh M, Geroliminis N (2016) Clustering of heterogeneous networks with directional flows based on 'snake' similarities. Transp Res Part B Methodol 91:250–269
60. Behrisch M, Bieker L, Erdmann J, Krajzewicz D (2011) SUMO-simulation of urban mobility-an overview. In: SIMUL. The third international conference on advances in system simulation 2011, pp 55–60
61. Krauss S, Wagner P, Gawron C (1997) Metastable states in a microscopic model of traffic flow. Phys Rev E 55(5):5597
62. Van Aerde M, Rakha H (1995) Multivariate calibration of single regime speed-flow-density relationships. In: Proceedings of the 6th 1995 vehicle navigation and information systems conference, vol 334, p 341
63. Van Aerde M, Rakha H (1995) Multivariate calibration of single regime speed-flow-density relationships. In: Proceedings of the 6th 1995 vehicle navigation and information systems conference, pp 334–341
64. Polak J, Jones P, Vythoulkas P, Sheldon R, Wofinden D (1993) Travellers' choice of time of travel under road pricing. Transport studies unit oxford university-publications-TSU-all series-
65. Elokda E, Cendese C, Zhang K, Censi A, Lygeros J, Frazzoli E (2023) Karma priority lanes for fair and efficient bottleneck congestion management. In: 2023 31st Mediterranean conference on control and automation (MED), pp 458–463
66. Jenelius E, Mattsson L-G, Levinson D (2011) Traveler delay costs and value of time with trip chains, flexible activity scheduling and information. Transp Res Part B Methodol 45(5):789–807

67. Ben-Akiva ME, Lerman SR, Lerman SR (1985) Discrete choice analysis: theory and application to travel demand, vol 9. MIT press
68. Noland RB, Polak JW (2002) Travel time variability: a review of theoretical and empirical issues. Transp Rev 22(1):39–54
69. Hess S, Polak JW, Daly A, Hyman G (2007) Flexible substitution patterns in models of mode and time of day choice: new evidence from the UK and the Netherlands. Transportation 34(2):213–238

Chapter 6
Highlights of Lane-Free Automated Vehicle Traffic with Nudging

Markos Papageorgiou, Panagiotis Typaldos, Dionysios Theodosis, Georgios Chalkiadakis, Iason Chrysomallis, Niloufar Dabestani, Iasson Karafyllis, Milad Malekzadeh, Mehdi Naderi, Ioannis Papamichail, Georgios Titakis, Dimitrios Troullinos, and Venkata Karteek Yanumula

Abstract Connected Automated Vehicles (CAVs) have superb capabilities compared to human drivers, which calls for revisiting of established conventional road traffic principles. Recently, the *TrafficFluid* concept, a novel paradigm for vehicular traffic at high levels of vehicle automation, was proposed, relying on two combined principles: (a) *Lane-free* traffic, whereby vehicles are not bound to fixed traffic lanes and (b) Vehicle *nudging*, whereby vehicles influence other vehicles in front or aside of them. This chapter provides highlights of related research: three methodological approaches to lane-free vehicle movement strategies with nudging, namely, optimal control, nonlinear feedback control, multiagent reinforcement learning; a specialized microscopic simulation environment; macroscopic traffic flow models emerging from lane-free vehicle driving with nudging, using a theoretical and an empirical approach; a new, highly efficient traffic control measure for lane-free traffic; a challenging case study of a large-scale complex roundabout; and joint path optimization of multiple vehicles in a lane-free environment.

M. Papageorgiou (✉) · P. Typaldos · D. Theodosis · N. Dabestani · M. Malekzadeh · M. Naderi · I. Papamichail · G. Titakis · D. Troullinos · V. K. Yanumula
Dynamic Systems & Simulation Laboratory, School of Production Engineering and Management, Technical University of Crete, 73100 Chania, Greece
e-mail: mpapageorgiou@tuc.gr

P. Typaldos
e-mail: ptypaldos@tuc.gr

D. Theodosis
e-mail: dtheodosis@tuc.gr

N. Dabestani
e-mail: ndabestani@tuc.gr

M. Malekzadeh
e-mail: mmalekzadehkebria@tuc.gr

M. Naderi
e-mail: mnaderi@tuc.gr

I. Papamichail
e-mail: ipapamichail@tuc.gr

P. Ioannou and A. A. Malikopoulos (eds.), *Transportation Mobility in Smart Cities*, Springer Tracts on Transportation and Traffic 21,
https://doi.org/10.1007/978-3-031-64769-7_6

6.1 Motivation

Despite huge technological progress in many domains, daily traffic congestion continues to reign daily on the roads around the globe causing serious travel delays, environmental degradation, and decreased traffic safety. Remarkably, this is in contrast with even simple species which manage their transport problems efficiently, see, e.g., [10]. Various traffic management methods [51] have been developed and partly employed in the past decades with different levels of success. More recently, a wide variety of vehicle automation and communication systems have been developed that tremendously improve vehicles' individual capabilities, enabling a new generation of potential traffic management tools [15, 55]. This tendency is continuing with the emergence of high-automation vehicles which are tried out in real traffic environments, see, e.g., [2]. In the not-too-far future, vehicles may communicate with each other and with the infrastructure, and drive automatically, based on own sensors, communications, and appropriate movement control strategies.

Recently, the *TrafficFluid* concept, a novel paradigm for vehicular traffic at high levels of vehicle automation, was proposed [54], relying on two combined principles: (a) *Lane-free* traffic, whereby vehicles are not bound to fixed traffic lanes, as in conventional traffic, but may drive anywhere on the 2D surface of the road and (b) vehicle *nudging*, whereby vehicles influence other vehicles in front or aside of them. These two principles are absent in human driving due to a significant perception limitation. Specifically, human vision is aligned with one direction at a time, which is naturally the direction of movement, and this has two influential implications:

(i) Traffic lanes were introduced in the 1950s to render human driving safer, as driving on a lane reduces to following the vehicle in front at a safe distance. On the other hand, lanes must be wide enough to accommodate the widest circulating vehicles, hence they lead to lower lateral occupancy and, as a consequence, to

G. Titakis
e-mail: gtitakis1@tuc.gr

D. Troullinos
e-mail: dtroullinos@tuc.gr

V. K. Yanumula
e-mail: vyanumula@tuc.gr

G. Chalkiadakis · I. Chrysomallis
Intelligent Systems Laboratory ("InteLLigence"), School of Electrical and Computer Engineering, Technical University of Crete, 73100 Chania, Greece
e-mail: gchalkiadakis@tuc.gr

I. Chrysomallis
e-mail: ichrysomallis@tuc.gr

I. Karafyllis
Department of Mathematics, National Technical University of Athens, 15780 Athens, Greece
e-mail: iasonkar@central.ntua.gr

lower flow and capacity. Also, traffic lanes necessitate occasional lane changing, which is a difficult and accident-prone maneuver.

(ii) Driving is only influenced by traffic conditions downstream, which leads to structurally anisotropic flow and lack of pressure at the macroscopic level, thus impacting traffic flow capacity and stability.

In summary, traffic lanes and anisotropy, stemming from human perception limitations, are impacting traffic safety and efficiency. Particles in other fluids (water, gas) are isotropic and do not form lanes. Even in living nature (bird flocks, fish schools) and in some human activities (crowd or bicycle-avenue movement), lanes are not encountered.

Forthcoming Connected Automated Vehicles (CAVs) have superb capabilities: (i) fast and reliable all-round perception; (ii) communication with other CAVs and the infrastructure; (iii) fast and efficient (algorithmic) decision-making. Thus, CAVs are different objects, compared to human-driven vehicles, which suggests that the current road traffic organization, in particular, traffic lanes and anisotropy, should be revisited, as they appear like fossils from the era of human driving. In absence of lanes, CAVs may drive smoothly in lateral direction, as abrupt lane changing becomes obsolete. Also, higher (lateral) occupancy may lead to higher capacity. Furthermore, in absence of lanes, flow and capacity increase proportionally with road width (Sect. 6.7), which, among others, enables a novel and highly efficient traffic control measure, internal boundary control (Sect. 6.8). More generally, design of vehicle movement strategies under the TrafficFluid concept may be performed so as to inherit opportune properties to the emerging traffic flow. In other words, instead of modeling traffic flow as a physical process, we have the opportunity to actively design it, via appropriate vehicle movement strategies, as an *artificial fluid*.

The TrafficFluid concept is investigated within an ERC (European Research Council) project by the same name, as well as several national research projects by various groups, and first overviews of related works have already appeared [64]. The present chapter provides highlights of research carried out within the ERC TrafficFluid grant. Sections 6.2, 6.3, 6.4 outline three respective methodological approaches for the development of lane-free vehicle movement strategies with nudging, each with its advantages and limitations, namely, optimal control, nonlinear feedback control, and multiagent reinforcement learning. Section 6.5 presents a specialized microscopic simulation environment for lane-free vehicle traffic with nudging. Sections 6.6 and 6.7 present macroscopic traffic flow models emerging from lane-free vehicle driving with nudging, using a theoretical and an empirical approach, respectively. Section 6.8 presents a new, highly efficient traffic control measure for lane-free traffic, which is internal boundary control. Section 6.9 presents a challenging case study of a large-scale complex roundabout, while Sect. 6.10 presents a first step toward joint path optimization of multiple vehicles in a lane-free environment.

6.2 Optimal Vehicle Trajectory Planning

Optimal-control-based planning of individual vehicle trajectories is one of the employed approaches for Connected and Automated Vehicles (CAVs) in a lane-free traffic environment with vehicle nudging. Specifically, a nonlinear constrained Optimal Control Problem (OCP) is developed and solved via a very efficient, real-time feasible numerical solution algorithm. The OCP is employed in a Model Predictive Control (MPC) mode for each CAV independently, with a rolling horizon of 8 s and event-triggered re-planning. Each CAV broadcasts every new path planning to other CAVs in its vicinity to enable more reliable prediction of their short-term movement by them.

This approach is distinguished by high flexibility regarding the application infrastructure and scenario considered. On the other hand, care is required, e.g., with the tuning of multiple sub-objectives, as the entailed vehicle behavior depends largely on appropriate balancing of competing requirements for collision avoidance, passenger comfort, driving efficiency, and more.

6.2.1 Optimal Control Problem (OCP)

An OCP consists of a state-space dynamic model, control constraints, and an objective function to be minimized. The present OCP is formulated for each individual vehicle based on a discrete-time vehicle model. In a first infrastructure scenario, vehicles are moving on an unfolded straight ring road, with small lateral speed compared to longitudinal speed. Therefore, the employed model consists of two decoupled double-integrators (for position and speed), one for each of the longitudinal and lateral movements [80], with the longitudinal and lateral accelerations being the control inputs. Clearly, the lateral movement uses real-valued states and control thanks to lane-free road driving, which allows for the vehicles to be positioned within the road boundaries continuously in the lateral direction, in contrast to the introduction of discrete variables engaged in most lane-based approaches.

In view of limited vehicle capabilities, passenger comfort, and to respect the road boundaries, the control inputs (longitudinal and lateral acceleration) are constrained appropriately:

- A state-dependent lower limit on the longitudinal acceleration accounts for non-negative speeds.
- A constant upper bound reflects the vehicle's longitudinal acceleration capabilities. In addition, a state-dependent upper bound on the same input is used only in rare emergency situations, when following of an obstacle vehicle downstream at a pre-specified distance is required.
- In the lateral direction, the vehicle must stay within the (stationary and straight) road boundary lines, which gives rise, via an intermediary feedback boundary controller, to state-dependent upper and lower bounds of lateral acceleration.

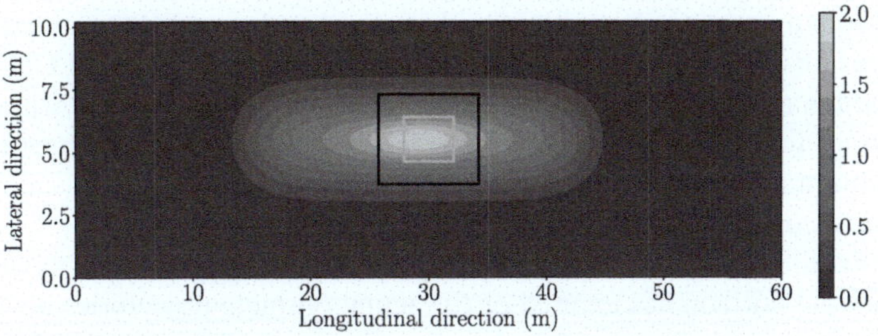

Fig. 6.1 Heat diagram of an obstacle ellipsoid for collision avoidance. Cyan rectangle reflects the obstacle dimensions, augmented with the EV dimensions in the black rectangle

Turning to the objective function, various weighted sub-objectives are combined, and minimization leads to efficient and comfortable vehicle advancement, while avoiding obstacles. Specifically, the sub-objectives are as follows:

- Fuel efficiency and passenger comfort in the longitudinal direction are achieved by a quadratic cost term of longitudinal acceleration. In addition, a quadratic cost term of the lateral acceleration further improves the passenger comfort and suppresses unnecessary lateral movements.
- Deviation of longitudinal and lateral speeds from their respective desired values is penalized via respective quadratic cost terms.
- Obstacle avoidance is achieved via an imaginary positive-valued, smooth, longitudinally, and laterally symmetric ellipsoid hemisphere around each obstacle in the vicinity of the ego vehicle (EV), see Fig. 6.1. The ellipsoid extends beyond the physical vehicle dimensions, including time-gap-like margins in longitudinal and lateral directions. The ellipsoid extends all around the obstacle, even beyond its front part, which is useful for preventing the EV from cutting-in dangerously close in front of an obstacle and enables nudging of the EV by obstacles behind it that gives rise to structurally new traffic flow characteristics [28]. Nudging is created also by the lateral extent of the ellipsoid if the two vehicles approach each other laterally.
- To account for infeasible lateral movements that might result due to the uncoupled model equations in longitudinal and lateral directions, a quadratic term penalizes the rate of longitudinal over lateral speed below a threshold value.

6.2.2 Numerical Solution and Model Predictive Control (MPC)

The OCP is defined as the minimization of the objective function discussed above, subject to the equality constraints (model) and the inequality constraints. An efficient Feasible Direction Algorithm (FDA) [52], using the Polak–Ribiere method, is employed to solve the OCP problem within an event-based MPC framework. The trajectories of 8-s duration (time horizon) are computed in 12.4 ms of CPU time on average, which allows for application in real time. MPC calls the FDA in an event-based manner for each vehicle, and this results in asynchronous path computation and application in emulated real time for each of the vehicles. Each vehicle broadcasts its latest computed trajectories to all other vehicles in its vicinity for use in their own path planning. MPC of each vehicle triggers an update of its path planning if: (i) half of the time horizon has expired or (ii) actual obstacle trajectories deviate from the predicted ones or (iii) new obstacles enter the scene. At each path planning, the pre-set longitudinal desired speed of each vehicle may be lowered, depending on the initial vehicle speed and the average speed of surrounding vehicles, so as to suppress unnecessary bursts of longitudinal acceleration.

6.2.3 *Simulation Results*

Simulations with TrafficFluid-Sim (Sect. 6.5) were carried out on an unfolded lane-free ring road of 1 km length and 10.2 m width at different traffic densities, mixing vehicles of different sizes and desired speeds.

At traffic level, a fundamental diagram (FD—flow versus density) with a typical inverse-V shape emerges, as shown in Fig. 6.2. It is interesting to note that no backward-moving jams are observed at any density value, i.e., the emerging traffic flow is stable. At vehicle level, it was observed that the vehicles move efficiently, smoothly, and safely over the full range of possible vehicle densities, while maintaining passenger convenience.

As a highlight of lane-free driving, Fig. 6.3 shows the lateral position of a vehicle within the road width over time, for a density of 100 veh/km. In fact, lateral vehicle movement is flexible and differs at different vehicle densities. In under-critical densities, the vehicle is seen in Fig. 6.3 to leverage the complete available lateral space (road width) in its attempt to attain or maintain the desired longitudinal speed. It is interesting to note that lateral maneuvering is more intensive at periods when

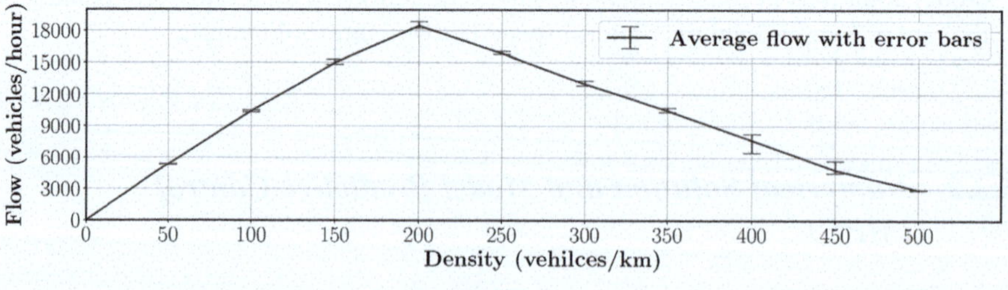

Fig. 6.2 Fundamental diagram. Error bars reflect maximum and minimum deviations among five simulation replications

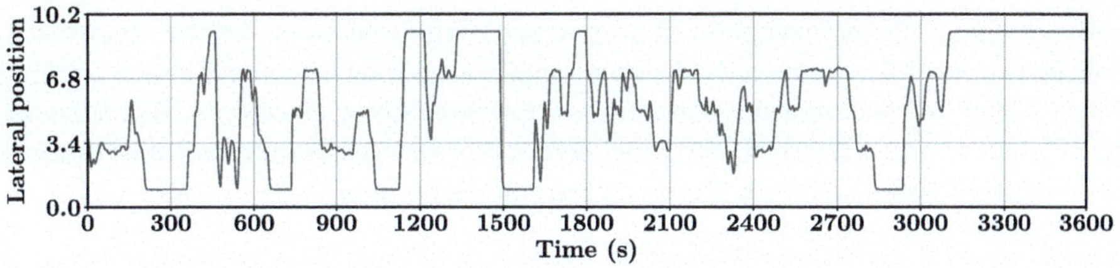

Fig. 6.3 Lateral position over time for a vehicle at density of 100 veh/km

the longitudinal speed falls short of the desired speed, which happens due to slower vehicles ahead. In such circumstances, the stronger employed lateral displacements witness the striving of the vehicle to find appropriate opportunities to accelerate longitudinally toward the desired speed. It may also be seen that, at periods, the vehicle is moving exactly on the left or right road boundaries (its center being half-width aside), after having reached those asymptotically, and without ever crossing the boundary. Such efficient vehicle maneuvering is not only beneficial for vehicle advancement, but also improves the overall traffic efficiency, contributing to the high flows of the FD in Fig. 6.2. In critical or over-critical densities, the lateral movement is more limited due to lack of lateral space, which is occupied by other vehicles.

Videos showing the vehicle movement in different densities can be viewed at https://bit.ly/TF-OPP.

6.2.4 Future Work

Future work is focused on introducing on-ramps and off-ramps with appropriate merging and exiting extensions of the present vehicle movement strategy to allow testing of the performance of the approach at bottlenecks that may result from high inflow from an on-ramp. Checking the improvements in the flow due to lateral positioning of vehicles according to their longitudinal desired speeds was preliminarily addressed in [17].

6.3 Nonlinear Feedback Controllers

Development of purely reactive (no predictions) Nonlinear Feedback Controllers (NLFCs) for real-time vehicle movement control is one of the employed approaches for Connected and Automated Vehicles (CAVs) in a lane-free traffic environment with vehicle nudging. This approach results in individual (decentralized) controllers for each vehicle with limited real-time information requirements about other neighboring vehicles. The approach is distinguished by rigorous results that mathematically guarantee the achievement of all pursued main objectives. On the other hand, consideration of fundamentally different application scenarios may, under circumstances, call for the rigorous derivation of corresponding extensions. The derived NLFCs allow for a variety of different choices that entail accordingly different vehicle behaviors which all enjoy the same guaranteed qualitative properties.

6.3.1 Introduction

This approach considers the development of cruise controllers (CC) for CAVs that must guarantee the following objectives for the CAV motion:

1. The vehicles must not collide with each other and must always remain within the road boundaries.
2. The vehicles must move in a specific direction of the road and must not be led to a halt.
3. The speeds of all vehicles must always be lower than the road speed limit and, if possible, converge asymptotically to a desired speed.

In addition to the above objectives, the CC design procedure takes into account certain technical constraints, such as the technical characteristics of each vehicle (which limits the acceleration that can be applied) and the road characteristics (which limits the vehicle orientation angle). Finally, the CC of each vehicle can only use real-time measurements (e.g., about other vehicles or about the road boundaries) from a neighborhood of the given vehicle. As a consequence, the CC is fully decentralized and the vehicles can be deemed autonomous.

The CC design problem is essentially a feedback controller design problem. The current speed and the distances from other nearby vehicles and from the road boundaries are measured and are "fed-back" to the CC, which in turn determines the appropriate vehicle acceleration and steering angle. Nonlinear feedback design is a well-established branch of mathematical control theory [65] and provides the necessary tools for the achievement of the aforementioned objectives.

On the other hand, the movement of CAVs on a road reminds strongly of a well-known problem in Mathematical Physics: the movement of molecules of a fluid. Indeed, CAVs can be seen as self-driven particles moving within a pre-specified domain. The mathematical description of a set of particles leads to a well-known topic in Mathematics: the n-body problems [71]. The use of the analogy between vehicles and fluid particles has been criticized in the past for vehicles with human drivers [13] due to a specific property that appears in vehicle systems with human drivers: anisotropy, i.e., the fact that decisions of the human driver are affected mainly (if not exclusively) by downstream vehicles. For CAVs, this is not necessarily true anymore. Indeed, if nudging and lane-free movement are allowed, as in the TrafficFluid paradigm [54], then many characteristics of compressible fluid flow (including isotropy) can be restored. Therefore, the literature for n-body problems can also provide important tools for the CC design.

In what follows, we review available results concerning the CC design for lane-free CAVs. The results have been obtained by utilizing ideas and notions from four different areas in Applied Mathematics: (i) mathematical control theory (nonlinear feedback design), (ii) dynamical systems (Hamiltonian systems and n-body problems), (iii) mathematical physics (mechanics of particles and systems), and (iv) fluid mechanics (compressible fluid flow).

6.3.2 *Cruise Controller (CC) Design*

The geometry of the road plays an important role for the CC design. This is particularly true for the case of lane-free movement of vehicles. For roads of constant width with no intersections, there are two basic geometries (and every road with a different geometry can be transformed to a road with one of the following geometries):

(a) the straight road and
(b) the ring road.

In addition to the geometry of the road, another issue that plays an important role for the CC design is that vehicles are (in general) not identical but have different sizes and different desired speeds. Note also that, in contrast to particle models in Physics, the double-integrator model may, under circumstances, be less accurate in the description of the movement of a vehicle, e.g., when strong turning maneuvers are involved. Therefore, for CC design we opt for the bicycle kinematic model with inputs the vehicle acceleration and the steering angle of the front wheels relative to the vehicle orientation, see Fig. 6.4. From a mathematical point of view, an additional difficulty comes from the fact that the incorporation of all objectives leads to the definition of the overall control system on an open set which is not homeomorphic to a linear space. This case is rarely studied in the literature of nonlinear control theory.

The first paper that provided CC for the simplest case, i.e., the case of a straight road with constant width and identical vehicles, was [29]. The main result of [29] guarantees all the objectives that were described in Sect. 6.3.1, as well as the following properties: (i) the speeds of all vehicles tend asymptotically to a given set point, (ii) the orientations of all vehicles tend asymptotically to be parallel to the road, and (iii) the accelerations and the steering angles tend asymptotically to zero. The avoidance of collisions and road boundary violation were achieved by using potentials. However, in contrast with particles from models in Mathematical Physics, the potentials in [29] vanish identically when the distance between two vehicles is greater than a pre-set threshold, the so-called interaction distance. This particular characteristic entails that the CC needs real-time information about only those surrounding vehicles that are located within the area specified by the interaction distance. Note that distance is defined such that iso-distance curves are ellipses, so that the road width be better exploited in lane-free traffic, see Fig. 6.5. Note also that the extent of the interaction zone around the vehicle implies that, beyond repulsion by front obstacles, nudging forces are also exerted to the vehicle due to obstacles at its sides or behind it. The

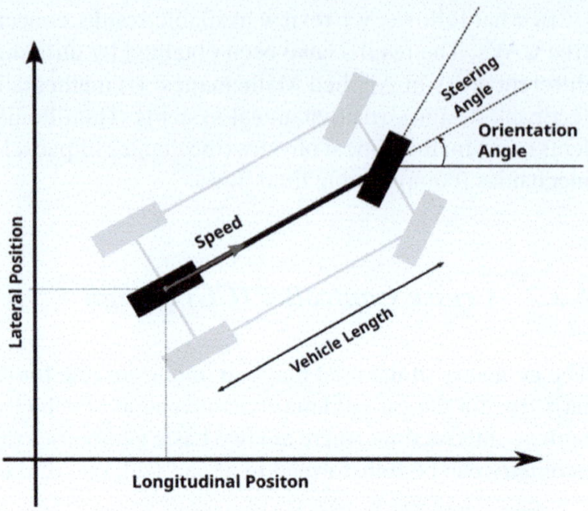

Fig. 6.4 The bicycle kinematic model

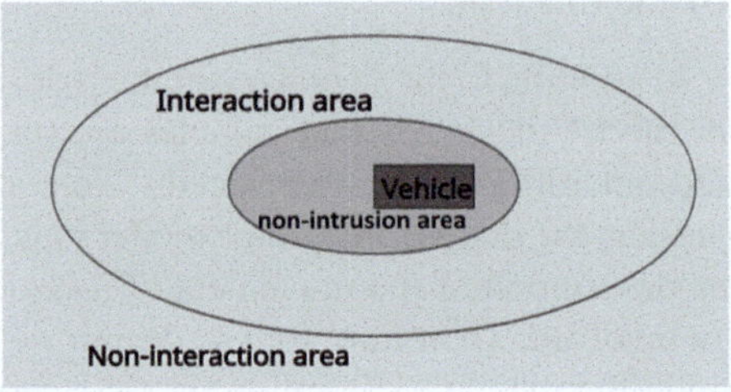

Fig. 6.5 Interaction area, non-interaction area, and non-intrusion area around a vehicle

family of CCs proposed in [29] was based on a Lyapunov function constructed by using the Newtonian mechanical energy of the vehicles/particles and additional penalty terms that guarantee that all vehicle orientation angles remain within certain bounds.

It is important to note that the CCs proposed in [29] do not guarantee that the vehicles tend to a specific "final" arrangement on the road. The vehicles are self-organized and their "ultimate" arrangement depends on the initial condition. The speed constraints are achieved by means of nonlinear controller gains, while the achieved convergence is not exponential. Moreover, the CCs do not require speed measurements but only the measurement of relative positions of neighboring vehicles.

The scope of the CCs in [29] was extended in [28], still for the case of a straight road with constant width. The CCs in [28] guarantee all features of [29] but, as an extension, the vehicles are allowed to have different sizes. Moreover, the CCs in [28] have additional terms that imitate the action of viscosity in fluid flow. The viscous terms can give faster convergence of the speeds and smaller accelerations. When the viscous terms are present, the controllers require real-time measurements of relative speeds of neighboring vehicles. Finally, the results in [28] provide two different families of CCs: one family based (as in [29]) on the Newtonian mechanical energy of the vehicles/particles and another family based on a pseudo-relativistic mechanical energy, where the speed limits in relativistic mechanics ($\pm$ the speed of light) are replaced by the two speed limits for vehicles (zero and the speed limit of the road, respectively) and allow the use of constant controller gains.

All the results in [28] were re-established in [69] for the case of a ring roads with constant width. There is a difference between the ring road and the straight road: in the case of the ring road, the angular speeds rather than the actual speeds of all vehicles tend asymptotically to a given set point. This is the effect of the major topological difference between the straight road and the ring road.

The case of variable-width roads with possible on-ramps and off-ramps was studied in [31]. The results in [31] utilize the inviscid, pseudo-relativistic Lyapunov setting and allow each vehicle to move within its own (curved) road boundaries, which may or may not coincide with the actual road boundaries. Moreover, each vehicle has its own desired speed. The controllers in [31] guarantee all features of the controllers in [28, 29] in the constant-width case, but can also work in the variable-width case. Each vehicle's own boundaries do not need to be known to the vehicle controller a priori to their full extent but are only required to be known in real time in a small neighborhood of the vehicle. It should be emphasized that, despite having own boundaries, the vehicle controllers can handle obstacle avoidance with obstacles located out of their own boundaries (but within their interaction area) as in previous cases. The extensions in [31] allow for consideration of a variety of practical situations, e.g., when vehicles may have to be positioned laterally on the road according to their size (e.g., trucks) or desired speeds as in [17].

Videos showing the movement of AVs under the CCs that were given in [28, 29, 31, 69] are provided in: https://bit.ly/3OW6H4F (straight road), https://bit.ly/45sAIFP (ring road NCC and PRCC), https://youtu.be/h4nIUz2nl9s (Off-ramps).

6.3.3 Open Problems

As mentioned above, the CCs in [31] can handle most of the cases that may appear in practice. However, there are special cases of variable-width roads where some of the vehicles are asymptotically led to a halt. Simulations show that the set of initial conditions that can lead the vehicles to a halt is of zero Lebesgue measure and thus the possibility of having this outcome is zero. Criteria that determine whether or not the vehicles will be led to a halt are needed.

The performance of the proposed (Newtonian or pseudo-relativistic) CCs depends heavily on the potential functions and size of interaction area used. Extensive numerical experiments must be performed to determine appropriate potential functions and distance thresholds that yield desired performance.

Finally, the stability properties of the system of vehicles under the proposed CCs need to be better understood. Some progress in this direction has been achieved when the vehicles are constrained to 1D motion (or when their movement is lane based), see [30].

6.4 Reinforcement Learning and Multiagent Coordination Strategies

6.4.1 Introduction

Typical approaches for automated driving primarily involve rule-based movement strategies, use of optimization techniques [4, 26, 44, 78, 80], or feedback control methods [24, 28, 32]. However, there is an emergence of novel techniques that tackle automated driving with machine learning-based principles [3, 14, 35]. This is in light of the recent computational capabilities acquired by "smart" vehicles, and advances in the field of deep reinforcement learning (DRL) where classical reinforcement learning [68] is combined with deep neural networks. DRL algorithms manage to solve complex problems and learn through patterns on the observed data. These capabilities, combined with access to simulation infrastructures that allow for agent-based training, give rise to novel approaches of DRL in lane-based traffic [3, 14, 35]. By contrast, learning-based solutions tied with DRL in lane-free traffic have only recently been established. At the same time, the emergence of communication capabilities among automated vehicles [35, 81] allows also for the use of multiagent coordination techniques [36, 79] in order to improve driving behavior.

Against this background, in this section, we first outline the application of DRL algorithms for the challenging two-dimensional lane-free traffic environment from the perspective of a vehicle that needs to learn a driving policy in a populated road. Moreover, we provide a novel methodology that properly captures the *state information* that a vehicle receives when surrounded by other lane-free vehicles, and, importantly, makes use of a carefully designed *modular reward function* that can incorporate the effect of nudging, together with collision avoidance and desired speed pursuit [33, 34]. In addition, we discuss the integration within the lane-free traffic domain of the popular in DRL research "imitation learning" paradigm [1, 50, 58, 72]. In this line of work, one can make use of trajectories sampled from rule-based movement strategies, and consider these as "experts" or "mentors" that help accelerate the learning process of a "mentee" vehicle—and allow the mentee to surpass its mentors in performance [9]. Second, as already mentioned, another line of research in lane-free traffic considers the exploitation of communicated information across vehicles, in order to improve their movement strategies. Indeed, current and future advancements in communication infrastructures for automated vehicles conjecture new approaches, where vehicles can rely on accurate real-time communication with nearby traffic. This assumption has led to novel movement strategies that explicitly incorporate communication within the vehicles' decision-making [35]. As such, we investigate multiagent coordination for lane-free movement with a strong focus on scalability and dynamic interactions among vehicles [76, 77].

6.4.2 Deep Reinforcement Learning

In reinforcement learning (RL) [68], agents improve their strategies over time by making use of appropriate (positive or negative) reinforcement signals or *rewards* received for their actions. Rewards are delivered via a *reward function* that effectively maps the desirability of each action taken to a specific reward. The RL algorithms themselves then make use of the rewards received in order to estimate the long-term desirability (or value) of the actions taken at particular world states and devise successful policies that appropriately match world states to actions. Therefore, the design (or inference, in case it is not provided) of the reward function is of utmost importance to a learning agent, as it must properly reflect each objective set within a learning domain.

Several key objectives can be considered for the lane-free traffic paradigm [33, 34]. Starting with the safety of the controlled vehicle and the surrounding agents in mind, at each training step, a negative reward is yielded for each collision occurred in the simulation. Additionally, the concept of nudging between vehicles is employed as a metric. Based on the degree to which a surrounding vehicle is positioned within an ellipsoid function, a negative impact is reflected on the reward value, which addresses collision avoidance as well. Another important goal is desired speed pursuit. The reward function integrates this objective by granting higher rewards to speeds close to the predetermined desired speed value of each vehicle. Incorporating these three primary goals of safety, desired speed pursuit, and nudging, the learning agent is provided with essential underlying guidance in an RL manner.

An extension of this DRL approach involves the contribution of a single or multiple mentors' experience. Intuitively, observing a highly experienced driver naturally provides valuable insights to a learning agent. However, due to data privacy restrictions, sensor limitations, or the impracticality of using expensive tools, the observation of the expert's actions is not, in most cases, an option. By putting forward a *deep implicit imitation reinforcement learning* (DIIRL) framework, as seen in Fig. 6.6, we can enable the utilization of such incomplete expert knowledge, essentially exploiting only state-transition information. As explained in [9], this approach accelerates the early learning stages, i.e., the stages in which, commonly, redundant exploration of possible actions takes place. Experimental evaluation showcased a drastic reduction in required computational power and learning time. Moreover, DIIRL is shown to allow the "mentee" self-driving vehicle to surpass the performance of a non-optimal mentor.

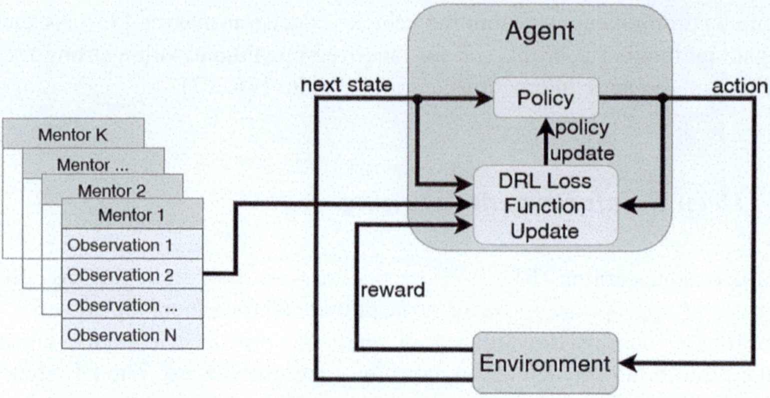

Fig. 6.6 High-level representation of the DIIRL framework

6.4.3 Multiagent Coordination

When one attempts to eliminate the restriction of lanes, the traffic environment and its presentation is fundamentally altered. Instead of pre-defined lateral positions, now we need to consider the continuous space of lateral placements, resulting in much more diverse formations of vehicles, e.g., as the one shown in Fig. 6.7. As such, this constitutes a different domain altogether for decision-making, when compared to lane-based approaches with multiagent coordination [14, 57]. In addition, vehicles in the examined environment operate at different desired speeds, resulting in further conflicts when they need to concurrently make a movement decision. To address this, a factor graph representation [38] with vehicles in the roles of agents controlling acceleration-related variables is natural for this challenging dynamic coordination domain, and allows the use of distributed constraint optimization algorithms [19] such as the one we examine, the Max-Sum message-passing algorithm [16].

The factor graph formulation, as shown in Fig. 6.7, contains two types of factors and is formed dynamically based on the vehicles' proximity and speed. The first type F_i accounts for the individual goal of each vehicle which is the desired speed objective, while the second type F_{ij} is pairwise, connecting two nearby vehicles to capture the risk of collision, again with the use of an ellipsoid function. The fundamental difference when compared with single-agent movement strategies is that through the use of the Max-Sum algorithm, vehicles send and receive messages from the vehicles nearby that share a pairwise factor. With this additional information from the messages, each vehicle's movement decisions are affected by the surrounding traffic's goals.

One challenge and potential bottleneck with this approach is that one is required to discretize the control of the vehicles, since Max-Sum relies on a pre-specified discretization of the domain. To tackle this issue, we provide a novel extension to Max-Sum, where the incorporation of an additional component, namely, Quadtrees [18],

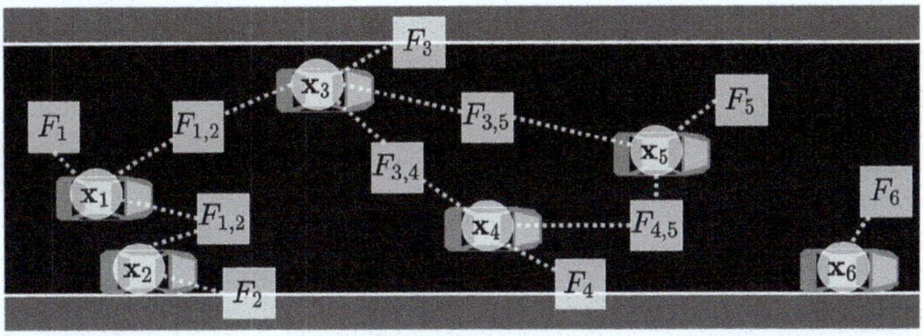

Fig. 6.7 An example of a factor graph of vehicles in an instance of the lane-free traffic environment

allows for dynamic discretization over the continuous (and two-dimensional) control space [77]. Essentially a tree structure is dynamically created and embedded within the message-passing process of Max-Sum, where vehicles automatically adjust the selection of discrete points depending on the current observed state.

6.4.4 Future Endeavors

Our DIIRL approach can be extended to continuous action space environments. As such, ongoing and future work includes the integration and evaluation of additional DRL algorithms, such as PPO [63] and NAF [22], into our framework. More generally, we are working toward enhancing our learning and coordination approaches so that they move closer to real-world applicability. To this end, we are in the process of incorporating rule-based subsystems into our framework, assigning them the task of maintaining the safety and comfort of passengers. Essentially, this will constitute a hybrid approach, where DRL or factor graph formulations handle the strategic decision-making of the vehicles (e.g., their preferred lateral alignment and so on), instead of directly affecting acceleration values.

6.5 Microscopic Simulation Environment

Research endeavors regarding the TrafficFluid paradigm necessitate a microscopic simulation tool that can model vehicle movement strategies in lane-free traffic environments involving nudging, with flexibility in its design in order to accommodate emerging needs. Existing microscopic simulators (e.g., SUMO [43], Aimsun [8]) are built from the ground-up considering conventional lane-based traffic and car-following driving, with their principles being hard-coded into their underlying structure (leader–follower schemes and lane-changing behavior). While these are no longer relevant for lane-free movement with vehicle nudging, many aspects of exist-

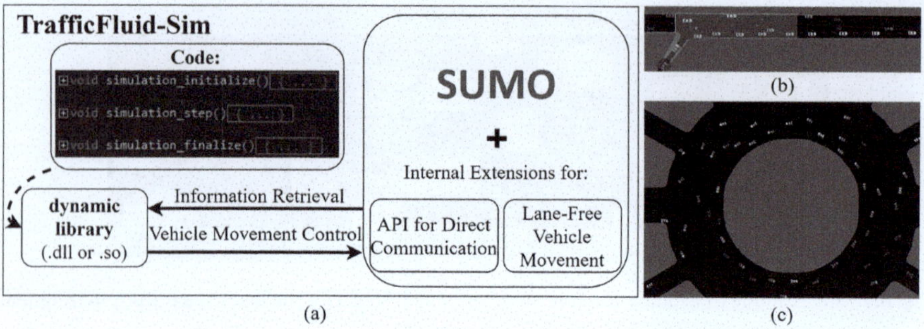

Fig. 6.8 **a** High-level overview of TrafficFluid-Sim, **b** Observing traffic downstream, **c** Roundabout scenario with bicycle model

ing simulators can be of use (e.g., graphical interface, road network structure, loop detectors, etc.). Therefore, we utilize the source code of SUMO and complement it with several adjustments and additional developments, rendering it compatible with TrafficFluid requirements.

The result is *TrafficFluid-Sim* [74, 75], a microscopic simulator tool where users can build and examine vehicle movement strategies under the TrafficFluid paradigm through a provided structure for code development in C/C++, used to generate a dynamic library that is combined with the simulator application. An Application Programming Interface (API) is provided through this dynamic library, which allows users to exchange information with the simulator in an online manner without time-related bottlenecks. Figure 6.8a illustrates the main components of TrafficFluid-Sim. With the API, one can monitor the traffic environment through requests regarding: the vehicles' current state (position, speed, orientation) and properties (type, length, width, desired speed); information about nearby traffic from the perspective of a specific vehicle (see Fig. 6.8b); properties of road sections; online measurements of flow from loop detectors; and online measurements of density and mean speed at specific road sections. This information can be directly incorporated in movement strategies that rely on communication schemes either among vehicles (vehicle to vehicle) or with the infrastructure (vehicle to infrastructure). Most importantly, the API also provides the capability to control each vehicle with an appropriate control input depending on the selected vehicle dynamics.

Two different movement dynamics are supported in the simulator, namely, the double double-integrator and the bicycle model. The former considers that longitudinal and lateral control of the vehicles are decoupled, and users need to supply the simulator (through the API) with the control input of longitudinal and lateral acceleration (in m/s^2) at every discrete time step. The orientation of the vehicle is not controlled by the user and is always parallel with the road axis. This assumption is suitable for highway scenarios, where vehicles typically operate with high longitudinal speeds and minor lateral maneuvers (Fig. 6.8b). However, refined control over orientation is desirable for certain scenarios with stronger turning maneuvers, as

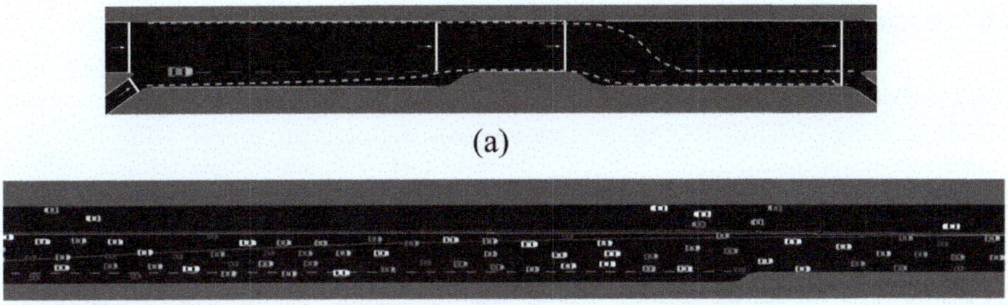

(a)

(b)

Fig. 6.9 **a** Smooth boundaries guiding vehicles according to their routes; **b** Real-time moving boundaries for Internal Boundary Control application

encountered in urban road networks, or more generally for a more accurate depiction of the vehicle movement dynamics. We address this by integrating the bicycle model (see Fig. 6.4 in Sect. 6.3.2) as an alternative, where the front wheels of the vehicle are steerable, and users need to yield a scalar acceleration value (in m/s^2) and a steering angle (in *rad*) as control input through the API. The bicycle model is, for instance, utilized for roundabout scenarios as shown in Fig. 6.8c, where the turning operation is inherent to the required vehicle movement, see also Sect. 6.9.

Additional developments on the tool, specific to lane-free environments, are related to demand handling at origins and design of smooth lateral boundaries. For the former, we have devised two different insertion policies that determine how vehicles are inserted at origin points of a road network according to given demand flow values, taking into account the available longitudinal and lateral road space. Initial vehicle speed depending on traffic conditions near the entry point and initial lateral placement are specified for each vehicle that enters the road.[1] Furthermore, the design of lateral boundaries aims at smoothly guiding vehicles according to their routing scheme. For instance, in Fig. 6.9a, the vehicles entering from the on-ramp or exiting through the off-ramp follow paths within the corridors delineated by the shown smooth lateral boundaries. Finally, real-time shifting of lateral boundaries is a required feature for the application of Internal Boundary Control (IBC) (see Sect. 6.8), where the internal lateral boundary of the road partitions a bidirectional highway to two directions. Using IBC, we can give more lateral space to one direction for congestion mitigation, as shown in Fig. 6.9b.

TrafficFluid-Sim is continuously expanding and evolving to accommodate emerging research needs related to the TrafficFluid concept. For example, latest developments enable signal-free lane-free crossing at urban junctions.

[1] The interested reader may refer to https://bit.ly/3tf53jD for illustrative videos.

6.6 Emerging Macroscopic Traffic Flow I: Theoretical

6.6.1 Introduction

Macroscopic models address traffic flow as a fluid with particular characteristics expressed in aggregate variables, such as vehicle density and mean speed, and reflect its evolution via partial differential equations (PDEs). The first and most influential traffic flow model for human drivers was independently proposed by Lighthill and Whitham [42] and Richards [59] (LWR model). The well-known LWR model is a first-order model governed by one scalar hyperbolic PDE that expresses the conservation of vehicles. First-order traffic models, while well studied, present several limitations, as they do not capture significant traffic flow dynamics. Therefore, researchers proposed second-order models for traffic flow with human drivers (like the ARZ model, see [5, 6, 40, 82]) which improve the modeling accuracy and some of the deficiencies of first-order models at the expense of higher complexity (an additional PDE).

There are three main issues with macroscopic traffic flow models: (i) their derivation from microscopic models, (ii) the theoretical properties of their solutions, and (iii) the numerical schemes that allow the approximation of their solutions. While the theoretical properties and the numerical analysis of macroscopic traffic flow models are extensively studied in the literature, their rigorous derivation from a microscopic model remains an issue. The rigorous derivation of the LWR model from a microscopic Follow-the-Leader model was performed in the last decade (see [21, 23]), and the connection of the ARZ model with a microscopic Follow-the-Leader model was shown in [5].

Although some researchers have proposed macroscopic models for the traffic flow of connected automated vehicles (CAVs), there are very few macroscopic models for lane-free CAV traffic with nudging, and, in what follows, we review some available related results that have been obtained by utilizing ideas and notions from three different areas in applied mathematics: (i) analysis of PDEs, (ii) numerical analysis of PDEs and Navier–Stokes equations, and (iii) fluid mechanics (compressible fluid flow).

6.6.2 Macroscopic Models for Lane-Free CAV Traffic

Since nudging and lane-free movement are allowed by the cruise controllers (CCs) proposed in [28, 29] (see Sect. 6.3), it should be expected that many characteristics of compressible fluid flow (including isotropy) will be present in the corresponding macroscopic models. The formal derivation of the macroscopic models for the CCs that were proposed in [28, 29] was given in [28, 30]. The macroscopic models involve the continuity equation and an additional PDE for the mean speed. The speed PDE is strongly reminiscent of the momentum equation for 1D polytropic compressible fluid flow: it contains the inertial term, a viscosity term with density-dependent viscosity, a pressure term and a friction term (the analog of the relaxation term in macroscopic traffic models for human drivers). Therefore, it is justified to say that the macroscopic models in [28, 30] are "fluid-like" models. For the CCs that are based on the Newtonian mechanical energy, the macroscopic model is highly nonlinear. Some properties (convergence to a traveling wave for small densities) of this model are given in [30]. On the other hand, for the CCs that are based on a pseudo-relativistic mechanical energy (pseudo-relativistic cruise controllers—PRCCs) the macroscopic model is quasi-linear and more manageable. We next focus on the quasi-linear macroscopic model for lane-free CAV traffic with PRCCs, which is called CAV Macroscopic Model (CAVMM).

It should be noted that there are some crucial differences between the CAVMM and the 1D compressible Navier–Stokes (NS) equations:

(a) In compressible fluid flow, the density is required to be positive, while in CAVMM the density is required to be positive and less than a maximum density.
(b) In compressible fluid flow, the speed can take arbitrary real values, while in CAVMM the speed is required to be positive and less than the speed limit of the road.
(c) In compressible fluid flow, the friction term tends to bring the fluid speed to zero, while in CAVMM it tends to bring the fluid speed to the desired speed.
(d) In compressible fluid flow, the pressure and viscosity terms are always present, while in CAVMM these terms vanish when the density is lower than a specific value (the so-called interaction density).
(e) In CAVMM, the acceleration is multiplied by a factor that blows up when the speed tends to zero or to the road speed limit. This factor is analogous to factors appearing in relativistic fluid mechanics and has to do with the pseudo-relativistic character of the mechanical energy that was used for the derivation of the PRCCs.

The most important implication shown in [28] is the fact that, by changing the functions or the parameter values of the CCs, we can directly determine the physical properties of the "traffic fluid." To understand how far the implications of the relations between the CCs parameters and the characteristics of the traffic fluid go, it is important to notice that for polytropic flow of gases, the dynamic viscosity and the pressure are always increasing functions of the fluid density, while for a traffic fluid that emerges from the use of PRCCs it is possible to obtain a traffic fluid with dynamic viscosity and pressure, which are non-monotone functions of the fluid density (and can have local minima). Thus, the traffic fluid can have very different physical properties from those of real compressible fluids. Since such properties may be established or tuned via appropriate choices at the underlying CC level, we may talk about designed or "artificial traffic fluids."

A model reduction is possible for the CAVMM when we make a certain selection for the potential function involved in the PRCCs, which leads to a particular selection for the pressure function in the macroscopic model, yielding the Special CAV Macroscopic Model (SAVMM). It was shown in [70] that (at least formally) the solutions of the SAVMM converge exponentially to the solutions of a nonlinear heat

Table 6.1 Macroscopic Models. Density and speed are denoted by ρ and v, respectively

	Human drivers	SAVMM	1D NS for a polytropic compressible fluid
	Continuity Equation: $\rho_t + (\rho v)_x = 0$		
Reduced Model	LWR: $v = f(\rho)$, $\rho > 0, v > 0$	$v = \beta^{-1}\left(-\rho^{-2}\mu(\rho)\rho_x\right)$, $\rho \in (0, \rho_{\max}), v \in (0, v_{\max})$	–
Speed Equation	ARZ: $v_t + (v + \rho f'(\rho))\, v_x = -k\,(v - f(\rho))$ $\rho > 0, v \in \mathbb{R}$	$q(v)\,(v_t + v v_x) + k\rho^{-2}\mu(\rho)\rho_x = \rho^{-1}\,(\mu(\rho)v_x)_x - k\beta(v)$ $\rho \in (0, \rho_{\max}), v \in (0, v_{\max})$	$v_t + v v_x + A\rho^{\gamma-2}\rho_x = \rho^{-1}(\mu(\rho)v_x)_x - \kappa(\rho, v)v$ $\rho > 0, v \in \mathbb{R}$
Derived equation	$s_t + v s_x = -ks$, $s = v - f(\rho)$	$s_t + v s_x = -ks$, $s = \beta(v) + \rho^{-2}\mu(\rho)\rho_x$	–
Constants	$k > 0$	$k, \rho_{\max}, v_{\max} > 0$, $v^* \in (0, v_{\max})$, $\bar{\rho} \in (0, \rho_{\max})$	$A > 0, \gamma \in (1, 2)$
Functions	$f : (0, +\infty) \to (0, +\infty)$ decreasing	$\beta(v) = \int_{v^*}^{v} q(s)ds$, $q(v) = \frac{v_{\max}^3(v+v^*)-2v_{\max}^2 v^* v}{2(v_{\max}-v)^2 v^2}$ $\mu : (0, \rho_{\max}) \to [0, +\infty)$, $\lim_{\rho\to\rho_{\max}^-} (\mu(\rho)) = +\infty$, $\mu(\rho) = 0$ for $0 < \rho \leq \bar{\rho}$, $\mu(\rho) > 0$ for $\bar{\rho} < \rho < \rho_{\max}$	$\mu : (0, +\infty) \to (0, +\infty)$ increasing, $\kappa : (0, +\infty) \times \mathbb{R} \to [0, +\infty)$

equation. The situation is completely analogous to the human-drivers case where the ARZ model reduces to the LWR model [40]. The numerical study and derivation of some properties of this particular nonlinear heat equation were performed in [70].

All macroscopic models mentioned above are presented in Table 6.1.

6.6.3 *Open Problems*

The rigorous derivation of the macroscopic model for lane-free CAV traffic with PRCCs remains an open problem. This is expected since the rigorous derivation of the NS equations from the microscopic motion of fluid particles is also an open problem (Hilbert's 6th problem). A novel particle method was proposed in [27] that can be used both as numerical scheme and as a tool to show existence of solutions for the 1D NS equations of a polytropic compressible fluid. Since the CAVMM has been constructed based on the consideration of vehicles as "self-driven" particles of a fluid [28], it is expected that the particle method in [27], modified appropriately, may be suitable for macroscopic traffic flow models for CAVs. In this way, both numerical and existence-of-solutions issues can be addressed. A rigorous and extensive study of the properties of the solutions and appropriate numerical schemes for the macroscopic model for lane-free CAV traffic with PRCCs is also needed.

6.7 Emerging Macroscopic Traffic Flow II: Empirical

As already discussed in Sects. 6.2–6.4, various methodological approaches can be explored to design a safe and efficient vehicle movement strategy in lane-free traffic, including optimal control [80], nonlinear feedback control [29], as well as multiagent systems [76]. For quick verification and demonstration of the TrafficFluid concept, an ad hoc vehicle movement strategy for lane-free traffic with nudging was suggested at an early stage [53], while an improved and strongly modified ad hoc strategy followed [45] and was implemented in TrafficFluid-Sim [74, 75], i.e., the simulation environment discussed in Sect. 6.5.

All these vehicle movement strategies can be employed for microscopic traffic simulation that produces macroscopic data necessary to investigate empirically emerging macroscopic properties of the new TrafficFluid concept. As for conventional traffic, microscopic simulation requires demanding computations, while macroscopic traffic flow models are less demanding for simulation purposes and may provide more direct insights in some macroscopic properties of lane-free traffic with nudging. They also have an analytical form, which allows for their usage for various significant traffic engineering tasks (estimation, traffic control strategy design) beyond simulation. Before employing a traffic flow model in practical applications, it is crucial to first calibrate it using traffic data. The calibration procedure aims at appropriately specifying the macroscopic model parameter values, so that the representation of the traffic flow is as accurate as the model structure allows. The calibration method typically involves the minimization of the discrepancy between the macroscopic model and the traffic data, by use of proper optimization tools [67, 73].

In this section, an empirical approach is presented, employing a space-time discrete macroscopic traffic flow model structure, which requires low computational effort [56]. The model is an extended version of the well-known Cell Transmission Model (CTM) [12]. Specifically, CTM does not reproduce the capacity drop phenomenon, i.e., the empirical observation that, at the head of congestion, the discharge flow is reduced compared to the capacity of the road. For this reason, CTM was extended in several possible ways to enable the reproduction of capacity drop, see [37] for an overview and comparison. The same model has been used for the formulation of optimal control problems, feedback control design, as well as macroscopic simulations for Internal Boundary Control (IBC) in lane-free automated vehicle traffic [46, 47], see Sect. 6.8.

The extended CTM includes a few parameters, whose values may depend on factors such as highway geometry, driver behavior (in case of manually driven vehi-

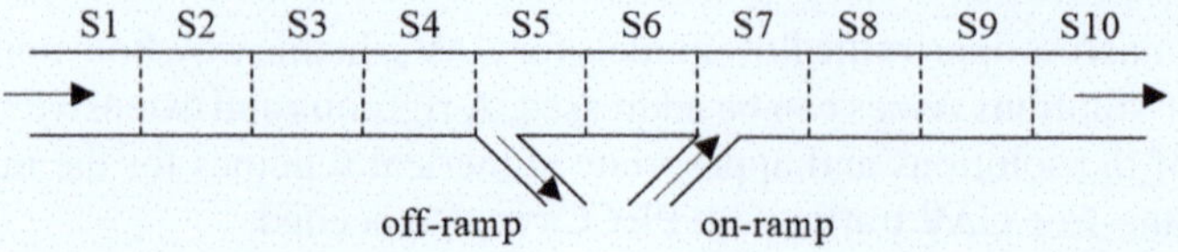

Fig. 6.10 A 5-km highway stretch subdivided in 10 sections for macroscopic modeling

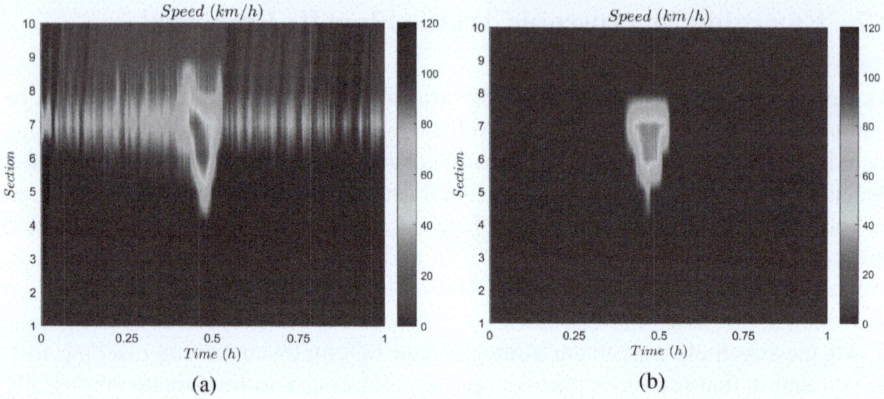

Fig. 6.11 Space-time speed evolution: **a** Microscopic simulation; **b** Macroscopic simulation with calibrated extended CTM

cles) or moving strategy used (in case of automated traffic), distribution of vehicle sizes, weather conditions, etc. The reliability and accuracy of the model rely on the appropriate specification of its parameter values. The extended CTM was calibrated in [56], by use of the CALISTO software tool [66], to fit lane-free traffic data, and optimal model parameters were determined for different highway widths and different vehicle nudging levels to investigate their impact on macroscopic characteristics, including capacity, critical density, free speed, and capacity drop.

Numerous microscopic simulations were designed on a highway stretch of 5 km in length and 10.2 m in width that includes an on-ramp and an off-ramp (see Fig. 6.10) by use of the ad hoc vehicle movement strategy [45], implemented in TrafficFluid-Sim, to generate the traffic data necessary to calibrate the macroscopic traffic flow model. The mainstream and on-ramp demands are designed so that congestion is created at the on-ramp section, propagates upstream and eventually dissolves due to lower demand (Fig. 6.11a). This is necessary in order to have calibration data that span the whole spectrum of possible traffic conditions, including free flow, critical, and congested traffic conditions.

After generating the traffic data via microscopic simulation, CALISTO was used for calibration, and Fig. 6.11b displays the macroscopic simulation results produced with the calibrated extended CTM. The obtained fit is reasonably good and the capacity drop, observed in the data of the microscopic simulation at the head of the congestion, is reproduced by the macroscopic model.

Lane-free traffic is deemed to imply that incremental changes of the road width may lead to corresponding incremental changes of capacity. This is because lane-free driving vehicles may spread and exploit any portion of additional road width, even if it is much less than a whole conventional traffic lane. To investigate microscopically the validity of that assumption, i.e., to investigate the impact of the highway width on the parameters of the macroscopic model, separate simulations were performed for different widths, starting from the nominal width of 10.2 m (corresponding to the width of 3 conventional lanes) and widening or narrowing it by multiples of 0.85 m (corresponding to a quarter of a conventional lane). Furthermore, it is interesting to investigate and demonstrate the impact of different values of the coefficient that regulates the influence of nudging forces in the ad hoc movement strategy.

The calibrated capacity values obtained for different highway widths and different nudging strengths are presented in Fig. 6.12 (upper). It is clear that capacity increases with increasing highway width. It is also clear that capacity increases with increasing vehicle nudging strength for any width. The same figure displays the lines produced via linear regression without the intercept term, i.e., assuming that the capacity is proportional to the width, for each nudging case. It is also interesting to note that the capacities reached are about 2–2.5 times higher than the capacity observed on a highway with conventional lanes.

On the other hand, the calibrated free speed does not depend on the width of the highway, something that is physically comprehensible, since road widening does not affect the desired speeds of vehicles nor the way vehicles interact with each other at near-critical densities. Thus, the free speed is found to be almost constant in dependence of the highway width. Regarding nudging, the calibrated free speed has an average value of 96.6 km/h, 94.3 km/h, and 92.1 km/h for the full nudging, weak nudging, and no nudging cases, respectively. This indicates a slight dependence of free speed on nudging strength, which can be physically justified by the fact the nudging increases the speed of interacting vehicles at near-critical densities. Because of linearly increasing capacity and virtually constant free speed, in dependence of the highway width, in a triangular fundamental diagram, the critical density, calculated from the calibrated values of capacity and free speed, must also depend linearly on the highway width, and increase for increased levels of nudging, as is indeed shown in Fig. 6.12 (lower).

The obtained results with respect to changing road width deliver full justification and a solid basis for IBC in lane-free automated vehicle traffic [46, 47], see Sect. 6.8.

The discharge rate at the head of congestion was found in the microscopic simulations to be independent of the nudging strength; as a consequence, capacity drop is found to increase with vehicle nudging. Despite some uncertainties in the values and role of some macroscopic model parameters, at a phenomenological level the calibrated macroscopic model matched well the discharge flow and capacity drop produced in the microscopic simulation. More investigations are necessary for better understanding of some issues related to discharge flow and capacity drop. Future investigations will also consider calibration and comparison of other discrete-time macroscopic model structures, both known from conventional traffic and new ones that may be better suited for lane-free automated vehicle traffic with nudging.

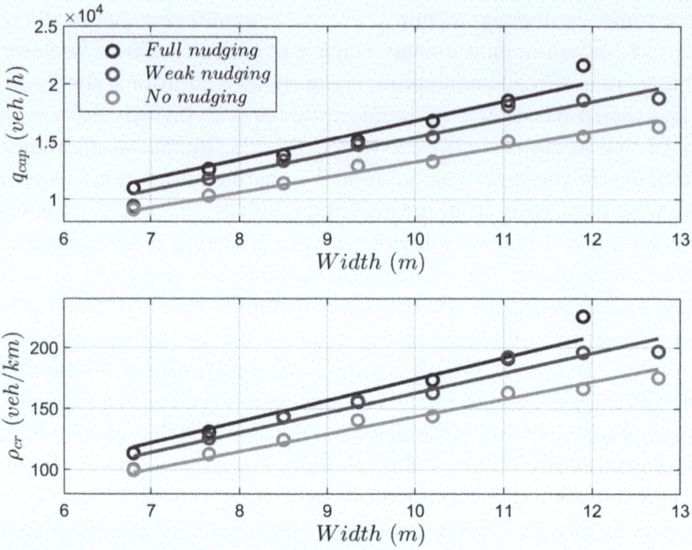

Fig. 6.12 Capacity (upper) and critical density (lower) versus highway width for full, weak, and no nudging

6.8 Internal Boundary Control (IBC)

Elevating the throughput of the road infrastructure by reversing the direction of one or more lanes at special occasions (holidays, road works, evacuation, special events) is one way to reduce traffic congestion in traditional traffic management. However, lane-based tidal flow faces strong limitations [47] that prevent real-time operation. In contrast, in lane-free traffic [54], it is possible to introduce real-time Internal Boundary Control (IBC) [47], a novel traffic control measure, which relies on the fact that the traffic flow and capacity exhibit incremental changes in accordance with corresponding incremental changes of the road width, as demonstrated in [56] (see also Sect. 6.7). In fact, IBC may lead to tremendous improvements and be one of the strongest arguments for introducing the TrafficFluid concept in the era of CAV traffic.

Consider a road with two opposite traffic directions serving CAVs. The overall capacity of the cross-road (for both directions) can be shared flexibly in space and time among the two directions by virtual shifting of the internal boundary (IB) according to the current demands in either direction, enabling more efficient usage of the available infrastructure. The current IB location may be communicated to CAVs so that they drive accordingly.

Figure 6.13 illustrates a road stretch with two directions of traffic: direction a from left to right and direction b from right to left. For macroscopic modeling and control, rather than using a space-continuous separation curve as in Fig. 6.13, the road stretch is partitioned into n sections, with respective lengths L_i, and the total road width

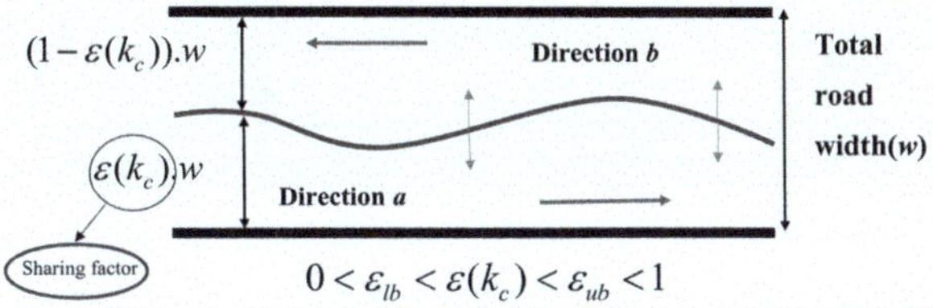

Fig. 6.13 Space-time flexible internal road boundary

(both directions) w, which is assumed constant over all sections for simplicity, can be shared in real time, independently for each section, between the two traffic directions. As a result, each direction is assigned a corresponding road width $w_i^a = \epsilon_i \cdot w$ and $w_i^b = (1 - \epsilon_i) \cdot w$, where $0 \leq \epsilon_i \leq 1$ is the sharing factor per section $i = 1, 2, \ldots, n$, to be specified in real time as an IBC control input, i.e., we have $\epsilon_i(k_c)$, where k_c is the discrete-time index of control, with control interval T_c in the order of 1 min. The total (both directions) section capacity q_{cap} as well as the total critical density ρ_{cr} and the total jam density ρ_{max} are accordingly shared at each section between the two traffic directions a and b. Based on [47], these quantities are given by direction from the following relations:

$$\begin{aligned} q_{i,cap}^a(\epsilon_i) &= \epsilon_i \cdot q_{cap}, \quad q_{i,cap}^b(\epsilon_i) = (1 - \epsilon_i) \cdot q_{cap} \\ \rho_{i,cr}^a(\epsilon_i) &= \epsilon_i \cdot \rho_{cr}, \quad \rho_{i,cr}^b(\epsilon_i) = (1 - \epsilon_i) \cdot \rho_{cr} \\ \rho_{i,max}^a(\epsilon_i) &= \epsilon_i \cdot \rho_{max}, \quad \rho_{i,max}^b(\epsilon_i) = (1 - \epsilon_i) \cdot \rho_{max}. \end{aligned} \tag{6.1}$$

While applying IBC, it is not permissible to fully close any of the two directions; hence, the designated road width in any direction must never be narrower than the greatest width of vehicles traveling on the road, leading to stricter constraints on the sharing factors, as follows:

$$0 < \epsilon_{lb} \leq \epsilon_i \leq \epsilon_{ub} < 1, \tag{6.2}$$

where $\epsilon_{lb} \cdot w$ and $(1 - \epsilon_{ub}) \cdot w$ are the minimum admissible widths to be assigned to directions a and b, respectively.

In conventional traffic management, the traffic density (in veh/km) gives a clear indication of the state of traffic, depending on its value compared to the critical density. However, IBC has the critical density determined by the sharing factor (see (6.1)); hence altering dynamically according to the control action. To tackle this problem, the following relations define the relative densities (dimensionless) per section and per direction:

$$\tilde{\rho}_i^a(k) = \frac{\rho_i^a(k)}{\epsilon_i(k-1)\rho_{cr}}, \; \tilde{\rho}_i^b(k) = \frac{\rho_i^b(k)}{(1 - \epsilon_i(k-1))\rho_{cr}}. \tag{6.3}$$

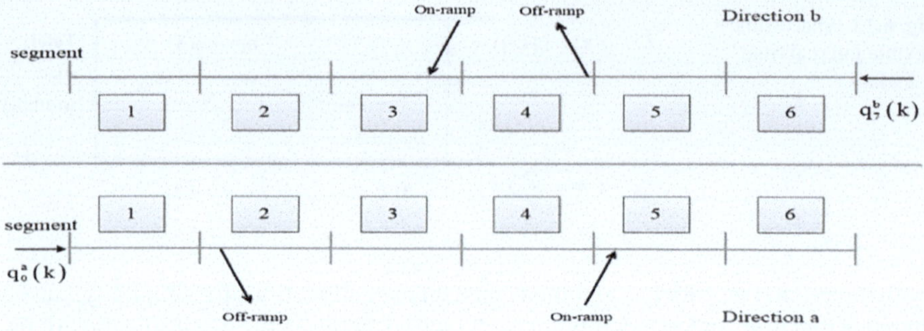

Fig. 6.14 The considered highway stretch

The relative densities enable consistent assessment of the traffic conditions in the IBC context. Specifically, if the relative density of a section and direction is less than 1, it reflects under-critical (free-flow) traffic conditions; if it is around 1, it reflects capacity flow; and if it is greater than 1, it reflects over-critical (congested) traffic conditions.

The characteristics of IBC are analyzed in [47], where its high improvement potential is demonstrated by formulating and solving an open-loop optimal control problem, in the form of a convex Quadratic Programming (QP) problem, aiming at minimizing a physical criterion, the total time spent by all vehicles. However, simpler real-time approaches with similar efficiency, but without the need for online demand prediction, are preferable. Therefore, a feedback-based Linear-Quadratic Regulator (LQR) was developed in [46] for IBC, aiming at balancing the relative densities in the two directions, and has been demonstrated to be robust and similarly efficient as the open-loop optimal control solution, without the need for accurate modeling and external demand prediction.

The performance of the IBC is investigated using the bidirectional highway stretch depicted in Fig. 6.14 with 3 km length. Using the demand flows presented in Fig. 6.15 with constant sharing factor at $\epsilon_i = 0.5$ for all sections in the nonlinear extended CTM, we obtain the simulation results of the no-control case that Fig. 6.16 displays (spatio-temporal evolution of the relative density), while by applying IBC, Fig. 6.17 demonstrates that the congestion is utterly avoided in both directions.

A video of the macroscopic simulation without and with IBC (LQR) may be viewed in https://bit.ly/45uVnn4. It is noted that the IBC concept was also tested, verified, and demonstrated by use of microscopic simulation, see Fig. 6.9b for a snapshot. Also, the integrated use of IBC and ramp metering was investigated in [25], and further integration of IBC with variable speed limits (on highways) or with traffic signals (at urban intersections) is currently under investigation.

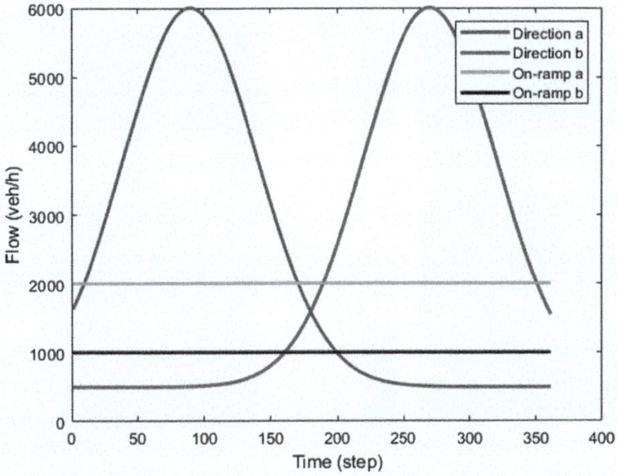

Fig. 6.15 Demand flows per direction and on-ramp

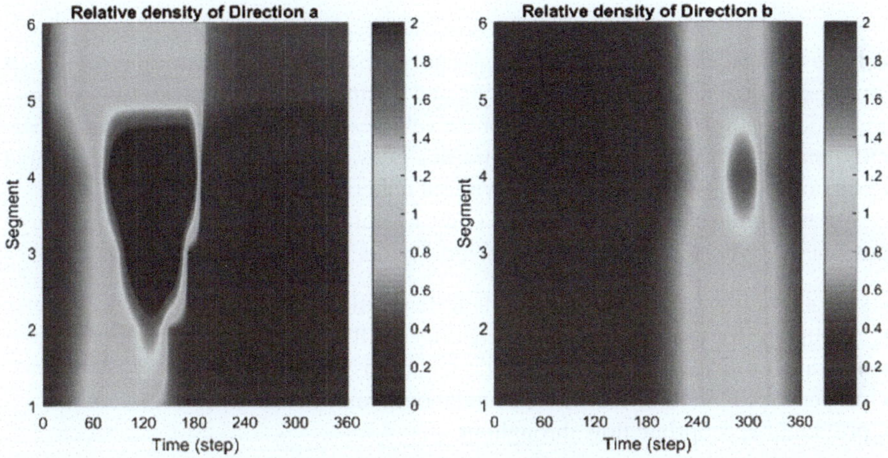

Fig. 6.16 Relative density for the two directions in the no-control case

6.9 Large-Scale Lane-Free Roundabouts

Roundabouts are a key element of urban traffic infrastructure, allowing for more efficient flow at light traffic; however, they may be a bottleneck when demand is high [20]. Consequently, if roundabouts are managed efficaciously, traffic flow in their vicinity may be improved. Although several works have focused on controlling connected automated vehicles (CAVs) moving on roundabouts, they mostly consider simple roundabout cases, like single or double lane ones with a limited number of radial streets, see a short survey in [49], while there are many large and complicated roundabouts around the world. In particular, the Place Charles de Gaulle roundabout

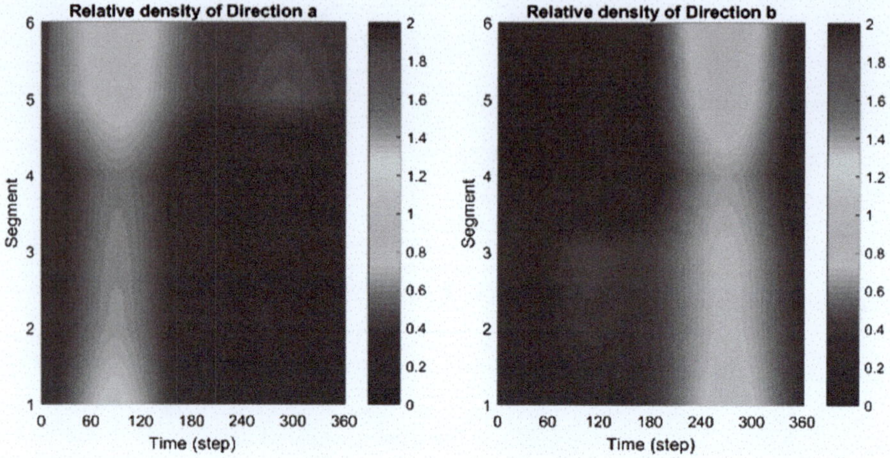

Fig. 6.17 Relative density for the two directions in the control case

in Paris is intricate, with 38 m width, 84 m outer radius, and 46 m inner radius. It comprises a dozen bidirectional (entering and exiting) radial streets, i.e., 144 distinct origin-destination (OD) movements for the vehicles. Such big roundabouts are inherently lane-free even for conventional traffic. Thus, once on the roundabout, human drivers or CAVs must find their way without adhering to traffic lanes.

We aim to develop a decentralized control strategy for CAVs moving on a large and lane-free roundabout according to some goals mentioned below:

- smooth motion of vehicles toward their destination;
- accounting for trip distance and mitigating conflicts among vehicles;
- avoiding collisions, boundary violation, and missing exit;
- proper exploitation of the whole roundabout area;
- providing appropriate flow throughout.

Due to frequent and significant steering of vehicles while moving on the roundabout, the kinematic bicycle model [29], rather than the double double-integrator, is used to represent their dynamics, where each vehicle has four state variables, namely, longitudinal and lateral position, orientation, and speed; and two control inputs: acceleration and steering angle (Fig. 6.4). Then, to ease analysis and controller design for the circular movement, the dynamics are transformed into polar coordinates, where the vehicle position is determined by a radius and an angle. Also, a new state variable, the deviation from the circular angle, is introduced to replace orientation. Specifically, if this variable is zero, a vehicle has a circular motion. Consequently, smaller values of deviation result in fewer and weaker conflicts among vehicles on the roundabout.

The nonlinear feedback controllers (NLFCs) mentioned in Sect. 6.3 and presented in [29, 69] are employed as the kernel of the control system for real-time decision-making by each vehicle. To control vehicles driving in the connected branches with

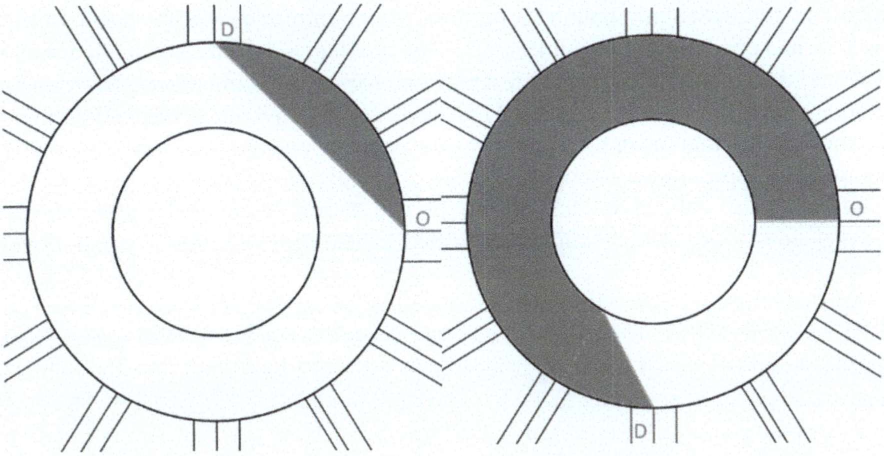

Fig. 6.18 Defined movement corridor: **a** visible ODs, **b** invisible ODs

different orientations, the NLFC developed for straight roads in [29] is modified such that non-zero desired orientation, i.e., the entrance or exit branch orientation, can be used. Another NLFC developed for circular roads in [69] is employed to navigate vehicles within the roundabout. Using this controller, vehicles' deviations from circular angle are intended to converge to zero; thus it was also modified to accommodate non-zero desired deviation for entering or exiting vehicles.

In view of the large number of OD couples in big roundabouts, it is sensible to define respective corridors which delineate the admissible movement zones of corresponding OD vehicles and help to decrease the number of conflicts among vehicles on the roundabout. For instance, if a vehicle's destination is close to its entrance branch, it seems logical to avoid excursions to the inner roundabout boundary. The roundabout's outer boundary is considered as the outer boundary of all OD corridors. On the other hand, to achieve better infrastructure utilization, the corridors' interior boundaries are differently defined for two OD corridor types: (1) The destination is visible from the origin, in which case the straight line connecting the leftmost point of the origin branch with the leftmost point of the destination branch is the inner boundary (Fig. 6.18a) and (2) the destination is not visible from the origin. Under this circumstance, a vehicle may utilize the full width of the roundabout until it is sufficiently close to its destination. Then, to avoid missing the exit or moving perpendicularly toward the exit, a line connecting the inner boundary to the leftmost point of the exit branch is chosen as the left boundary of the corridor as shown in Fig. 6.18b.

By appropriate determination of position-dependent desired orientations within each OD corridor, vehicles are properly guided toward their destination along their trip. In [49], a heuristic approach is suggested to determine desired orientation, while in [48] an optimal control method is proposed that minimizes a weighted sum of trip distance and deviation from circular angle. The problem is solved analytically for two extreme cases, namely, for the shortest path and for the minimum deviation from the circular angle. Moreover, a backward Dijkstra's algorithm is suggested to determine the optimal orientation in a discretized grid of the roundabout surface. A simpler option is to approximate the optimal desired orientation by locally combining the two orientations, related to the shortest path and the minimum angle deviation, via a weighted average, which can be performed in real time with low computational effort.

Special boundary controllers are introduced to ensure that vehicles do not exceed their designated OD corridors. The boundary controllers apply to both straight and circular boundaries and are used to derive lower and upper limits for the vehicle steering angle, corresponding to the left and right boundaries, respectively. The goal of a boundary controller is to guide the vehicle's lateral position and orientation asymptotically to the road boundary's position and orientation (straight or circular), respectively. The design of the boundary controller considers a subsystem of the bicycle model, which includes only the two relevant state variables, and the steering angle as the only control variable. A linear state-feedback controller is designed for the linearized subsystem around the desired equilibrium point. The steering angles generated by two boundary controllers, one for each of the left (inner) and right (outer) boundaries, are utilized to saturate the steering angle determined by the NLFCs [49].

Because of several changes applied to the NLFCs, such as implementation in discrete time, practical limits for control signals, and varying desired orientation, their original features, like collision avoidance, cannot be fully guaranteed. Therefore, a sort of safety filter, limiting vehicle acceleration is designed. Furthermore, to suppress aggressive behavior of vehicles in crowded situations, an adaptive desired speed policy is suggested. Specifically, each vehicle determines its desired speed as a decreasing function of the local density in its close vicinity. Also, vehicles moving on the entrance branches take into account the local density of a sector of the roundabout in front of the entrance and reduce their desired speed as density in the sector increases.

Employing the outlined movement strategy, a microscopic simulation with 1600 vehicles moving on Place Charles de Gaulle roundabout is conducted in TrafficFluid-Sim platform [75] (Sect. 6.5) whose result can be watched at https://bit.ly/36exR42 that demonstrates the effectiveness of the proposed scheme. A snapshot of the simulation is provided in Fig. 6.19. Indeed, vehicles move toward their destinations, do not collide with each other, do not exceed the boundaries, and do not miss their exit. However, they may apply accelerations and decelerations to avoid collisions, especially in high-density conditions.

Fig. 6.19 Snapshot of microscopic simulation for the Charles de Gaulle roundabout

6.10 Joint Trajectory Optimization of Automated Vehicles

While the vast majority of proposed CAV (connected automated vehicles) driving methods address each vehicle individually, an interesting question refers to possible performance gains achievable if multiple vehicle trajectories would be optimized jointly. This is indeed considered in research addressing signal-free urban junction crossing or automated highway merging [60] but is less common for highway driving vehicles. More specifically, works on the subject of optimal 2D joint trajectory planning for multiple vehicles in lane-based traffic are limited, because lateral vehicle positions are discrete (on lanes), not real valued, except for special circumstances, such as during lane changing. Solution of discrete optimization problems is computationally expensive even for a single vehicle [7, 39, 41, 62], and complexity scales exponentially.

Joint trajectory specification for a (big or small) group of vehicles is an interesting and appealing topic also in the lane-free context, where all involved variables are real valued, and appropriate numerical solution algorithms like the Feasible Direction Algorithm (FDA) (see Sect. 6.2, for the decentralized approach) exhibit polynomial complexity. As a first step in this direction, [11] delivers a generalization of the decentralized problem of Sect. 6.2 (see also [80]), by formulating a centralized nonlinear constrained Optimal Control Problem (OCP), for joint trajectory planning for a number of CAVs in a lane-free traffic environment with vehicle nudging. For the vehicle dynamics, a double double-integrator model is utilized, which is quite common for the problem at hand. However, as a precaution, a specific term is considered in the objective function to discourage infeasible heading angles. Compared to the

Fig. 6.20 Initial vehicle arrangement of eight vehicles and three obstacles (red)

decentralized problem, a number of vehicles are jointly considered in [11], and their goals are accumulated in one common objective function. This leads to optimal decisions for each vehicle that are no more "egoistic" as in decentralized approach, but decisions that improve the average performance of all vehicles as one entity.

The objective function consists of several sub-objectives for each vehicle, i.e., increase of passenger comfort, reduction of fuel consumption, vehicle advancing (reaching desired speeds), collision avoidance, and mitigation of infeasible maneuvers. Fixed and state-dependent control input bounds account for various technical limitations as well as for road boundary respect. As mentioned earlier, the solution of the formulated OCP is computed by use of the very efficient FDA [52], which exploits the structure of the state equations to map the OCP into a nonlinear programming problem in the reduced space of control variables. Note that FDA iterations can be stopped at any time, even before convergence, delivering a control trajectory that is feasible, i.e., satisfies all state equations and constraints, but may be sub-optimal. It should be noted that the OCP at hand is non-convex, hence the delivered minimum can be a local minimum. However, if this local minimum serves the main purposes of the application, then it is acceptable.

To demonstrate the efficiency of the proposed approach, a challenging scenario is considered with a group of eight vehicles driving lane-free on a straight road. Initially, the eight vehicles are located longitudinally in inverse order of their respective longitudinal desired speeds; therefore, vehicles need to overtake one another, via appropriate maneuvers, and reach their desired speeds. In addition, there are three vehicles (see Fig. 6.20) in front, which are not optimized, but drive independently with constant longitudinal speed and zero lateral speed during the whole optimization time horizon [11].

Figure 6.21 displays the open-loop optimal longitudinal and lateral movements of all eight vehicles of the group. It can be seen that, at the end of the time horizon, all the vehicles have successfully reached their respective desired speeds without any collision, and vehicles' longitudinal positions are ordered according to their desired speeds, meaning that the fastest vehicle (with the highest desired speed) is the first vehicle, the second-fast vehicle is second in positioning order, and so forth. The time horizon was taken sufficiently long to cover this re-ordering of longitudinal vehicle positions [11].

The proposed joint trajectory specification approach is very efficient regarding the collective goals of the vehicles, but, equally importantly, also in terms of computation times (see [11]). Specifically, the controlled vehicles move efficiently and safely in a coordinated way, achieving comfortable, fuel-efficient, and collision-free maneuvers, while, at the same time, they respect the road boundaries. Videos showing the movement of CAVs guided by the centralized approach are provided in https://bit.ly/3OS3rZl.

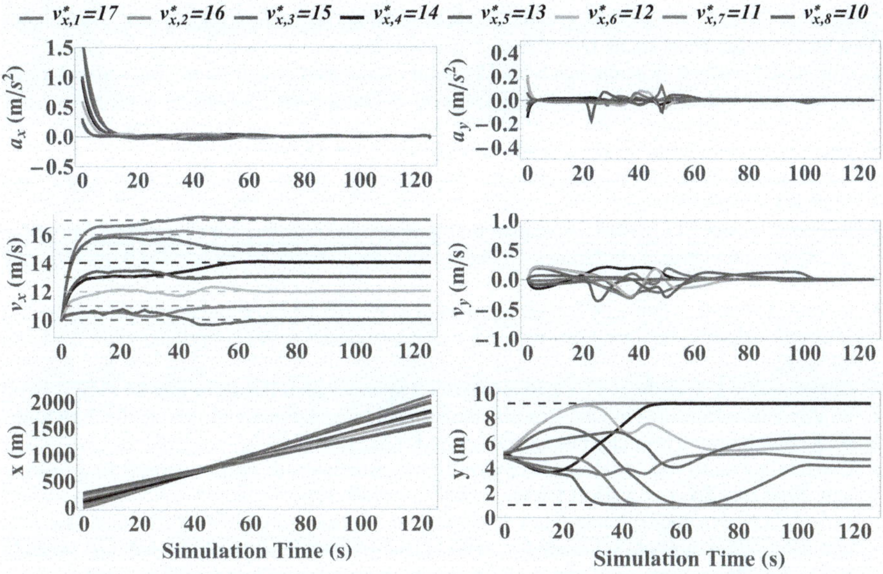

Fig. 6.21 Optimal longitudinal and lateral trajectories (accelerations, speeds, positions) for eight CAVs

In a nutshell, [11] describes the development of a methodological and computational tool for intelligent control of a group of vehicles in a coordinated way in lane-free traffic. There are multiple potential applications of this tool. In particular, while [80] considered independent MPC-based driving for individual vehicles (Sect. 6.2), the present tool would allow for a variety of different entities to be considered in lane-free traffic, each of which is controlled in an independent way using MPC. Such entities may comprise individual vehicles, groups of vehicles, vehicle platoons in different formations (e.g., vehicle flocks, 2D snake-like platoons), and more. Vehicle flocks in lane-free traffic were also considered in [61].

Acknowledgements The research leading to these results has received funding form the European Research Council under the European Union's Horizon 2020 Research and Innovation programme/ERC Grant Agreement no. [833915], project TrafficFluid.

References

1. Abbeel P, Ng AY (2004) Apprenticeship learning via inverse reinforcement learning. In: Proceedings of the twenty-first international conference on machine learning, p 1. ACM Press
2. Aeberhard M, Rauch S, Bahram M, Tanzmeister G, Thomas J, Pilat Y, Homm F, Huber W, Kaempchen N (2015) Experience, results and lessons learned from automated driving on Germany's highways. IEEE Intell Transp Syst Mag 7(1):42–57

3. Aradi S (2022) Survey of deep reinforcement learning for motion planning of autonomous vehicles. IEEE Trans Intell Transp Syst 23(2):740–759
4. Ardelt M, Coester C, Kaempchen N (2012) Highly automated driving on freeways in real traffic using a probabilistic framework. IEEE Trans Intell Transp Syst 13(4):1576–1585
5. Aw A, Klar A, Materne T, Rascle M (2002) Derivation of continuum traffic flow models from microscopic follow-the-leader models. SIAM J Appl Math 63(1):259–278
6. Aw A, Rascle M (2000) Resurrection of "second order" models of traffic flow. SIAM J Appl Math 60(3):916–938
7. Burger C, Lauer M (2018) Cooperative multiple vehicle trajectory planning using MIQP. In: 2018 21st international conference on intelligent transportation systems (ITSC), pp 602–607 (2018)
8. Casas J, Ferrer JL, Garcia D, Perarnau J, Torday A (2010) Traffic simulation with aimsun. In: International series in operations research and management science. Springer, New York, pp 173–232
9. Chrysomallis I, Troullinos D, Chalkiadakis G, Papamichail I, Papageorgiou M (2023) Deep reinforcement learning with implicit imitation for lane-free autonomous driving. In: 26th European conference on artificial intelligence (ECAI) (2023)
10. Couzin ID, Franks NR (2003) Self-organized lane formation and optimized traffic flow in army ants. Proc Roy Soc Lond Ser B Biol Sci 270(1511):139–146
11. Dabestani N, Typaldos P, Yanumula VK, Papamichail I, Papageorgiou M (2023) Joint trajectory optimization for multiple automated vehicles in lane-free traffic with vehicle nudging. In: 2023 IEEE 26th international conference on intelligent transportation systems (ITSC). IEEE
12. Daganzo CF (1994) The cell transmission model: a dynamic representation of highway traffic consistent with the hydrodynamic theory. Transp Res Part B: Methodol 28(4):269–287
13. Daganzo CF (1995) Requiem for second-order fluid approximations of traffic flow. Transp Res Part B: Methodol 29(4):277–286
14. Di X, Shi R (2021) A survey on autonomous vehicle control in the era of mixed-autonomy: From physics-based to AI-guided driving policy learning. Transp Res Part C: Emerg Technol 125:103008
15. Diakaki C, Papageorgiou M, Papamichail I, Nikolos I (2015) Overview and analysis of vehicle automation and communication systems from a motorway traffic management perspective. Transp Res Part A: Policy Pract 75:147–165
16. Farinelli A, Rogers A, Petcu A, Jennings N (2008) Decentralised coordination of low-power embedded devices using the max-sum algorithm. In: Padgham, Parkes, Müller and Parsons (eds.) International conference on autonomous agents and multi-agent systems (AAMAS), pp 639–646
17. Faros I, Yanumula VK, Typaldos P, Papamichail I, Papageorgiou M (2022) A lateral positioning strategy for connected and automated vehicles in lane-free traffic. In: Wang M, Jaekel B, Lehnert M, Zhou R, Li Z (eds) 4th symposium on management of future motorway and urban traffic systems (MFTS 2022), Dresden, Germany, pp 20–28
18. Finkel RA, Bentley JL (1974) Quad trees a data structure for retrieval on composite keys. Acta Informatica 4(1):1–9
19. Fioretto F, Pontelli E, Yeoh W (2018) Distributed constraint optimization problems and applications: a survey. J Artif Intell Res 61:623–698
20. Flannery A, Datta T (1997) Operational performance measures of American roundabouts. Transp Res Record: J Transp Res Board 1572(1):68–75
21. Francesco MD, Rosini M (2015) Rigorous derivation of nonlinear scalar conservation laws from follow-the-leader type models via many particle limit. Arch Ration Mech Anal 217(3):831–871
22. Gu S, Lillicrap T, Sutskever I, Levine S (2016) Continuous deep q-learning with model-based acceleration. In: International conference on machine learning, PMLR, pp 2829–2838
23. Holden H, Risebro NH (2018) The continuum limit of follow-the-leader models — a short proof. Discret Contin Dyn Syst - A 38(2):715–722
24. Ioannou PA, Chien CC (1993) Autonomous intelligent cruise control. IEEE Trans Veh Technol 42(4):657–672

25. Jin X, Yu X, Hu Y, Wang Y, Papageorgiou M, Papamichail I, Malekzadeh M, Guo J (2022) Integrated control of internal boundary and ramp inflows for lane-free traffic of automated vehicles on freeways. In: 2022 IEEE 25th international conference on intelligent transportation systems (ITSC). IEEE, pp 1234–1239
26. Kamal MAS, Taguchi S, Yoshimura T (2016) Efficient driving on multilane roads under a connected vehicle environment. IEEE Trans Intell Transp Syst 17(9):2541–2551
27. Karafyllis I, Papageorgiou M (2023) A particle method for 1-d compressible fluid flow. Stud Appl Math
28. Karafyllis I, Theodosis D, Papageorgiou M (2022) Constructing artificial traffic fluids by designing cruise controllers. Syst Control Lett 167:105317
29. Karafyllis I, Theodosis D, Papageorgiou M (2022) Lyapunov-based two-dimensional cruise control of autonomous vehicles on lane-free roads. Automatica 145:110517
30. Karafyllis I, Theodosis D, Papageorgiou M (2022) Stability analysis of nonlinear inviscid microscopic and macroscopic traffic flow models of bidirectional cruise-controlled vehicles. IMA J Math Control Inf 39(2):609–642
31. Karafyllis I, Theodosis D, Papageorgiou M (2023) Forward completeness and applications to control of automated vehicles. arXiv:2307.11515
32. Karafyllis I, Theodosis D, Papageorgiou M (2023) Nonlinear adaptive cruise control of vehicular platoons. Int J Control 96(1):147–169
33. Karalakou A, Troullinos D, Chalkiadakis G, Papageorgiou M (2022) Deep RL reward function design for lane-free autonomous driving. In: Lecture notes in computer science. Springer International Publishing, pp 254–266
34. Karalakou A, Troullinos D, Chalkiadakis G, Papageorgiou M (2023) Deep reinforcement learning reward function design for autonomous driving in lane-free traffic. Systems 11(3):134
35. Kiran BR, Sobh I, Talpaert V, Mannion P, Sallab AAA, Yogamani S, Perez P (2022) Deep reinforcement learning for autonomous driving: a survey. IEEE Trans Intell Transp Syst 23(6):4909–4926
36. Kok JR, Vlassis N (2006) Collaborative multiagent reinforcement learning by payoff propagation. J Mach Learn Res 7:1789–1828
37. Kontorinaki M, Spiliopoulou A, Roncoli C, Papageorgiou M (2017) First-order traffic flow models incorporating capacity drop: overview and real-data validation. Transp Res Part B: Methodol 106:52–75
38. Kschischang F, Frey B, Loeliger HA (2001) Factor graphs and the sum-product algorithm. IEEE Trans Inf Theory 47(2):498–519
39. Levy R, Haddad J (2022) Cooperative path and trajectory planning for autonomous vehicles on roads without lanes: a laboratory experimental demonstration. Transp Res Part C: Emerg Technol 144:103813
40. Li T (2003) Global solutions of nonconcave hyperbolic conservation laws with relaxation arising from traffic flow. J Differ Equ 190(1):131–149
41. Li T, Wu J, Chan CY, Liu M, Zhu C, Lu W, Hu K (2020) A cooperative lane change model for connected and automated vehicles. IEEE Access 8:54940–54951
42. Lighthill MJ, Whitham GB (1955) On kinematic waves ii. a theory of traffic flow on long crowded roads. Proc Roy Soc Lond. Ser A. Math Phys Sci 229(1178):317–345
43. Lopez PA, Behrisch M, Bieker-Walz L, Erdmann J, Flötteröd YP, Hilbrich R, Lücken L, Rummel J, Wagner P, Wiessner E (2018) Microscopic traffic simulation using sumo. In: 2018 21st international conference on intelligent transportation systems (ITSC), pp 2575–2582
44. Makantasis K, Papageorgiou M (2018) Motorway path planning for automated road vehicles based on optimal control methods. Transp Res Record: J Transp Res Board 2672(19):112–123
45. Malekzadeh M, Manolis D, Papamichail I, Papageorgiou M (2022) Empirical investigation of properties of lane-free automated vehicle traffic. In: 2022 IEEE 25th international conference on intelligent transportation systems (ITSC), pp 2393–2400
46. Malekzadeh M, Papamichail I, Papageorgiou M (2021) Linear–quadratic regulators for internal boundary control of lane-free automated vehicle traffic. Control Eng Pract 115:104912

47. Malekzadeh M, Papamichail I, Papageorgiou M, Bogenberger K (2021) Optimal internal boundary control of lane-free automated vehicle traffic. Transp Res Part C: Emerg Technol 126:103060
48. Naderi M, Mavroeidi M, Papamichail I, Papageorgiou M (2023) Optimal orientation for automated vehicles on large lane-free roundabouts. In: IEEE conference on decision and control (CDC). IEEE
49. Naderi M, Papageorgiou M, Karafyllis I, Papamichail I (2022) Automated vehicle driving on large lane-free roundabouts. In: 2022 IEEE 25th international conference on intelligent transportation systems (ITSC), pp 1528–1535
50. Nair A, McGrew B, Andrychowicz M, Zaremba W, Abbeel P (2018) Overcoming exploration in reinforcement learning with demonstrations. In: 2018 ieee international conference on robotics and automation (ICRA). IEEE, pp 6292–6299
51. Papageorgiou M, Ben-Akiva M, Bottom J, Bovy PH, Hoogendoorn SP, Hounsell NB, Kotsialos A, McDonald M (2007) ITS and traffic management. Handbooks Oper Res Management Sci 14:715–774
52. Papageorgiou M, Marinaki M, Typaldos P, Makantasis K (2016) A feasible direction algorithm for the numerical solution of optimal control problems–extended version. Internal Report no. 26, Dynamics Systems and Simulations Laboratory, Technical University of Crete, Chania, Greece
53. Papageorgiou M, Mountakis KS, Karafyllis I, Papamichail I (2019) Lane-free artificial-fluid concept for vehicular traffic. arXiv:1905.11642
54. Papageorgiou M, Mountakis KS, Karafyllis I, Papamichail I, Wang Y (2021) Lane-free artificial-fluid concept for vehicular traffic. Proc IEEE 109(2):114–121
55. Papamichail I, Bekiaris-Liberis N, Delis AI, Manolis D, Mountakis KS, Nikolos IK, Roncoli C, Papageorgiou M (2019) Motorway traffic flow modelling, estimation and control with vehicle automation and communication systems. Annu Rev Control 48:325–346
56. Papamichail I, Schoenn-Anchling N, Malekzadeh M, Markantonakis V, Papageorgiou M (2023) Macroscopic traffic flow model calibration for lane-free automated vehicle traffic. In: 2023 IEEE 26th international conference on intelligent transportation systems (ITSC). IEEE
57. Pendleton S, Andersen H, Du X, Shen X, Meghjani M, Eng Y, Rus D, Ang M (2017) Perception, planning, control, and coordination for autonomous vehicles. Machines 5(1):6
58. Price B, Boutilier C (2003) Accelerating reinforcement learning through implicit imitation. J Artif Intell Res 19:569–629
59. Richards PI (1956) Shock waves on the highway. Oper Res 4(1):42–51
60. Rios-Torres J, Malikopoulos AA (2016) A survey on the coordination of connected and automated vehicles at intersections and merging at highway on-ramps. IEEE Trans Intell Transp Syst 18(5):1066–1077
61. Rostami-Shahrbabaki M, Weikl S, Niels T, Bogenberger K (2023) Modeling vehicle flocking in lane-free automated traffic. Transp Res Record p. 03611981231159405
62. Sakaguchi Y, Bakibillah ASM, Kamal MAS, Yamada K (2023) A cyber-physical framework for optimal coordination of connected and automated vehicles on multi-lane freeways. Sensors 23(2):611
63. Schulman J, Wolski F, Dhariwal P, Radford A, Klimov O (2017) Proximal policy optimization algorithms. arXiv:1707.06347
64. Sekeran M, Rostami-Shahrbabaki M, Syed AA, Margreiter M, Bogenberger K (2022) Lane-free traffic: History and state of the art. In: 2022 IEEE 25th international conference on intelligent transportation systems (ITSC). IEEE, pp 1037–1042
65. Sontag ED (1990) Mathematical control theory. Springer, US
66. Spiliopoulou A, Papamichail I, Papageorgiou M, Chrysoulakis J (2014) Calisto user's manual. Deliverable 4.1, Technical Report for the Project SMOOTH, ARCHIMEDES III
67. Spiliopoulou A, Papamichail I, Papageorgiou M, Tyrinopoulos Y, Chrysoulakis J (2017) Macroscopic traffic flow model calibration using different optimization algorithms. Oper Res Int J 17:145–164
68. Sutton RS, Barto AG (2018) Reinforcement learning: an introduction. MIT Press

69. Theodosis D, Karafyllis I, Papageorgiou M (2023) Cruise controllers for lane-free ring-roads based on control lyapunov functions. J Frankl Inst 360(9):6131–6161
70. Theodosis D, Karafyllis I, Titakis G, Papamichail I, Papageorgiou M (2024) A nonlinear heat equation arising from automated-vehicle traffic flow models. J Comput Appl Math 437:115443
71. Thirring W (1997) Classical mathematical physics. Dynamical systems and field theories. Springer, New York
72. Torabi F, Warnell G, Stone P (2019) Recent advances in imitation learning from observation. In: Proceedings of the twenty-eighth international joint conference on artificial intelligence. In: International joint conferences on artificial intelligence organization, pp 6325–6331
73. Treiber M, Kesting A (2013) Traffic flow dynamics. Data, models and simulation. Traffic flow dynamics. Springer, Berlin, Heidelberg,
74. Troullinos D, Chalkiadakis G, Manolis D, Papamichail I, Papageorgiou M (2021) Lane-free microscopic simulation for connected and automated vehicles. In: 2021 IEEE international intelligent transportation systems conference (ITSC). IEEE, pp 3292–3299
75. Troullinos D, Chalkiadakis G, Manolis D, Papamichail I, Papageorgiou M (2022) Extending SUMO for lane-free microscopic simulation of connected and automated vehicles. SUMO Conf Proc 3:95–103
76. Troullinos D, Chalkiadakis G, Papamichail I, Papageorgiou M (2021) Collaborative multiagent decision making for lane-free autonomous driving. In: Proceedings of the 20th international conference on autonomous agents and multiagent systems, pp 1335–1343
77. Troullinos D, Chalkiadakis G, Samoladas V, Papageorgiou M (2022) Max-sum with quadtrees for decentralized coordination in continuous domains. In: Proceedings of the thirty-first international joint conference on artificial intelligence, IJCAI-22, Vienna, Austria, pp. 518–256
78. Typaldos P, Papageorgiou M, Papamichail I (2022) Optimization-based path-planning for connected and non-connected automated vehicles. Transp Res Part C: Emerg Technol 134:103487
79. Wooldridge M (2009) An introduction to multiagent systems. Wiley
80. Yanumula VK, Typaldos P, Troullinos D, Malekzadeh M, Papamichail I, Papageorgiou M (2023) Optimal trajectory planning for connected and automated vehicles in lane-free traffic with vehicle nudging. IEEE Trans Intell Veh 8(3):2385–2399
81. Yu C, Wang X, Xu X, Zhang M, Ge H, Ren J, Sun L, Chen B, Tan G (2019) Distributed multiagent coordinated learning for autonomous driving in highways based on dynamic coordination graphs. IEEE Trans Intell Transp Syst 21(2):735–748
82. Zhang H (2002) A non-equilibrium traffic model devoid of gas-like behavior. Transp Res Part B: Methodol 36(3):275–290

Chapter 7
Models and Control Algorithms for Electric Automated Buses in Smart Cities

Silvia Siri, Stefano Bracco, Cecilia Pasquale, and Simona Sacone

Abstract Electric mobility is one of the key components of the smart mobility transition that most of the cities worldwide are living nowadays. Besides the electrification of private cars, a great contribution in this direction is given by public transport, since a growing number of electric buses is appearing in most of the cities. At the same time, another important innovation is completely changing the mobility concept, that is, the technology allowing increasing levels of automation on board of vehicles. The mobility of public transport of the future smart cities will be then represented by fleets of electric automated vehicles. In order to correctly plan and control such vehicles, it is necessary, first of all, to develop appropriate models able to represent the dynamic behavior of buses when interacting with other vehicles on the road and able to estimate their energy consumption when traveling in the city roads. On the basis of such models, proper control algorithms can be designed in order to regulate the behavior of buses which have to follow a specific route and respect a fixed timetable. In this chapter, we provide an overview on electric buses, by highlighting the main modeling and control challenges, and by providing some possible traffic and consumption models, as well as some traffic control algorithms for optimizing bus speeds, dwell and charging times.

S. Siri (✉) · C. Pasquale · S. Sacone
Department of Informatics, Bioengineering, Robotics and Systems Engineering, University of Genova, Genova, Italy
e-mail: silvia.siri@unige.it

C. Pasquale
e-mail: cecilia.pasquale@unige.it

S. Sacone
e-mail: simona.sacone@unige.it

S. Bracco
Electrical, Electronics and Telecommunication Engineering and Naval Architecture Department, University of Genova, Genova, Italy
e-mail: stefano.bracco@unige.it

P. Ioannou and A. A. Malikopoulos (eds.), *Transportation Mobility in Smart Cities*, Springer Tracts on Transportation and Traffic 21,
https://doi.org/10.1007/978-3-031-64769-7_7

7.1 Introduction

July 2023 has been recognized by the World Meteorological Organization as the hottest month ever measured in human history. According to the United Nations, this record ends the era of global warming and begins the era of global boiling. This shocking warning calls for the urgent implementation of policies to limit global temperature rise in order to prevent the occurrence of catastrophic climate events. All human activities have an impact on the health of the planet, however the most damaging ones are those involving the use of fossil fuels for their execution. Prominent among these are transport activities, particularly road transport, which are some of the main sources of greenhouse gases [1].

Several actions can be implemented to achieve a more sustainable road transport. These actions involve the implementation of policies to manage and regulate both demand and supply of transport. On the transport demand side, a first step toward sustainability is to promote the use of collective transport modes, such as public transport, car sharing, car pooling, and so on. At the same time, on the transport supply side, there is the need for a transition to less impactful propulsion systems and the development of effective and possibly better performing public transport services compared to private transport. The combination of these needs poses major challenges that may require a complete revision of the mobility paradigms implemented today. A public transport system that is both high performing and sustainable is the subject of this chapter, which focuses primarily on the control of electric automated buses.

In recent years, we have witnessed a massive conversion of fossil-fueled bus fleets to electric-powered buses [2], which has led to increasing attention by researchers in analyzing the technological requirements [3] and in designing planning approaches [4] for electric public transport systems (see Fig. 7.1 showing an electric bus used for testing at the Savona Campus, University of Genova). Yet, it is worth noting that the technological advances concerning bus fleets are not only involving the transition to electrification, but also the adoption of sophisticated communication systems, e.g., Vehicle to Vehicle (V2V) and Vehicle to Infrastructure (V2I). The development of smart strategies for managing electric mobility, such as Vehicle to Grid (V2G) and Vehicle to Building (V2B), making vehicles directly connected with the national power grid must also be considered. At the same time, the technological progress is pushing toward increasing levels of automation on board of buses, leading to the era of automated vehicles, for which some prototypes and case studies are under testing in many cities worldwide [5]. Therefore, it is reasonable to assume that the public transportation systems of the future will be made of buses that are, at the same time, electric, automated, and connected [6].

Aim of this chapter is to address some important issues in the design of electric automated buses of the near future, which regard modeling and control methods for the definition of optimal speed profiles, optimal dwell and charging times for buses. In order to provide some backgrounds to the reader in the area of electric mobility, Sect. 7.2 describes the major trends in the electric mobility sector and some specific technological aspects referred to electric buses. Section 7.3 provides, then,

Fig. 7.1 Full electric bus at Savona Campus, University of Genova, May 2021

a state of the art of research works for planning and controlling electric buses in smart city road networks. Section 7.4 introduces the main challenges for modeling the behavior of electric buses through the definition of proper energy consumption models and traffic models. Section 7.5 describes some important control approaches for the optimal timing of buses, in terms of speed profiles and dwell/charging times. Finally, some conclusive remarks for automated electric public transport systems in smart cities are highlighted in Sect. 7.6.

7.2 Electric Buses in Smart Cities

In the last years, most of the countries worldwide have accelerated the development of electric mobility, which is seen as one of the main means to achieve decarbonization, through the adoption of important directives and legislative acts. One of the areas in which electric mobility has seen a higher spread is surely public transport in the urban context. In this section, a general overview of electric mobility is firstly provided, then the development of electric buses, their associated technologies, and their impact on the power grid are analyzed more in detail.

7.2.1 *Electric Mobility in Smart Cities*

Among directives and legislative acts for the development of electric mobility, referring specifically to the European context, it is worth mentioning the Directive 2014/94/EU on the implementation of an alternative fuels infrastructure (DAFI), the Green Deal [7] and the measures underlying the Fit for 55 plan, the Next-Generation EU [8], and the Strategy and Action Plan for Sustainable and Intelligent Mobility [9], which envisages at least 30 million zero-emission vehicles on European roads by 2030. A boost to electric mobility can also be attributed to the EU Regulation/2019/631, which defines the performance levels of new vehicles in terms of CO_2 emissions and encourages the market introduction of increasingly green vehicles.

To allow this electric mobility transition, different incentives have been devised in different countries for the purchase of low-emission vehicles and recharging infrastructures, important measures in terms of exemptions from road taxes have also been conceived, as well as local concessions such as the possibility to access restricted traffic zones and free parking areas in some cities [3]. Despite the many advantages of electric mobility, first and foremost the reduction of noise and emissions in cities, there are still some obstacles that slow down its diffusion, as highlighted in [10]: the cost of purchasing an electric vehicle and its charging infrastructure, the lower autonomy compared to traditional vehicles, the poor distribution of charging infrastructures in some areas, and the need for more charging infrastructures on motorways. There has certainly been a great development of charging infrastructures in Europe in recent years, let us mention, for instance, the network of Ionity ultra-fast chargers installed in 24 countries [11] and the CEUC (Central European Ultra Charging) project in Eastern Europe [12]. Another factor that will undoubtedly influence the uptake of electric vehicles is the cost of the charging service, which is strongly linked to the evolution of electricity prices, which have risen sharply in the last 2 years.

As reported by the European Alternative Fuels Observatory (EAFO) [13], the DAFI directive recognizes different types of alternative fuels to be used in order to decarbonize the transport sector. In particular, electricity, hydrogen, biofuels, synthetic and paraffin fuels, natural gas, including biomethane, in gaseous form (compressed natural gas—CNG), liquefied natural gas (LNG), and liquefied petroleum gas (LPG) can all be seen as a valuable solution to mitigate the environmental impact of vehicles. On July 2023 in the European Union (EU27), alternative fuel cars and vans represented 5.39% of the total fleet, and the highest percentages were recorded in Poland (12.09%, where 11.86% is represented by LPG), Sweden (9.85%), and Italy (9.15%). If the attention is only focused on electric vehicles, namely, Plug-in Hybrid Electric Vehicles (PHEVs) and Battery Electric Vehicles (BEV), also called "pure electric vehicles," at the end of 2022 there were more than 3.2 million BEV cars and vans and more than 2.7 million PHEVs, respectively, accounting for 1.15% and 0.94% of the total fleet. Sweden, Denmark, Luxembourg, and the Netherlands constitute the countries where the percentage of electric cars and vans exceeds 5%. It is also very interesting to analyze the data related to new annual registrations. In Sweden and Finland, BEVs and PHEVs cover more than 50% of new registrations, while in Belgium, Denmark, Malta, and the Netherlands they represent a percentage in the range between 30% and 50%. If we focus on charging infrastructures in Europe, provisional data reported by EAFO in July 2023 [13] assess that there are more than 500,000 charging points for electric vehicles, 11.4% represented by direct current (DC) chargers, and the remaining part by alternating current (AC) chargers.

7.2.2 *Spread and Technical Features of Electric Buses*

The number of alternative fueled buses on the road in EU27 from 2008 to 2022 is shown in Fig. 7.2, where buses are distinguished in Battery Electric Buses (BEBs), Plug-in Hybrid Electric Buses (PHEBs), and hydrogen, LPG, CNG and LNG buses. The majority of alternative fueled buses is represented by those fed by CNG, but Battery Electric Buses (BEBs) are those which have seen the greatest increase in recent years, mainly due to the EU policies and incentive mechanisms toward the electrification of the urban public transport. This is well described by the graph reported in Fig. 7.3, where annual registrations of alternative fueled buses are shown and where it can be observed that, in 2022, BEBs accounted for 67.7% of new alternative fueled bus registrations. The total number of alternative fueled buses in 2022 is instead reported in Fig. 7.4, specified for each EU country. France, Germany, and the Netherlands have the largest number of electric buses, while in Denmark and Finland BEBs represent almost the entirety of the alternative fueled buses. More detailed data on the diffusion of electric buses in Europe and worldwide are reported in [14].

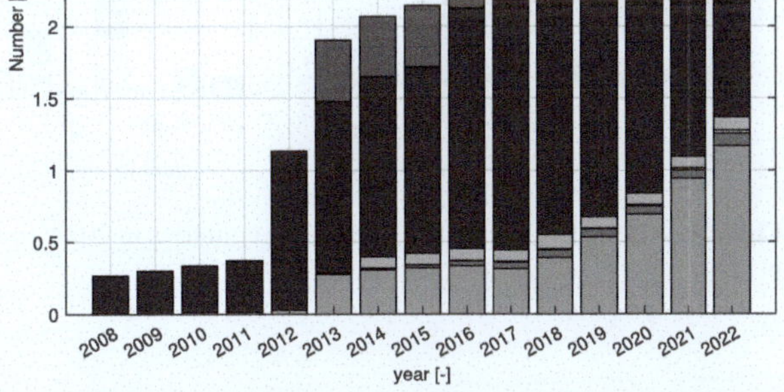

Fig. 7.2 Alternative fueled bus fleet in EU27 (own elaboration from data available in [13])

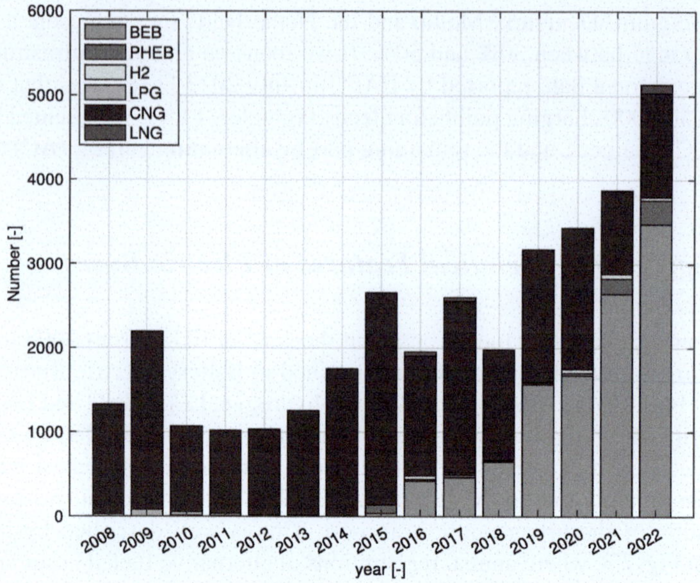

Fig. 7.3 Number of newly registered alternative fueled buses in EU27 (own elaboration from data available in [13])

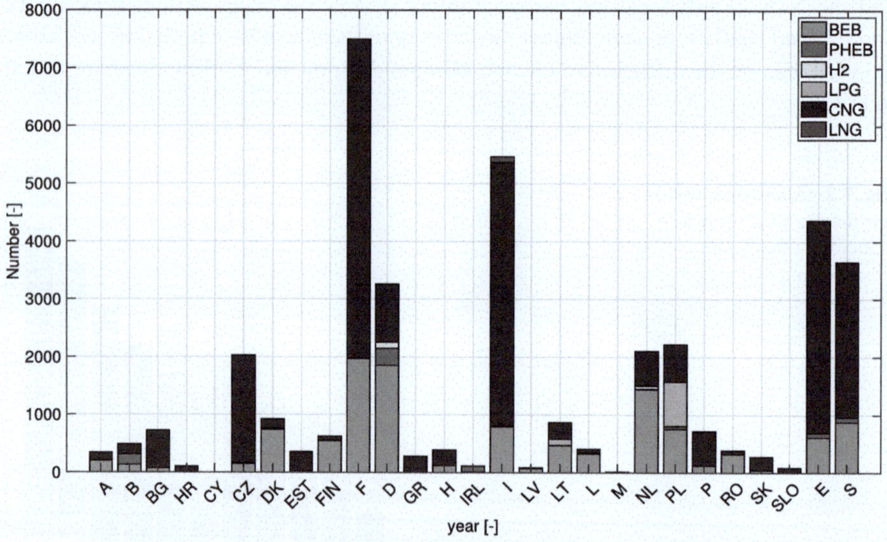

Fig. 7.4 Alternative fueled bus fleet in EU countries in 2022 (own elaboration from data available in [13])

As cited by the Directive 2019/1161/EU on the promotion of clean and energy-efficient road transport vehicles, which sets minimum procurement targets by 2025 and 2030 for zero-emission vehicles, the markets for low- and zero-emission urban buses are characterized by increased market maturity, whereas markets for low- and zero-emission trucks are at an earlier stage of market development. It is worth noting that BEBs and buses fed by hydrogen can be considered as low- or zero-emission buses when both the electricity to feed BEBs and the hydrogen stored in H_2 buses come from renewable energy sources, such as wind, solar, and hydro. This is particularly true in countries with a high share of renewables in total electricity generation and when hydrogen is produced by electrolyzers fed by green electricity (for this reason it is called "green H2").

As reported in [15], the main advantages of BEBs compared to buses fed by fossil fuels (mainly diesel) can be found in lower environmental impact (almost negligible if charged electricity comes from renewable sources), lower noise and vibration for passengers and people nearby, higher efficiency also due to regenerative braking, higher reliability, and lower maintenance costs [3]. Moreover, the electrification of urban public transport fleets can enhance the citizen awareness toward sustainability and promote the supply chain of companies specialized in the maintenance of electrical and electronic components and batteries. Electric buses can be well integrated with micro-mobility and e-sharing services, and in some cases their charging stations can also be used by other types of electric vehicles, for instance, last-mile logistic vans and trucks. Finally, as it will be better clarified in Sect. 7.2.4, BEBs could also provide services to the electric power system through their smart charging infrastructures.

One of the main drawbacks of BEBs is the capital cost: compared to a diesel bus having the same size and performance, the purchase price of an electric bus is about double. But, due to lower maintenance costs and higher efficiency, several studies report a lower Total Cost of Ownership (TCO) for BEBs [15, 17]. Obviously, a low TCO can be reached if the operational costs related to the charging can benefit from flexible electricity prices and the application of smart charging strategies. On the other hand, capital costs can be considerably reduced by benefitting from purchase incentives and tax exemptions. It is important to highlight that, when urban public transport companies decide to renovate their bus fleets, they must not think of simply replacing diesel vehicles with electric ones, but develop an organic plan for the size of the electric vehicle fleet, taking into account the transport needs and technical characteristics of these vehicles and their charging systems, which could also be coupled with renewable power plants installed in the proximity of depots. Furthermore, to mitigate the initial capital costs, different options can be adopted such as the purchase of batteries or of the bus on lease, the participation in joint purchase agreements involving several municipalities, and the use of public–private partnerships [15]. All these aspects must be dealt with, together with suitable strategic planning optimization algorithms, as better detailed in Sect. 7.3.

BEBs are available on the market in several different lengths (5 [m], 12 [m], 18 [m], 24 [m]) and with different types of batteries, among which Nickel–Manganese–Cobalt (NMC) and Lithium–Ion–Phosphate (LFP) are considered the best ones. It is common to divide BEBs into two groups: long-haul buses and short-haul buses. In the first case, they have a battery with a rated capacity in the range 250–660 [kWh], a driving range higher than 200 [km], and they are typically charged twice a day during long stops. In the second group, there are buses having a smaller battery (50–250 [kWh]) that is charged not only during long stops but also during the journey. Average energy consumptions of BEBs stay in the range between 0.76 and 2.79 [kWh/km] with an average value of 1.65 [kWh/km] [18]. Apart from external conditions, the energy consumption of a BEB strongly depends on the energy consumed by the internal air conditioning system, the lights, and all the other auxiliary systems [19]. The typical lifespan of batteries is between 7 and 10 years and a great influence on that derives from the ambient temperature and the cycling operation. Once they reach the end of life, they could be used as second-life batteries in order to store the surplus energy coming from renewable power plants in small and large size applications in different sectors (residential, tertiary, and industrial).

7.2.3 Charging Systems and Modes for Electric Buses

Since BEBs have larger batteries and higher energy consumptions than electric cars and vans, together with less time spent parked, they are usually charged in DC with power levels higher than 50 [kW]. The classical way is the so-called *depot charging* when the bus is charged overnight at 25–150 [kW] for at least 5 h. This can be done either by a cable connected to an AC/DC charging station or by a pantograph mounted on the top of the bus. Since the number of charging points in the depot may be smaller than the number of electric buses in the fleet, they can be managed by optimizing the charging schedule according to their timetable and battery state of charge when they arrive at the depot. Depot charging requires space for the installation of charging stations. In addition, with a large number of charging stations, the total committed electrical power of the depot becomes significant, with the need for dedicated grid-connected substations. Then, especially when buses are charged by a cable, the presence of technical personnel is required for the connection and disconnection of vehicles. It has been proved that buses with NMC batteries show a good battery performance during depot charging.

Opportunity charging systems partially recharge buses at the terminus using a pantograph. It takes only 10–15 minutes to re-energize the vehicle and it allows for a shorter depot recharge at the end of the day with lower power than with overnight systems. The system, however, runs the risk of breaking down during peak hours, when dwell times are shorter. The terminus intervals, in fact, serve not only to give drivers a rest but also to provide a margin of flexibility in the event of delays. If recharging is not fast enough, the timetable may be compromised. With opportunity charging systems the typical charged power is in the range between 150 [kW] and 600 [kW] and the most suitable buses are those equipped with Lithium–Titanate–Oxide (LTO) batteries [15]. By recharging several times during the day, buses can be equipped with smaller batteries; this reduces weight and increases space for occupants but, obviously, terminals must be equipped with all the necessary infrastructure (power substation, charging infrastructure, etc.).

Flash charging refers to the possibility of charging the buses with short "flashes" of energy along the line, at certain stops, just long enough for passengers to get on and off. In this way, the vehicle normally arrives at the terminus with a battery charge of around 40–50%, allowing the full recharging time to be shortened to just 3 to 5 min. At the end of the service, when returning to the depot, a similar situation occurs: the battery allows the vehicle to be fully recharged with less power than required by the overnight and opportunity systems. With flash charging very high charging power, around 500–600 [kW], is transferred to the bus by a pantograph system; as a result, charging times become comparable with refueling times of diesel buses. Buses can present a hybrid configuration with lithium–ion batteries and super-capacitors.

Finally, among the possible charging modes for electric buses it is worth mentioning also *wireless charging* and *overhead conductive charging in motion*. In the former case, inductive charging occurs when the bus places its wireless receiver (located at the bottom of the vehicle) above a charging coil positioned below the road surface. This can happen at bus terminals or when the bus stops for passengers to get on and off. The transferred power is around 50–250 [kW]. This solution has a reduced visual impact but also requires structural and maintenance work on the road surface. In the latter case, there are in-motion charging trolleybuses that use overhead catenary wires installed on only 20–40% of the bus route and otherwise use battery power as BEBs.

In all the aforesaid examined cases, it is important to note that the charging time not only depends on the charged power but also on the battery state of charge, since the maximum charging power is strongly affected by the battery state of charge. DC charging curves usually present a concave shape with maximum charging power with state of charge between 30% and 80%; so, to benefit from very high charging power, it is convenient to keep the state of charge of the battery within the aforesaid limits.

7.2.4 Smart Charging of Electric Buses

Charging infrastructures for electric mobility are part of the national electric power system, within which they constitute "electrical loads" connected at low or medium voltage, this last in the specific case of charging hubs equipped with multiple charging stations or single high-power charging points. The spread of electric vehicles and the associated charging infrastructure will increasingly have a significant impact on the demand curve and power flow management in electricity networks. Indeed, uncontrolled and simultaneous charging of large numbers of vehicles in confined urban areas can lead to problems in the electricity network, such as congestions and degradation of service quality. To address these issues, controlled and optimized charging logics need to be implemented to manage the charging of vehicle fleets in a way that maximizes the potential of the electricity grid and charging systems. It is certainly advantageous to charge electric vehicles at times when there is a surplus of electricity from renewable sources, at times of day when the grid is less congested, or when electricity prices are low [16]. Within these new strategies for managing electric mobility, V2G and V2B technologies will increasingly find their place, making electric vehicles active rather than passive players in the national electricity system. Fleets of electric vehicles capable of operating in V2G mode will be able to provide services to the grid, smoothing the demand curve and reducing absorption peaks, just as the use of V2B technology can lead to optimized management of energy flows in the increasingly sustainable and "prosumer" building sector. The integration of electric mobility systems, renewable energy production, storage systems, and smart buildings will be one of the main pillars of the smart cities of the future.

The term *smart charging* refers to the possibility of implementing charging strategies for the individual electric vehicle that allow the charging power to be modulated or delayed over time based on external signals, such as price indications or technical information from the electric power grid management, while still meeting the needs of the end user of the vehicle [16]. Such strategies can provide both economic benefits to the end user and better management of the electricity distribution network, avoiding new investments in network components and overloads or interruptions of the energy supply service due to excessive peaks in demand. Smart charging therefore means recharging a vehicle or, more generally, a fleet of vehicles at times of low demand (e.g., at night), when electricity prices are low or when local renewable energy production is high.

Through the use of the V2G technology, vehicles act as true storage systems, able to absorb energy from the grid when charging and release it when discharging. Fleets of grid-connected electric vehicles, such as electric buses when parked overnight, can be charged (G2V, grid to vehicle) at times of low demand, low prices, or when excess energy production from non-programmable renewable sources is expected. The same buses are then discharged through V2G technology to provide energy to the grid to meet an increase in demand or an unplanned decrease in renewable production.

The application of V2G technologies to the electric bus sector can lead to economic benefits in terms of revenues from the sale of services in energy markets. This can help reduce the TCO of electric buses, making them more competitive with diesel buses. Unlike private electric cars, which are difficult to predict in terms of dwell time, electric bus fleets are more predictable and therefore more suitable for participation in energy markets acting as virtual power stations.

7.3 State of the Art of Planning and Control Methods for Automated Electric Bus Fleets

The adoption of new generation buses in smart cities not only provides challenges related to the technological equipment, but also implies new problems related to both planning and online control, that is the main goal of this chapter. In fact, the classic planning and control problems widely studied for traditional buses (see the survey paper [20] for more details) need to be revised to consider the recharging requirements [21] and the automated or semi-automated features of these vehicles [22]. The process of planning a bus fleet can be distinguished into three levels, i.e., strategic, tactical, and operational, corresponding to long, medium, short time horizons [20]. In addition, online control schemes must be considered for making decisions on bus fleets in real time depending on the real conditions of the system. As highlighted in [21], planning and control problems for conventional buses must be extended to consider the needs of electric buses.

Specifically, at the *strategic level*, besides the decisions related to transit lines design, other long-term decisions regard the choice of the equipment to be installed both on board of the vehicle and in the charging infrastructure (as done, for instance, in [23]), the location of charging stations [24], and the sizing of the bus fleet. In several works, some of these decisions are made jointly. For example, in [25] a linear programming model is proposed to find the optimal number of electric buses according to specific choices in terms of technological equipments both in the vehicle and in the infrastructure, while in [26, 27] the decisions related to technological choices and the location of charging points are made jointly. At the *tactical level*, decisions mainly concern the definition of frequency and timetables. These problems can be extended to consider the energy needs of electric buses taking into account, for instance, more flexible timetables; an approach of this type can be found in [28]. The review work [29] analyzes all the strategic and tactical planning problems involving electric buses (in particular, charging station location problems, bus timetabling, and scheduling problems) and highlight their impact on the distribution grid.

At the *operational level*, most of the research works found in the literature deal with vehicle and crew scheduling [20]. In the presence of electric buses, the problem of vehicle scheduling is particularly crucial since it is related with the availability of the charging infrastructure. An interesting work on electric bus scheduling can be found, for instance, in [30], where the presence of a single depot is assumed, and in [31], where the presence of multiple depots is considered. In [32], a charging facility sharing strategy is proposed to allow charging stations of electric buses to be utilized by electric cars for a given fee during certain time windows. In [33], a single-line vehicle scheduling problem is solved by considering the use of skip-stop operation strategies in order to minimize both operator costs and passenger costs. In [34], an optimization tool is proposed for the charging scheduling and management of a fast-charging system for electric buses. When considering electric vehicles, the influence of the surrounding traffic becomes particularly relevant for the effective energy consumption and thus for the actual ability to complete the tasks assigned to a vehicle. In [35], an electric bus schedule in presence of stochastic traffic conditions is proposed.

Regarding *online control*, the main control strategies for traditional buses fall mainly into two categories, as suggested in [20]:

- *Station control strategies*, in which the control actions are implemented at bus stops. These actions may be based on holding buses at the stop to avoid bus bunching and to increase the regularity of service [36] or on the application of stop-skipping strategies to increase the bus speed [37] or to minimize passenger waiting times [38].

- *Inter-station control strategies* that are taken while the bus is traveling between two stops. These actions can be based on bus speed control [39, 40], or by implementing traffic signal priority strategies in which the timing of traffic lights is appropriately regulated to prioritize buses [41].

Combined approaches, such as the one presented in [42], involve the implementation of multiple control actions at the same time. Indeed, in [42], the implemented actions are related with holding the bus at stops, bus speed control, and traffic lights control. In case electric buses are considered, aspects related to driving conditions, traffic state, and the presence of intersections equipped with traffic lights become very important. In fact, most approaches related to the control of electric buses fall into the category of inter-station control, integrating the concepts already developed in eco-driving strategies for electric vehicles, i.e., finding the optimal speed profile of electric vehicles in order to reduce energy consumption, considering aspects of vehicle movement, road characteristics and signs (see [43, 44]), with the need to ensure efficient public transport service. An example of an eco-driving approach applied to the control of electric buses is presented in [45]. Other works, such as [46–48], combine the typical purposes of eco-driving control with the goal of limiting bus bunching.

The approaches mentioned above have in common the fact that the influence of traffic on bus trajectories is neglected. This is surely reasonable in case reserved lanes are present for public transport, as it happens in many cities. However, in case reserved lanes are not present in urban lines or for intercity buses, this assumption can no longer apply and therefore control strategies that integrate the prediction of traffic conditions are needed. In this line of research it is possible to mention [49–52], in which the objective is to optimize speed, dwell times, and/or charging times of electric automated buses in real time, taking into account the measured traffic state and the predicted traffic conditions that the bus will encounter along its path.

Finally, it is worth concluding by indicating that further extensions can be made to these control schemes in order to take into account additional features related to the motion of automated vehicles. For example, in [53], a hierarchical control scheme is proposed for a platoon of fully autonomous buses, in which the state and trajectory of the buses are defined by a high level of control and are used as a reference for a low level of control to constrain lateral and longitudinal movements of each bus.

7.4 Models for Electric Automated Buses in Smart Cities

Both for the design phase and for the real-time control of electric automated buses in smart cities, it is necessary to apply suitable models. Such models must be able to represent at least two important aspects:

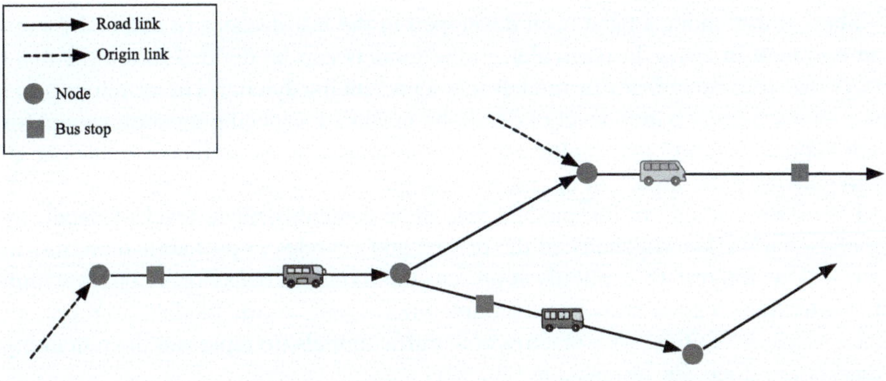

Fig. 7.5 The considered network representation

- the dynamic behavior of buses when interacting with other vehicles on the road, by means of appropriate *traffic models*;
- the energy consumption of buses, depending on their trips and the presence of charging stations, by means of appropriate *energy consumption models*.

Let us consider electric automated buses which have to follow specific routes with fixed stops and fixed timetables in a road network. The considered charging modes are opportunity charging at the terminus and flash charging along the line in case the stops are equipped with a charging infrastructure.

The road network can be described by means of a directed graph that is composed of M road links, O origin links, and N nodes. $\mathcal{S}$ represents the set of bus stops in the network, while $\mathcal{B}$ is the set of considered buses. For each node $n \in N$, let us define with O_n the set of road exiting links, and with I_n, $\bar{I}_n$ the set of entering road links and entering origin links, respectively. Buses can stay in the bus stops, not only for allowing passengers to get off or get on the bus, but also for charging the battery via the electric charging infrastructure (in case the sop is provided with a flash charging infrastructure). In Fig. 7.5, the main components of the considered network are reported.

7.4.1 Traffic Models

As previously introduced, in order to plan and control electric automated buses, it is very important to represent their interaction with the other vehicles present in the road. Indeed, such interaction is always very relevant, except for the cases in which dedicated lanes are reserved for public transport. In particular, there is a double interaction of buses with traffic, since they influence the surrounding traffic and, at the same time, they are affected by the traffic conditions around.

Many traffic models have been developed in the last decades (a detailed survey can be found in [54]). In particular, traffic models can be divided into three main categories: *microscopic* traffic models that represent the dynamics of all vehicles and their interactions at a high level of detail, by means of a specific dynamic model for each vehicle; *macroscopic* traffic models that represent the traffic dynamics at an aggregate level, by means of aggregate variables, specifically density, mean speed, and flow; *mesoscopic* traffic models that show an intermediate level of detail, by representing the heterogeneity of the drivers and vehicles in probabilistic terms. In case it is of interest to explicitly represent the dynamic behavior of buses within the traffic flow, a good choice is to couple microscopic traffic models (to represent buses as single entities) with macroscopic traffic models (to represent the remaining vehicles), as done, for instance, in [49, 55].

Let us now describe the traffic model adopted in [49] for studying electric buses, taking inspiration from the *Micro-Macro METANET* model proposed for the first time in [56]. In this model, each road link m is divided into N_m sections of length L_m and with λ_m lanes. The time horizon is discretized into K time steps, with sampling time interval T [h].

The main aggregate variables of the Micro-Macro METANET model are the following:

- $\rho_{m,i}(k)$ is the traffic density in section i of road link m at time step k [veh/km/lane];
- $\rho^{\text{tot}}_{m,i}(k)$ is the traffic density including the presence of buses in section i of road link m at time step k [veh/km/lane];
- $v_{m,i}(k)$ is the mean traffic speed in section i of road link m at time step k [km/h];
- $q_{m,i}(k)$ is the traffic flow in section i of road link m at time step k [veh/h];
- $q_o(k)$ is the traffic flow that enters the network from origin link o at time step k [veh/h];
- $Q_n(k)$ is the traffic flow entering node n at time step k [veh/h];
- $\beta_{m,n}(k)$ is the splitting rate from node n to link m at time step k (the splitting rates must guarantee that $\sum_{\mu \in O_n} \beta_{\mu,n}(k) = 1, n = 1, \ldots, N, k = 0, \ldots, K-1$).

For each bus the following variables are defined:

- $v^b(k)$ is the speed of bus b at time step k [km/h];
- $p^b(k)$ is the position of bus b at time step k [km];
- $\delta^b_{m,i}(k)$ is equal to 1 if bus b is in section i of link m at time step k, 0 otherwise;
- $w^b_s(k)$ is equal to 1 if bus b is in bus stop s at time step k, 0 otherwise.

Let us now consider the *macroscopic behavior* of the traffic flow. As in the METANET model [57], the evolution of the traffic density in each section i of each road link m is given by

$$\rho_{m,i}(k+1) = \rho_{m,i}(k) + \frac{T}{L_m \lambda_m}\Big[q_{m,i-1}(k) - (k) q_{m,i}(k)\Big] \tag{7.1}$$

with the traffic flow expressed as

$$q_{m,i}(k) = \rho_{m,i}(k) v_{m,i}(k) \lambda_m. \tag{7.2}$$

The density dynamics defined by (7.1) does not consider the presence of buses which are instead included in the total traffic density $\rho_{m,i}^{\text{tot}}(k)$, calculated as follows:

$$\rho_{m,i}^{\text{tot}}(k) = \rho_{m,i}(k) + \sum_{b=1}^{B} \delta_{m,i}^{b}(k) o^{b}, \tag{7.3}$$

where o^b is the occupancy of bus b [veh/km/lane].

The speed dynamics in each section i of each road link m is computed considering the total density $\rho_{m,i}^{\text{tot}}(k)$ according to the following relation:

$$\begin{aligned} v_{m,i}(k+1) = v_{m,i}(k) + \frac{T}{\tau}\Big[V\left(\rho_{m,i}^{\text{tot}}(k)\right) - v_{m,i}(k)\Big] + \frac{T}{L_m} v_{m,i}(k)\Big[v_{m,i-1}(k) - v_{m,i}(k)\Big] \\ - \frac{\nu T\left[\rho_{m,i+1}^{\text{tot}}(k) - \rho_{m,i}^{\text{tot}}(k)\right]}{\tau L_m\left[\rho_{m,i}^{\text{tot}}(k) + \chi\right]} - \delta^{\text{on}} T \frac{v_{m,1}(k) q_o(k)}{L_m \lambda_m \left[\rho_{m,1}^{\text{tot}}(k) + \chi\right]}, \end{aligned} \tag{7.4}$$

where τ, ν, χ, and δ^{on} are model parameters.

The steady-state speed–density relation in (7.4) is given by

$$V\left(\bar{\rho}_{m,i}(k)\right) = v_{m,i}^{\text{f}} \cdot \left[1 - \left(\frac{\rho_{m,i}^{\text{tot}}(k)}{\rho_{m,i}^{\max}}\right)^{l}\right]^{\kappa}, \tag{7.5}$$

where $v_{m,i}^{\text{f}}$ is the free-flow speed in section i of road link m [km/h], $\rho_{m,i}^{\max}$ is the jam density in section i of road link m [veh/km/lane], while l, κ are other model parameters.

The inflow of traffic into node n is computed as

$$Q_n(k) = \sum_{\mu \in I_n} q_{\mu, N_\mu}(k) + \sum_{o \in \bar{I}_n} q_o(k) \tag{7.6}$$

while the traffic flows entering the first sections of the exiting links from node n are given by

$$q_{m,0}(k) = \beta_{m,n}(k) \cdot Q_n(k). \tag{7.7}$$

Let us now represent the *microscopic behavior* of buses traveling in the traffic flow. For each bus b, the covered distance is calculated on the basis of its speed, with the following dynamic equation:

$$p^b(k+1) = p^b(k) + v^b(k) T. \tag{7.8}$$

When the position of the bus is in the proximity of the stop, the corresponding binary variables $w_s^b(k)$ must be fixed equal to 1 for T_s^b time steps, with T_s^b representing the number of time steps in which bus b must wait at stop s (according to the timetable). This corresponds to the following relation:

$$w_s^b(h) = 1,\ h = k, \dots, k + T_s^b - 1,\ w_s^b(k + T_s^b) = 0 \ \text{ if } \pi_s - \Delta \leq p^b(k) \leq \pi_s + \Delta, \tag{7.9}$$

where π_s is the position of bus stop s [km], and Δ is a given tolerance.

Moreover, the bus speed $v^b(k)$ must be null if the bus is waiting at a stop, i.e.:

$$v^b(k) = 0 \ \text{ if } \sum_{s=1}^{S} y_s^b(k) = 1, \tag{7.10}$$

while, when the bus is traveling, the bus speed $v^b(k)$ is computed as the minimum between the design speed $v^{b,\text{sch}}$ due to the timetable [km/h] and the average traffic speed in the section in which the bus is traveling, i.e.,

$$v^b(k) = \min\left\{ v^{b,\text{sch}}, \sum_{m=1}^{M} \sum_{i=1}^{N_m} \delta_{m,i}^b(k) \cdot v_{m,i}(k) \right\} \ \text{ if } \sum_{s=1}^{S} y_s^b(k) = 0. \tag{7.11}$$

7.4.2 Consumption Models

Energy consumption of electric buses depends on many aspects, related with the technical characteristics of the vehicle, weather conditions, number of passengers, driving style of the driver, road slope, just to mention the most important ones. Many consumption models for electric buses have been studied by researchers in the last decade. For example, in [58], the developed microscopic energy consumption tool simulates the instantaneous energy consumption as a function of the line and vehicle characteristics, and of other conditions such as the ambient temperature, the bus load, and the traffic conditions. In [59], bus speed profiles are generated and then inputted to a consumption model which is based on the route topography, the auxiliary loads, and the weather conditions. In [60], the consumption model of an electric bus is integrated with a microscopic traffic model which provides the bus driving cycles in different traffic conditions and with different driver behaviors. In [61], the same consumption model is adopted with the final goal of evaluating different sizes and locations for the fast-charging infrastructure for the electric public transport system.

Let us now describe the main features of the energy consumption model adopted in [60, 61], where the reader can find more details. The main inputs of the model can be divided into:

- Bus rated data, such as the width and height, the curb weight, the tire size and nominal pressure, the gear ratio, the maximum power absorbable by the auxiliary

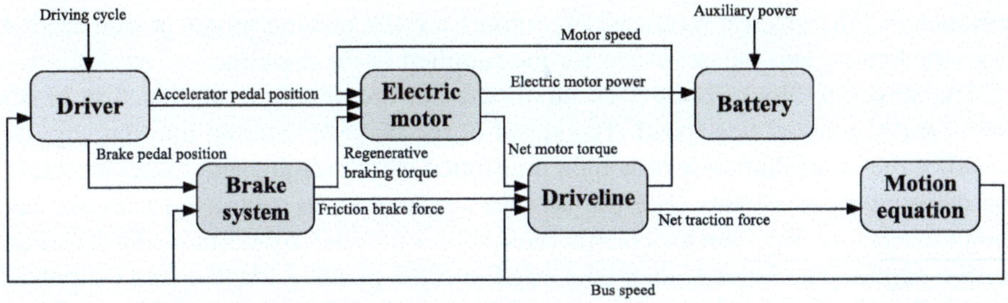

Fig. 7.6 Information diagram of the energy consumption model

systems, the drag coefficient, the rated values of motor power and torque, the maximum braking force and regenerative braking efficiency, the transmission spin loss, the rated capacity, the open-circuit voltage, and internal resistance of the battery.
- Time-dependent operating data, such as the bus driving cycle, the road slope, the ambient conditions, the number of onboard passengers, the power absorbed by the auxiliary systems, and the tire pressure.

It is worth noting that this model, as most of the energy consumption models found in the literature, represents time as a continuous variable. Figure 7.6 shows the main subsystems of the model and their connections, which have been implemented in MATLAB® and Simulink® and are now briefly described.

The *driver* block receives as inputs the bus driving cycle (to be used as a reference speed profile) and the actual speed of the bus. It works as a proportional–integral–derivative (PID) controller by computing the error between the reference speed and the actual speed and, then, by regulating the accelerator and brake pedal position in order to follow the speed profile of the driving cycle.

The *brake system* block receives as inputs the brake pedal position and the actual bus speed. The brake pedal position is transformed into a desired brake force, on the basis of the maximum brake force and the road adhesion coefficient. The desired brake force is then split into the regenerative brake force, properly converted into the corresponding regenerative braking torque (in input to the electric motor block), and the friction brake force (in input to the driveline block).

The *electric motor* block receives as inputs the motor speed, the accelerator pedal position, and the regenerative braking torque. The accelerator pedal position is converted into a corresponding accelerating torque, while the regenerative braking torque at the crankshaft is converted into the actual regenerative torque. Then, the accelerating torque and the regenerative torque are used to compute the net motor torque, which is used to calculate the mechanical output power of the motor on the basis of the motor angular speed. The electric input power of the motor is computed by summing the losses to the mechanical output power of the motor, with the losses depending on the net torque of the motor, the angular speed of the motor, and other parameters. The electric motor block outputs are the electric motor power needed from the battery and the net motor torque supplied to the driveline.

The *driveline* block receives as inputs the net motor torque, the friction brake force, and the actual bus speed. The speed of the bus is converted into the angular speed of the crankshaft, which is then multiplied by the torque spin losses to obtain the driveline power losses. The torque spin losses are also subtracted from the net motor torque coming from the electric motor block in order to determine the driveline torque output. The latter is then converted into the positive tractive force supplied by the driveline to the tires by means of the ratio between the tire radius and the gear ratio. The net tractive force acting on the vehicle is obtained as the difference between the positive tractive force and the friction brake force coming from the brake system block.

The *battery* block receives as inputs the electric motor power from the electric motor and the power absorbed by the auxiliary systems. Their sum represents the total discharge power of the battery. The battery is modeled using an equivalent electric circuit with an ideal voltage source and one resistance in series, representing the internal resistance of the battery.

The *motion equation* block receives as input the net traction force acting on the vehicle. The output of this block is the speed of the bus $v_b(t)$ at time t, which is an input to the driver block, according to a closed-loop logic, as explained before. The actual bus speed $v_b(t)$ is calculated by integrating the acceleration $a_b(t)$ determined by applying Newton's second law of motion to the bus, i.e.,

$$F^{\mathrm{t}}(t) - R^{\mathrm{air}}(t) - R^{\mathrm{r}}(t) - R^{\mathrm{g}}(t) = m_b^{\mathrm{eq}} \cdot a_b(t), \tag{7.12}$$

where $F^{\mathrm{t}}(t)$ is the net traction force, $R^{\mathrm{air}}(t)$ is the aerodynamic resistance, $R^{\mathrm{r}}(t)$ is the rolling resistance, and $R^{\mathrm{g}}(t)$ represents the resistance due to the grade, at time t. Moreover, in (7.12) m_b^{eq} is the equivalent mass of the bass obtained as

$$m_b^{\mathrm{eq}} = m_b(1 + \beta_b) \tag{7.13}$$

with m_b representing the static mass of the bus (depending on the bus characteristics and the number of passengers onboard) and β_b a suitable increasing factor, which varies from vehicle to vehicle. The concept of equivalent mass in (7.13) is used to account for the inertial force related to all the rotating components inside the vehicle.

Note that the net tractive force $F^{\mathrm{t}}(t)$ can assume positive values, in that case the bus is said in powering mode, negative values, which means that the bus is braking, and finally null values, causing the bus natural deceleration (coasting mode).

The aerodynamic resistance is computed as

$$R^{\mathrm{air}}(t) = \frac{1}{2}\rho^{\mathrm{air}} C^{\mathrm{air}} A_b^{\mathrm{f}} v_b(t)^2, \tag{7.14}$$

where ρ^{air} is the air density (depending on ambient conditions), C^{air} is the drag coefficient, and A_b^{f} is the bus frontal area.

The rolling resistance and the resistance due to the grade are given, respectively, by

$$R^{\mathrm{r}}(t) = K^{\mathrm{r}}(t) m_b g \cos\alpha(t) \tag{7.15}$$

$$F_c = m_b g \sin\alpha(t), \tag{7.16}$$

where $K^{\mathrm{r}}(t)$ represents the rolling resistance coefficient at time t (which depends on the vehicle speed $v_b(t)$ and the tire pressure), g is the acceleration of gravity, and $\alpha(t)$ indicates the angle representing the road slope at time t.

To summarize, the main outputs of this consumption model are

- the real speed profile of the vehicle and its acceleration;
- the battery current and voltage profiles during charging and discharging phases;
- the power exchanged between the battery and the motor;
- the battery state of charge;
- the electric consumption per kilometer of the bus.

7.5 Control Algorithms for Electric Automated Buses in Smart Cities

As described in Sect. 7.3, the control strategies for electric buses can account for different aspects and can be based on different assumptions. In this section, we focus on control algorithms for electric buses traveling in roads without reserved lanes, in which, then, the interaction with traffic is relevant. Moreover, referring to the classification reported in Sect. 7.3, we are interested to consider both station and inter-station control strategies, since the decisions to be made regard the optimal speed along the route, as well as the dwell times and charging times at stops. In the following, a multi-objective problem is reported for each bus present in the road network. It is worth noting that, in case more buses have to share the same charging infrastructure in some stops, it could be useful to plan their trajectories and timing jointly, either with centralized or distributed approaches. However, in the next section, the multi-objective problem for a single bus is stated.

7.5.1 The Multi-objective Problem for Bus Optimal Charging and Speed Trajectories

Let us consider a generic bus b traveling in a road stretch which starts at the initial stop and finishing at the last stop. The stretch is divided into N sections, each one having length L_i [km], $i = 1, \dots, N$. Let us denote the set of road sections as $\mathcal{I} = \{i : i = 1, \dots, N\}$. The time horizon to be considered starts at the scheduled departure time of bus b from the first stop and finishes at the scheduled last time at the terminus. The planning horizon resulting from the bus timetable is discretized into K time steps, with sample time indicated with T. For simplifying the notation, let us denote with $\mathcal{K} = \{k : k = 0, \ldots, K-1\}$ the set of time steps of the considered planning horizon.

The following state variables are defined for each time step $k \in \mathcal{K} \cup \{K\}$:

- $p^b(k)$ is the position of bus b at time step k [km];
- $e^b(k)$ is the energy stored in the battery of bus b at time step k [kWh].

The following control variables are defined for each time step $k \in \mathcal{K}$:

- $v^b(k)$ is the speed of bus b at time step k [km/h];
- $w_s^b(k) \in \{0, 1\}$ is equal to 1 if bus b is in stop $s \in \mathcal{S}$ at time step k, 0 otherwise;
- $c_s^b(k) \in \{0, 1\}$ is equal to 1 if bus b is charging its battery in stop $s \in \mathcal{S}$ at time step k, 0 otherwise.

On the basis of the considered state variables, it is possible to write the following state equations:

$$p^b(k+1) = p^b(k) + T v^b(k) \qquad k \in \mathcal{K} \tag{7.17}$$

$$e^b(k+1) = e^b(k) + T \sum_{s \in \mathcal{S}} \varsigma_s c_s^b(k) - \delta^b \left[p^b(k+1) - p^b(k) \right] - T\beta^b \qquad k \in \mathcal{K}. \tag{7.18}$$

The state equation given by (7.17) computes the distance covered by the bus on the basis of its speed. On the contrary, (7.18) updates the energy stored in the bus battery by increasing it considering the charging power ς_s associated with stop s [kW] and by decreasing it taking into account the electric consumption δ^b of the bus when traveling [kWh/km] and the electric consumption β^b of the bus due to the auxiliary equipment [kW].

An important aspect of this problem is that the traveling behavior of buses is influenced by traffic. Hence, when planning the optimal speeds and dwell/charging times of the bus, it is necessary to account for the predicted traffic conditions that the bus will encounter along the route, in particular, it is necessary to know $v_i^{\text{pred}}(k)$, i.e., the predicted mean traffic speed in section i at time kT [km/h], $i \in \mathcal{I}$, $k \in \mathcal{K}$. This prediction can be obtained from historical traffic data or on the basis of a prediction traffic model, as done, for instance, in [51].

Some auxiliary variables are needed in the optimization problem in order to determine in which road section the bus is, according to its position $p^b(k)$ and the position p_i of the beginning of each section i. In particular, the following sets of auxiliary binary variables are introduced for $i \in \mathcal{I}$, $k \in \mathcal{K}$

$$z_i^b(k) = \begin{cases} 1 \text{ if } p^b(k) \geq p_i \\ 0 \text{ otherwise} \end{cases} \tag{7.19}$$

$$y_i^b(k) = \begin{cases} 1 \text{ if } p^b(k) \le p_{i+1} \\ 0 \text{ otherwise.} \end{cases} \tag{7.20}$$

In order to represent the logic conditions in (7.19) and (7.20) through linear relations, the following inequalities must be written, respectively,

$$p^b(k) - p_i + M\big(1 - z_i^b(k)\big) \ge \sigma \qquad i \in \mathcal{I},\ k \in \mathcal{K} \tag{7.21}$$

$$p_i - p^b(k) + M z_i^b(k) \ge 0 \qquad i \in \mathcal{I},\ k \in \mathcal{K} \tag{7.22}$$

$$p_{i+1} - p^b(k) + M\big(1 - y_i^b(k)\big) \ge 0 \qquad i \in \mathcal{I},\ k \in \mathcal{K} \tag{7.23}$$

$$p^b(k) - p_{i+1} + M y_i^b(k) \ge \sigma \qquad i \in \mathcal{I},\ k \in \mathcal{K}, \tag{7.24}$$

where σ is a small quantity and M is a large quantity arbitrarily chosen.

After defining $z_i^b(k)$ and $y_i^b(k)$, it is still necessary to introduce another set of auxiliary binary variables, as follows:

$$\lambda_i^b(k) = \begin{cases} 1 \text{ if } z_i(k) = y_i(k) = 1 \\ 0 \text{ otherwise} \end{cases} \tag{7.25}$$

meaning that $\lambda_i^b(k)$ is equal to 1 if the bus is in section i at time step k.

In order to represent (7.25) through linear relations, the following inequalities must be considered:

$$\lambda_i(k) \le z_i(k) \qquad i \in \mathcal{I},\ k \in \mathcal{K} \tag{7.26}$$

$$\lambda_i(k) \le y_i(k) \qquad i \in \mathcal{I},\ k \in \mathcal{K} \tag{7.27}$$

$$\lambda_i(k) \ge z_i(k) + y_i(k) - 1 \qquad i \in \mathcal{I},\ k \in \mathcal{K}. \tag{7.28}$$

In the considered optimization problem, three objectives are introduced, that are

- the minimization of the deviations from the bus timetable;
- the minimization of the final energy lack in the bus battery;
- the minimization of the oscillations in the speed profile.

The first objective F_1 regards the quadratic deviations from the bus timetable and can be written as

$$F_1 = \sum_{s \in \mathcal{S}} \sum_{k \in \mathcal{K}} \alpha_s(k) \Big(\bar{w}_s^b(k) - w_s^b(k) \Big)^2, \tag{7.29}$$

where $\alpha_s(k)$ is a weight used to properly penalize early arrivals and late departures from stops, and $\bar{w}_s^b(k) \in \{0, 1\}$ is equal to 1 if the bus must be in bus stop s at time step k, 0 otherwise (according to the pre-defined timetable of the bus).

The second objective F_2 on the final energy lack in the bus battery can be expressed as

$$F_2 = \zeta^{b,\text{fin}}, \tag{7.30}$$

where $\zeta^{b,\text{fin}}$ is the final lack of energy in the bus compared with the desired level $\bar{e}^{b,\text{fin}}$ [kWh], i.e.,

$$\zeta^{b,\text{fin}} = \max\{\bar{e}^{b,\text{fin}} - e^b(K), 0\}. \tag{7.31}$$

In order to represent (7.31) in a linear way, the following inequalities should be introduced:

$$\zeta^{b,\text{fin}} \geq \bar{e}^{b,\text{fin}} - e^b(K) \tag{7.32}$$

$$\zeta^{b,\text{fin}} \geq 0. \tag{7.33}$$

The third objective function F_3 regards the quadratic deviations of the speed in consecutive time steps, i.e.,

$$F_3 = \left(v^b(k+1) - v^b(k)\right)^2. \tag{7.34}$$

The multi-objective optimization problem involving bus b can be stated with the following mixed-integer linear quadratic formulation.

Problem 1

$$\min J = \left(F_1, F_2, F_3\right) \tag{7.35}$$

with F_1, F_2, F_3 expressed as (7.29), (7.30), (7.34), subject to the state equations given by (7.17) and (7.18), constraints (7.21)–(7.24), (7.26)–(7.28), (7.32)–(7.33) and the following other constraints:

$$p^b(k) - (\pi_s - \Delta) \geq \sigma - (L + \sigma)\left(1 - w_s^b(k)\right) \quad s \in \mathcal{S}, k \in \mathcal{K} \tag{7.36}$$

$$(\pi_s + \Delta) - p^b(k) \geq \sigma - (L + \sigma)\left(1 - w_s^b(k)\right) \quad s \in \mathcal{S}, k \in \mathcal{K} \tag{7.37}$$

$$\sum_{s \in \mathcal{S}} w_s^b(k) \leq 1 \qquad k \in \mathcal{K} \tag{7.38}$$

$$\sum_{k \in \mathcal{K}} w_s^b(k) \geq \gamma_s \qquad s \in \mathcal{S} \tag{7.39}$$

$$c_s^b(k) \leq w_s^b(k) \qquad s \in \mathcal{S},\ k \in \mathcal{K} \tag{7.40}$$

$$c_s^b(k) \leq \eta_s \qquad s \in \mathcal{S},\ k \in \mathcal{K} \tag{7.41}$$

$$0 \leq v^b(k) \leq \sum_{i \in \mathcal{I}} \lambda_i(k) v_i^{\text{pred}}(k) \qquad k \in \mathcal{K} \tag{7.42}$$

$$v(k)^b \leq \left(1 - \sum_{s \in \mathcal{S}} w_s^b(k)\right) \cdot v^{b,\max} \qquad k \in \mathcal{K} \tag{7.43}$$

$$e^{b,\min} \leq e^b(k) \leq e^{b,\max} \qquad k \in \mathcal{K}, \tag{7.44}$$

where L is computed as $L = \sum_{i \in \mathcal{I}} L_i$. ■

Constraints (7.36)–(7.37) impose that $w_s^b(k)$ is equal to 0 if the bus position is lower than $\pi_s - \Delta$ or greater than $\pi_s + \Delta$, i.e., when the bus is not in the proximity of stops, with π_s indicating the position of bus stop s [km] along the road stretch and Δ representing a tolerance.

Constraints (7.38) ensure that each bus can wait at most in one stop in each time step, while (7.39) guarantee that each bus waits in a stop for at least a minimum number of time steps γ_s given by the bus timetable. Constraints (7.40) and (7.41) impose, respectively, that the bus can be charged in a bus stop only if it is waiting there and if the stop is provided with a charging infrastructure, with $\eta_s \in \{0, 1\}$ being equal to 1 if bus stop s is provided with a charging infrastructure, and 0 otherwise.

Constraints (7.42) impose that the speed cannot be negative and cannot be higher than the predicted traffic flow speed in the section in which the bus is traveling. Moreover, constraints (7.43) impose that the bus speed cannot exceed the maximum allowed speed $v^{b,\max}$ [km/h] if it is traveling and, at the same time, they impose that the bus speed is null if the bus is waiting at a stop.

Finally, (7.44) fix lower and upper bounds for the energy stored in the batteries of the bus, respectively, given by $e^{b,\min}$ and $e^{b,\max}$ [kWh].

Problem 1 is a multi-objective optimization problem, since it is characterized by three objectives that are, in general, in conflict. Such a problem can be transformed into a single-objective optimization problem by applying, for instance, the weighted sum method, as done in [50], in which the objective function is a weighted sum of the three objectives. In other cases, multi-objective optimization techniques can be applied in order to find the Pareto frontier, i.e., the set of efficient solutions or, in other words, the set of alternatives with different trade-offs among the objectives (see [51] for an example of Pareto frontier in case Problem 1 is rewritten considering only objectives F_1 and F_2).

7.5.2 Control Algorithms for Bus Optimal Charging and Speed Profiles

The multi-objective optimization problem described in Sect. 7.5.1 can be solved offline, for instance, before a bus leaves the first stop, in order to optimally plan the speed profiles and the dwell/charging times at sops, on the basis of the timetable and the traffic conditions predicted before the bus leaves. However, it is well known that some unpredicted situations can occur along the path, e.g., longer waiting times at stops due to unforeseen larger number of passengers, failures to the charging infrastructure causing battery energy levels lower than the expected ones, road accidents provoking traffic breakdowns, and so on. In order to be able to react to unexpected events, it is beneficial to adopt *online control schemes* to be applied in real time on the basis of real state measurements.

A general control scheme is shown in Fig. 7.7, where a bus controller exploits data coming from road sensors (collecting traffic data, e.g., density, mean speed, flow) and from onboard sensors (collecting data on the bus, e.g., battery state of charge, position, speed) to determine the optimal speed profile and the optimal dwell/charging times at stops. The control algorithm to be used can be based on the iterative solution of Problem 1 at different time steps.

In [50], for example, the proposed control scheme is composed of two main modules, both based on real-time measurements, i.e., the traffic prediction model, which allows to make a prediction of the traffic state in the road, and the optimal control module, consisting in solving a problem similar to Problem 1 rewritten in a single-objective form. The control scheme is run at each time step of the considered planning horizon, according to a sort of receding-horizon logic.

The control scheme proposed in [52] is again composed of a traffic prediction module and an optimal control module but this latter is, in this case, based on a multi-objective algorithm based on Problem 1. In particular, the optimal control

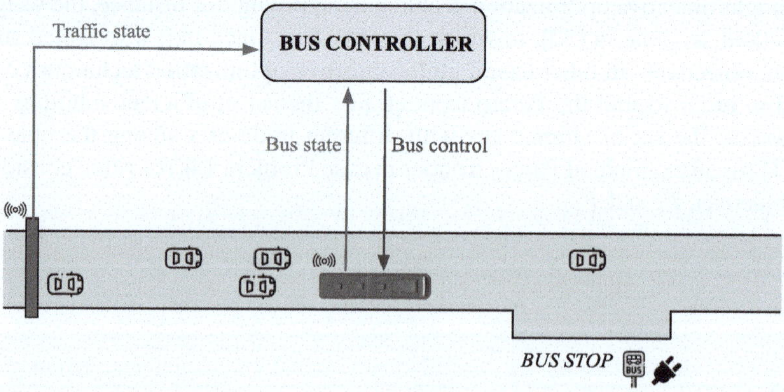

Fig. 7.7 Bus control scheme

module can be applied with two different lexicographic approaches, giving priority to the timetable or to the final energy level, on the basis of a condition which verifies, depending on the state measured in real time, if the bus is expected to be late at the next stop or not. Such control scheme is re-run every time the bus is ready to leave a stop, and the considered control horizon covers all the time necessary for the bus to reach the last stop.

As aforementioned, in case more buses traveling in the road network have to share some charging equipment or, in general, need to be planned jointly, the bus controller shown in Fig. 7.7 can be integrated with the controller of other buses. This corresponds to the definition of *distributed control schemes* in which the different controllers can communicate each other some relevant information or can take some decisions jointly. The definition of such schemes, which must at the same time provide effective solutions for the bus trajectories and be computationally efficient, is an important challenge to be addressed in future research works.

7.6 Conclusion

Electric and automated buses have the potential to transform public transport in smart cities, providing new opportunities for operators to improve regional connectivity and making the transport service easier, safer, and more sustainable. The challenges to be faced are many, not only on the technological side, but also on the social, policy, and economic sides. From the point of view of researchers working in the area of modeling and control, all these technological, social, and economic developments should be captured in specific models and control algorithms. In particular, as for modeling issues, an important effort should be made to integrate traffic models and energy consumption models, that so far have been developed almost separately. Such models should try to capture as much as possible all the traffic-related issues (e.g., traffic breakdowns, accidents, merging phases) and all the energy aspects (V2G/V2B logics, smart charging, attitudes of drivers). With a similar logic, the control algorithms should be used to manage buses in real time, considering the possibility of V2V communication among vehicles, the integration with other mobility planning systems, energy management systems, and all the other data sources that are present in the smart cities, in order to guarantee accessibility, connectivity, inclusiveness, and, in general, a better quality of life for citizens.

References

1. Transport (2019) Increasing oil consumption and greenhouse gas emissions hamper EU progress towards environment and climate objectives, Report of the European Environment Agency, vol 12 (2019)
2. Borén S (2020) Electric buses' sustainability effects, noise, energy use, and costs. Int J Sustain Transp 14:956–971
3. Manzolli JA, Trovao JP, Antunes CH (2022) A review of electric bus vehicles research topics - Methods and trends. Renew Sustain Energy Rev 159:112211
4. Lim LK, Ab Muis Z, Ho WS, Hashim H, Chien CBP (2022) Review of the energy forecasting and scheduling model for electric buses. Energy 125773
5. Jia Q-S, Panetto H, Macchi M, Siri S, Weichhart G, Xu Z (2022) Control for smart systems: challenges and trends in smart cities. Ann Rev Control 53:358–369
6. Azad M, Hoseinzadeh N, Brakewood C, Cherry CR, Han LD (2019) Fully autonomous buses: a literature review and future research directions. J Adv Transp 4603548
7. European Green Deal, European Council, https://www.consilium.europa.eu/en/policies/green-deal/
8. Recovery plan for Europe, European Commission, https://commission.europa.eu/strategy-and-policy/recovery-plan-europe_en
9. Mobility Sytrategy, European Commission, https://transport.ec.europa.eu/transport-themes/mobility-strategy_en
10. Energy & Strategy Group (2022) Smart mobility report 2022, Politecnico di Milano
11. Ionity. https://ionity.eu/
12. CEUC: New fast-charge corridor for Europe (2018). https://www.electrive.com/2018/04/26/ceuc-new-fast-charge-corridor-for-europe/
13. European Alternative Fuels Observatory, European Commission, https://alternative-fuels-observatory.ec.europa.eu/general-information/alternative-fuels
14. Electric bus, main fleets and projects around the world, Sustainable bus. https://www.sustainable-bus.com/electric-bus/electric-bus-public-transport-main-fleets-projects-around-world/
15. Electric buses in public transport. A vade mecum, MOTUS-E Report (in Italian) (2022). https://www.motus-e.org/wp-content/uploads/2022/10/Autobus-elettrici-nel-trasporto-pubblico.-Un-vademecum-1.pdf
16. Bracco S, Delfino F, Piazza G (2022) The role of electric mobility in future grids (in Italian). AEIT J 11(12):6–12
17. Mahmoud M, Garnett R, Ferguson M, Kanaroglou P (2016) Electric buses: a review of alternative powertrains. Renew Sustain Energy Rev 62:673–684
18. Rodrigues ALP, Seixas SRC (2022) Battery-electric buses and their implementation barriers: analysis and prospects for sustainability. Sustain Energy Technol Assess 51:101896
19. Papa G, Santo Zarnik M, Vukasinovic V (2022) Electric-bus routes in hilly urban areas: overview and challenges. Renew Sustain Energy Rev 165:112555
20. Ibarra-Rojas OJ, Delgado F, Giesen R, Muñoz JC (2015) Planning, operation, and control of bus transport systems: a literature review. Transp Res Part B: Methodol 77:38–75
21. Perumal SS, Lusby RM, Larsen J (2022) Electric bus planning & scheduling: a review of related problems and methodologies. Eur J Oper Res 301:395–413
22. Tian Q, Wang DZ, Lin YH (2022) Optimal deployment of autonomous buses into a transit service network. Transp Res Part E: Logist Transp Rev 165:102865
23. Bi Z, Keoleian GA, Ersal T (2018) Wireless charger deployment for an electric bus network: a multi-objective life cycle optimization. Appl Energy 225:1090–1101
24. Ferro G, Minciardi R, Parodi L, Robba M (2023) Optimal location and line assignment for electric bus charging stations. IEEE Syst J 17:1950–1961
25. Pelletier S, Jabali O, Mendoza JE, Laporte G (2019) The electric bus fleet transition problem. Transp Res Part C: Emerg Technol 109:174–193

26. Battaïa O, Dolgui A, Guschinsky N, Rozin B (2019) MILP model for fleet and charging infrastructure decisions for fast-charging city electric bus services. Comput Ind Eng 109336 (2019)
27. Battaïa O, Dolgui A, Guschinsky N, Kovalyov MY (2023) Designing fast-charge urban electric bus services: an integer linear programming model. Transp Res Part E: Logist Transp Rev 171:103065
28. Duan M, Liao F, Qi G, Guan W (2023) Integrated optimization of electric bus scheduling and charging planning incorporating flexible charging and timetable shifting strategies. Transp Res Part C: Emerg Technol 152:104175
29. Alamatsaz K, Hussain S, Lai C, Eicker U (2022) Electric bus scheduling and timetabling, fast charging infrastructure planning, and their impact on the grid: a review. Energies 15:7919
30. Jahic A, Plenz M, Eskander M, Schulz D (2021) Route scheduling for centralized electric bus depots. IEEE Open J Intell Transp Syst 2:149–159
31. Gkiotsalitis K, Iliopoulou C, Kepaptsoglou K (2023) An exact approach for the multi-depot electric bus scheduling problem with time windows. Eur J Oper Res 306:189–206
32. Ji J, Bie Y, Wang L (2023) Optimal electric bus fleet scheduling for a route with charging facility sharing. Transp Res Part C: Emerg Technol 147:104010
33. Tang C, Shi H, Liu T (2023) Optimization of single-line electric bus scheduling with skip-stop operation. Transp Res Part D: Transp Environ 117:103652
34. He Y, Liu Z, Song Z (2020) Optimal charging scheduling and management for a fast-charging battery electric bus system. Transp Res Part E: Logist Transp Rev 142:102056
35. Tang X, Lin X, He F (2019) Robust scheduling strategies of electric buses under stochastic traffic conditions. Transp Res Part C: Emerg Technol 105:163–182
36. Daganzo CF (2009) A headway-based approach to eliminate bus bunching: systematic analysis and comparisons. Transp Res Part B: Methodol 43:913–921
37. Fu L, Liu Q, Calamai P (2003) Real-time optimization model for dynamic scheduling of transit operations. Transp Res Rec 1857:48–55
38. Sun A, Hickman M (2005) The real-time stop-skipping problem. J Intell Transp Syst 9:91–109
39. Sirmatel II, Geroliminis N (2018) Mixed logical dynamical modeling and hybrid model predictive control of public transport operations. Transp Res Part B: Methodol 114:325–345
40. Varga B, Tettamanti T, Kulcsár B (2018) Optimally combined headway and timetable reliable public transport system. Transp Res Part C: Emerg Technol 92:1–26
41. Chow AH, Li S, Zhong R (2017) Multi-objective optimal control formulations for bus service reliability with traffic signals. Transp Res Part B: Methodol 103:248–268
42. Zimmermann L, Coelho LC, Kraus W, Carlson RC, Koehler LA (2021) Bus trajectory optimization with holding, speed and traffic signal actuation in controlled transit systems. IEEE Access 9:143284–143294
43. Sciarretta A, De Nunzio G, Ojeda LL (2015) Optimal ecodriving control: energy-efficient driving of road vehicles as an optimal control problem. IEEE Control Syst Mag 35:71–90
44. Padilla GP, Weiland S, Donkers MCF (2018) A global optimal solution to the eco-driving problem. IEEE Control Syst Lett 2:599–604
45. Flores Paredes J, Padilla Cazar GP, Donkers MCF, Implementation and experimental results (2019) A shrinking horizon approach to eco-driving for electric city buses. IFAC-PapersOnLine 52:556–561
46. Varga B, Tettamanti T, Kulcsár B (2019) Energy-aware predictive control for electrified bus networks. Appl Energy 252:113477
47. Lacombe R, Gros S, Murgovski N, Kulcsár B (2020) Hierarchical control of electric bus lines. IFAC-PapersOnLine 53:14179–14184
48. Lacombe R, Gros S, Murgovski N, Kulcsár B (2021) Bilevel optimization for bunching mitigation and eco-driving of electric bus lines. IEEE Trans Intell Transp Syst 23:10662–10679
49. Pasquale C, Sacone S, Siri S, Ferrara A (2022) Optimal charging and speed control of electric buses based on traffic flow predictions. In: Proceedings of IEEE 6th IEEE conference on control technology and applications, pp 1011–1016

50. Pasquale C, Sacone S, Siri S, Ferrara A (2022) Traffic-prediction-based optimal control of electric and autonomous buses. IEEE Control Syst Lett 6:3331–3336
51. Pasquale C, Sacone S, Siri S, Ferrara A (2023) Multi-objective optimization of electric autonomous bus trajectories based on the *ε-constraint method*. In: Proceedings of the 31st mediterranean conference on control and automation, pp 472–477
52. Pasquale C, Sacone S, Siri S, Ferrara A (2023) Decentralized control of intercity electric automated buses via time-varying objective prioritization. In: Proceedings of the 62nd IEEE conference on decision and control, pp 4130–4135
53. Yu L, Li K, Kuang Z, Wang Z (2023) Decentralized motion planning for intelligent bus platoon based on hierarchical optimization framework. Transp Res Part C: Emerg Technol, vol 148, 104025
54. Siri S, Pasquale C, Sacone S, Ferrara A (2021) Freeway traffic control: a survey. Automatica 130:109655
55. Piacentini G, Goatin P, Ferrara A (2020) A macroscopic model for platooning in highway traffic. SIAM J Appl Math 80:620–638
56. Pasquale C, Sacone S, Siri S, Ferrara A (2018) A new Micro-Macro METANET model for platoon control in freeway traffic networks. In: Proceedings of IEEE 21st international conference on intelligent transportation systems, pp 1481–1486
57. Kotsialos A, Papageorgiou M, Diakaki C, Pavlis Y, Middelham F (2002) Traffic flow modeling of large-scale motorway networks using the macroscopic modeling tool METANET. IEEE Trans Intell Transp Syst 3:282–292
58. Fiori C, Montanino M, Nielsen S, Seredynski M, Viti F (2021) Microscopic energy consumption modelling of electric buses: model development, calibration, and validation. Transp Res Part D: Transp Environ 98:102978
59. El-Taweel NA, Zidan A, Farag HEZ (2021) Novel electric bus energy consumption model based on probabilistic synthetic speed profile integrated With HVAC. IEEE Trans Intell Transp Syst 22:1517–1531
60. Bracco S, Barbagelata C, Bianco G, Casati C, Siri S, Siri E (2021) Simulation models for the evaluation of energy consumptions of electric buses in different urban traffic scenarios. In: Proceedings of the 60th international conference on ecological vehicles and renewable energies
61. Leone C, Piazza G, Longo M, Bracco S (2021) Electrification of LPT in algeciras bay: a new methodology to assess the consumption of an equivalent e-bus. Energies 14:1–24

Chapter 8
E-Scooter Riding Behaviors and Risks from Naturalistic Driving Study and Crash Data Analysis

Renran Tian, Lingxi Li, Stanley Chien, Yaobin Chen, and Rini Sherony

Abstract E-scooters are becoming increasingly popular as a convenient, fun, and environment-friendly micro-mobility option, especially among younger generations. As the numbers of e-scooter riders increase across cities and towns, related crashes and injuries increase at the same time. This chapter presents a summary of the major results and findings obtained from a multi-year, large-scale naturalistic study that addresses the following important issues: (1) Baseline moving patterns of e-scooters in diverse road environments and location; (2) interaction of e-scooter riders with vehicles and other road users in different scenarios; and (3) the common scenarios for crashes or near-misses involving e-scooter riders. The collected data and analytical results can be used to develop behavior prediction models for e-scooter riders which will help support the development of automated driving systems in challenging urban environments to improve road safety.

8.1 Introduction

8.1.1 Background

With the increased awareness of environmental sustainability, advances of technologies, and changes in lifestyles and city designs, micro-mobility has rapidly proliferated in the US cities. More and more people are accepting and using the new mobility options, such as e-scooters and bikes, especially among the younger generations. The flexible micro-mobility options are best recognized to address first- and last-mile

R. Tian
North Carolina State University, 915 Partners Way, Room 4121, Raleigh, NC 27606, USA

L. Li · S. Chien · Y. Chen (✉)
Purdue University, 723 West Michigan St., SL 174, Indianapolis, IN 46202, USA
e-mail: chen62@purdue.edu

R. Sherony
Toyota Motor North America R&D, 1555 Woodridge Ave, Ann Arbor, MI 48105, USA

P. Ioannou and A. A. Malikopoulos (eds.), *Transportation Mobility in Smart Cities*, Springer Tracts on Transportation and Traffic 21,
https://doi.org/10.1007/978-3-031-64769-7_8

problem, fill the gap between the rider's home or destination and public transport stops, or complete short trips. They are also usually considered as fun and interesting activities. Compared with bikes, a more traditional two-wheel transportation option, electric scooters (e-scooters) have quickly spread in many towns and cities in the U.S. and across the world as a convenient, cost-effective, and clean mobility option. As early as 2017, 85 cities worldwide adopted e-scooter-sharing systems [1]. Since then, e-scooter systems have widely been used more than bike-sharing systems in the U.S., accounting for 63% of a total of 136 million shared micro-mobility trips in 2019 [2].

In a survey of 7,000 people conducted by Populus in July 2018, 70% of the respondents had a favorable view of e-scooters. This result complies with a recent study published in January 2021 [3], in which researchers found that 73% of survey respondents would or may use e-scooters. In another Qualtrics survey completed in August 2018, over 500 adults across the US cities from Boston to Los Angeles were asked about e-scooters. All of the respondents were familiar with various e-scooter share companies, while 176 respondents had used the programs personally. In addition, 55% of all respondents and 72% of those who have used e-scooters believed that the e-scooter sharing programs would be a lasting innovation. Among those who have taken rides with e-scooters, 88% of people were satisfied or extremely satisfied with the experience, 85% said they were happy with the availability of e-scooters, and 87% planned to continue using e-scooters. Based on existing studies [3–9], potential factors affecting the attitudes toward the acceptance of e-scooters include gender, age, race, education level, income level, perceived hedonic quality, prior experiences of ride-hailing services, and familiarity with e-scooters. While economic benefits, locational benefits, and sustainability are driving e-scooter usage [4], infrastructure, weather, and safety might be the top concerns [3].

Rental e-scooters can usually run at a maximum speed of about 10–20 mph. In general, e-scooters are suggested to be ridden in bike lanes or share the road with cars. However, people may also ride e-scooters on sidewalks. Currently, e-scooters are required to follow all traffic rules, but there are no specified regulations established for this new type of transportation tool. This poses various safety concerns for both e-scooter riders and other road users.

Compared with the systematic crash data collection and analysis for cyclists, crash data analysis for e-scooter riders are still relatively new. There is no national crash data for e-scooters yet, but many studies have tried to investigate the problem with various data sources. For every 100 k e-scooter trips, it is estimated that 20–25 injuries result in emergency room (ER) visits [10, 11]. Based on this ratio, there might be around 17,000–21,000 ER-level injuries in 2019 nationwide. By summarizing news reports, a reporter found that 29 people worldwide have been killed and reported in e-scooter crashes from the summer of 2018 to the end of 2019, mostly in the U.S. Though most e-scooter crash victims are e-scooter riders, there are two pedestrians (including a 90-year-old woman) and one cyclist killed during the crashes too [12]. 169 e-scooter-involved crashes were identified between 2017 and 2019 from U.S. media reports [13]. However, media reports may not cover all cases and can be biased. According to FHWA [14], 65 e-scooter-related injuries at Virginia Beach were documented by emergency services during the research period. However, none of these cases were reported in the media. So, the number of severe e-scooter crashes is likely much more than the listed numbers. Based on a study in an emergency department [15], the researchers found that among the randomly sampled crashes causing the ER visits, the e-scooter-related crashes jumped from 0 to 25% among all the vehicle-related injuries from 2018 to 2019, when all other sources like cars, bicycles, and motorbikes remained unchanged. In an earlier study [16], the researchers also found a significant increase of 500% in e-scooter-related ER visits for 2 consecutive years after the launch of e-scooter rental services.

From limited e-scooter crash data presented above, it is obvious that e-scooter usage poses potential injury risks to riders and other vulnerable road users such as pedestrians. This includes the scenarios where the e-scooter operator hits the pedestrians [17, 18] or tripped them when the e-scooter was parked inappropriately [8]. The study in [17, 18] argues that reduced visibility and quiet movement may contribute to crashes between e-scooters and pedestrians. The researchers in [8] found that the sidewalk is very often blocked from improperly parked dockless e-scooters. Road infrastructures like street-side attributes, sidewalk attributes, and non-street-side attributes have various effects on the e-scooter parking behaviors.

8.1.2 Related Studies

A naturalistic study was conducted to investigate the interaction of e-scooters and the riding environments, such as surrounding objects and potholes, via mobile sensing data [19]. The researchers developed a mobile sensing system, including four low-cost sensors: (1) GPS, (2) IMU, (3) 2D LiDAR (12 m max distance, 5.5 Hz scan frequency), and (4) portable camera. All the collected data were from the sensors facing forward. The research used proximity to objects as the measurement of riding circumstances. Instantaneous proximity maps were aggregated for each trip as the riding complexity measurements. Riding complexity, vibration, and velocity were surrogate riding safety metrics. The main goal was to find the relationships between these safety metrics, riding locations, and surface conditions. Due to the data collection and analysis limitations, this research did not focus on risky interactions or e-scooter riders' behaviors. The study was not able to recognize surrounding objects, their movements, and their detailed interactions. In addition, the study did not record the behavior of the e-scooter riders. All data were aggregated to the whole riding duration for a coarse-level data analysis. The hardware also limited the data quality.

Although there has been very limited naturalistic study on e-scooters, there were some studies that were conducted on e-cyclists, mainly in Europe. Using e-bicycles instrumented with GPS, video cameras, IMUs, pedal sensors, and brake force sensors, the study in [20] collected 88 self-reported safety–critical events from 410 total trips (86 h of data). Then each event was analyzed using video data about the environmental scenario and threats. The results showed that pedestrians (31%), light vehicles (21%), and bicycles (18%) are the main threats to e-cyclists in critical events. By comparing with the baseline data, the only significant factor affecting the existence of critical events is road location. The intersection is associated with more self-reported safety concerns.

8.1.3 Research Objectives and Chapter Organization

As discussed above, due to limited data sources the e-scooter-related safety issues have been investigated primarily in prior studies based on hospital data [15], survey data, road-side observations [10], and video recording and media reports [13]. These heterogeneous, biased, and scrappy data prevent researchers from understanding a complete picture of e-scooter micro-level behaviors on the roads, depicting the details about crashes, and predicting risks. Unlike vehicle crashes, there is no standard crash database available so far for e-scooter-related crashes. The naturalistic study can help to address the deficiencies with the capability to collect detailed quantitative behavior data in the naturalistic matter. With appropriately designed naturalistic driving experiments, the effects of individual differences can also be modeled and investigated. The naturalist study collects detailed quantitative behavior data in a naturalistic riding setting. In addition, the effects of individual differences are also modeled and studied.

To our best knowledge, there has not been any systematic research that focuses on the moving behavior and crash scenarios related to e-scooters for public use in an open-road environment. This chapter presents a summary of the major results and findings obtained from a multi-year, large-scale naturalistic study that addresses the following important issues:

- Baseline moving patterns of e-scooters in diverse road environments and locations.
- Interaction of e-scooter riders with vehicles and other road users in different scenarios.
- The common scenarios for crashes or near-misses involving e-scooter riders at different risk levels.

The collected data and analytical results can be used to develop behavior prediction models for e-scooter riders which will support the development of automated driving systems in challenging urban environment to improve road safety. The remaining of this chapter is organized as follows. Section 8.2 presents the proposed naturalistic data collection systems, process, and methods. A summary of the collected data is given. Section 8.3 discusses data analytics methodology and results. Section 8.4 presents e-scooter-related crash data analysis. Finally, conclusions are given in Sect. 8.5.

8.2 Naturalistic Data Collection

As described above, it is important to address the safety concerns related to e-scooter riders and their interaction with vehicles and other road users. Such research efforts require a large amount of behavior and interaction data of e-scooter riders in the naturalistic road environment, especially the micro-level movement and interaction data. There are three typical approaches to collecting such data. The first approach is to use a vehicle-based data collection system equipped with cameras and other sensors to collect e-scooter rider's behavior data via naturalistic driving. This method can be less efficient for the ego vehicle to encounter a large number of e-scooters during the driving on the road, especially in less busy roads or different days and times on a day. The second approach uses road-side-installed cameras and other sensors to collect data around the location. However, this method may miss details which are easily visible from the vehicle view angles or shorter distances. More importantly, road-side cameras can only be installed in limited predetermined road locations and are difficult to capture the different behaviors of e-scooters from all road locations.

The third approach is to directly collect the data from e-scooter by an e-scooter-based data collection system. Compared with the previous two methods, this approach has several advantages: (1) more efficient to encounter a large number of vehicles on the road; (2) capable of collecting data in all road locations; (3) easy to obtain direct measurements of the e-scooters' micro-level behavior; (4) accurate recording of surrounding objects; and (5) possible to record the riders' demographic backgrounds via controlled human subject experiments to better understand individual differences. The collected movement data of surrounding road users (vehicles, pedestrians, and other road users) can be then used to reconstruct the interactions between the e-scooter and vehicles and detect potential conflicts. Hence, in this study, we proposed to use the vehicle-based and e-scooter-based data collection methods to achieve the proposed goal.

8.2.1 Design of Data Collection Systems

To ensure the data quality and improve the richness of the information contained in the naturalistic data collection, vehicle- and wearable e-scooter-based data acquisition systems were designed and implemented. Each system consists of three types of sensors: cameras for video data; LiDAR for distance and IMU information; and GPS for latitude, longitude, and altitude data. For portability, the e-scooter-based system uses smaller sized cameras and a computing platform compared to the vehicle-based system. Both systems run on the Ubuntu/ROS platform.

The vehicle-based system includes six FLIR cameras which are strategically placed to cover 360° view of the field. The system is equipped with a 64-beam 360° Ouster LiDAR sensor for collecting detailed information about the distance and position of objects in the environment. These data are used to create highly accurate 3D models of the area. It also includes a Reach Emlid GPS for accurate determination of the location of the vehicle as it moves through the environment. Finally, all the data is processed by a high-powered desktop computer that stores and analyzes the data. The combination of these different sensors and the computer's processing power allows for the creation of highly detailed and accurate models of the environment.

The data collection system must cover 360° of view around the vehicle, which requires proper placement of the sensors on the vehicle. Four cameras are placed at the corners on the top of the vehicle, and two cameras are placed at the center, one front-facing and one back-facing. The cameras in the corners are equipped with wide-angle lenses, and the cameras at the center are equipped with 40° lenses. The LiDAR is placed at the center of the vehicle's roof to provide maximum coverage around the vehicle. The GPS antenna is required to be at least 50 cm away from any electronic disturbance and has a metallic base of at least 100 cm in diameter. The best position to place the GPS is at the back of the vehicle, on the roof. The complete system was described in detail in [21] by the authors. Figure 8.1 shows the vehicle-based data collection system used in this study. Figure 8.2 shows the placement of sensors on the data collection vehicle.

The e-scooter-based system is required to:

- Collect accurate motion dynamics of the e-scooter.
- Fit into different current and future e-scooter designs.
- Collect movement data of surrounding road users like cars and pedestrians to reconstruct the interactions and detect potential conflicts.

There are several constraints for selecting sensors and other devices to be mounted on an e-scooter such as the weight and duration of operation of the system. Since the system was designed to collect data for the surrounding objects, the final system consists of three USB cameras, a 3D LiDAR with integrated IMU, an RTK GPS unit, and an NVIDIA Jetson T $\times$ 2 development computer board for data collection. To hold and protect the system, all system components, except the sensors, are enclosed

Fig. 8.1 Vehicle-based data collection system

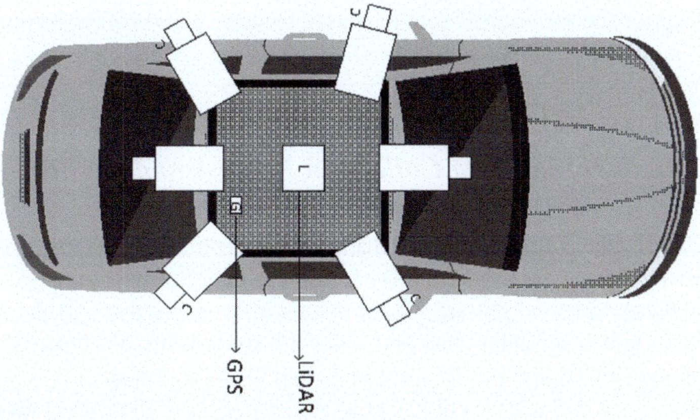

Fig. 8.2 Placement of sensors on the data collection vehicle

in a wearable case. The integrated system can collect data continuously for hours, meeting all the requirements. A detailed explanation of the wearable e-scooter-based data collection can be found in [22]. Figure 8.3 shows the e-scooter-based system.

To make the data acquisition systems to operate more efficiently and reliably, we performed pilot driving experiments under different driving scenarios and road environments. The data collected from the pilot study were analyzed and used to address critical issues related to data quality and accuracies including LiDAR-camera calibration accuracy, LiDAR resolution for the vehicle-based system, camera coverage

Fig. 8.3 Wearable e-scooter-based data collection system

and recording for the wearable e-scooter-based system, and time synchronization of the camera image and LiDAR point cloud.

8.2.2 Vehicle-Based Data Collection Process and Results

For vehicle-based data collection, an IRB approval was obtained for subject recruiting and naturalistic driving experiments. By following the approved process, a total of 55 subjects were recruited among 72 applicants after going through the screening process based on demographic data and criteria for eligibility. Of those 55 subjects between the ages of 18 and 66, 35 were male and 20 were female.

In this study, the subjects were asked to self-report how they view their driving style on a scale of 1–10, with 1 being very conservative, and 10 being very aggressive. About 80% of subjects drive conservatively. 68 data collection sessions were completed. Each session was designed for 20–40 min, depending on traffic and road conditions. The research team evaluated the roads of downtown Indianapolis to determine the areas preferred for the data collection. The subjects can choose their own routes among the destinations given to them. The subjects also had options to follow predetermined routes if they were unfamiliar with downtown Indianapolis. We conducted naturalistic data collection on different days and different times during a day to increase scene variabilities.

A summary of the data collected is as follows:

- Total time: 35.24 h—an average of 31 min per session.
- Total data size: 7.52 TB of raw data.
- Total number of frames: 7.6 million (each frame has 6 camera images, LiDAR point cloud, and GPS).
- Estimated total number of e-scooters: 400.

8.2.3 E-Scooter-Based Data Collection Process and Results

The e-scooter-centered data collection uses the wearable system developed in this study. It was observed that riding an e-scooter can be challenging for people who lacks familiarity with e-scooters or other products that generate translatable skills. Due to the high speed of surrounding vehicles on the road and the possibility of riding the e-scooter with vehicles on the road, inexperienced riders may be at an elevated risk of a crash and/or injury. After a careful evaluation of the e-scooter-centered data collection process, it was felt that this process may pose safety concerns if we ask the subjects to ride the e-scooter with the wearable data acquisition system in the city center. With these restrictions in mind, we completed the e-scooter-centered data collection with experienced team members riding the e-scooter alongside vehicles on the road, as per the rule. This data collection was conducted in the same areas as in the vehicle-based data collection. A total of 14 e-scooter-centered data collection sessions were conducted with 20–60 min long for each session.

Below is a summary of the data collected by the e-scooter-based system:

- Total time: 7.5 h.
- Total data size: 1 T.B. of raw data.
- Total number of frames: 900,000 (each frame has 3 camera images, LiDAR point cloud, and GPS).
- The estimated nearby vehicles: >300.

8.2.4 Nationwide Naturalistic Driving Data Collection

As was described in Sects. 8.2.2 and 8.2.3, the naturalist driving data collection was carried out in the Indianapolis downtown area with recruited subjects and research team members. The data size is relatively small and less representative in terms of locations, drivers, e-scooter riders, road conditions, infrastructure design, and driving environments. To enhance the prior naturalistic data collection, we conducted a national naturalistic driving study in several major geographical regions in the U.S. The same vehicle-based data collection system described in Sect. 8.2.1 was used with our research team members as drivers.

8.2.4.1 Selection of Locations

Several cities were proposed for data collection in the South, West, East, and Central regions of the U.S. The fundamental requirements are (1) the population of e-scooter riders is large and (2) accurate GPS information can be obtained in these regions for data collection (ideally with centimeter-level accuracy). GNSS (Global Navigation Satellite System) stations can provide a centimeter-level accuracy. Thus, we investigated data collection locations with GNSS station information at the following link: https://geodesy.noaa.gov/CORS/. The data collection locations were chosen based on the e-scooter population and availability of nearby GNSS stations. Also, the locations were chosen to provide a good representation of different regions in the U.S., so that the collected data would help to understand how the e-scooters are used in different environments.

When selecting locations for data collection, the number of e-scooters already present in an area is an important consideration. A high number of existing scooters in a location can indicate a high demand for the service in that area. This means that collecting data in that location would be more likely to be successful, as there is already a built-in user base for the service. Additionally, if the area already has a high density of e-scooters, it may be an indication that the infrastructure and regulations are already in place to support the use of e-scooters, which can make it easier to launch and operate the services.

Table 8.1 Detailed breakdown of the nationwide naturalistic data collection

	Total	Austin	San Diego	Boston	D.C	Indy
Hours of data collected	200	50	30	35	45	30
Days spent on data collection	53	14	17	7	7	6
Total number of e-scooters tagged	8,750	3,920	937	136	3,246	511
Total number of bikes tagged	8,160	1,787	1,627	1,049	3,296	401
Total number of interesting objects tagged	2,143	1021	255	256	548	63
Total raw data collected (Terabytes)	30.6	7.6	6.8	6.6	7.1	2.5

In summary, four major cities that satisfy these requirements were selected for the nationwide naturalistic data collection, which includes Austin, Texas (South), San Diego, California (West), Boston, Massachusetts (East), and Washington D.C. (East).

8.3 Summary of Data Collection Results

For the nationwide naturalistic data collection, a total of 200 h of naturalistic driving data were collected. Data collection was completed between May and August 2022. 8,750 e-scooters were tagged for further data analysis. 8,160 bikes and 2,143 of not commonly seen objects (such as unmanned delivery vehicles, horse wagons, etc.) on the road were also tagged. The detailed breakdown of the data collected at each geographical location is summarized in Table 8.1.

8.4 Naturalistic Data Analysis

The collected naturalistic driving data, from both the car-centered and e-scooter-centered perspectives, are analyzed through a multi-step process to study the behavior patterns and associated risks of e-scooter riders. This section describes the two main aspects of data analysis: preprocessing pipeline and data analysis method. The analysis results are provided.

8.4.1 Preprocessing Pipeline

The primary objective of data preprocessing is to assign motion and trajectory labels to all the visually captured e-scooter riders throughout the encountering cases to reconstruct the driving scenes. The whole process is fully automated with the goals to perform multi-modal data fusions for LiDAR, Camera, and GPS data efficiently and accurately.

8.4.1.1 Multi-modal Data Fusion

The detailed pipeline for multi-modal data fusion is described in one of the earlier papers [23], and this section provides a brief review of the process, as shown in Fig. 8.4. The pipeline contains four main steps. The first step is LiDAR-camera sensor fusion for object detection and tracking. This fusion process leverages system calibration outcomes to facilitate automated object detection. Computer vision algorithms are employed to detect and track vehicles and pedestrians. Consistency is maintained across consecutive frames to establish object continuity throughout the video sequence. Concurrently, we compute the 3D coordinates of all identified and tracked objects relative to the data collection platform at each time frame.

Subsequent to the first step, the second step is to compute local trajectories of the objects of interest. Object positions in the local coordinate system of the data collection platform are determined frame by frame. This information undergoes outlier removal, interpolation for data gaps, and noise reduction through filtering. Ultimately, local trajectories for these objects within the coordinate systems of the corresponding vehicles (e.g., cars or e-scooters) are established.

The third step of the preprocessing pipeline pertains to the fusion of GPS and IMU data for ego vehicle trajectory generation. GPS coordinates are used to geolocate the global positions of the vehicle (e.g., car or e-scooter). The accuracy of GPS-derived coordinates is further enhanced by incorporating data from the Inertial Measurement Unit (IMU). The final output encompasses the global trajectories of the data collection platform in the world coordinate system.

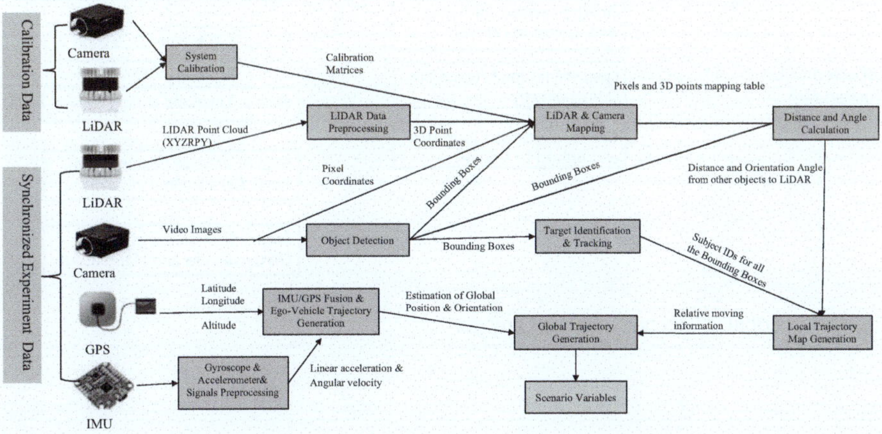

Fig. 8.4 Multi-modal sensor fusion pipeline for preprocessing raw inputs

The fourth step involves global scene reconstruction. Through a series of coordinate transformations, we amalgamate the global trajectories of the data collection platform with the local trajectories of all surrounding objects relative to the platform. The reconstructed scene encompasses the global trajectories of the ego vehicle (car or e-scooter) and all adjacent road users. Additionally, it facilitates the calculation of relative positions and distances among these objects across different time frames.

8.4.1.2 Distributed Data Preprocessing Process

To enhance the efficiency of data preprocessing, we optimized the preprocessing pipeline described above using a hybrid approach that leverages both local computing resources and the Google Cloud computing platform. Figure 8.5 illustrates the distribution of computational tasks between these two resources. All the raw data collected are initially stored in the form of ".rosbag" files. These files undergo an initial processing phase conducted locally within our laboratory environment, aimed at extracting images, LiDAR point cloud data, and GPS data separately. This step is referred to as "data preparation." The primary reason for conducting this initial step locally is the sheer volume of raw files, which are more efficiently handled locally. All the extracted files from the data preparation are compressed to further reduce their file sizes before being uploaded to Google Drive cloud storage.

Once all the files have been successfully uploaded, the remaining computational tasks are executed using the cloud computing platform to increase scalability and integration of heterogeneous computing frameworks. Consequently, data preprocessing speed is much improved through the optimization of the entire process.

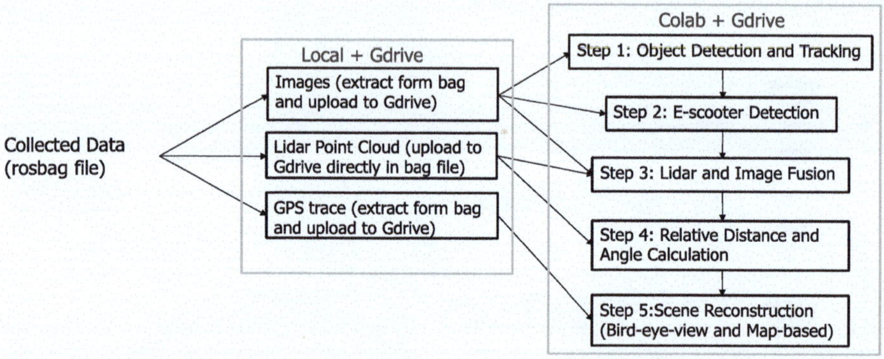

Fig. 8.5 Distributed data preprocessing process

8.4.2 *Data Analysis Method*

For each preprocessed encountering cases, trajectories of the e-scooter riders and vehicles are generated. Based on the trajectories, a list of behavior variables is calculated, including gap time and time to collision, shortest distance, and median speeds of the car and the e-scooter rider during the countering process.

8.4.2.1 Trajectory Generation

The trajectory generation method involves the following steps:

1. Smoothing relative distances: Initially, relative distances between the ego system (e.g., a vehicle) and detected objects are smoothed under the assumption that these relative distances do not change abruptly. The Relative Nearest Neighbor Consistency (RANSAC) algorithm is used to fit the best curve for the X, Y, and Z relative distances. This smoothing process not only improves data consistency but also allows for synchronization with GPS signals by estimating relative distances based on GPS coordinates over time.
2. Conversion to local Cartesian coordinates: The method assumes that we have GPS-based trajectories for the ego vehicle. Simultaneously, LiDAR points provide relative distances (X, Y, Z) to detected objects. Given the LiDAR's attachment to the ego vehicle and knowledge of the ego vehicle's GPS location, the location of detected objects (e.g., e-scooters) is inferred. This is done by converting GPS coordinates into local Cartesian coordinates.
3. Calculation of heading angles: Heading angles are calculated for each coordinate using information from the current point and the point at the next timestamp. The assumption is that the heading angle of the ego vehicle matches that of the LiDAR sensor.
4. Projection onto local Cartesian coordinate system: Using the heading angle of the LiDAR sensor and the relative location of the detected object (e-scooter), the method projects the location of the object onto the local Cartesian coordinate system. This step ensures that both the ego vehicle's trajectory and the detected object's trajectory are in the same local coordinate system.

8.4.2.2 Calculation of Gap Time and Estimated Time to Collision

Gap time and estimated time to collision (TTC) (there were no crashes in the data) are defined based on the positions, speeds, and heading angles of both an E-scooter and a vehicle. These two entities are considered to have potential conflicts if their coast trajectories (projected paths if they maintain their current courses) intersect. The gap time is the time difference between their expected arrivals at the point of intersection. If this time difference is less than three seconds, it raises concern for a potential collision, and the TTC is calculated as the average arrival time of both entities. Inaccurate TTCs resulting from imprecise GPS coordinates are manually excluded.

8.4.2.3 Estimation of the Shortest Distance and Median Speeds During the Encountering Process

The determination of the closest distance between the vehicle and the e-scooter is a straightforward process. With synchronized GPS coordinates available for both the e-scooter and the vehicle from prior preprocessing, we calculate the difference in their distances at each timestamp. Subsequently, we arrange these distance differences in ascending order and identify the closest distance within the dataset.

As mentioned previously, we possess GPS coordinates and timestamps for both the vehicle and the e-scooter. Leveraging this information, we estimate the speed of either the vehicle or the e-scooter by initially computing the difference in distance between consecutive GPS coordinates and then dividing it by the corresponding time difference. This calculation provides us with the estimated speed for each timestamp. Subsequently, we determine the median speed for the entire trajectory sequence. The rationale behind using the median value, as opposed to the average, to represent the trajectory lies in its robustness. The median is less susceptible to the influence of outliers or noisy data points, ensuring a more reliable representation of the trajectory.

8.4.3 Results

After scenario variables are calculated for all cases, the results are summarized and reported in this subsection from vehicle-centered and e-scooter-centered perspectives, respectively.

8.4.3.1 Results from Vehicle's Perspective

Table 8.2 presents a comprehensive overview of scenario variables derived from the analysis of 203 vehicle-e-scooter-rider encounters, with a focus on vehicle-centered data. Several noteworthy observations emerge. The average shortest distances between e-scooter riders and the ego vehicle are typically ranged from 15 to 16 m. The minimum recorded distance during data collection exceeds 3 m. Throughout these encounters, the average vehicle speed hovers around 12 mph, while the average e-scooter speed is approximately 8.5 mph. These parameter values offer insights into traffic conditions within downtown areas of urban centers. Furthermore, they underscore the potential of micro-mobility solutions to provide faster and more convenient urban transportation alternatives. Instances of encounters are also captured when vehicles and e-scooters operate at significantly higher speeds, approximately 30 and

20 mph, respectively. This highlights the potential for encounters at elevated and potentially hazardous velocities.

The average minimum time to collision (TTC) computed for all potential conflict cases is 3.77 s, with the lowest TTC recorded at approximately 1 s. Some cases involve close proximity and carry heightened risks, which will be subject to further analysis in subsequent sections. The minimum gap time, averaged across all cases, stands at 3.09 s. This value provides insight into the typical time difference between cars and e-scooter riders as they navigate shared zones where their trajectories intersect in city road environments. Given the average speeds of these two entities, it can be inferred that they tend to maintain approximately 10–15 m when crossing paths.

After observing the distributions of key scenario variables based on the analysis of the 203 encounters using vehicle-centered data (shown in Fig. 8.6), we can reveal that vehicle speeds predominantly cluster at very low values (indicating stationary vehicles) or within the range of 10–20 mph when vehicles are in motion within urban settings. In contrast, e-scooter speeds exhibit a more dispersed distribution covering a wider range. The distribution of minimum distances follows a normal-like pattern, with a right-skewed profile.

Out of the 203 analyzed encounter cases from the vehicle-centered dataset, 26.1% of them, totaling 53 cases, qualify as potential conflict situations. These encompass 10 cases with TTC falling within the 0–2 s range, 28 cases with TTC in the 2–4 s range,

Table 8.2 Scenario variables for all vehicle-centered cases

	Average	Minimum	Maximum
Minimum distance (meter)	15.29	3.14	61.82
Ego vehicle median speed (m/s)	5. 3	0	14.61
Ego vehicle median speed (mph)	11.86	0	32.68
E-scooter median speed (m/s)	3.79	0	8.99
E-scooter median speed (mph)	8.48	0	20.11
mTTC (s) (minimum TTC)	3.77	0.96	13.85
Minimum gap time (s)	3.09	0.03	16.91

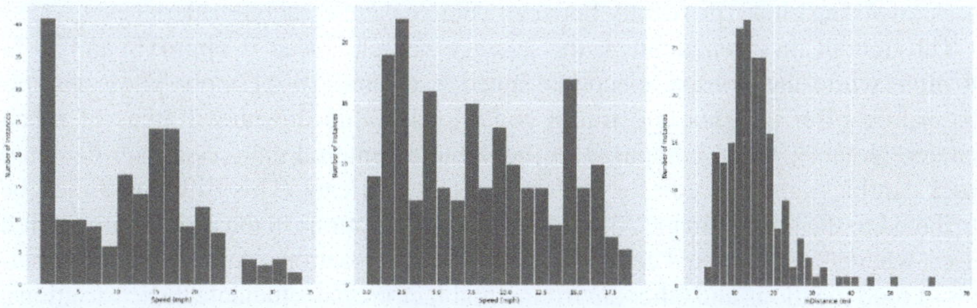

Ego-Vehicle Speed **E-Scooter Speed** **Shortest Distance**

Fig. 8.6 Distributions of scenario variables from vehicle-centered data analysis

Table 8.3 Comparison of scenario variables between potential conflict and baseline cases using vehicle-centered data

	Potential conflict cases	Baseline cases
Average minimum distance	16.22 m	14.94 m
Average ego vehicle median speed	18.25 mph	9.51 mph
Average e-scooter median speed	8.43 mph	8.61 mph
Average minimum gap time	0.77 s	5.54 s

and the remaining cases with TTC exceeding 4 s. Table 8.3 presents a comparative analysis of key scenario variables between potential conflict cases and baseline (non-potential conflict) cases. The findings indicate that the average minimum distance and the average e-scooter median speed remain similar across both categories of cases. This suggests that, regardless of whether a situation is potentially conflicting or not, e-scooter speeds exhibit minimal variation, and the relative distances between objects remain consistent.

Furthermore, the table reveals a significant disparity in average vehicle speeds between potential conflict cases, registering at 18.25 mph, and baseline cases, which record an average speed of 9.51 mph. This observation suggests that higher vehicle speeds are associated with an increased likelihood of potential conflicts. It also implies that potential conflict scenarios tend to occur when vehicles are traveling at relatively higher speeds. A more detailed analysis of specific road locations, such as mid-block or intersections, may provide further insights to elucidate these results.

8.4.3.2 Results from E-Scooter Rider's Perspective

The comprehensive analysis results of the scenario variables from 182 vehicle-e-scooter-rider encounter cases are given in Table 8.4 with a focus on e-scooter-centered data. Several noteworthy findings emerge. The average shortest distances between e-scooter riders and the ego vehicle are typically ranged from 7 to 8 m. The minimum recorded distance during data collection slightly exceeds 2 m. Compared to the vehicle-centered data, the e-scooter-centered dataset captures a higher frequency of cases involving closer proximity between vehicles and e-scooter riders.

Throughout these encounters, the average vehicle speed is approximately 13–14 mph, while the average e-scooter speed is around 10–11 mph. These parameter values offer insights into traffic conditions within downtown areas of urban centers. Notably, when compared to the vehicle-centered data, e-scooter-centered cases exhibit higher speeds for both e-scooters and cars. This difference may be attributed to the predominance of mid-block road locations in the e-scooter-centered cases, whereas car-centered cases often occur at intersections. A more detailed analysis of road locations and other environmental factors may elucidate these variations. Similar to vehicle-centered data, encounters also occur at higher speeds, with some

Table 8.4 Scenario variables for all e-scooter-centered cases

	Average	Minimum	Maximum
Minimum distance (meters)	7.27	2.35	21.68
Ego vehicle median speed (m/s)	6.1	0	18.4
Ego vehicle median speed (mph)	13.65	0	41.16
E-scooter median speed (m/s)	4.8	0	10.60
E-scooter median speed (mph)	10.74	0	23.71
mTTC (s) (minimum TTC)	3.99	1.05	12.51
Minimum gap time (s)	2.26	0.01	12.76

instances involving cars traveling at around 40 mph and e-scooters at approximately 20 mph.

The average minimum TTC computed for all potential conflict cases is approximately 3.99 s, with the lowest TTC recorded at about 1 s. These values closely resemble the TTC values observed in the vehicle-centered analysis.

The average minimum gap time, considering all cases, is 2.26 s. This illustrates the average time difference between cars and e-scooter riders as they traverse shared zones, as reflected in the e-scooter-centered data. Notably, given the higher average speeds of both entities in the e-scooter-centered data, it can be inferred that they tend to maintain a distance of approximately 10–15 m when their paths intersect. This finding aligns with the observations from the car-centered data.

After checking the distributions of crucial scenario variables extracted from the analysis of 182 encounters within the e-scooter-centered dataset, the distribution of vehicle speeds exhibits a normal-like pattern, which differs from the vehicle-centered data. This pattern strongly suggests that the majority of e-scooter-centered cases were recorded during periods of vehicle movement in the vicinity.

Similarly, the distribution of e-scooter speeds displays a more normal-like distribution when compared to the vehicle-centered data. This indicates a higher degree of data homogeneity. An intriguing observation is the presence of a peak in e-scooter speed, occurring at approximately 15–20 mph in both histograms derived from the vehicle-centered and e-scooter-centered datasets.

Furthermore, the distribution of minimum distances demonstrates a similar shape between e-scooter-centered and vehicle-centered data. However, it is noteworthy that for e-scooter-centered cases, this distribution skews to the left, indicating smaller distances between e-scooter riders and vehicles in these encounters.

Among all the 182 encountering cases analyzed from the e-scooter-centered data, 45.1% of them are 82 potential conflict cases. Table 8.5 presents a comparative analysis of key scenario variables between potential conflict cases and baseline (non-potential conflict) cases, following a similar pattern to the findings observed in the vehicle-centered data.

Firstly, the results reveal that the average minimum distance and the average e-scooter median speed remain similar across both categories of cases. This consistency

Table 8.5 Comparison of scenario variables between potential conflict and baseline cases using e-scooter-centered data

	Potential conflict cases	Baseline cases
Average minimum distance	7.67 m	6.95 m
Average ego vehicle median speed	13.60 mph	13.65 mph
Average e-scooter median speed	11.63 mph	10.07 mph
Average minimum gap time	0.84 s	3.97 s

indicates that regardless of the potential for conflict, e-scooter speeds exhibit minimal variation, and the relative distances between e-scooter riders and vehicles remain comparable.

Interestingly, in contrast to the vehicle-centered data, the average vehicle speed is also comparable between potential conflict and non-potential conflict cases in the e-scooter-centered dataset. This observation suggests a higher degree of homogeneity in the sampled cases within the e-scooter-centered data.

As expected, the average minimum gap time in potential conflict cases is significantly shorter than that in baseline cases, aligning with the results observed in the vehicle-centered analysis. This finding underscores the reduced temporal separation between vehicles and e-scooter riders in potential conflict scenarios, which is consistent with the vehicle-centered data.

8.4.3.3 Behavior Patterns in Different Encountering Geometries

In our analysis, we classify each encounter between the ego vehicle and the e-scooter rider into two fundamental categories: "crossing" and "parallel" scenarios. Further subcategorization divides "crossing" scenarios into "e-scooter crossing from right" and "e-scooter crossing from left," while "parallel" scenarios are divided into cases with "same moving directions" and cases with "opposite moving directions." When comparing scenario variables across these distinct geometric categories, several significant findings are obtained:

- Cases involving e-scooters crossing from the left exhibit longer distances and higher e-scooter speeds compared to those crossing from the right. This aligns with the intuitive expectation that e-scooter riders tend to accelerate more swiftly toward the vehicle's path when afforded greater distances to adjust their actions. Additionally, the "crossing from left" category displays a higher incidence of potential conflict cases, indicative of increased dynamic interactions.
- Generally, "parallel" scenarios tend to feature higher vehicle speeds in comparison to "crossing" scenarios. This difference may be attributed to the specific road locations where these categories are more prevalent. Specifically, "parallel" cases often occur at mid-block locations, while "crossing" cases are frequently found at intersections. Interestingly, both "crossing" geometries exhibit similar

vehicle speeds, while "parallel" scenarios involving vehicles moving in opposite directions record higher vehicle speeds. This variance may be attributed to the influence of e-scooters on vehicle speeds in "parallel" cases with the same direction of movement.

- Among potential conflict cases, those within the "parallel with opposite direction" category and the two "crossing" categories display shorter minimum TTC (mTTC) values compared to potential conflict cases within the "parallel with same direction" category. This observation is rationalized by the increase in relative speed when the vehicle and the e-scooter move toward each other, thereby reducing the available time to avoid a potential collision.
- E-scooter speeds exhibit a distinct ranking across the geometric categories, ascending from "crossing from right" to "crossing from left," followed by "parallel with same direction," and culminating in "parallel with opposite direction." This ranking reflects a gradual increase in e-scooter speed across these categories, from low to high.

8.5 Crash Data Analysis

In Sects. 8.2 and 8.3, we discussed the processes of naturalistic data collection and analysis. Naturalistic driving studies remain a valuable tool for gaining insights into real-world driving behavior and contributing to developing effective road safety interventions. However, naturalistic driving alone cannot obtain the complete picture of real crash scenarios. Hence, it is worthwhile to conduct crash data analysis to fully understand the e-scooter-related safety issues.

In this section, we aim to investigate the crash databases to identify and analyze crash cases between vehicles and e-scooter riders. Representative crash scenarios can be reconstructed from the data analysis with a set of scenario variables. Two databases were investigated, including the Enhanced Tennessee Roadway Information System (ETRIMS) and Tennessee Integrated Traffic Analysis Network (TITAN). ETRIMS is a query-based system that delivers information on roadways and crashes. TITAN is an online system for collecting and storing crash data reported by law enforcement authorities. Each crash data in TITAN is accompanied by crash reports filled out by the reporting officer. The crash reports also describe the crashes using drawings and narratives that specify the exact location, direction of travel, and maneuvers of those involved. The analysis of e-scooter-related crash characteristics was accomplished by extracting information from both sources for 3 years: 2019, 2020, and 2021.

E-scooter crashes are reported as non-motorist or pedestrian crashes in the crash databases. Therefore, all potential e-scooter crashes were extracted from ETRIMS using the crash descriptions "Other Type Non-Motorist" and "Pedestrian crashes." Crashes containing either one of these descriptions were subsets from crashes for the 3 years. Crash characteristics such as the number of lanes and posted speed limit were also obtained. Then, all crashes obtained from the previous step were looked up

Table 8.6 Number of crashes under the two descriptions

Year	Other type non-motorist	Pedestrian crashes
2019	157 (14)	1,714 (6)
2020	63 (3)	1,359 (4)
2021	251 (23)	1,475 (13)

in TITAN using their unique IDs to verify if they were e-scooter crashes. The crash reports were searched for the keyword "*scooter*." All reports with matching keywords were reviewed in detail to ascertain if they were truly e-scooter crashes. Specifically, the crash narratives and diagrams were studied. Crashes that could be identified with certainty as e-scooter crashes were retained for further analysis. Lastly, crash typing was completed on the retained crashes to obtain maneuvers for the motorist and non-motorist involved in the crashes. Crash typing was completed using the Pedestrian and Bicycle Crash Analysis Tool (PBCAT) version 3. PBCAT is an open-access crash typing application developed by the Federal Highway Administration.

A total of 63 e-scooter crashes were identified between 2019 and 2021. A breakdown of the number of crashes under the two crash descriptions is presented in Table 8.6. The number of e-scooter crashes identified from a review of crash reports is presented in parenthesis.

A summary of the main crash variables obtained from PBCAT is presented below, which can help us better understand the crash characteristics. A description of variables obtained from PBCAT is also provided.

8.5.1 Crash Locations

The summary of crash location distribution is shown in Fig. 8.7. It is not difficult to see that most crashes occurred at intersections.

8.5.2 Motorist and E-Scooter Maneuver

The motorist maneuver includes going straight, turning right, turning left, unknown maneuvers, and other maneuvers. The summary of motorist maneuvers before crashes is shown in Fig. 8.8. The relative direction of travel for e-scooters relative to motorists just before the crashes is shown in Fig. 8.9. As we can observe, going straight is the most frequent motorist maneuver before crashes. Crossing path from motorist's right is the most frequent direction of travel for e-scooters relative to motorists just before the crashes.

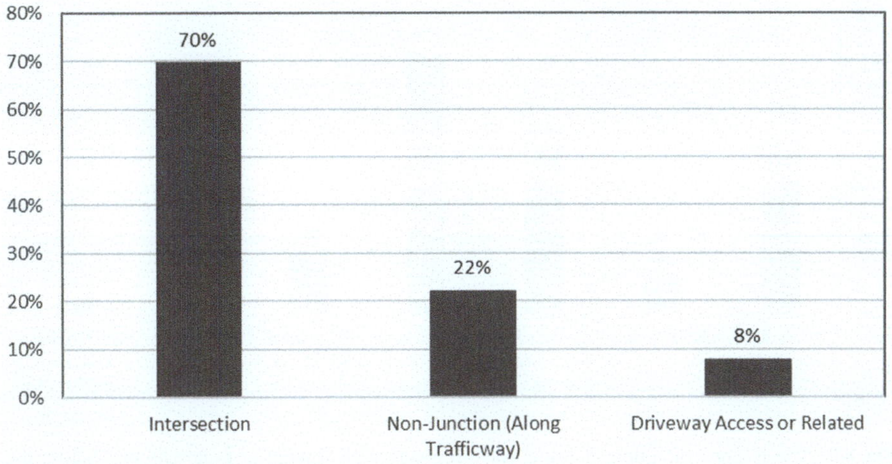

Fig. 8.7 Location of e-scooter crashes

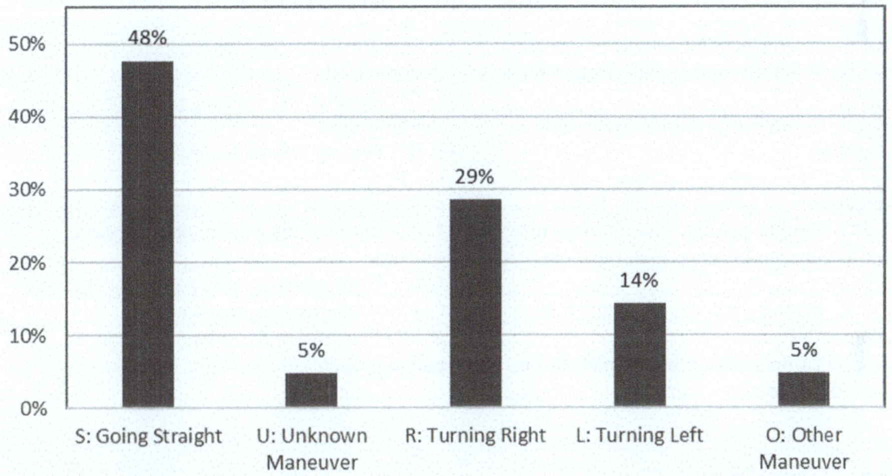

Fig. 8.8 Motorist maneuvers just before crashes

8.5.3 *Crash Type*

Crash types between motorists and non-motorists are summarized in Table 8.7. The first letter denotes the motorist maneuver, and the last two letters denote the relative traffic direction of the e-scooters.

The distribution of different crash types is shown in Fig. 8.10. S-CR is the most frequent crash type, followed by S-PS, S-CL, and R-CR.

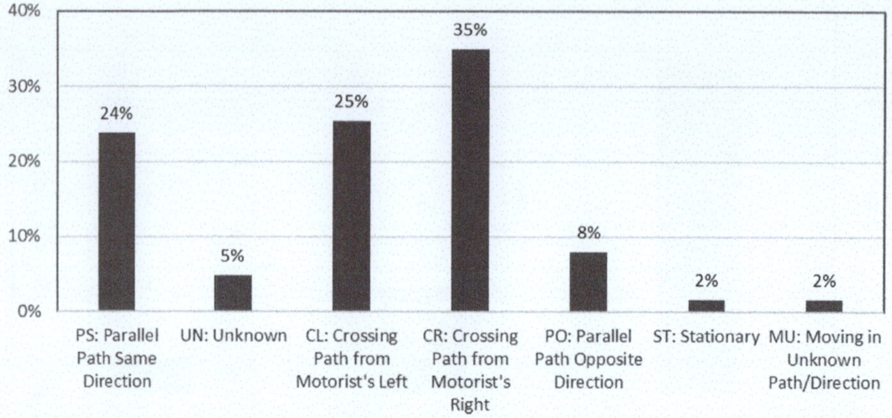

Fig. 8.9 The relative direction of travel for e-scooters with respect to motorists just before the crashes

Table 8.7 Summary of crash types between motorists and non-motorists

S-PS: Straight-parallel path same direction	R-CL: Turning right-crossing path from motorist's left
S-CR: Straight-crossing path from motorist's right	L-CL: Turning left-crossing path from motorist's left
R-PS: Turning right-parallel path same direction	L-ST: Turning left-stationary
O-PS: Other maneuver-parallel path same direction	O-CR: Other maneuver-crossing path from motorist's right
S-PO: Straight-parallel path opposite direction	L-PO: Turning left-parallel path opposite direction
S-CL: Straight-crossing path from motorist's left	L-PS: Turning left-parallel path same direction
L-CR: Turning left-crossing path from motorist's right	O-UN: Other maneuver-unknown
R-CR: Turning right-crossing path from motorist's right	R-MU: Turning right-moving in unknown path/direction

8.6 Conclusions

In this chapter, two data acquisition systems were developed to collect naturalistic data for vehicle and e-scooter–rider interactions. After carefully designing the routes and areas for vehicle-centered and e-scooter-centered data collection activities in the Indianapolis downtown area, 55 subjects with diverse demographic backgrounds were recruited to participate in the data collection experiments. 68 vehicle-centered data collection sessions and 14 e-scooter-centered data collection sessions were conducted. About 35 h of vehicle-centered data and 8 h of e-scooter-centered data were collected, resulting in more than 8 TB of raw data with more than 8-million frames. Each frame has synchronized camera images, LiDAR point cloud, and GPS. More than 700 encounters between e-scooter riders and cars are captured in the

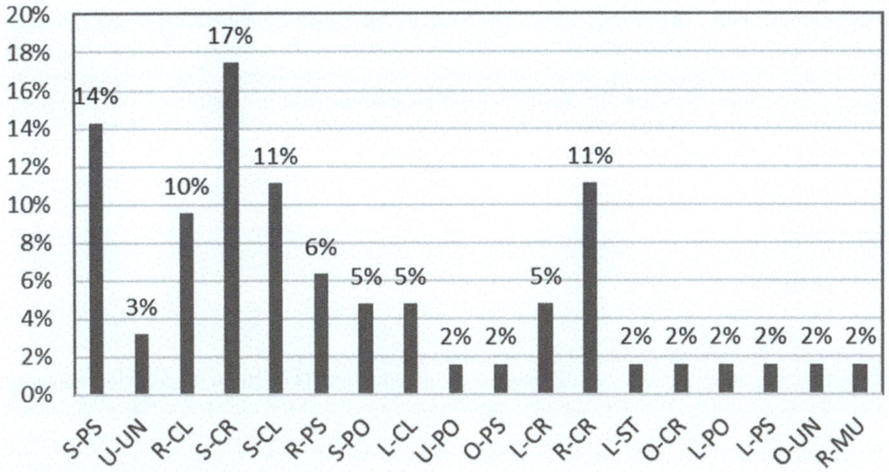

Fig. 8.10 Crash types obtained from PBCAT

dataset. To further enhance our prior naturalistic data collection, a nationwide naturalistic driving study was conducted to collect broader representative naturalistic driving data involving interactions of e-scooters with vehicles and other road users. The processed and notated datasets from these NDD studies are very valuable for understating e-scooter's behavior interacting with vehicles and other road users.

To process the large-scale data collection, an efficient data preprocessing pipeline was developed. The data processing speed has been significantly increased by implementing a hybrid process that is composed of local computers and Google cloud computing platform. The upgraded pipeline also removes most of the manual work to save time and increase scalability. A state-of-the-art algorithm was adopted to detect and track vehicles and humans from the collected video data. Furthermore, a computer-vision approach was developed with a new benchmark dataset to detect e-scooter riders efficiently and automatically.

Recognizing the limitations of the naturalistic driving data to obtain the complete picture of real crash scenarios, we have performed crash data analysis to gain more insights into real-world e-scooter-related crashes.

The research results and findings presented in this chapter have shown the behavior metrics of e-scooter riders in different vehicle and e-scooter interactions. E-scooter rider detection benchmark dataset and classifier are also released publicly. The results of this study will help to fill the gap in quantitative empirical data about e-scooter rider behaviors, which can potentially support the development and testing of vehicle–e-scooter collision avoidance mitigation systems. In the follow-up effort, the results can be further integrated with crash data to better model e-scooter crash risks and guide the training of e-scooter behavior prediction algorithms with more labels to the dataset.

Acknowledgements The authors would like to thank the Toyota Collaborative Safety Research Center (CSRC) for providing financial support. The authors are very grateful to the following graduate students for their significant contributions to this project: Avinash Prabu, Dan Shen, Zhengming Zhang, Kumar Apurv, Siddhant Betrabet, Atharva Kulkarni, Nitya Ranjan, and Taryn Spisak. Thanks go to Diwas Thapa, Syed Mohammad Adil, and Sabyasachee Mishra of University of Memphis for their technical contributions to e-scooter crash data analysis.

References

1. Howe E, Bock B (2018) Global Scooter Sharing Market Report
2. National Association of City Transportation Officials (NACTO) (2020) Shared Micromobility in the U.S.: 2019. Available online: https://nacto.org/shared-micromobility-2019/. Accessed Feb 2021
3. Almannaa MH, Alsahhaf FA, Ashqar HI, Elhenawy M, Masoud M, Rakotonirainy A (2021) Perception analysis of e-scooter riders and non-riders in Riyadh, Saudi Arabia: survey outputs. Sustainability 13(2):863
4. Ching-Fu C, Cheng-Chien K, Yi-Ju L (2017) Investigating barriers and facilitators of attitude and intention to use e-scooter sharing system. In: 22nd international conference of Hong Kong Society for transportation studies: transport and society, HKSTS 2017. Hong Kong Society for Transportation Studies Limited, pp 399–406
5. Aguilera-García Á, Gomez J, Sobrino N (2020) Exploring the adoption of moped scooter-sharing systems in Spanish urban areas. Cities 96:102424
6. Fitt H, Curl A (2019) Perceptions and experiences of Lime scooters: summary survey results
7. Huang FH, Lin SR (2018) A survey of user experience of two-wheeler users in long-term interactions. In: Congress of the international ergonomics association. Springer, Cham, pp 1465–1472
8. James O, Swiderski JI, Hicks J, Teoman D, Buehler R (2019) Pedestrians and e-scooters: an initial look at e-scooter parking and perceptions by riders and non-riders. Sustainability 11(20):5591
9. Sanders RL, Branion-Calles M, Nelson TA (2020) To scoot or not to scoot: findings from a recent survey about the benefits and barriers of using E-scooters for riders and non-riders. Transp Res Part A: Policy Practice 139:217–227
10. Austin Public Health (2018) Dockless Electric Scooter-Related Injuries Study. Report. Austin Public Health (APH), Available online via. Accessed Feb 2021. https://www.austintexas.gov/sites/default/files/files/Health/Epidemiology/APH_Dockless_Electric_Scooter_Study_5-2-19.pdf
11. Multnomah County Health Department (2019) Scooter-related Injuries in Multnomah County July–November 2018. Online via. Accessed Feb 2021: https://www.portlandoregon.gov/transportation/article/709715.
12. Griswold A (2020) At least 29 people have died in electric scooter crashes since 2018, Quartz report, accessed 02.2021. https://qz.com/1793164/at-least-29-people-have-died-in-electric-scooter-crashes/
13. Yang H, Ma Q, Wang Z, Cai Q, Xie K, Yang D (2020) Safety of micro-mobility: analysis of E-Scooter crashes by mining news reports. Accid Anal Prev 143:105608
14. FHWA (2017) National Household Travel Survey. Available online at: https://nhts.ornl.gov/vehicle-trips. Accessed Feb 2021
15. Beck S, Barker L, Chan A, Stanbridge S (2020) Emergency department impact following the introduction of an electric scooter sharing service. Emerg Med Australas 32(3):409–415
16. Badeau A, Carman C, Newman M, Steenblik J, Carlson M, Madsen T (2019) Emergency department visits for electric scooter-related injuries after introduction of an urban rental program. Am J Emerg Med 37(8):1531–1533

17. Nisson PL, Ley E, Chu R (2020) Electric scooters: case reports indicate a growing public health concern
18. Sikka N, Vila C, Stratton M, Ghassemi M, Pourmand A (2019) Sharing the sidewalk: a case of E-scooter related pedestrian injury. Am J Emerg Med 37(9):1807-e5
19. Ma Q, Yang H, Mayhue A, Sun Y, Huang Z, Ma Y (2021) E-scooter safety: the riding risk analysis based on mobile sensing data. Accid Anal Prev 151:105954
20. Dozza M, Piccinini GFB, Werneke J (2016) Using naturalistic data to assess e-cyclist behavior. Transp Res F: Traffic Psychol Behav 41:217–226
21. Prabu A, Ranjan N, Li L, Tian R, Chien S, Chen Y, Sherony R (2022) SceNDD: a Scenario-based naturalistic driving dataset. In: IEEE 25th International Conference on Intelligent Transportation Systems (ITSC), pp 4363–4368. https://doi.org/10.1109/ITSC55140.2022.9921953
22. Prabu A, Shen Dan, Tian R, Chien S, Li Lingxi, Chen Y, Sherony R (2021) A wearable data collection system for studying micro-level e-scooter behavior in naturalistic road environment. In: Proceedings of Fast-zero'21, Kanazawa, Japan
23. Prabu A, Tian R, Chien S, Li L, Chen Y, Sherony R (2023) Risk assessment and mitigation of e-scooter crashes with naturalistic driving data. In: 27th international technical conference on the Enhanced Safety of Vehicles (ESV). Yokohama, Japan

Chapter 9
Challenges and Algorithms for Connected and Autonomous Vehicle Safety and Mobility in Smart Cities

Azim Eskandarian, Goodarz Mehr, Anshul Nayak, and Xu Shang

Abstract Connected vehicles can enhance safety by enabling real-time communication between vehicles and infrastructure, alerting drivers to potential hazards, and improving situational awareness. This technology also benefits vulnerable road users and reduces the economic burden of traffic incidents. In this chapter, we take a look at various challenges facing connected autonomous vehicles and proposed algorithms to address those challenges and enhance vehicle safety and mobility in smart cities. Specifically, cooperative adaptive cruise control subject to temporary communication or target detection loss; connected lane-change advance warning systems for improving traffic flow; probabilistic trajectory forecasting for vulnerable road users, especially pedestrians; and, finally, approaches to collision avoidance and mitigation for autonomous vehicles.

9.1 Cooperative Adaptive Cruise Control Subject to Temporary Communication or Target Detection Loss

Connected vehicle technology is crucial in addressing the high number of motor vehicle crashes and their economic impact. In the United States alone, 38,680 people died and 2.36 million were injured in traffic incidents in 2020, incurring an economic

A. Eskandarian (✉)
College of Engineering, Virginia Commonwealth University, 601 West Main Street, Suite 331, Richmond, VA 23284-3068, USA
e-mail: eskandariana@vcu.edu

G. Mehr · A. Nayak · X. Shang
Virginia Tech, Blacksburg, VA 24061, USA
e-mail: goodarzm@vt.edu

A. Nayak
e-mail: anshulnayak@vt.edu

X. Shang
e-mail: shangxu@vt.edu

P. Ioannou and A. A. Malikopoulos (eds.), *Transportation Mobility in Smart Cities*, Springer Tracts on Transportation and Traffic 21,
https://doi.org/10.1007/978-3-031-64769-7_9

damage of over $55 billion. Globally, traffic incidents claim 1.35 million lives annually, with pedestrians, motorcyclists, and cyclists among those most affected. Connected vehicles can enhance safety by enabling real-time communication between vehicles and infrastructure, alerting drivers to potential hazards, and improving situational awareness. They also benefit vulnerable road users and reduces the economic burden of traffic incidents, and can help achieve the UN target of reducing road traffic deaths and injuries in half by 2030.

Cooperative Adaptive Cruise Control (CACC) is an evolution of Adaptive Cruise Control (ACC) that enhances longitudinal control of a vehicle by utilizing inter-vehicle distance measurements from a radar and vehicle state information obtained through wireless communication [14, 19]. CACC plays a crucial role in enhancing safety by reducing variability in vehicle behavior, a major contributing factor to crashes [71]. Furthermore, implementing CACC in a signalized corridor can reduce fuel consumption by over a third [58]. Additionally, mass deployment of CACC can increase traffic throughput from 2000 vehicles per hour per lane in manual driving to 4200 vehicles per hour per lane. These benefits are made possible by CACC's ability to maintain shorter and smoother gaps between vehicles, resulting in improved efficiency and performance overall [19].

Compared to ACC, CACC enables vehicle following with smaller inter-vehicle distances and maintains string stability [55]. The most common CACC information flow topology is predecessor-following, where the preceding vehicle sends its desired acceleration to the following vehicle [60, 62, 67]. The preceding vehicle's desired acceleration is used as the feedforward term in the CACC controller of the ego vehicle and dominantly determines the resulting control action, especially during transient behavior [40, 60, 74]. For example, when using a PD controller, the feedforward term determines up to 80% of the resulting control action [60]. Hence, the preceding vehicle's desired acceleration is extremely important for CACC controllers.

However, communication impairments, such as communication loss [57, 61, 74], latency [11, 12], and malicious interference [9, 10, 13], can compromise the performance of CACC. The study of communication loss in CACC has attracted a lot of attention in recent years. There are two kinds of communication loss scenarios in CACC, one is random communication loss during a short period and the other is communication loss during an extended period. To address random communication loss during a short period, many researchers have implemented a consensus-based controller to utilize vehicle state information from the available communication links [20, 32, 64]. During extended communication loss; however, the CACC controller cannot receive the preceding vehicle's desired acceleration and has to degrade to ACC to ensure safety. Hence, it is essential to improve the CACC controller performance during extended communication loss.

To that end, three approaches have been proposed to deal with extended communication loss. The first is fallback to ACC; when communication is lost, CACC falls back to ACC to maintain string stability, which increases the inter-vehicle distance and degrades the controller's performance [3]. The second approach is to use a better wireless communication protocol. Lee and Chanson proposed a wireless communication protocol in which a lost packet was resent until the elapsed time it exceeded the maximum time delay requirement [37]. This method can induce a high communication delay, which may compromise the string stability of CACC [44]. The third approach is utilizing an advanced CACC control algorithm. Nunen et al. proposed a model predictive control (MPC) method for CACC, in which vehicles share their own predictions of the desired acceleration in the MPC controller using wireless communication and each following vehicle saves this information for its preceding vehicle [74]. The following vehicle uses this saved information as a feedforward signal in the CACC controller when communication loss occurs. However, this method is only effective when communication loss duration is within the prediction horizon of the MPC controller, which is usually around 1 s [22]. Another study [57] proposed a dynamic spacing policy during communication loss, which introduced the risk of collision during braking. Ploeg et al. utilized the estimated preceding vehicle's acceleration as a feedforward term in the CACC controller during communication loss [61]. This method can be implemented for a few seconds during communication loss. Hence, compared to the methods above, the estimation of the preceding vehicle's acceleration is a better solution to improve the inter-vehicle distance tracking performance of CACC during extended communication loss.

ACC systems may also suffer from target detection loss due to sensor and environmental issues. A radar sensor has limited field of view and can miss the target, i.e., the preceding vehicle, on curved roads, especially in urban driving environments for full speed range ACC [4, 23, 68]. During unfavorable weather conditions such as fog, object reflectance is low and can result in the loss of target [8, 21]. For these situations with short-term target detection loss, it has been suggested that the acceleration term be suppressed until the sensor detects the target again [4, 23, 68, 78]. It's possible, however, that the preceding vehicle has taken an off-ramp which can also lead to target detection loss. Thus, for target detection loss on curved roads, it has been proposed that the following vehicle travels through the curve for target re-acquisition to determine whether to use ACC again [68, 78].

To evaluate the performance of CACC, we first implemented a CACC controller using three robots to form a platoon and compared it to an ACC controller. Our results indicated that the CACC system practically maintained the desired time gap during constant speed, acceleration, and deceleration periods. The ACC system, however, resulted in significantly larger and smaller actual time gaps during the acceleration and deceleration periods, respectively [41].

Next, we proposed a solution to enable ACC vehicle following during target detection loss. In our proposed solution, a following vehicle can compute the inter-vehicle distance using the front vehicle's odometry obtained through wireless communication. The system design is tested using our autonomous mobile robot platform, where we see that the designed system enables vehicle following for both cases where distance sensor measurement is and is not available. As inter-vehicle distance can be computed using accurate odometry data during target detection loss, the same vehicle following performance could be achieved as when distance sensor measurement is available [42].

Finally, we proposed an innovative CACC control strategy with an adaptive Kalman filter and a new method of calculating the measurement vector of the preced-

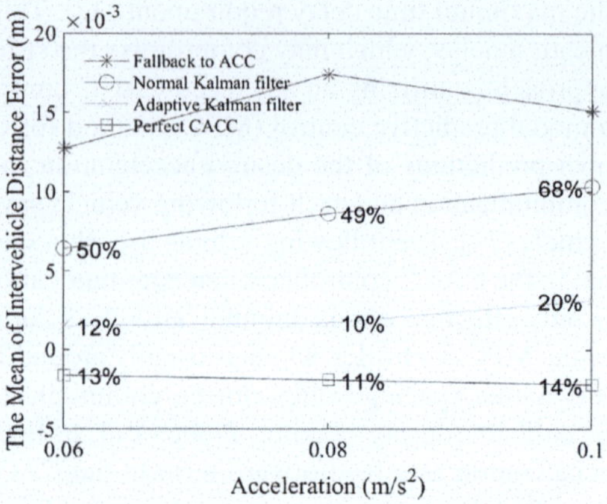

Fig. 9.1 The mean of inter-vehicle distance error with different decelerations (the percentage means the corresponding mean divided by that for the fallback to ACC)

ing vehicle, which is based on the inter-vehicle distance, the relative velocity, and the position and velocity of the ego vehicle. This calculation method can avoid the oscillatory feedforward control input. It effectively improves the performance of CACC during extended communication loss. The adaptive Kalman filter can estimate the acceleration of the preceding vehicle and improve the performance of CACC during communication loss, which reduces the mean and RMS of inter-vehicle distance error significantly, compared to falling back to ACC. In our simulations and experiments, the results show that the mean of inter-vehicle distance error in the adaptive Kalman filter is less than the normal Kalman filter based on the Singer model, as shown in Fig. 9.1. Hence, implementing the adaptive Kalman filter is a better solution than the normal Kalman filter during communication loss [76].

9.2 Connected Lane Change Advance Warning System for Improved Traffic Flow

Lane changes play a vital role in driving, requiring careful consideration of various factors such as the purpose and urgency of the lane change, the presence of nearby vehicles, and the behavior of other drivers [15]. A successful maneuver necessitates the driver or autonomous vehicle to identify a suitable gap in the target lane, adjust their speed, maintain a proper position in relation to surrounding vehicles, and navigate to the desired lane while avoiding potential collisions [35]. A single error or unsafe driving conduct during any stage of this process can lead to an accident. In the United States, approximately four to ten percent of all reported motor vehicle crashes can be attributed to unsafe lane-changing maneuvers. Moreover, these incidents not only result in fatalities but also incur economic loss by causing traffic congestion and delays [39, 66, 72]. In fact, it was estimated that congestion cost the U.S. around $166 billion in 2019, with traffic incidents accounting for approximately a quarter of total delays on American roads [54, 65]. To mitigate such incidents and minimize delays, it is crucial for vehicles to receive accurate and timely information to facilitate safe lane changes.

Several previous studies have produced algorithms aimed at assisting autonomous vehicles in executing safe lane changes [7, 16, 17, 31, 43, 45, 70, 75]. These studies primarily focused on developing rule-based or machine learning-based methods and control algorithms that determine the optimal trajectory for the maneuver while avoiding collisions. Although these methods are capable of completing a safe lane change from start to finish, they generally lack the ability to determine the ideal timing for initiating a lane change to reach a target position on another (perhaps non-adjacent) lane while minimizing traffic disruption. For instance, consider an autonomous vehicle traveling in the leftmost lane of a four-lane highway that needs to exit using an off-ramp. Existing methods struggle to identify the optimal moment to commence changing lanes to ensure that the vehicle reaches the rightmost lane by the time it reaches the off-ramp. If the lane change is initiated too early, the vehicle will spend a significant amount of time traveling at a potentially reduced speed in the rightmost lane. On the other hand, if the maneuver is initiated too late, the vehicle may miss the exit entirely or be compelled to perform multiple rushed lane changes, leading to potential slowdowns in nearby traffic and an increased likelihood of incidents [27].

Prior research on human drivers suggests that implementing advance warning systems could be a viable solution to address the issue of rushed and unsafe lane changes in autonomous vehicles [27, 29, 30]. The key concept behind this approach is creating a model capable of assessing the probability of successfully executing one or multiple lane changes to reach a desired position ahead. Although some studies have produced such probability models, they all have certain limitations [27, 30, 46]. For instance, some models are applicable only to specific traffic types, such as those involving an exponential headway distribution, while others cannot be deployed in real time. Most importantly, all existing models are restricted to scenarios involving only two lanes, which is impractical for many situations, particularly in highway driving scenarios.

Our work takes advantage of vehicular communication to address the limitations of previous studies. Specifically, we formulate a model that can accurately estimate the probability of reaching a certain goal state in a different lane using one or multiple lane changes under general traffic conditions [48, 49], and also show how it can be incorporated in advance warning systems to reduce traffic delay during recurrent and non-recurrent congestion [50–52, 63, 73, 79, 81]. In what follows, Sect. 9.2.1 formulates the problem, Sect. 9.2.2 discusses our methodology for developing and validating the model, and Sect. 9.2.3 demonstrates some of the applications of the developed model.

9.2.1 Problem Formulation

Our goal is to answer the following question: what is the probability that a vehicle can reach a position in a different lane at a certain distance ahead using one or multiple lane changes? On its own, this question is not well-defined because of all the environmental factors involved. Therefore, we make several simplifying assumptions to mathematically formulate the problem and develop the probability model.

Without loss of generality, we focus on a highway road segment under free-flow traffic conditions. The road segment has n lanes ($n \geq 2$) numbered from left to right by 1 to n, i.e., the leftmost lane is lane 1 and the rightmost lane is lane n. We assume that vehicles in lane i, $1 \leq i \leq n$, are all passenger cars represented by a point at their center of gravity and all have the same velocity v_i that does not change over time, but may be different from velocity v_j where $i \neq j$. Furthermore, we assume that headway (front bumper to front bumper) distances in lane i are independent identically distributed (i.i.d.) random variables $H_{i,k}$ that have a common cumulative distribution function $F_i(h)$. Since our focus is on highway driving, we assume that headway distances on lane i are from a log-normal probability distribution defined by parameters (μ_i, σ_i), but other probability distributions can be used as well [53, 77].

We further assume that the lane-changing behavior of the vehicle under study (the ego vehicle) follows a Gipps gap acceptance lane-change model [26], where the vehicle changes lanes only if the headway distance between its leading and trailing vehicles in the adjacent target lane j, $1 \leq j \leq n$, is larger than a minimum acceptable (critical) gap g_j. As is the case for velocity, we assume that g_j does not change over time but can be different for different lanes. Additionally, we assume that once the vehicle finds such acceptable gap, it instantly starts changing lanes and completes it in a set time t_j, after which its velocity matches the velocity of vehicles in the target lane v_j. Here we assume that only the ego vehicle changes lanes and none of the other vehicles do so, though statistically speaking at any instant the number of vehicles entering a lane is the same as those exiting that lane [28], so our assumption should not have a noticeable impact on the results.

A near-term goal state is defined as a point beyond the line of sight of the driver or perception sensors, but not so far that reaching it requires maneuvers beyond multiple lane changes. Denoting by d the longitudinal distance (distance along the road) from the current position to the goal state, here we assume that $0.1\,\text{km} \leq d \leq 5\,\text{km}$.

Using these assumptions, the problem can be formulated as follows: what is the success probability of reaching a point on lane j a distance d ahead of the current position on lane i, $j \neq i$, assuming that for vehicles on lane k, $1 \leq k \leq n$, all velocities are equal to v_k, headway distances are i.i.d. random variables from a log-normal distribution defined by parameters (μ_k, σ_k), and the ego vehicle changes lanes according to the Gipps gap acceptance lane-change model with critical gap g_k?

9.2.2 *Methodology*

We first consider the case where $n = 2$ and then expand the model to include cases where $n > 2$ using a recursive approach. Without loss of generality, assume that the ego vehicle is currently on lane 1 and the goal state is a distance d ahead on lane 2. Our objective is to determine the probability of finding an acceptable gap in lane 2 and completing a lane change before the ego vehicle travels distance d. Based on the assumptions in Sect. 9.2.1, this probability, denoted by $P(S)$, is a function of $d, v_1, v_2, \mu_2, \sigma_2, g_2$, and t_2. That is, $P(S) = f_2(d, v_1, v_2, \mu_2, \sigma_2, g_2, t_2)$ (index 2 of function f refers to the number of lanes).

The success probability $P(S)$ of reaching the goal state a distance d ahead is equivalent to the probability of finding an acceptable gap in lane 2 during the time the ego vehicle travels a total distance $d_i = d - t_2 v_1$, given that it takes t_2 seconds to change lanes. Denote by d_r the distance traveled by the ego vehicle relative to lane 2 while traveling a total distance d_i. That is,

$$d_r = d_i | \left(1 - \frac{v_2}{v_1}\right) |. \tag{9.1}$$

The ego vehicle travels distance d_r relative to lane 2 while traveling a total distance d_i and searching for an acceptable gap, so if we freeze the motion of vehicles on lane 2 and only consider the relative motion of the ego vehicle, $P(S)$ is equal to the probability of finding a point on lane 2 along distance d_r ahead of (or behind if $v_2 > v_1$) the current position that is at least $g_2/2$ away from each of its two nearest vehicles on that lane. The probability of finding such a point is equivalent to the probability of finding a gap no smaller than g_2 along distance $d_e = d_r + g_2$ in lane 2 starting from a distance $g_2/2$ behind (or ahead of if $v_2 > v_1$) the ego vehicle position.

The problem of finding $P(S)$ can now be formulated in a more abstract way. Assume that the real line $\mathbb{R}$ is populated with points such that the inter-arrival distances are i.i.d. random variables from a log-normal probability distribution defined by parameter pair (μ_2, σ_2). These points represent vehicles on lane 2 whose headway distances are i.i.d. random variables from a log-normal distribution. Here, $P(S)$ is equal to the probability of finding a gap no smaller than g_2 in the interval $[0, d_e]$. Before moving on, the following lemma is needed to calculate $P(S)$.

Lemma 9.1 *If X is a log-normal random variable from a distribution with parameters (μ, σ) and k is a positive real number, then $Y = X/k$ is a log-normal random variable from a distribution with parameters $(\mu - \ln(k), \sigma)$.*

Proof For $y > 0$, we have

$$\begin{aligned} P[Y \leq y] &= P[\frac{X}{k} \leq y] = P[X \leq ky] \\ &= \frac{1}{2} + \frac{1}{2}\text{erf}[\frac{\ln(ky) - \mu}{\sqrt{2}\sigma}] = \frac{1}{2} + \frac{1}{2}\text{erf}[\frac{\ln(y) - (\mu - \ln(k))}{\sqrt{2}\sigma}], \end{aligned} \tag{9.2}$$

where erf is the error function. According to (9.2), Y is a log-normal random variable from a distribution with parameters $(\mu - \ln(k), \sigma)$. □

Up to this point, we have reduced the number of parameters from the original 7 down to d_e, g_2, μ_2, and σ_2. The next step is scaling the abstract formulation by a factor of $1/d_e$. This way, using Lemma 9.1 we can deduce that $P(S)$ is equal to the probability of finding a gap no smaller than $g = g_2/d_e$ in the interval [0, 1], assuming $\mathbb{R}$ is populated with points such that the inter-arrival distances are i.i.d. random variables from a log-normal distribution defined by parameter pair $(\mu = \mu_2 - \ln(d_e), \sigma = \sigma_2)$. This final step shows that probability $P(S)$ can be calculated based on three parameters, g, μ, and σ. In other words, $P(S) = f_2(d, v_1, v_2, \mu_2, \sigma_2, g_2, t_2) = q(g, \mu, \sigma)$. Function q can be seen as the joint distribution of the success probability over the three parameters g, μ, and σ, and can be calculated using numerical Monte Carlo simulations.

Now that we know the value of $P(S)$ for $n = 2$, i.e., $f_2(d, v_1, v_2, \mu_2, \sigma_2, g_2, t_2)$, its value for cases where $n > 2$ can be calculated using the following theorem.

Theorem 9.1 *For $n \geq 3$, probability $P(S) = f_n(d, v_{1:n}, \mu_{2:n}, \sigma_{2:n}, g_{2:n}, t_{2:n})$ (where $w_{l:m}$ means $w_1, w_2, \ldots, w_m$ for any parameter w and indices $l \leq m$) is calculated by the following equation:*

$$\begin{aligned} &f_n(d, v_{1:n}, \mu_{2:n}, \sigma_{2:n}, g_{2:n}, t_{2:n}) \\ &\quad = \int_0^d f_2(d - x, v_{n-1:n}, \mu_n, \sigma_n, g_n, t_n) \\ &\quad \times \frac{\partial}{\partial x} f_{n-1}(x, v_{1:n-1}, \mu_{2:n-1}, \sigma_{2:n-1}, g_{2:n-1}, t_{2:n-1}) \mathrm{d}x \\ &\quad = \frac{\partial}{\partial x} \int_0^d f_2(d - x, v_{n-1:n}, \mu_n, \sigma_n, g_n, t_n) \\ &\quad \times f_{n-1}(x, v_{1:n-1}, \mu_{2:n-1}, \sigma_{2:n-1}, g_{2:n-1}, t_{2:n-1}) \mathrm{d}x. \end{aligned} \tag{9.3}$$

Proof Using the continuous form of the law of total probability [38], we can write

$$P(S) = \int_{-\infty}^{\infty} P(S|X = x) f_X(x) \mathrm{d}x. \tag{9.4}$$

Here, conditioning the probability on location x where the vehicle changes lanes from lane $n - 2$ to lane $n - 1$, $0 \leq x \leq d$, it is easy to see that $P(S|X = x)$ is nothing but $f_2(d - x, v_{n-1:n}, \mu_n, \sigma_n, g_n, t_n)$, because it is the probability of successfully reaching the goal state on lane n a distance $d - x$ ahead from the current position x on lane $n - 1$. As for $f_X(x)$, note that it is the derivative of $F_X(x)$, the cumulative distribution function of the probability that at some point on or before x the vehicle reached lane $n - 1$ using $n - 2$ lane changes. This probability, in turn, is equal to the success probability for a case where the goal state is located a distance x ahead on lane $n - 1$. Therefore, $f_X(x) = \frac{\partial}{\partial x} f_{n-1}(x, v_{1:n-1}, \mu_{2:n-1}, \sigma_{2:n-1}, g_{2:n-1}, t_{2:n-1})$.

Substituting these values in (9.4) and limiting the integration bounds to 0 and d result in (9.3), completing the proof. □

Because of the assumptions that were made when developing the probability model in Sect. 9.2.1, it is essential to validate the model and evaluate its performance under general traffic conditions. The validation scheme we used was to define a target position on a different lane of a (simulated or real) highway a distance d ahead of a fixed starting point and count the number of vehicles that initiate a lane-changing maneuver at the starting point and successfully reach the target position. Dividing that number by the number of vehicles that initiate the maneuver determines the success probability which can be compared with model predictions to validate it. This validation approach was implemented through VISSIM traffic simulations, where we also varied different parameters to understand how they impacted the probability.

9.2.3 Applications

As mentioned previously, the developed probability model can be used to design connected advance warning systems that can reduce traffic delay during recurrent and non-recurrent congestion. To that end, we devised three case studies to better analyze the impact of such systems.

In our first study, we used the probability model to develop a connected advance warning system that used driver and traffic information to advise vehicles on when to change lanes to reach an off-ramp. Traffic simulations were carried out on a two-lane segment of the westbound I-66 inter-state highway, and we studied how traffic flow and the ratio of vehicles trying to exit through the off-ramp affected system performance. The results showed that the system was able to reduce average delay by up to 83%—wiping out the congestion in some cases—and had the greatest impact when the ratio of vehicles trying to take the off-ramp was high [52].

In our second study, we used the probability model to develop a connected advance warning system that used driver and traffic information to advise vehicles on when to change lanes to avoid lane drops. Traffic simulations were carried out on a segment of the southbound I-81 inter-state highway with two-lane drops—transitioning from four lanes to two lanes—and we studied how traffic flow and the ratio of vehicles equipped with our system affected overall performance. The results showed that the system was able to reduce average delay by up to 50% and had the greatest impact when the ratio of equipped vehicles was high [50].

In our final study, we used the probability model to develop Sentinel, a connected advance warning system that used driver and real-time traffic information to advise vehicles on when to change lanes during a traffic incident. Traffic simulations were carried out on a four-lane segment of the westbound I-66 inter-state highway where the rightmost lane got blocked due to a traffic incident for a set period of time, and we studied how traffic flow, incident duration, and the ratio of vehicles equipped with our system affected overall performance. The results indicated that the system was able to reduce average delay by up to 37% and had the greatest impact when the ratio of equipped vehicles was high and the incident took longer to clear up [51].

9.3 Probabilistic Trajectory Forecasting

9.3.1 Motivation

In this section, we explore the significance of awareness of the surrounding environment for self-driving cars in ensuring correct and safe maneuvering. Particularly, the study focuses on the accurate estimation of future states of vulnerable road users (VRUs), such as pedestrians and bicyclists, during complex maneuvers. While considerable advancements have been made in predicting the motion of VRUs, most existing prediction models rely on deterministic approaches that provide single point estimates of future states (Fig. 9.2a). However, in dynamic environments with multiple interactions, deterministic predictions may be prone to inaccuracy. Since, humans tend to change directions swiftly, deterministic predictions may fail to capture this randomness in pedestrian trajectory. Therefore, providing a probability distribution associated with pedestrian's future predicted states will be more robust compared to a single point estimate (Fig. 9.2b). The risk-aware region around the future state captures all the probable locations the pedestrian can be present at a future time with certain confidence. It enables planning algorithms to either completely avoid the region or plan intelligently to execute uncertainty-aware motion planning that is more robust and less prone to collision [34]. Therefore, the current study proposes a risk-aware motion prediction model for pedestrians which can be combined with intelligent planning algorithms downstream. The uncertainty during future motion prediction is estimated as a distribution of trajectories with a mean predicted path and associated variance such that the ground truth trajectory mostly lies within 2σ confidence interval of the predictions.

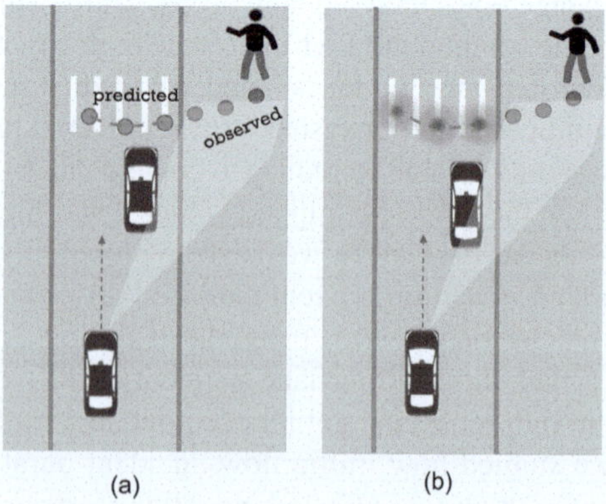

Fig. 9.2 **a** Deterministic trajectory forecasting with point prediction **b** Probabilistic trajectory forecasting with associated uncertainty

In recent years, the extensive utilization of Deep Neural Networks (DNNs) has been observed in pedestrian trajectory forecasting. These networks predominantly employ deep learning architectures such as recurrent neural networks (RNN) [69] and long short-term memory (LSTM) networks [6] to capture the temporal dynamics of sequential data. Notable studies such as scene-LSTM [47] and social-LSTM [5] have leveraged LSTM network architecture to enhance trajectory forecasting. However, the primary focus of prior work has been on improving prediction accuracy, with limited emphasis on quantifying uncertainty.

On the contrary, DNN models like the Bayesian Neural Networks (BNN) [80] have been used recently to capture the uncertainty in time series. Typically, a Bayesian neural network introduces stochastic weights or activation functions into the model and is trained using Bayes rule:

$$P(\theta|D) = \frac{P(D|\theta)\, P(\theta)}{P(D)}. \tag{9.5}$$

The probability $P(D|\theta)$ is called likelihood while $P(\theta)$ represents the prior distribution over weights. $P(\theta|D)$ is the posterior distribution which gets statistically updated using the prior and likelihood. Consequently, the posterior distribution also becomes stochastic and samples can be drawn during inference for computing the distribution of outputs rather than deterministic estimates. A BNN has several advantages over traditional neural networks such as estimating uncertainty, classifying the uncertainty into epistemic and aleatoric uncertainty as well as integrating the knowledge of prior into the model. For instance, provided a set of training input states, X = $x_1, x_2, ..., x_T$ and outputs, Y = $y_1, y_2, ..., y_T$ for a pedestrian, the distribution of its future predicted states, y^* can be computed by marginalizing the posterior, $p(\theta|X, Y)$ over some new input data point, x^*:

$$p(y^*|x^*, X, Y) = \int_\theta p(y^*|x^*, \theta')p(\theta'|X, Y)\, d\theta', \tag{9.6}$$

where θ' represents weights and $p(\theta')$ refers to the probability of sampling from prior weight distribution. During Bayesian prediction, computing the posterior, $p(\theta'|X, Y)$ can be quite challenging and many sampling algorithms like Monte Carlo Markov Chain (MCMC) and variational inference [33] have been used to approximate the posterior. But, most of the sampling methods are either computationally expensive or introduce a lot of parameters into the model. This often requires the incorporation of variational methods to approximate the posterior during inference. Hence, we use the Monte Carlo (MC) dropout method [25] which has been used to accurately approximate a Bayesian neural network without significantly changing the model. The MC dropout method assumes that weights are stochastically dropped during inference. This process is then repeated for N forward passes to generate a random distribution of predicted values. The current study applies Monte Carlo dropout to an encoder–decoder NN architecture for uncertainty quantification during predictions.

9.3.2 Methods

9.3.2.1 Encoder–Decoder Model

An encoder–decoder-based LSTM architecture was developed in this study to forecast future trajectories of pedestrians across multiple time horizons. The architecture consists of an encoder and a decoder, where the encoder generates an embedded representation of important features as a sequence of encoded space vectors. The decoder utilizes these encoded features to estimate the outputs. Let $\{x_t\}_T$ represent the x-position up to T time steps as $\{x_1, x_2, .., x_T\}$. The encoder maps the data into an encoded space using a nonlinear function denoted as $e = g(x)$. Subsequently, the decoder employs the encoded features to predict the future positions for F forward time steps, denoted as $\{x_{T+1}, x_{T+2}, ..., x_{T+F}\}$. The current architecture comprises two LSTM cells for the encoder and one LSTM cell for the decoder (Fig. 9.3).

The ablation study revealed that encoding both the position and velocity of the pedestrian significantly improved prediction accuracy compared to encoding only position information. Consequently, the input data for the LSTM layers consists of a multivariate time series with four features: $\{x, y, u, v\}$ representing the pedestrian's x- and y-positions and velocity, respectively. The output of the neural network retains the same features, and the predicted values $\hat{x}$ and $\hat{y}$ correspond to the pedestrian's future position. To incorporate regularization, weight dropout with a certain probability (p) is applied, followed by the **tanh** activation function after each LSTM layer.

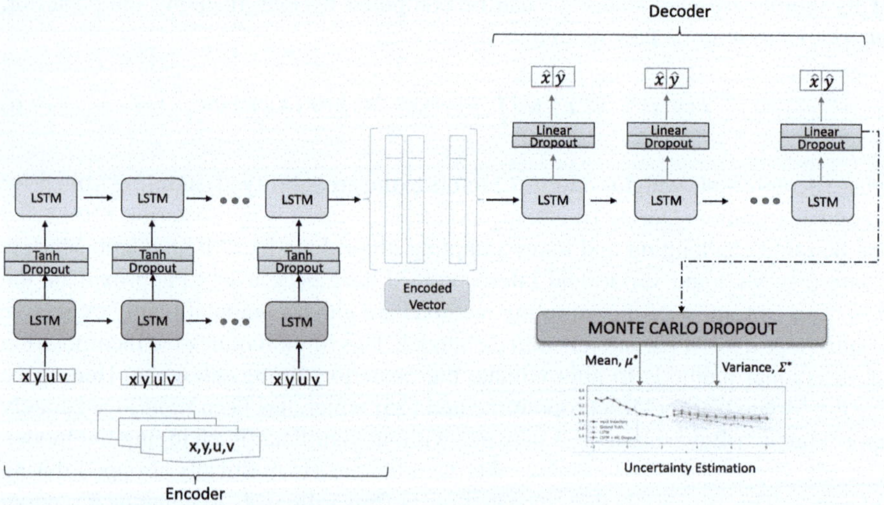

Fig. 9.3 LSTM architecture with Monte Carlo dropout

9.3.2.2 MC Dropout

Monte Carlo dropout (MC dropout) has been implemented during inference to quantify uncertainty (Fig. 9.3). The MC dropout model incorporates stochastic dropout during test time, randomly dropping weights of the neural network at each layer. Thus, for any new input set x^*, the inference is computed by performing random dropout at each layer of the model. The dropout probability is denoted as p, and the model is inferred N times to obtain a set of outputs $\{y_1^*, y_2^*, ..., y_N^*\}$. From tis output distribution, the mean $\bar{y}^*$ and variance $\bar{\Sigma}_{y^*}$ of the marginal distribution can be estimated. The variance represents the model uncertainty, also known as epistemic uncertainty indicating the variation in model predictions (Eq. 9.7).

$$
\begin{aligned}
\text{Mean}, \bar{y}^* &= \frac{1}{N}\sum_{n=1}^{N} {y^*}_{(n)} \\
\text{Var}, \bar{\Sigma}_{y^*} &= \frac{1}{N}\sum_{n=1}^{N} ({y^*}_{(n)} - \bar{y}^*)^2.
\end{aligned}
\tag{9.7}
$$

9.3.3 *Results*

In this section, we provide an overview of the datasets, implementation details, performance metrics, and prediction results for the encoder–decoder network. Our models were trained using publicly available pedestrian datasets, including the ETH dataset [59] comprising the ETH and HOTEL scenes, as well as the UCY dataset [36] encompassing the UNIV, ZARA1, and ZARA2 scenes. Following the approach of previous works [56], we considered 8 historical steps (equivalent to 3.2 s) to predict 12 future steps (equivalent to 4.8 s).

Initially, our model was trained on the ETH dataset, and to augment the number of trajectories, we employed a sliding window approach. A sliding window with a length of T = 1 step was utilized to generate multiple small trajectories from a single large trajectory. For example, a pedestrian's original trajectory of 29 steps resulted in 10 small trajectory pairs (x, y) with 20 steps each. In total, 1597 multivariate time series trajectory sequences were generated, which were then split into 1260 training sequences and 337 testing sequences.

9.3.3.1 Implementation Details

All the neural networks are trained end to end using TensorFlow. Adam optimizer with a learning rate of $1e - 3$ was used to compute the mean-squared error (MSE) loss.

Each model was trained for 100 epochs with a batch size of 32. The LSTM encoder–decoder model was trained with "tanh" activation while the other two models had "ReLU" activation. 10% of the training data was used for validation. The MSE error was monitored on the validation loss with callback functions like EarlyStopping and ReduceLRonPlateau. The model was compiled and fit using train and test data.

9.3.3.2 Performance Metrics

The trained model is then used to predict the future position of the pedestrian. By default, the model predicts deterministic future states. However, probabilistic predictions can be inferred using Monte Carlo dropout. We can run the stochastic inference using MC dropout repeatedly to generate a distribution of trajectories (Fig. 9.4). The mean of the distribution represents the predicted path while the associated variance quantifies uncertainty. We adopt the widely used performance metrics [5], namely, average displacement error (ADE) and final displacement error (FDE) for prediction comparison between the deterministic and probabilistic models.

(a) Average Displacement Error (ADE): It refers to the mean of Euclidean distance over all estimated points of every trajectory with the corresponding points in ground truth.

$$\text{ADE} = \frac{1}{T}\sum_{t=t_0}^{t_f} ||\hat{Y}_{(t)} - Y_{(t)}||. \tag{9.8}$$

(b) Final Displacement Error (FDE): The mean of Euclidean distance between the predicted final destination and true final destination across all trajectories.

$$\text{FDE} = ||\hat{Y}_{(t_f)} - Y_{(t_f)}||, \tag{9.9}$$

where $\hat{Y}_t$ is the predicted location at timestamp t and Y_t is the ground truth position.

9.3.3.3 Trajectory Prediction

For predicting uncertainty and evaluating its trustworthiness, we generated a distribution of trajectories using MC dropout. We have shown the uncertainty estimation of a single pedestrian trajectory from the ETH dataset using the 1D CNN network (Fig. 9.4). For the current scenario, we predict 12 steps or 4.8 s into the future based on 3.2 s of past trajectory data. A single trajectory (red) is generated by deterministic prediction and provides point estimates of future states. However, we generate probabilistic predictions by applying MC dropout to each neural network model. For instance, the encoder–decoder model with MC dropout model was sampled N = 30 times during inference with a stochastic dropout, p = 0.2 (20% of the weights are

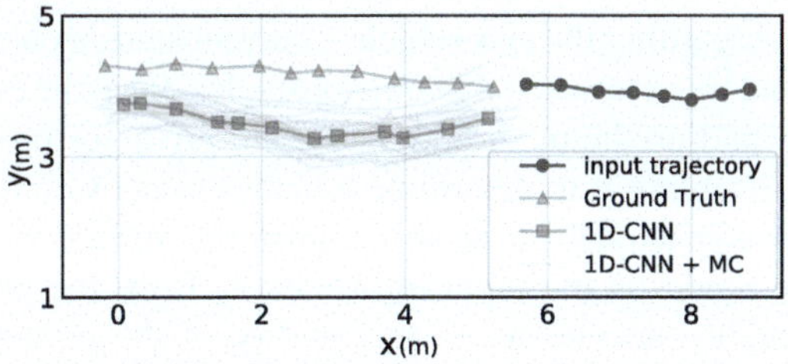

Fig. 9.4 Trajectory prediction using Monte Carlo dropout

randomly dropped) to generate a distribution of N different trajectories. The variance of the distribution quantifies the uncertainty during trajectory prediction.

9.4 Collision Avoidance and Mitigation

Every day, approximately 3,700 individuals lose their lives worldwide due to road accidents, as stated by a report from the CDC [2]. These accidents can occur due to various factors, including human errors, drowsy driving, or driving under influence. Additionally, vehicle malfunctions such as flat tires or brake failures can also lead to collisions. To address these issues, autonomous driving has emerged as a potential solution [24]. Unlike humans, autonomous vehicles do not suffer from lapses in concentration and can respond more effectively in emergencies caused by other vehicles malfunctioning because of their high computational speed. Moreover, with the development of communication technologies, the connectivity of autonomous vehicles could improve the performance of the collision avoidance algorithm and make it possible to redesign the collision avoidance algorithm with more information.

Connectivity could improve the performance of the collision avoidance algorithm. It is one of the most important parts in autonomous driving and is realized by the safe controller. Figure 9.5 illustrates the workflow for autonomous driving. The planner

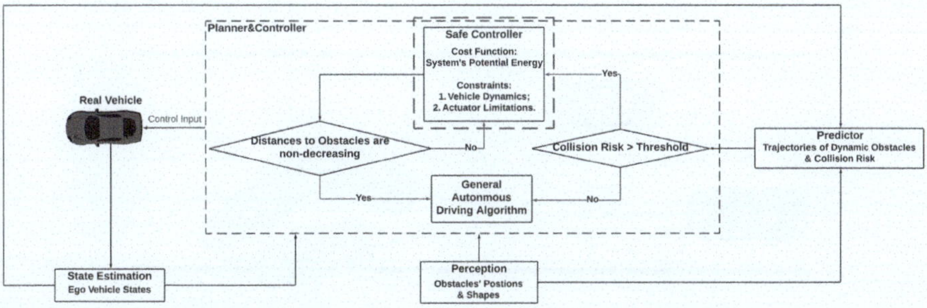

Fig. 9.5 Strategy illustration

obtains information regarding the ego vehicle's current state, as well as the positions of static obstacles and trajectories of dynamic obstacles, from the estimation module, perception module, and predictor. Additionally, the predictor provides the planner with collision risk, enabling it to determine whether to activate the safe controller. Once the safe controller is enabled, it will control the vehicle to move away from the dangerous state based on the input information. Thus, the performance of the safe controller is highly dependent on the quality of the perception, estimation, and localization data. As we have discussed in the previous section, connectivity could significantly increase the quality of these data which leads to improvements in collision avoidance performance. Moreover, with the development of $V2V$ and $V2X$ technology, the safe controller is able to know the accurate states of all dynamic obstacles and static obstacles which makes it possible to develop a generalized safe controller for all the environment.

Moreover, connectivity could transfer more information than the data mentioned above which provides the possibility for us to redesign the safe controller. In situations where emergencies arise, it is crucial for the safe controller to possess the capability to handle complex environments and realize collision mitigation, as this problem may have limited or even no feasible solutions. However, collision mitigation is often overlooked in existing methods. To address this, we developed an MPC+APF (model predictive control + artificial potential field) approach, utilizing connectivity, that can be applied in diverse environments to achieve collision mitigation. Taking into account the statistic indicating an average vehicle occupancy of 1.5 people in 2018 [18], we assume that the connectivity can provide the safe controller with occupancy information from other vehicles. With this information, our collision mitigation strategy focuses on protecting the areas occupied by passengers and redirecting potential collisions to occur with the vehicle's empty spaces. For example, consider the scenario depicted in Fig. 9.6, where an unavoidable collision arises. The red obstacle vehicle loses control at high speed and veers into the right lane. The blue ego vehicle has the option to either continue turning right along the blue dashed line or switch lanes by turning left along the green dashed line. Given the assumption that there is only one driver in the blue car and a driver with a passenger in the co-pilot seat of the red car, our method will tend to select the green dashed line to prioritize the safety of the occupants.

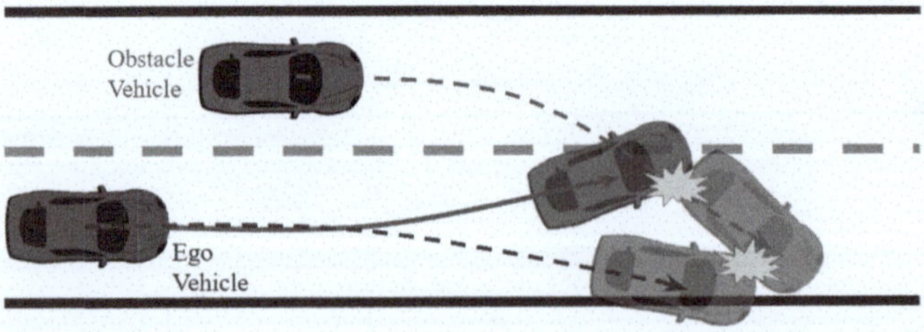

Fig. 9.6 Strategy illustration

We design a safe controller which combines the artificial potential function and model predictive control in this research. The controller is built on two important facts that are provided by connectivity: 1. The connectivity could provide accurate environment information; 2. the connectivity could provide occupancy information for other vehicles. In situations where a collision is unavoidable, the safe controller will try to identify a safe trajectory or protect the part of the vehicle occupied by humans. Throughout the remainder of this section, "ego vehicle" refers to the autonomous vehicle under our control, whereas "obstacle vehicle" represents a vehicle exhibiting abnormal behavior that causes a potential threat or hazard.

9.4.1 *Methodology*

9.4.1.1 Potential Function

We have devised a novel potential function for our problem, drawing inspiration from the electrical potential field generated by a line charge. The potential function has two fundamental components: finite-long line charge and infinite-long line charge.

The potential field of the point P generated by the finite-long line charge can be determined as indicated in [1]:

$$V_F = \int_{-a}^{b} \frac{k\lambda_1 dx}{r} = \int_{-a}^{b} \frac{k\lambda_1 dx}{\sqrt{x^2 + d^2}} = k\lambda_1 ln(\frac{b + \sqrt{b^2 + d^2}}{-a + \sqrt{a^2 + d^2}}), \tag{9.10}$$

where k represents the Coulomb constant, λ_1 denotes the line density of the charger, a corresponds to the lower bound of the line, b represents the upper bound of the line, and d represents the distance between point P and the charged finite line, as illustrated in Fig. 9.7.

Similarly, the potential field of the point P generated by an infinite-long line charge can be derived following [1]:

$$V_{Inf} = 2k\lambda_1 ln(\frac{d_0}{d}), \tag{9.11}$$

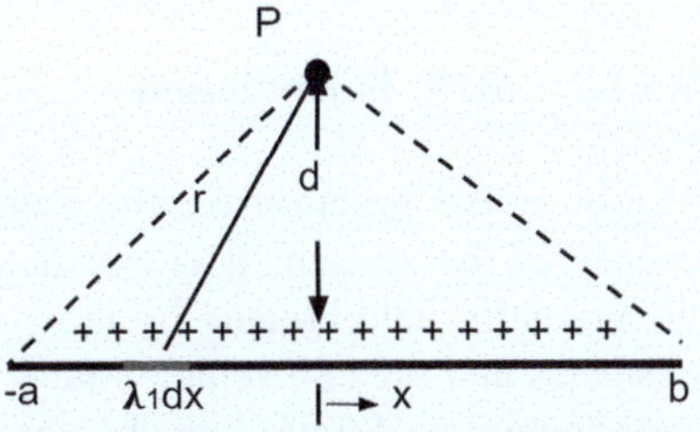

Fig. 9.7 Finite-long line charge

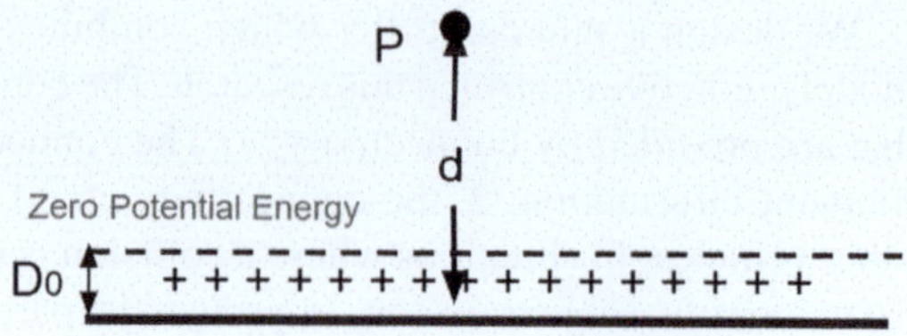

Fig. 9.8 Infinite-long line charge

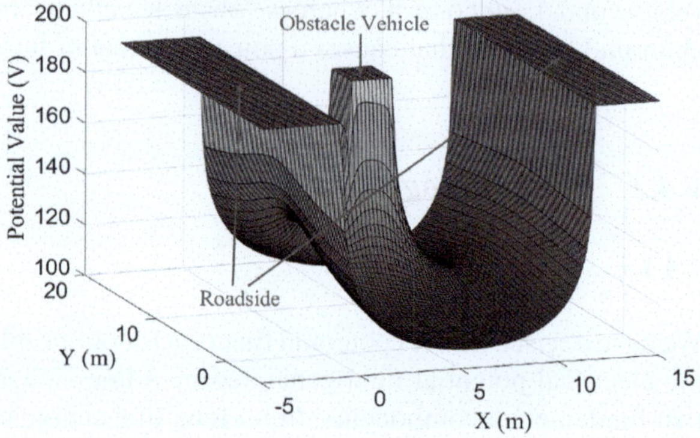

Fig. 9.9 Potential field in highway

where k represents the Coulomb constant, λ_1 denotes the line density of the charger, d represents the distance between point P and the infinite-long line charge, and d_0 corresponds to the distance at which the potential energy is defined as zero which is shown in Fig. 9.8.

For the obstacle vehicle, we consider its shape as a rectangle in a two-dimensional space. The potential field is created by superimposing the potential fields of four finite-long line charges. In the context of a highway or a normal road, the road structure is regarded as an infinite-long line charge, while in an intersection, it is treated as a semi-infinite-long charge. The overall potential field is obtained by superimposing the potential fields of all surrounding obstacles. Mathematically, it can be expressed as $V = \sum_{i=1:n} w_i V_{Obs_i}$, where w_i represents the weight and V_{Obs_i} represents the potential function of the ith obstacle. The potential fields for the highway and the intersection with a single obstacle vehicle are illustrated in Figs. 9.9 and 9.10, respectively. □

9.4.1.2 MPC Formulation

The objective function for the model predictive control problem is the potential energy of the system. It is calculated by placing the charged ego vehicle within the potential field created by the surrounding obstacles. To approximate different obstacles and the ego vehicle, they can be represented as polygons and modeled as a combination of finite line charges and infinite line charges. Additionally, specific

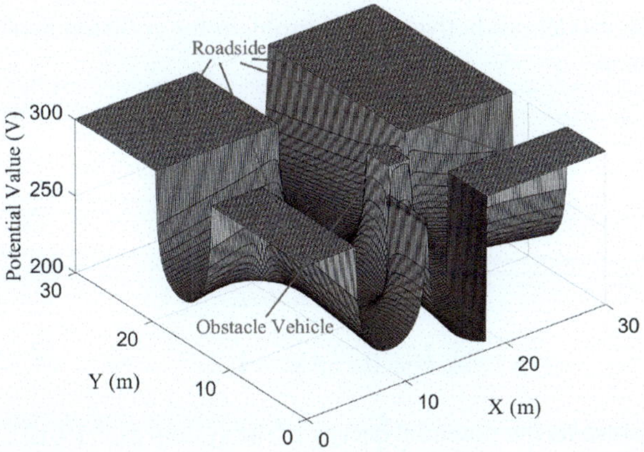

Fig. 9.10 Potential field in intersection

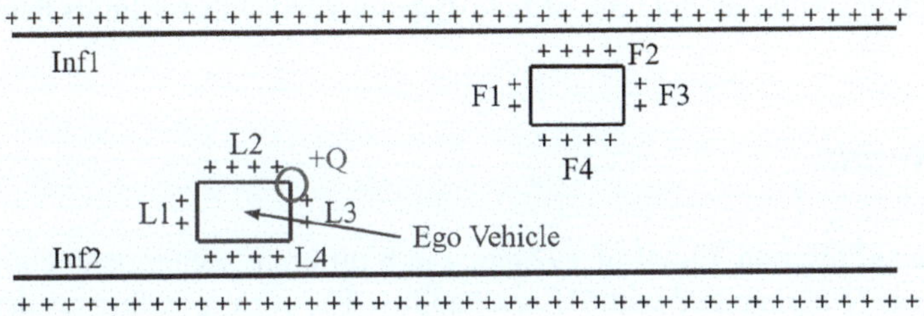

Fig. 9.11 Charged system

part protection can be implemented by introducing an additional point charge to the designated area. The penalty for the proximity of this specific part to obstacles is higher compared to other parts, as influenced by the presence of the point charge. This design reduces the likelihood of collisions occurring at the specific part. Figure 9.11 illustrates this concept, where the ego vehicle is represented on the left side and constructed using four finite-long line charges. An additional point charge is placed at the driver's position to provide specific part protection.

The potential energy of the system could be calculated as

$$E = E_R + E_V + E_{pR} + E_{pV}, \tag{9.12}$$

where E_R and E_V represent the potential energy between the ego vehicle and road structures and obstacle vehicles, respectively. E_{pR} and E_{pV} represent the potential energy between the extra point charge and road structures and the obstacle vehicles, respectively.

The nonlinear model predictive control problem for collision avoidance and collision mitigation is

$$\min_{z_k, u_k} \sum_{k=1}^{N} E(z_k) \tag{9.13}$$

$$s.t.\ \ z_{k+1} = z_k + f(z_k, u_k)T$$

$$u_{min} \leq u_k \leq u_{max},$$

$$k = 0, ..., N-1,$$

where $E(z_k)$ represents the potential energy of the system as in (9.12), N denotes the prediction horizon, and T is the time step. In the constraints, z_k, u_k, and $f(z_k, u_k)$ are vehicle states, vehicle inputs, and vehicle models. We try to minimize the system's potential energy without violating vehicle dynamics and actuator limits listed above.

9.4.2 Results

We conducted tests on a total of 192 cases in 8 different scenarios to compare the effectiveness of our method with HJ-reachability. The corresponding results are presented in Table 9.1. For each test, the ego vehicle initiates movement from its initial state until the distances between the ego vehicle and surrounding obstacles continue to non-decrease. If no collision occurs during this process, we consider it a successful avoidance of collision by the ego vehicle. When the protection mode is activated, the average success rate for APF+MPC is 56.71%, while with the protection mode off, the success rate is 57.81%. Notably, both of these rates surpass the success rate of HJ-reachability with system decomposition, which stands at 35.94%—an improvement of approximately 20%.

Table 9.1 Success rate comparison

Scenario	APF + MPC (protection on) (%)	APF + MPC (protection off) (%)	HJ-reachability (%)
1	50	41.67	25
2	41.67	62.5	20.83
3	33	50	29.17
4	75	75	75
5	87.5	70.83	66.67
6	33	41.67	8.33
7	87.75	75	33.33
8	45.83	45.83	29.17

References

1. Electric field of line charge. http://hyperphysics.phy-astr.gsu.edu/hbase/electric/elelin.html
2. Global road safety (2022). https://www.cdc.gov/transportationsafety/global/index.html
3. Abou Harfouch Y, Yuan S, Baldi S (2017) An adaptive switched control approach to heterogeneous platooning with intervehicle communication losses. IEEE Trans Control Netw Syst 5(3):1434–1444
4. Ahmed-Zaid F, Engelman GH, Haney PR (2005) Object detection in adaptive cruise control. US Patent 6,968,266
5. Alahi A, Goel K, Ramanathan V, Robicquet A, Fei-Fei L, Savarese S (2016) Social lstm: human trajectory prediction in crowded spaces. In: Proceedings of the IEEE conference on computer vision and pattern recognition, pp 961–971
6. Altché F, de La Fortelle A (2017) An lstm network for highway trajectory prediction. In: 2017 IEEE 20th international conference on intelligent transportation systems (ITSC). IEEE, pp 353–359
7. Atagoziyev M, Schmidt KW, Schmidt EG (2016) Lane change scheduling for autonomous vehicles. IFAC-PapersOnLine 49(3):61–66
8. Austin PM (1987) Relation between measured radar reflectivity and surface rainfall. Mon Weather Rev 115(5):1053–1070
9. Blum J, Eskandarian A (2006) CARAVAN: A communications architecture for reliable adaptive vehicular adhoc networks. In: SAE transactions, pp 736–746
10. Blum JJ, Eskandarian A (2005) Adaptive space division multiplexing: an improved link layer protocol for inter-vehicle communications. In: Proceedings of the 2005 IEEE intelligent transportation systems conference (ITSC). IEEE, pp 455–460
11. Blum JJ, Eskandarian A (2009) Avoiding timeslot boundary synchronization for multihop message broadcast in vehicular networks. In: Proceedings of the 2009 IEEE vehicular technology conference. IEEE, pp 1–5
12. Blum JJ, Natarajan R, Eskandarian A (2011) Synchronization challenges in media access coordination for vehicular ad hoc networks. Wirel Commun Mob Comput 11(7):954–963
13. Blum JJ, Neiswender A, Eskandarian A (2008) Denial of service attacks on inter-vehicle communication networks. In: Proceedings of the 2008 IEEE conference on intelligent transportation systems (ITSC). IEEE, pp 797–802
14. Blum JJ, Tararakin A, Eskandarian A (2008) Efficient certificate distribution for vehicle heartbeat messages. In: Proceedings of the 2008 IEEE vehicular technology conference. IEEE, pp 1–5
15. Brackstone M, McDonald M, Wu J (1998) Lane changing on the motorway: factors affecting its occurrence, and their implications
16. Cesari G, Schildbach G, Carvalho A, Borrelli F (2017) Scenario model predictive control for lane change assistance and autonomous driving on highways. IEEE Intell Transp Syst Mag 9(3):23–35
17. Chandra R, Selvaraj Y, Brännström M, Kianfar R, Murgovski N (2017) Safe autonomous lane changes in dense traffic. In: Proceedings of the 2017 IEEE international conference on intelligent transportation systems (ITSC). IEEE, pp 1–6
18. Davis SC, Boundy RG (2021) Transportation energy data book: Edition 39. Tech. rep., Oak Ridge National Lab.(ORNL), Oak Ridge, TN (United States)
19. Dey KC, Yan L, Wang X, Wang Y, Shen H, Chowdhury M, Yu L, Qiu C, Soundararaj V (2015) A review of communication, driver characteristics, and controls aspects of cooperative adaptive cruise control (CACC). IEEE Intell Transp Syst Mag 17(2):491–509
20. Di Bernardo M, Salvi A, Santini S (2014) Distributed consensus strategy for platooning of vehicles in the presence of time-varying heterogeneous communication delays. IEEE Trans Intell Transp Syst 16(1):102–112
21. DiFranco JV, Rubin WL (1968) Radar detection
22. Dunbar WB, Caveney DS (2011) Distributed receding horizon control of vehicle platoons: stability and string stability. IEEE Trans Autom Control 57(3):620–633

23. Engelman GH, Richardson MJ, Barber PA, King PJ (2001) Adaptive vehicle cruise control system and methodology. US Patent 6,233,515
24. Eskandarian A, Wu C, Sun C (2019) Research advances and challenges of autonomous and connected ground vehicles. IEEE Trans Intell Transp Syst 22(2):683–711
25. Gal Y, Ghahramani Z (2016) Dropout as a Bayesian approximation: representing model uncertainty in deep learning. In: International conference on machine learning. PMLR, pp 1050–1059
26. Gipps PG (1986) A model for the structure of lane-changing decisions. Transp Res Part B Methodol 20(5):403–414
27. Gong S, Du L (2016) Optimal location of advance warning for mandatory lane change near a two-lane highway off-ramp. Transp Res Part B Methodol 84:1–30
28. Goswami V, Bham GH (2006) A study of lane change frequency on a multilane freeway. In: Applications of advanced technology in transportation, pp 792–797
29. Hang J, Yan X, Ma L, Duan K, Zhang Y (2018) Exploring the effects of the location of the lane-end sign and traffic volume on multistage lane-changing behaviors in work zone areas: A driving simulator-based study. Transp Res Part F Traffic Psychol Behav 58:980–993
30. He J, He Z, Fan B, Chen Y (2020) Optimal location of lane-changing warning point in a two-lane road considering different traffic flows. Phys Stat Mech Appl 540:123000
31. Iberraken D, Adouane L, Denis D (2018) Multi-level Bayesian decision-making for safe and flexible autonomous navigation in highway environment. In: Proceedings of the 2018 IEEE/RSJ international conference on intelligent robots and systems (IROS). IEEE, pp 3984–3990
32. Jia D, Ngoduy D (2016) Platoon based cooperative driving model with consideration of realistic inter-vehicle communication. Transp Res Part C Emerg Technol 68:245–264
33. Jospin LV, Laga H, Boussaid F, Buntine W, Bennamoun M (2022) Hands-on Bayesian neural networks-a tutorial for deep learning users. IEEE Comput Intell Mag 17(2):29–48
34. Kahn G, Villaflor A, Pong V, Abbeel P, Levine S (2017) Uncertainty-aware reinforcement learning for collision avoidance. arXiv:1702.01182
35. Kesting A, Treiber M, Helbing D (2007) General lane-changing model MOBIL for car-following models. Transp Res Rec 1999(1):86–94
36. Leal-Taixé L, Fenzi M, Kuznetsova A, Rosenhahn B, Savarese S (2014) Learning an image-based motion context for multiple people tracking. In: Proceedings of the IEEE conference on computer vision and pattern recognition, pp 3542–3549
37. Lee KK, Chanson ST (2002) Packet loss probability for real-time wireless communications. IEEE Trans Veh Technol 51(6):1569–1575
38. Leon-Garcia A (2017) Probability, statistics, and random processes for electrical engineering
39. Li-sheng J, Wen-ping F, Ying-nan Z, Shuang-bin Y, Hai-jing H (2009) Research on safety lane change model of driver assistant system on highway. In: Proceedings of the 2009 IEEE intelligent vehicles symposium (IV). IEEE, pp 1051–1056
40. Lidström K, Sjöberg K, Holmberg U, Andersson J, Bergh F, Bjäde M, Mak S (2012) A modular CACC system integration and design. IEEE Trans Intell Transp Syst 13(3):1050–1061
41. Lin Y, Eskandarian A (2017) Experimental evaluation of cooperative adaptive cruise control with autonomous mobile robots. In: Proceedings of the 2017 IEEE conference on control technology and applications (CCTA). IEEE, pp 281–286
42. Lin Y, Wu C, Eskandarian A (2018) Integrating odometry and inter-vehicular communication for adaptive cruise control with target detection loss. In: Proceedings of the 2018 IEEE intelligent vehicles symposium (IV). IEEE, pp 1848–1853
43. Liu F, Li Y, Zuo L, Xu X (2017) Research on lane change decision for autonomous vehicles based on multi-kernels least squares policy iteration. In: Proceedings of the 2017 IEEE international conference on intelligent human-machine systems and cybernetics (IHMSC), vol 1. IEEE, pp 385–389
44. Liu X, Goldsmith A, Mahal SS, Hedrick JK (2001) Effects of communication delay on string stability in vehicle platoons. In: Proceedings of the 2001 IEEE intelligent transportation systems conference (ITSC). IEEE, pp 625–630
45. Liu Y, Wang X, Li L, Cheng S, Chen Z (2019) A novel lane change decision-making model of autonomous vehicle based on support vector machine. IEEE Access 7:26543–26550

46. Luo Y, Turgut D, Bölöni L (2015) Modeling the strategic behavior of drivers for multi-lane highway driving. J Intell Transp Syst 19(1):45–62
47. Manh H, Alaghband G (2018) Scene-lstm: a model for human trajectory prediction. arXiv:1808.04018
48. Mehr G, Eskandarian A (2020) Estimating the likelihood of reaching a road target using multiple lane changes for driver assistance. In: Proceedings of the 2020 IEEE international conference on intelligent transportation systems (ITSC). IEEE, pp 1–6
49. Mehr G, Eskandarian A (2021) Estimating the probability that a vehicle reaches a near-term goal state using multiple lane changes. IEEE Trans Intell Transp Syst 23(6):5326–5337
50. Mehr G, Eskandarian A (2021) A probabilistic approach to driver assistance for delay reduction at congested highway lane drops. Int J Transp Sci Technol 10(4):353–365
51. Mehr G, Eskandarian A (2021) Sentinel: an onboard lane change advisory system for intelligent vehicles to reduce traffic delay during freeway incidents. IEEE Trans Intell Transp Syst 23(7):8906–8917
52. Mehr G, Eskandarian A (2021) Traffic-aware lane change advance warning system for delay reduction at congested freeway diverge areas. J Transp Eng Part Syst 147(9):04021052
53. Mei M, Bullen AGR (1993) Lognormal distribution for high traffic flows. Transp Res Rec (1398)
54. National Traffic Incident Management Coalition (NTIMC): Working together for improved safety, clearance, and communications. http://ntimc.transportation.org/Pages/NationalUnifiedGoal(NUG).aspx
55. Naus GJ, Vugts RP, Ploeg J, van De Molengraft MJ, Steinbuch M (2010) String-stable CACC design and experimental validation: a frequency-domain approach. IEEE Trans Veh Technol 59(9):4268–4279
56. Nikhil N, Tran Morris B (2018) Convolutional neural network for trajectory prediction. In: Proceedings of the European conference on computer vision (ECCV) workshops, pp 0–0
57. van Nunen E, Ploeg J, Medina AM, Nijmeijer H (2013) Fault tolerancy in cooperative adaptive cruise control. In: Proceedings of the 2013 IEEE conference on intelligent transportation systems (ITSC). IEEE, pp 1184–1189
58. Park BB, Malakorn K, Lee J et al (2011) Quantifying benefits of cooperative adaptive cruise control towards sustainable transportation system
59. Pellegrini S, Ess A, Van Gool L (2010) Improving data association by joint modeling of pedestrian trajectories and groupings. In: Computer vision–ECCV 2010: 11th European conference on computer vision, Heraklion, Crete, Greece, September 5–11, 2010, Proceedings, Part I 11. Springer, pp 452–465
60. Ploeg J, Scheepers BT, Van Nunen E, Van de Wouw N, Nijmeijer H (2011) Design and experimental evaluation of cooperative adaptive cruise control. In: Proceedings of the 2011 IEEE conference on intelligent transportation systems (ITSC). IEEE, pp 260–265
61. Ploeg J, Semsar-Kazerooni E, Lijster G, van de Wouw N, Nijmeijer H (2014) Graceful degradation of cooperative adaptive cruise control. IEEE Trans Intell Transp Syst 16(1):488–497
62. Ploeg J, Shukla DP, Van De Wouw N, Nijmeijer H (2013) Controller synthesis for string stability of vehicle platoons. IEEE Trans Intell Transp Syst 15(2):854–865
63. Roncoli C, Bekiaris-Liberis N, Papageorgiou M (2017) Lane-changing feedback control for efficient lane assignment at motorway bottlenecks. Transp Res Rec 2625(1):20–31
64. Santini S, Salvi A, Valente AS, Pescapé A, Segata M, Cigno RL (2016) A consensus-based approach for platooning with intervehicular communications and its validation in realistic scenarios. IEEE Trans Veh Technol 66(3):1985–1999
65. Schrank D, Eisele B, Lomax T (2019) Urban mobility report 2019. Tech. rep., Texas A&M Transportation Institute, The Texas A&M University
66. Sen B, Smith JD, Najm WG et al (2003) Analysis of lane change crashes. Tech. rep., United States Department of Transportation (DOT) National Highway Traffic Safety Administration (NHTSA)
67. Shladover SE, Nowakowski C, Lu XY, Ferlis R (2015) Cooperative adaptive cruise control: definitions and operating concepts. Transp Res Rec 2489(1):145–152

68. Sudou H, Hashizume T (2006) Adaptive cruise control system
69. Suo Y, Chen W, Claramunt C, Yang S (2020) A ship trajectory prediction framework based on a recurrent neural network. Sensors 20(18):5133
70. Vallon C, Ercan Z, Carvalho A, Borrelli F (2017) A machine learning approach for personalized autonomous lane change initiation and control. In: Proceedings of the 2017 IEEE intelligent vehicles symposium (IV). IEEE, pp 1590–1595
71. Van Arem B, Van Driel CJ, Visser R (2006) The impact of cooperative adaptive cruise control on traffic-flow characteristics. IEEE Trans Intell Transp Syst 7(4):429–436
72. Van Dijck T, van der Heijden GA (2005) VisionSense: an advanced lateral collision warning system. In: Proceedings of the 2005 IEEE intelligent vehicles symposium (IV). IEEE, pp 296–301
73. Van Driel CJ, Van Arem B (2010) The impact of a congestion assistant on traffic flow efficiency and safety in congested traffic caused by a lane drop. J Intell Transp Syst 14(4):197–208
74. Van Nunen E, Verhaegh J, Silvas E, Semsar-Kazerooni E, Van De Wouw N (2017) Robust model predictive cooperative adaptive cruise control subject to V2V impairments. In: Proceedings of the 2017 IEEE conference on intelligent transportation systems (ITSC). IEEE, pp 1–8
75. Wang M, Wang Z, Paudel S, Schwager M (2018) Safe distributed lane change maneuvers for multiple autonomous vehicles using buffered input cells. In: Proceedings of the 2018 IEEE international conference on robotics and automation (ICRA). IEEE, pp 1–7
76. Wu C, Lin Y, Eskandarian A (2019) Cooperative adaptive cruise control with adaptive Kalman filter subject to temporary communication loss. IEEE Access 7:93558–93568
77. Yin S, Li Z, Zhang Y, Yao D, Su Y, Li L (2009) Headway distribution modeling with regard to traffic status. In: Proceedings of the 2009 IEEE intelligent vehicles symposium (IV). IEEE, pp 1057–1062
78. Zanten A, Kost F Handbook of driver assistance systems: basic information, components and systems for active safety and comfort, vol 40, pp 919–967
79. Zhang C, Sabar NR, Chung E, Bhaskar A, Guo X (2019) Optimisation of lane-changing advisory at the motorway lane drop bottleneck. Transp Res Part C Emerg Technol 106:303–316
80. Zhang X, Mahadevan S (2020) Bayesian neural networks for flight trajectory prediction and safety assessment. Decis Support Syst 131:113246
81. Zhang Y, Ioannou PA (2016) Combined variable speed limit and lane change control for highway traffic. IEEE Trans Intell Transp Syst 18(7):1812–1823

Chapter 10
Connected Autonomous Driving Using Reconfigurable Intelligent Metasurfaces

Ehizogie Emoyon-Iredia, Taqwa Saeed, Nouman Ashraf, Christos Liaskos, Michele Segata, Paolo Casari, Sergi Abadal, Eduard Alarcon, Andreas Pitsillides, and Marios Lestas

Abstract Beyond 5G/6G communication systems promise to significantly impact the development of a New Generation of CCAM. Reconfigurable Intelligent Meta-

E. Emoyon-Iredia · M. Lestas (✉)
Department of Electrical and Computer Engineering and Informatics, Frederick University, Nicosia, Cyprus
e-mail: eng.lm@frederick.ac.cy

E. Emoyon-Iredia
e-mail: st015229@stud.frederick.ac.cy

Frederick Research Center, Nicosia, Cyprus

T. Saeed
Certec - Rehabilitation Engineering and Design, Lund University, Lund, Sweden

N. Ashraf
School of Electrical and Electronic Engineering, Technological University Dublin, Dublin, Ireland
e-mail: nouman.ashraf@tudublin.ie

C. Liaskos
Department of Computer Science and Engineering, University of Ioannina, Ioannina, Greece
e-mail: cliaskos@cse.uoi.gr

M. Segata · P. Casari
Department of Information Engineering and Computer Science, University of Trento, Trento, Italy
e-mail: michele.segata@unitn.it

P. Casari
e-mail: paolo.casari@unitn.it

S. Abadal · E. Alarcon
Department of Computer Architecture, Universitat Politècnica de Catalunya (UPC), Barcelona, Spain
e-mail: abadal@ac.upc.edu

E. Alarcon
e-mail: eduard.alarcon@upc.edu

A. Pitsillides
Department of Computer Science, University of Cyprus, Nicosia, Cyprus
e-mail: pitsillides.andreas@ucy.ac.cy

P. Ioannou and A. A. Malikopoulos (eds.), *Transportation Mobility in Smart Cities*, Springer Tracts on Transportation and Traffic 21,
https://doi.org/10.1007/978-3-031-64769-7_10

surfaces (RIM) are by now established as a key enabling technology for 6G Systems. They have been extensively investigated the last few years, as they possess exotic properties allowing for precise control over any aspect of an impinging wave. As such, they can be harnessed for the realization of Programmable Wireless Environments (PWE). Despite their investigation for deployment in a number of applications, their integration in Connected Autonomous Driving Applications has not been considered in literature with respect to cooperative driving performance. In this chapter, we consider the HyperSurface (HSF) as the enabling technology for RIM and present recent results demonstrating the feasibility and the associated performance gains achieved from the integration of RIM in Connected Automated Driving Applications. We discuss HSF design considerations which can affect CCAM performance with respect to the system architecture, the controller network and the associated routing protocols, the workload characterization and the closed loop beam steering design. We envision the chapter to serve as a synopsis of tools and methodologies that can be used for RIM-enabled CCAM system design.

10.1 Introduction

Cooperative, Connected, and Automated Mobility (CCAM) is a thriving sector with the promise to bring enormous improvements in road safety, traffic efficiency, and comfort in driving with significant social, economic, and environmental benefits. It has been well established by now that automated driving can greatly improve safety by eliminating human-error-related accidents and has tremendous social and economic benefits by alleviating congestion and reducing accident-related costs [1]. Moreover, the automotive industry is moving towards more automation and electrification, which both hold promise for better environmental practices [2]. Finally, automated driving systems, at their maturity, could increase mobility for seniors and people with disabilities and expand transportation options for underrepresented communities [3]. As a result of the aforementioned benefits, the autonomous (driverless) car market was valued at USD 22.22 billion in 2021, and it is expected to reach USD 75.95 billion by 2027 [4]. However, Cooperative Autonomous Vehicles (CAVs) require an underlying, highly dependable and high-performance wireless communication infrastructure. Applications that can benefit from such infrastructure include cooperative driving and perception, Vehicular Edge Computing (VEC), and distributed learning. In this context, Millimeter wave (mmWave) communications have been well established to serve cooperative vehicular applications due to their high-bandwidth performance [5] but are prone to strong signal attenuation in the presence of blocking bodies, e.g., around buildings and intersections. The unpredictable appearance of blockage and the temporary disconnection that ensue may thus prevent mmWave technology from delivering the dependable, high-bandwidth, and low-delay communications that cooperative vehicular applications require. Recently, a new wireless communication paradigm named programmable wireless environments (PWEs) has emerged, which strives to constitute the wireless propagation phenomenon into a fully controllable process, and thus mitigate previously unmanageable phenomena [6]. To realize this, PWEs coat the environment's planar objects with metasurfaces supporting well-defined networking and programming interfaces, denoted as Reconfigurable Intelligent Surfaces (RISs). RISs can passively alter the power, direction, polarization, and phase of any impinging wave by tuning the impedance of their reflecting elements [7]. In a popular RIS implementation, such tuning is achieved through varactors and varistors, or by constructing the reflecting elements from appropriate materials, e.g., liquid crystals. In either case, the objective is to provide a desired scattering diagram. Multiple deployed RISs can be dynamically orchestrated to realize custom end-to-end wireless propagation routes between vehicles, e.g., bypassing obstacles and offering a virtual line-of-sight link, thus enhancing the quality and reliability of the channel [8]. RISs have been considered in a number of recent works targeting vehicular communication scenarios [9], focusing mostly on preserving reliability under fast mobility. For example, a robust transmission scheme for RIS-assisted vehicular communication was proposed in [10], while the works in [11, 12] derived the performance of a vehicular RIS-assisted network in terms of outage probability and security. Moreover, the authors in [13] optimized RIS placement to maximize the received power for a communication scenario between a base station and an autonomous vehicle, while, in [14], the authors investigated communication aspects that can affect the vehicular network, such as configuration delays and power consumption. It is evident from the above that the integration of RIS technologies in vehicular networking has been investigated merely from a communications perspective, leaving the effects on CCAM applications and automated driving performance largely unexplored. One of the reasons obstructing such investigations, is the lack of validated performance evaluation tools for the considered settings: validated mathematical models, simulation environments, and above all, practical testbeds. So, addressing this gap in the field, we have recently developed a simulation framework based on the PLEXE/VEINS [15] platform which can be used for evaluation and design.

In this book chapter, we present recent results which demonstrate the feasibility and potential of integrating RIS technologies in CCAM applications. We focus on a specific RIS technology, namely the Hypersurface (HSF) and discuss its architecture and analysis tools that we have developed. Those tools can be used to characterize performance indicators relevant to CCAM applications such as the workload, reconfiguration delays, power consumption, fault tolerance, and feedback-based tracking stability. Finally, we show simulation results which demonstrate the ability of RIS technology to assist in the unobstructed realization of a merging maneuver on an intersection and also show the feasibility and limitations of the technology in scenarios with cascaded metasurfaces.

10.2 Metasurfaces Review

Metamaterials are based on the fundamental idea that the physical properties of matter stem from its atomic structure [16]. Therefore, one can create artificially structured materials (comprising sufficiently small elementary units known as meta-atoms) to yield any required energy-manipulating behavior, including types not found in natural materials [17]. A unique feature of metamaterials is that they have simple structures and therefore, offer a variety of techniques for generally low-cost and scalable fabrication. Techniques such as printed circuit boards, flexible materials such as Kapton, 3D printing, Large Area Electronics, bio-skins, and microfluidics, to name a few have been successfully employed for manufacturing [18]. For example, in order to create photonic metamaterials, building blocks of customized micro- or nanostructured metallo-dielectric sub-wavelength are densely packed into a functional material. The realization of novel, unusual, and occasionally even unheard-of optical properties, such as magnetism at optical frequencies, negative or large positive refractive indices, zero reflection via impedance matching, perfect absorption, enormous circular dichroism, or improved nonlinear optical properties, is made possible by this seemingly simple but groundbreaking concept. A few potential uses for metamaterials are compact polarization optics, cloaking technology, and ultrahigh-resolution imaging systems [19].

In order to achieve tunability, an impinging wave of any physical kind (such as electromagnetic, mechanical, acoustic, or thermal) triggers a metamaterial's surface elements, initiating a spatial distribution of energy over and within it; this phenomenon is known as surface plasmon. Surface currents are also induced over the material which can be modified by tuning the actuator elements within it (e.g., simple switches) accordingly. The metamaterial surface then organizes its adjustable components in a circuit to transform the exciting source into the producing source. The enabling physical phenomenon is the Huygens principle, which states that any electromagnetic (EM) wavefront departing from a surface can be traced back to an equivalent current distribution over a surface. This means metamaterials can produce any custom departing EM wave as a response to any impinging wave, just by tuning the state of embedded switches/actuators.

The relative size of a meta-atom compared to the wavelength of the excitation (impinging wave) defines the energy manipulation precision and efficiency of a metamaterial [18]. The 2D equivalent of metamaterials are metasurfaces (MSs), which are synthetic ultrathin sheet materials. They comprise of unit cells, which are sub-wavelength meta-atoms, arranged on a generally flat surface [20]. Their unit-cell response is described in terms of (local) complex reflection and transmission coefficients. Control over these complex coefficients, locally or globally, can give rise to a multitude of tunable functionalities and applications, such as perfect absorption, anomalous reflection, wavefront shaping, polarization control, and dispersion engineering. Historically, early metasurfaces were used to provide a fixed EM response for a specific incident wave that was assumed to be known beforehand. It was soon realized that the capability of dynamically tuning the metasurface properties, either to change the desired output functionality or to adapt it to an input wave of different characteristics, would greatly enhance the potential of metasurfaces for practical applications, giving rise to tunable metasurfaces. These have now been established as an important part of next-generation wireless systems because they have facilitated the realization of novel components with engineered and uneven, unnatural functionalities for the radio frequency (RF), terahertz (THz), and optical spectrum. Metasurfaces have tremendous significance in modern wireless vehicular communication technologies. Examples like the Dedicated Short-Range Communication (DSRC) and Long-Term Evolution (LTE) limit the communication speeds that can be achieved due to the limited communication resources availability [21]. As a result, the use of mmWave and THz communications to solve the bandwidth shortage problem in next-generation capacity-demanding networks is now being considered. At these frequencies, the communication technologies have more bandwidth available, resulting in higher possible data rates. This helps to improve the safety of autonomous driving even if the higher frequency suffers a higher propagation loss and a higher noise power [22]. Since mmWave signals rely heavily on directional multiple-input multiple-output (MIMO) antenna technologies which employ advanced beam forming techniques to achieve desired directionality and receiver alignment, they face numerous challenges in high-device mobility scenarios such as vehicular networks. One of such challenges is that mmWave communication is susceptible to blockages that can result in Non-Line-of-Sight (NLOS) channels, which degrade performance significantly. When direct communications have poor quality, such as in these NLOS channel cases, RISs can tune the wireless propagation environment to aid transmissions between the sender and receiver. RISs address this issue by providing virtual line of sight paths [23].

Metasurfaces are a key enabling technology for RISs due to their special properties, such as their ability to provide abrupt phase change, amplitude modulation, and polarization conversion of EM waves. Metasurfaces allow interactivity and interconnectivity since they can exchange information with their external environment and by sensing external effects they can effectively adjust their internal structure according to their goal. Surface waves can be manipulated by creating metasurface unit cells with the appropriate impedance which can then be patterned in ways that can guide or break waves in specific directions; these are used in scattering control applications [16]. Despite their exceptional qualities, the use of metamaterials and metasurfaces is constrained by accessibility issues since developing a metamaterial is seen as a task for specialized researchers, which limits its applicability to the more expansive engineering field [24]. Excellent reviews on metasurfaces have appeared in the literature, covering the underlying physics, the different applications and functionalities, the approaches to tunability, and the opportunities of nonlinearity and nonreciprocity [16–24].

10.3 Hypersurface Architecture

Long-term research on the subject of achieving reconfigurability in metamaterials culminated in the concept of Software-Defined Metamaterials (SDMs), the main idea of which was to incorporate a network of miniature controllers within the metamaterial structure [25]. The realization of HSF units, which may be incorporated in programs like PWEs, is made possible by adding a complete set of protocols and an Application Programming Interface (API) to the underlying metasurface hardware. This remarkable paradigm shift now allows for the control of the wireless channel [26].

For a definition, the HSF is a class of programmable metasurfaces which aims to embed intelligence within the metasurface to make it self-adaptive, and not entirely dependent on bulky and power-hungry external control devices such as a field-programmable gate array (FPGA) [27]. HSFs provide precise beamforming, efficient radiation pattern, and compact design with the cost of complexity. To this end, HSFs integrate a network of controller chips within the metasurface body. The controllers receive programmatic directives and perform simple alterations on the metasurface structure, adjusting its EM behavior globally, locally, upon request, or depending on the environment. This has several advantages. First, the SDM can host multiple functionalities concurrently and adaptively. Second, the HSF can now be connected to external devices or even other HSFs (forming inter-networks) to better adapt to the surroundings or increase the operative range. Last but not least, the HSF concept reduces the knowledge required to design a metamaterial for a given purpose [24]. The internal network of controller chips is central to the HSF approach and the main difference with respect to the rest of programmable metasurfaces. The controllers will not only incur significant fabrication and integration costs but also consume non-negligible power during operation. To optimize both costs and power, it is therefore crucial to understand how final application demands will affect the design of the HSFs. In the longer term, the HSF could incorporate new components such as embedded nanosensors, a full-integrated network, or an energy harvesting system, and exploit smaller and faster controllers to create devices capable of reacting microwave or THz signals in a truly autonomous manner, without having to rely on the constant intervention of an external controller.

Specifically, in vehicular applications the RIS is envisioned to be part of the road-side infrastructure enhancing the quality of Vehicle-to-Vehicle (V2V) and Vehicle-to-Infrastructure (V2I) communications when these are disrupted by blockages. CCAM applications rely heavily on the quality of the communications channels enabling cooperative actions and as such, any feedback control algorithms realizing automated maneuvers must account for the characteristics of the established communications. Beyond that, the reconfiguration delays of RISs are also relevant as these contribute to feedback delays which can compromise the system performance and stability. Both the RIS assisted communications and the reconfiguration delays together with the RIS power consumption are dependent on the adopted RIS architecture. In this context, implementing the HSF architecture can assist in disseminating reconfigura-

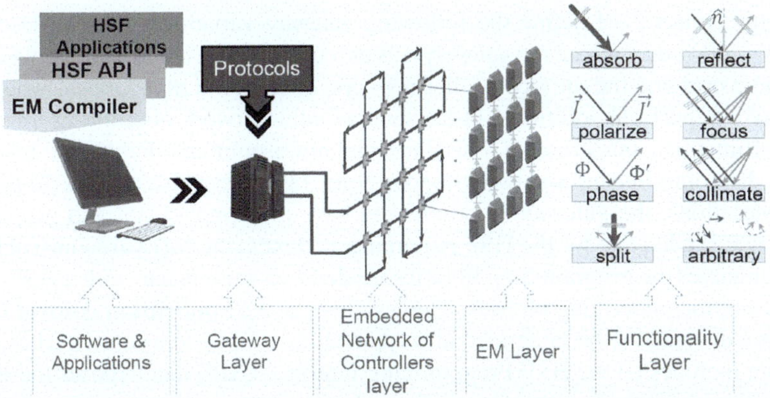

Fig. 10.1 Hypersurface structure and mode of operation. (Reproduced from [29] with permission from IEEE. © [2018] IEEE)

tion directives to each of the unit cells. Interconnected HSFs constitute a complete hardware and software platform that: (i) Express the physical phenomena behind MSs and their interactions in a callable software library form. (ii) Interconnects MSs to the Internet-of-Things (IoT) ecosystem. (iii) Enable using MSs in a conventional sense-input/adapt-output loop. The nodes within the controller network may have any set of integrated features, i.e., sensors or actuators. Sensors can exemplary sense the characteristics of the impinging waves (e.g., their direction of arrival). Actuators, such as semiconductor- or Microelectromechanical systems (MEMS)-based switches, are employed to control the current flow distribution over the surface of the HSF.

HSFs comprise four distinct functionality layers as shown in Fig. 10.1. First, the **intra-networking layer**. The HSF incorporates an array of networked controls and the aforementioned intra-network can be implemented in a variety of methods, including gateway-driven, wireless, and wired networks. In the HSF paradigm, a wired networking strategy was selected, which assumes that the controllers and the gateway are both connected over a wired means. The minimum threshold for enabling distributed intelligence in an HSF is wired networking, which enables smart meta-atoms to detect EM waves, communicate, and act locally on currents. This inevitably reduces the HSF's reliance on external computers while enabling the bypassing of faulty controllers and increasing the HSF's lifespan. Between the HSF and the Internet stands the **gateway layer**. It consists of the hardware and protocols for establishing a connection between HSFs and any other smart device (such as a cell phone or an IoT device). On the HSF digital board, this layer appears as a two-sided component which is able to locate peers and exchange data with them since it functions internally as a peer in the controller intra-network. By converting the intra-network data into a format that is compatible with widely used protocols (such WiFi, Ethernet, and Bluetooth), mainstream connectivity is made available externally.

In this context, we define the following gateway variations: (i) The described *pass-through mode*, where the gateway acts as a translator of commands between the external world and the intra-controller network, and (ii) the *smart mode*, where the gateway can perform additional tasks, such as intra-network monitoring for errors and malfunctions, intra-routing assistance, and multi-point intelligent data injection within the controller intra-network topology for faster response overall. In both of these variations, the gateway is assumed to have a computational and processing power that is adequate for the HSF requirement. Next is the **software control layer** which involves the programming libraries needed to communicate with a HSF while abstracting the underlying physics. The last layer is the **application control layer**. HSFs are designed to be integrated into applications and systems easily. Thus, their software includes (i) the HSF Programming Interface (HPI) that exposes the HSF's capabilities in software callbacks. (ii) The user-side interrupt service, i.e., a user-side daemon, that receives and handles interrupts, e.g., messages generated spontaneously by HSFs to denote hardware failures. (iii) The EM Compiler, an HSF-external or -internal software service that translates the HPI callbacks to corresponding spatio-temporal actuation directives for the HSF active elements [28].

10.4 Controller Network

As indicated in the previous section, the main characteristic of the hypersurface is the embedded controller network which facilitates the dissemination of the packetized directives from the input gateway to each of the unit cells. The nontrivial topology of this controller network necessitates routing protocols different from existing solutions which in turn affect the reconfiguration delays as well as the power consumption within the metasurface. In this section, we detail the characteristics of the controller network, the developed routing protocol as well as the resulting reconfiguration delays. The control layer may be as simple as directly connecting the state-altering components to the gateway, but, as the meta-atom density rises, a network of embedded data controllers is required to provide control within the HSF. The HSF-integrated Controller network (CN) faces severe resource constraints due to the global design requirements of a low-cost, low-power solution that conforms with the necessary downsizing without interfering with the delicate EM properties of the metasurface. This asynchronous design necessitates three pins per input/output port to support the unidirectional channels required by the employed four-way handshaking protocol [29]. For each controller node interfacing, 25 pins (shown in Fig. 10.2) have been allocated with this restriction imposed by cost and miniaturization considerations. Taking into account that a number of these pins are allocated for global metasurface reconfiguration functions, this allows for only four unidirectional input/output ports with each port reserving three pins. This results in the crucial restriction of two unidirectional input ports and two unidirectional output ports, which shapes to a great extent the uncommon, asymmetric *Manhattan type* topology (shown in Fig. 10.3) and presents the most severe design challenges for routing protocols.

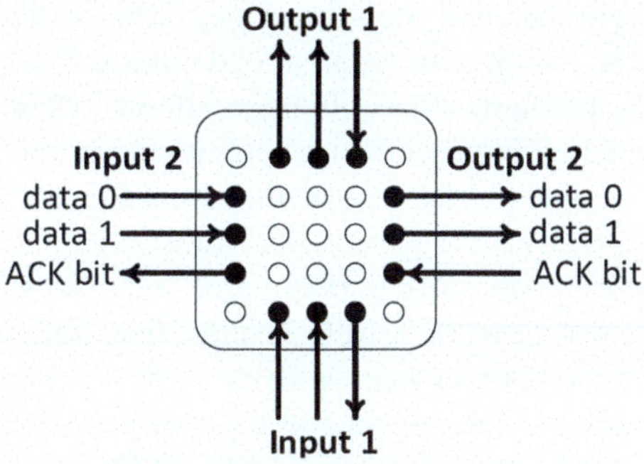

Fig. 10.2 Pin allocation of the controller node and the layout of unidirectional input/output channels

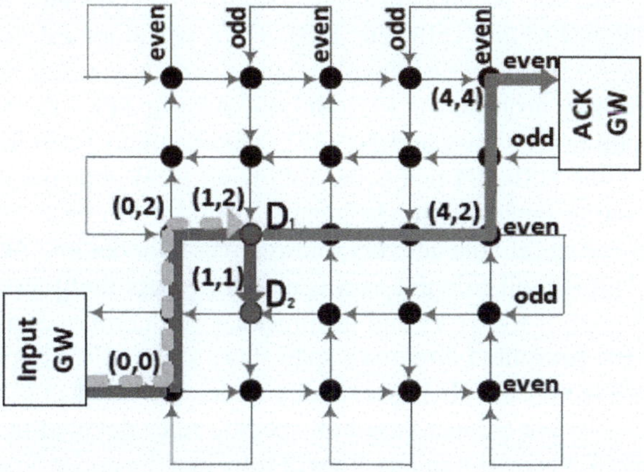

Fig. 10.3 A 5 × 5 HSF-embedded controller network topology with two packetized directives routing scenarios shown in solid blue and dotted orange links; the ACK path is shown in solid green. (Reproduced from [30] with permission from IEEE. © [2020] IEEE)

As depicted in Fig. 10.3, there is single directionality at each row and column that alternates from one row (or column) to the next. A mesh topology with only half of the unidirectional links present is analogous to this. In order to provide adaptability in the event of defects or malfunctions, controller nodes at the network edges have links accessible for further connectivity. Other factors that pose design challenges for the control layer are low power consumption, dense deployment, the small size of meta-atoms, avoiding EM interference, robustness against any form of fault, etc. There are similarities between the geometry of HSF-embedded CN and Network-on-Chip (NoC) interconnected multicore architectures in terms of miniaturization and general layout. However, due to the aforementioned resource constraints, the CN differs significantly from most conventional NoCs in terms of its operational features [30]. This necessitated the development of a routing mechanism that can suit these restrictions.

Dimension-Order Routing (DOR), also referred to as **XY** algorithm was first formulated because it enjoys widespread use in 2D mesh NoCs [31, 32] due to its inherent deadlock-avoidance property and implementation simplicity. Unfortunately, due to alternating directionality of successive network rows and columns in the CN, DOR quickly becomes unusable [30]. Thus, a topology-aware variation of the XY algorithm has been developed. This variant, referred to as topology agnostic XY–YX, avoids routing up at odd columns (which only route down) and routes through the previous even column as shown in Fig. 10.3. In the figure, nodes D_1 and D_2 are reached through the first column (column 0) instead of the second (column 1).

Agnostic XY–YX is distributed (i.e., executed locally at each node) and adhere correctly to the geometrical idiosyncrasy of the HSF CN topology. More importantly, since it does not perform any 360° turns, it is livelock- and deadlock-free as long as the network is fully healthy (has no faulty links or nodes). HSFs, however, can be complex devices with densely integrated tuning elements, unit cells, controller chips, etc., in non-standard packages in their embedded CN version. Such HSFs must provide protection against faults that may occur during construction, deployment, or use and be brought on by factors like connector misalignment, aging, physical damage to unit cells, etc. [33]. Thus, we adopt a design methodology that assumes the formation of disjointed rectangular regions called Faulty Blocks (FBs), which was initially proposed in [34] and utilizes principles from the Turn Model for adaptive routing by Glass and Ni [35]. The resultant Fault-Tolerant (FT) routing protocol for the HSF CN meets the following ideal objectives: (1) it is distributed, with all routing decisions made locally at nodes to conserve resources, (2) it is feasible, ensuring reliable delivery of directives to nodes and ACK messages to the ACK GW, (3) it is fault-tolerant as it is able to bypass faulty nodes contained in FBs, (4) it is devoid of livelocks and deadlocks, (5) it comprises a reasonable fault model using FBs, (6) routed directives span short routes when bypassing faulty nodes, and (7) it is adaptive as it bypasses faulty nodes and blocks [34].

FBs are created based on specific rules detailed in [30], that guarantee that each block has two boundary lines of fully healthy nodes on each side as shown in Fig. 10.4. Additionally, if two FB overlap in deadlock-inducing manner (also defined based on the said rules) they merge to form a super-FB. Packets are routed through the boundaries of FB as long as their destination is outside of FBs. Destinations inside FBs are considered faulty and do not receive packets. The packets are forwarded using what we refer to as South-Last routing where a packet that encounters an FB is routed around it in a clockwise direction. However, before encountering a boundary of an FB and after being forwarded away from an FB, a packet simply follows agnostic XY–YX routing. This is accomplished by configuring boundary nodes to perform predetermined routing commands and other nodes employing agnostic XY–YX. It is important to note also, that ACK packets are sent from the node prior to the destination node. That is, when the node before the destination forwards the last bit of the packet to the destination and receives ACK through the handshake protocol, the generation and transmission of an ACK packet is triggered. The ACK packet is forwarded using the same strategy to the ACK GW as depicted in Fig. 10.4.

Simulation results demonstrate the effectiveness of the proposed FT routing protocol under two fault distribution scenarios: Random Faults (RF) and spatially Correlated Faults (CF, or clustered) placements. A custom-coded simulator built with the AnyLogic Tool platform to accurately replicate the special features of the HSF CN system, namely its asynchronous operation, is used to evaluate the proposed FT routing algorithm. RFs entail a per-node probability of failure (p_f), while CFs are

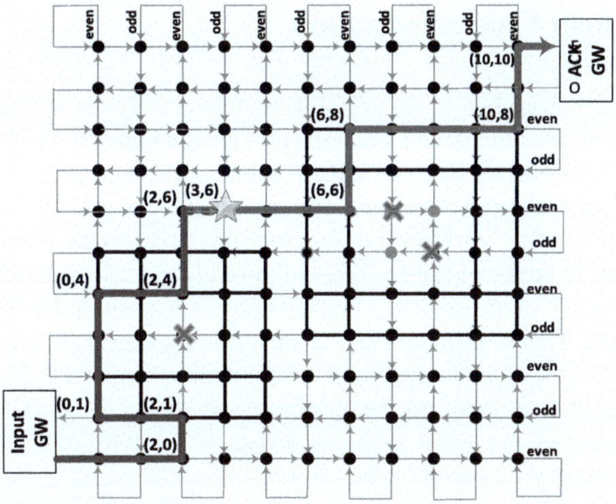

Fig. 10.4 Routing around faulty blocks: directives (red), ACKs (blue). (Reproduced from [30] with permission from IEEE. © [2020] IEEE)

represented using a spatially dependent failure model, where nodes adjacent to a defective node are more likely to also fail. Delivery reliability is measured in terms of the percentage of network nodes that are in a healthy state and, as a result, are assured to receive packetized directives sent to them by the proposed FT routing algorithm (RA). The average routing hop count as a function of the number of defective nodes and the size of the FB region is the subject of these measurements. The average per-packet hop count of directives delivered to healthy nodes under the RF and CF scenarios only rises, albeit slowly, as a higher proportion of network nodes fail, as seen in Fig. 10.5.

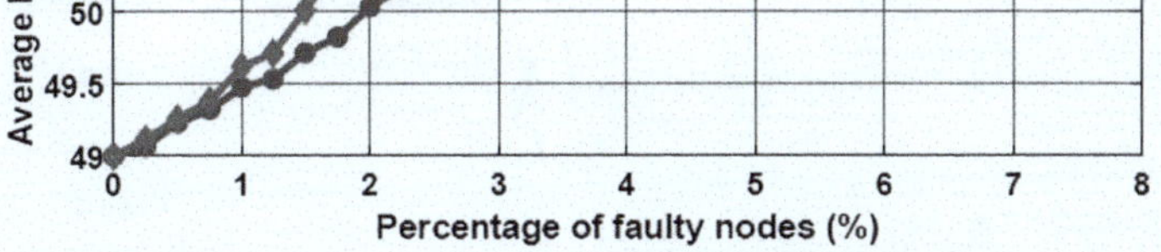

Fig. 10.5 Average per-packet hop count versus percentage of faulty nodes. (Reproduced from [30] with permission from IEEE. © [2020] IEEE)

10.5 Workload Characterization

The routing policy described in the previous section is only one of the factors which dictates the reconfiguration delays once the unit cells which need to be reconfigured at each reconfiguration step have been determined. However, the question is, which cells need to be reconfigured at each reconfiguration step to realize a particular functionality and what traffic/workload burden does this bring to the controller network. The traffic results from the packetized reconfiguration directives disseminated by the input gateway to each of the unit cells that need to be reconfigured using the proposed routing algorithm. This workload/traffic characterizing is crucial in determining the reconfiguration timings and delay which in turn dictate limitations in system performance. For vehicular applications, beam steering to track a mobile target is relevant, however, the speed of the target (vehicle, drone, vulnerable user) cannot exceed a particular value as the reconfiguration intervals might become less than the reconfiguration delay rendering the tracking infeasible. Below, we present the adopted methodology for workload characterization and we indicate performance limitations of the HSF which are relevant to vehicular applications.

The system of the HSF under investigation is shown in Fig. 10.6. It is made up of a bi-dimensional array of $M \times N$ unit cells coupled to a gateway device that connects the HSF tile to other tiles or to the external world. We assume that each unit cell has a square shape with sides of length d_u and can be configured in up to N_s states. These states represent the individual phases of cells that collectively create the phase gradient of the surface prompting the steering of the beam. Each cell can have one of discrete N_s phase values realized through tuning the varactor elements integrated into the unit cell structure [36]. In the case shown in Fig. 10.6, $N_s = 4$ which is the minimum number of states required to ensure high-quality beam steering performance as pointed out in [36]; each state is represented by a color.

Metasurface coding refers to the assignment of states to the unit cells of the HSF, which generally depends on the desired functionality and input parameters. In the particular case of anomalous reflection, the metasurface needs to implement linear phase gradients Φ'_x and Φ'_y in the x and y directions, respectively, as shown

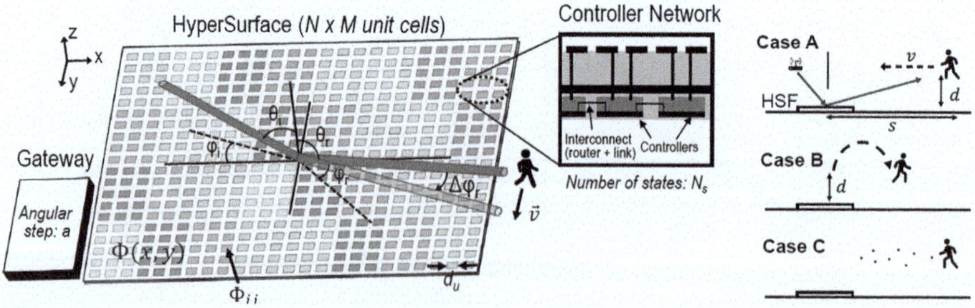

Fig. 10.6 Mobility system model. (Reproduced from [28] with permission from IEEE. © [2021] IEEE)

in Fig. 10.6 [37, 38]. In such a configuration, the metasurface would ideally present full reflectivity and a phase profile $\Phi(x, y) = \Phi'_x x + \Phi'_y y$ so that it complies with the required phase gradients. Such a phase profile allows to use the momentum conservation law for wave vectors at the air–metasurface interface, yielding

$$\begin{aligned} k_i \sin\theta_i \cos\phi_i + \Phi'_x &= k_r \sin\theta_r \cos\phi_r, \\ k_i \sin\theta_i \sin\phi_i + \Phi'_y &= k_r \sin\theta_r \sin\phi_r, \end{aligned} \tag{10.1}$$

where Φ'_x and Φ'_y are the phase gradients along the x-axis and the y-axis, respectively, whereas k_i and k_r represent the incidence and reflection wave vectors, where azimuth is ϕ and elevation is θ. The incidence wave vector, $k_i = \frac{2\pi}{\lambda_i}$ and the reflection wave vector, $k_r = \frac{2\pi}{\lambda_r}$ have λ_i and λ_r as their respective wavelengths. The phase gradients required to reflect a beam impinging from $\{\theta_i, \phi_i\}$ toward $\{\theta_r, \phi_r\}$ can be obtained by isolating Φ'_x and Φ'_y from Eq. (10.1).

Since the HSF is able to tune the phase on a per-cell basis, the phase Φ_{ij} of the unit cell at column i and row j can be expressed as

$$\Phi_{ij} = (\Phi'_x i + \Phi'_y j)d_u - \Phi_{00}, \tag{10.2}$$

where Φ_{00} is an arbitrary phase, assumed to be zero for simplicity. It should be noted that just a small set of discrete states are accounted for by unit cells S with cardinality $|S| = N_s$. Each unit cell is thus programmed to the state that is closer to the required reflection phase Φ_{ij} with simple nearest-neighbor mapping. Also, note that the unit cells are generally designed so that the phase difference between consecutive states is $\frac{2\pi}{N_s}$ [39].

Our framework demonstrates the proposed method to determine the communication workload within the HSF. In order to obtain the unit cell state matrices over time, the previously described mobility model and metasurface coding procedure are essentially implemented in MATLAB®. The positions of the HSF, illumination source, and moving target are utilized for the initial evaluation of the incidence and reflection angles. The phase gradients are then obtained using Eq. (10.1), and each unit cell's phase Φ_{ij} is determined using Eq. (10.2). Finally, Φ_{ij} is subjected to the closest neighbor mapping. This technique can be used to determine the matrix of unit cell states for any incidence and reflection conditions. By iterating such calculations, it is possible to produce successive unit cell state matrices that correspond to any angular granularity of movement. Only the unit cell state matrices in steps matching to the angular step parameter are obtained in order to model the gateway. According to our assumptions, which unit cells will receive packets from the gateway are described by a *diff* operation between adjacent unit cell state matrices [28]. Our methodology can determine the time instant at which packets are generated together with the intended destinations within the HSF. The main output is, thus, a packet injection trace for each simulated scenario. The communication traces are then parsed to obtain the relevant metrics such as the inter-arrival time, the percentage of updated cells between successive reconfigurations, or the destination matrix containing the ratio of packets delivered to a given node with respect to all the injected messages. The traces obtained from the methodology above are finally fed to an AnyLogic-based custom-made simulator [40], which evaluates the time consumed by the controllers network to deliver reconfiguration packets to the corresponding controllers. The delay of the network (to deliver packets relative to one reconfiguration of the HSF) is the time elapsed between the first data packet entering the network and the last ACK packet is received by the ACK gateway. The obtained reconfiguration delays, along with the information of the spatial distribution of packets, are finally utilized to characterize the patterns of the traffic within the network in terms of packet arrival times at the nodes. Statistical analysis is then performed on these patterns to define the appropriate models for the generated traffic [37].

We then extend this analysis by calculating the tolerance of the HSF, this is, the type and speed of movements that the HSF is capable of serving correctly. To this end, we compare the reconfiguration speed required by the application with the reconfiguration delay incurred by the controller network. For ease of exposition we assume that the HSF has N × M = 50 × 50 unit cells, with $N_s = 4$ possible states and an angular step of $\alpha = 5°$. For a specific reconfiguration delay, the HSF tolerance establishes the (maximum) application requirements that can be handled by a given HSF. Here, we describe the controller network's capacity to support a specific target motion. To this purpose, before the start of the subsequent reconfiguration cycle, we assess if the HSF is able to send all reconfiguration requests successfully. Thus, we compare the interval between modifications with the reconfiguration delay. In the considered beam steering application, the HSF must have a reconfiguration delay that is smaller than the time the target takes to change its location by $\alpha°$. This ability is known as the HSF tracking tolerance. Tolerant tracking can offer guidelines for the appropriate type of movement and speeds that the HSF is able to handle. Figure 10.7a shows the maximum speed the HSF can tolerate for an object following projectile motion. These curves are obtained through $\omega_{HSF} = \frac{\alpha}{D_{rec}}$, where ω_{HSF} is the HSF tolerance in units of deg/sec and D_{rec} is the reconfiguration delay obtained via controller network simulations. It is critical to note that we make the assumption that the unit cells' time needed to change state is negligible. The earlier results demonstrate that the reconfiguration delay of the surface depends on the changing angles (i.e., elevation, azimuth, incidence, and reflection) and the extent of change, which depend on the type of movement and the angular step. Thus, the tolerance results in Fig. 10.7a are specific for projectile motion. Different patterns of movements and different values of a would generate different graphs. As mentioned earlier, the HSF tolerance metric can be used as a guideline for the maximum speeds the HSF can tolerate. An example is shown in Fig. 10.7b where the HSF tolerance graph is compared with the angular speeds of different objects with different linear speeds (we consider $\alpha = 5°$). The objects are assumed to be moving in a straight line parallel to the surface and 5 m away from it. The linear speeds of the different objects are indicative of a human walking with a speed of 5 km/h, a human running with a speed of 13 km/h and a car moving with a speed of 120 km/h. The HSF is able to track any object as long as its angular speed is less than ω_{HSF}. However, the angular speed relies on both the linear speed of the object and its distance from the HSF. Since we

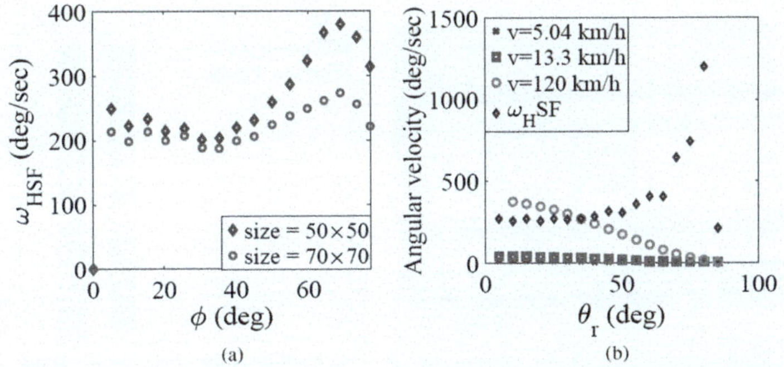

Fig. 10.7 **a** HSF tolerance for projectile movement for different sizes of the surface. **b** HSF tolerance versus the angular speeds of objects with different linear speeds. (Reproduced from [28] with permission from IEEE. © [2021] IEEE)

are not considering a motion in a circular path, but a straight line instead, the distance from the HSF changes with movement and thus the angular velocity is not constant throughout the full range of the considered motion. Therefore, one can observe that the HSF can track all movements, except for objects moving with 120 km/h when $\theta_r < 38$ (i.e., fails for smaller angles which is when the angular velocity of the object is greater than the HSF tolerance).

We illustrate the effectiveness of this strategy through the following example. Consider a 50 × 50 metasurface tracking a linearly moving target with an average speed of 50 km/h. We create clusters of controller nodes that report a high chance, $p > 0.7$, of obtaining the same state throughout the duration of the motion under consideration. These clusters are seen in Fig. 10.8a, where each square on the heatmap represents a node. Nodes that make up the cluster have the same color. It should be noticed that there are only a few color variants in the color scale because a significant section of the nodes have a high likelihood of having the same condition. This pattern is particularly common when the motion is in a straight line parallel to the surface. These findings support the potential application of multi-casting routing in the controller network in addition to the currently underutilized uni-casting routing. To evaluate the degree to which the communication burden may be reduced through multi-casting, we consider the following scenario. We take into consideration 1 km of straight road, along with 20 km/hr of average moving traffic. The injection rate into the road is set to 50 veh/h and the vehicle traces are once more generated on VISSIM. We switch between two users in order to take into account the situation when several users are being served via time division multiplexing of the metasurface resources. At the start of the simulation, the two users are placed 70 m apart. We take into account two cases. In Case 1, each meta-atom is reconfigured one at a time using routing, while in Case 2, each meta-atom along a row is reconfigured using the far-end meta-atom's configuration as the packet travels from source to destination. This can be done by using a variant of the XY algorithm, which reconfigures all nodes having the same Y coordinate as the destination, and adding a single bit to

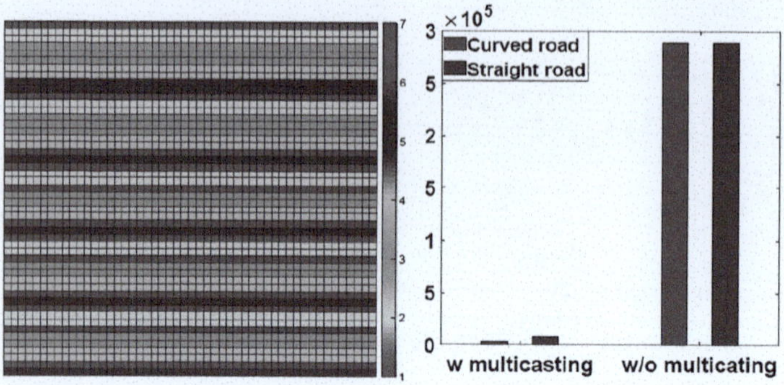

(a) (b)

Fig. 10.8 **a** Spatial correlation in cell states **b** Reduction in workload when multicasting is employed. (Reproduced from [14] with permission from IEEE. © [2021] IEEE)

the packet header to indicate multi-casting. The same experiment is repeated when the roadway is also slightly curved. The results are shown in Fig. 10.8b where a significant reduction in the number of messages is reported [14].

10.6 Communication Models

The introduction of metasurfaces on the roadside infrastructure is aimed at enhancing V2V and V2X communications when these are obstructed. The characteristics of the communication channel though, may critically affect the performance of the cooperative automated driving strategies thus necessitating an adaptive policy where the control parameters are regulated based on the state of the channel quality. Such adaptive control policies have not as of now been investigated in literature in the presence of metasurfaces. A significant part of such an investigation is the characterization of the metasurface-assisted communications channel. Despite a plethora of recent works [41], the problem is still open as the mathematical models need to be validated with experimental data from field deployments. In this chapter, we adopt a channel modeling approach which is based on obtaining the directivity using the metasurface coding procedure described in Sect. 10.5. This approach is attractive as, beyond the communications channel, it allows characterization of the reconfiguration delays and power consumption, thus offering a unified framework for cross-layer design and optimization.

Directivity of the metasurface is a property that measures the amount of energy radiated in a specific direction. Directivity hence depends on the surface's scattering diagram, which in turn is significantly affected by metasurface parameters like the size and the number of unit cells. We can determine the phase profile of the surface, which specifies the radiation pattern that is reflected (with respect to the receiver) in the directivity, from our earlier described concept of the metasurface coding (see Sect. 10.5). The communication channel gain can be defined as follows:

$$G_{total} = G_T G_{RIS} G_R, \tag{10.3}$$

where G_T, G_{RIS}, and G_R denote the transmitter gain, the RIS gain, and the receiver gain, respectively. The gain of the surface can then be expressed as follows:

$$G_{RIS} = \epsilon D, \tag{10.4}$$

where D represents the directivity and ϵ the efficiency of the surface. We can thus express the received power strength at the receiver through the following equation:

$$P_r = \frac{G_{total} P_t}{PL_{total}}, \tag{10.5}$$

where P_t denotes the transmitter power and PL_{total} the path loss, generally modeled as a free space path loss:

$$PL_{total} = PL_{T \to R \to D} = \left(\frac{4\pi f}{c}\right)^2 (d_1 + d_2)^n, \tag{10.6}$$

where f is the frequency, c the speed of light, n the path loss exponent, and d_1 and d_2 represent the distances between the transmitter and the RIS, and the RIS and the receiver, respectively. Notice that the path loss is defined over the sum of the two distances d_1 and d_2. This is only correct if we operate in the near field of the RIS, otherwise, we would need to consider the double path loss effect, i.e., the product of the distances.

Here we assume a large RIS enabling to work in the near field of the antenna. In our future work, we will consider a smaller RIS with a larger number of elements, enabling higher gains to compensate for double path loss effects. After computing the total path loss, the SNR at the receiver can be defined as usual:

$$SNR = \frac{P_r}{P_{noise}}. \tag{10.7}$$

Given the SNR, we compute the probability of reception using the standard bit error rate curves for OFDM available in Veins [38]. We test the implementation of the channel model in a simplified scenario. The channel model is comprised of the far field model, the path loss model on the incident and reflected paths, plus the model computing incidence and reflection angles required for the far-field model. We consider the intersection described in Fig. 10.15 with two vehicles only and without implementing any maneuver. We first run the simulation without communication, recording the positions of the vehicles over time. After 30 s of simulated time, we compute the incidence and reflection angles between the two vehicles considering the RIS to be located at the center of the intersection, facing south. We then code the RIS statically for such incidence and reflection angles and then re-run the simulation by enabling communication between the two vehicles. In particular, we have the vehicle coming from the south transmit periodic messages to the other, and record

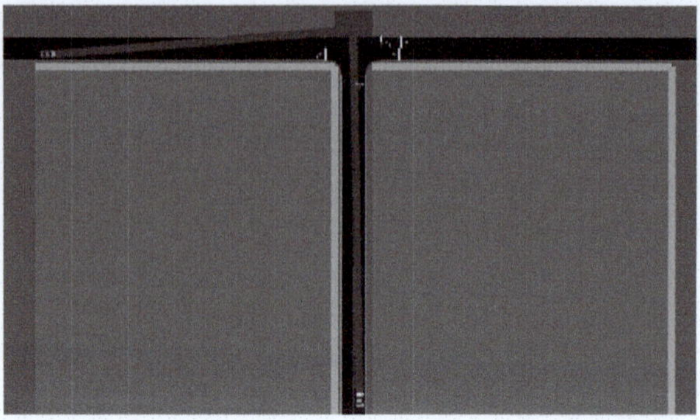

Fig. 10.9 Screenshot of the validation scenario in the SUMO GUI with the RIS drawn as a blue rectangle, together with the projection of the path for which the RIS has been configured. (Reproduced from [42] with permission from IFIP)

the gain of the RIS as a function of the position of the receiving vehicle. With respect to the model, we consider a center frequency of 25 GHz and a free-space path loss.

Figure 10.9 shows a screenshot of the simulation in SUMO at simulation time 30 s. Besides the intersection, the vehicles, and the buildings, our implementation draws the RIS in the scenario as a blue rectangle (not to scale) plus a projection of the path for which the RIS has been configured, enabling the user to visually inspect the simulation. Figure 10.10 shows the evolution of the gain, the total path loss, as well as the incidence and reflection angles as a function of the distance to the intersection (measured for the vehicle traveling west to east). With respect to the gain, the graph displays a step-like pattern which is due to the resolution of the far field pattern [42]. Qualitatively, the gain starts from a relatively high value but continues to increase till roughly 100 m to the intersection, i.e., 30 s into the simulation as expected. After passing the optimal point, the gain quickly drops due to the changes in the incidence and reflection angles, especially the azimuth. By looking at the total path loss, we observe that the RIS in its optimal range provides more than 20 dB of gain. On the other hand, vehicles closer to the intersection would experience lower path losses. This suggests that tracking algorithms might focus their attention on vehicles closer to the intersection, as intuition might also indicate.

10.7 Beam Steering

The use of metasurfaces in autonomous driving scenarios is often challenged by the mobile nature of the receiver and/or the transmitter. Although in some settings static metasurfaces can yield satisfactory performance, in most cases reconfiguration of the metasurface parameters in order for the reflected beam to adapt to user mobility is necessary. Even when the mobility patterns are known, an open loop or closed approach may be adopted for the metasurface to appropriately configure the state altering elements. Feedback-based methodologies are however more attractive as they are known to be more robust with respect to hardware imperfections or errors and model uncertainties.

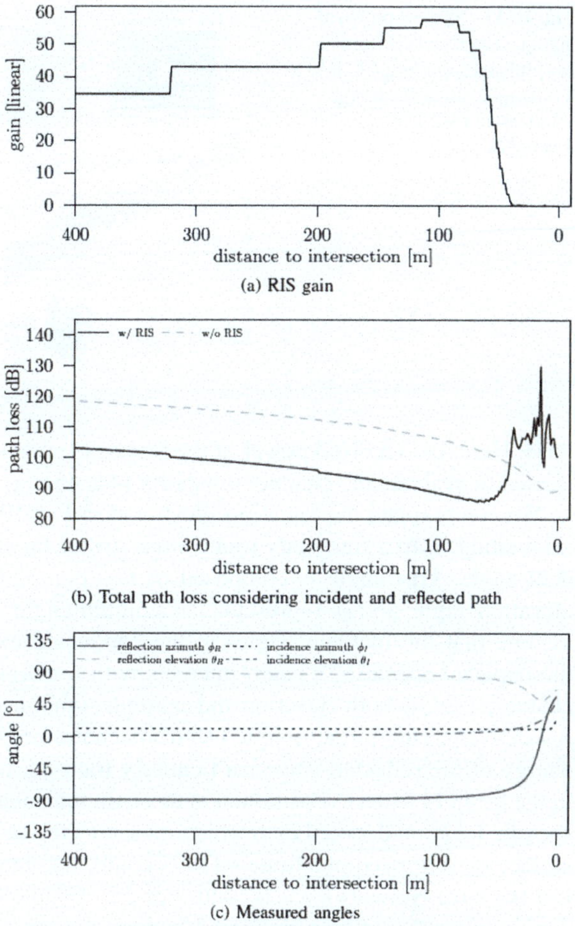

Fig. 10.10 Evolution of the **a** gain at the antenna, **b** the total path loss, and the **c** reflection and incidence angles as a function of the distance to the intersection for the vehicle traveling west to east (receiver). (Reproduced from [42] with permission from IFIP)

We consider an iterative beam steering approach reliant on feedback, which utilizes power measurements at the receiver. This method considers the necessary coding patterns for a multi-state metasurface and the procedure used for transmitting reconfiguration signals from the metasurface controller to the designated unit cells. The underlying framework relies on the HSF paradigm. We have chosen a widely recognized form of extremum seeking control (ESC) to act as the feedback-driven iterative control algorithm, as demonstrated in [43]. Its demonstrated ability to track mobile targets traveling at speeds exceeding 100 km/h makes it well-suited for V2X scenarios. The considered scenario requires a transmitter and a receiver establishing a reliable communication channel via a HSF acting as the RIS (**RIM**), due to the lack of LOS capability. The goal is to repeatedly adjust the controller states to point the impinging wave in the direction of the receiver after directing a beam toward the

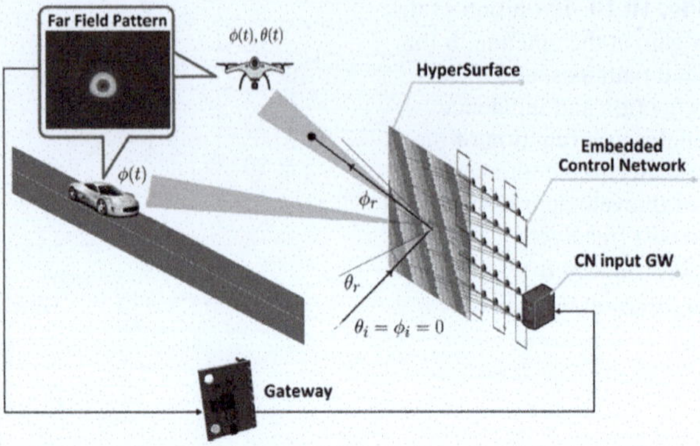

Fig. 10.11 System physical layout. (Reproduced from [44] released under a CC BY 4.0 license; https://doi.org/10.1109/TITS.2023.3241214)

RIM. Here, the received power at the receiver (which is intended to be maximized) is used to evaluate the channel reliability [44, 45].

We contemplate the situation illustrated in Fig. 10.11, wherein a solitary beam, originating from a stationary transmitter, strikes the metasurface at a constant incident angle. This angle is composed of two components, where θ_i represents the elevation angle, and ϕ_i represents the azimuth angle. For the sake of simplicity, we assume that the incident angles in the conducted experiments are set to zero. In the depicted scenario, a beam interacts with a metasurface, undergoing reflection characterized by both elevation and azimuth components denoted as $\theta_r(t)$ and $\phi_r(t)$, respectively. Due to the unique electromagnetic properties of the metasurface, the energy of the reflected wave isn't entirely focused, resulting in a side-lobe pattern in the far-field pattern. Positioned in three-dimensional space, there is a receiver at a specific location denoted as l, which measures the received power $P(t)$. The metasurface is comprised of a finite set of N controllers interconnected via the embedded CN (Control Network). Each controller can assume one of L finite discrete states. These controller states, represented as $c_1, c_2, \ldots, c_N$, are consolidated into a vector c = $[c_1, c_2, \ldots, c_N]$, with each entry taking on a value from the set S = $s_1, s_2, s_3 \ldots s_L$. The incident and reflection angles are also consolidated into vectors: $\psi_i = [\theta_i, \phi_i]$ and $\psi_r = [\theta_r, \phi_r]$, respectively. The reflection angle ψ_r is dependent on the incidence angle and the controller states, such that ψ_r = f(c, ψ_i). The function f(.) has been tabulated and is highly nonlinear.

The chosen controller states c result in a specific Far Field pattern, which in turn affects the received power $P(t)$. Therefore, $P(t)$ is also a function of the controller states c, the incidence angle ψ_i, and the receiver's location l, denoted as $P(t) = g(c, \psi_i, l)$.

The optimization problem at hand is to select the controller states to maximize the received power, as expressed by

$$\max_{c \epsilon C} P(t).$$

However, the complexity arises from the multidimensional input space and the highly nonlinear and uncertain nature of the function g(.). To simplify the problem, we reduce it to a single input variable, namely the reflection angle ψ_r, which relates to the controller states c through the function g(.). This adjustment in the input space ensures the concavity of $P(t)$ concerning the input variable ψ_r, which is crucial when applying Extremum Seeking Control (ESC). Thus, the optimization problem becomes

$$P: \max_{\psi_r \epsilon [-90,90]} P(t). \tag{10.8}$$

The described problem encounters several challenges:

1. Nonlinearity and Uncertainty: The functions f (.) and g (.) are highly nonlinear, and their specifications are uncertain due to construction inaccuracies and the potential for faulty behavior in both the metasurface and the CN.
2. Disturbances: There are disturbances to contend with, such as background noise affecting the measurements of received power.
3. Single Point of Measurement: Measurements are taken at a single point in 3D space, at the receiver's location.

To tackle these challenges, particularly the robustness issue concerning uncertainties, a closed-loop implementation has been favored over open-loop control, as discussed in [45–47]. In a closed-loop implementation, it is assumed that the power measured at the receiver is fed back to the input gateway of the HSF. The input gateway processes this measurement according to the control algorithm under development and generates the desired reflection angle ψ_r. This angle is then transformed into the desired controller states c using the inverse of the function f(.), assuming a one-to-one mapping. The desired controller states are then communicated to the corresponding controllers through the embedded controller network. Communication in this context is an active area of research, as addressed in [29]. In this scenario, we assume the use of the XY algorithm, which functions effectively in the absence of faults, and consider unicast routing one by one to reach the respective controllers. Once the controller messages are dispatched, the controllers adjust their states accordingly, thereby altering the Far Field pattern. This process is iteratively repeated.

Here, we assume the presence of a feedback controller in the following form:

$$\begin{cases} \dot{x} = h(x, P) \\ \psi_r = v(x, P). \end{cases} \tag{10.9}$$

In this context, ψ_r represents the desired angle of reflection, P(t) signifies the measured signal strength, x represents a vector of controller states, and h(.) and v(.) denote potentially nonlinear functions. The challenge at hand is to engineer the functions h(.) and v(.) in a manner that effectively addresses problem P. For a visual representation of the feedback system, please refer to the block diagram shown in Fig. 10.12.

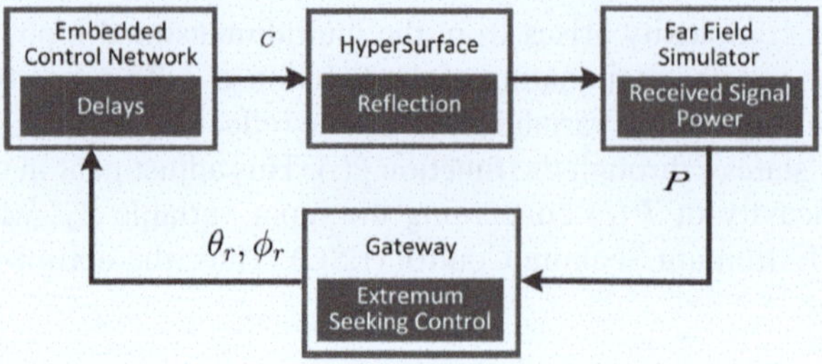

Fig. 10.12 Block diagram of closed loop feedback system. (Reproduced from [44] released under a CC BY 4.0 license; https://doi.org/10.1109/TITS.2023.3241214)

ESC is an adaptive, nonlinear, and model-independent control technique selected to address the complexity of the HSF's behavior, characterized by its intricate nonlinearity and substantial uncertainty. This approach involves applying sinusoidal perturbations to the input signal and measuring the resulting variations in the output signal in order to identify and track a local maximum of a measurable objective function. The ensuing discussion primarily focuses on the elevation angle, denoted as θ_r, yet it is equally applicable to the azimuth angle ϕ_r.

In the proposed algorithm, P(t) represents the system's output and thus serves as the input for the controller, while θ_r(t) acts as the controller's output, serving as the control variable. To introduce a perturbation to the best estimate of the input signal θ, a sinusoidal dither signal $\alpha sin(\omega t)$ is added, resulting in a perturbed input signal denoted as θ. This $\theta_r(t)$ is utilized to configure the HSF system, and the resultant P(t) is measured at the receiver. The measured signal then undergoes a filtration process, wherein low-frequency components are eliminated by passing it through a High-Pass Filter (HPF). The signal following filtration, referred to as P_h, is demodulated by multiplying it with the dither signal $\alpha sin(\omega t)$. Subsequently, this demodulated signal is integrated to obtain an updated value for θ. The continuous-time algorithm for the ESC scheme under consideration is as follows:

$$\begin{cases} P(t) = f(\theta_r(t)) \\ \theta_r(t) = \theta^*(t) + \alpha sin(\omega t) \\ \theta^*(t) = k\zeta(t) \\ \zeta(t) = -\omega_1\zeta(t) + \omega_1(P(t) - \eta(t))\alpha \sin(\omega t) \\ \eta(t) = -\omega_1\zeta(t) + \omega_1(P(t), \end{cases} \tag{10.10}$$

where k is gain of the gradient-based update law, ω is the perturbation frequency, ω_h is the cut-off frequency of the HPF, ω_l is the cut-off frequency of the Low-Pass Filter (LPF) and ζ is the demodulated signal. For the stability analysis of ESC, readers are referred to [48].

In this section, we assess the performance of the proposed control scheme by examining its effectiveness in directing the reflected beam toward a potentially mobile receiver to maximize received power. To carry out this evaluation, we employ two simulators: one developed in MATLAB® and another created using the Analogic simulation platform. The latter simulator simulates the process of transmitting packetized directives within the HSF-embedded CN, taking into account its unique features, such as asynchronous operation. The MATLAB® simulator consists of three key components: first, the implementation of the controller; second, the function that maps the reflection angle generated by the controller to the controller states; and third, the module that characterizes the far-field pattern resulting from the chosen controller states. The far-field pattern has a direct impact on the strength of the received signal. Throughout the evaluation process, we normalize received power values relative to the maximum value, ensuring that they fall within the range of $0 \leq P(t) \leq 1$.

In many applications, such as V2V (Vehicle-to-Vehicle), V2I (Vehicle-to-Infrastructure), and UAVs (Unmanned Aerial Vehicles), the target receiver often exhibits mobility. We investigate this phenomenon and its associated limitations. In the mobility model considered, a single mobile receiver travels in a straight line parallel to the surface. For simplicity, we assume that the target receiver is located at the same height as the HSF, resulting in $\phi_i = \phi_r = 0$. As the receiver moves in parallel to the HSF, its elevation angle θ continuously changes concerning the HSF. Consequently, the proposed algorithm must continually adapt the HSF parameters to ensure maximum received power. In the initial set of experiments, we assume a constant receiver speed of 5.04 km/h, equivalent to a walking scenario. Figure 10.13 illustrates graphical representations of the beam steering scheme's behavior and the corresponding received signal power. As mentioned earlier, the target's movement is strictly horizontal, keeping its azimuth fixed while only varying the elevation angle. Figure 10.13a illustrates the target's changing elevation angle, starting at 42° and gradually decreasing linearly to 36° over time. The initial deviation of the actual elevation angle from the one required to steer the beam toward the receiver results in the received signal power reaching only half of its maximum value, as depicted in Fig. 10.13b. However, the proposed beam steering algorithm quickly adjusts the beam's direction toward the target location. Within a very short time (0.25 s), the beam aligns closely with the receiver, leading to received signal power values close to their maximum. As the target continues to move, the algorithm demonstrates effective tracking, maintaining high power levels.

Figure 10.13c displays the spatial distribution of traffic within the HSF controller network, while the beam tracks the walking target. The figure highlights a bias toward the upper right and lower left parts of the metasurface, which bear a significant portion of the load. Additionally, Fig. 10.13d illustrates the delay incurred by the CN at each time instant during the tracking of the moving target. It's evident that the delays remain within a reasonable range and do not adversely affect the algorithm's performance.

In the second set of experiments, we continue to examine the linear motion of the receiver, but this time we increase its speed to 72 km/h. It's important to note

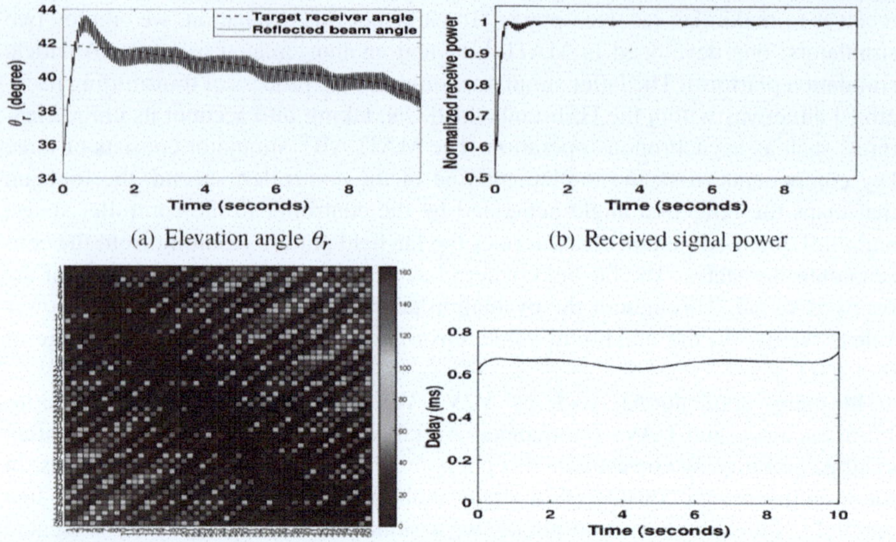

(a) Elevation angle θ_r (b) Received signal power

(c) Spatial distribution of the CN traffic (d) Delay introduced by the embedded control network

Fig. 10.13 Mobile receiver moving in parallel to the HSF with a constant speed of 5.04 km/h. (Reproduced from [44] released under a CC BY 4.0 license; https://doi.org/10.1109/TITS.2023.3241214)

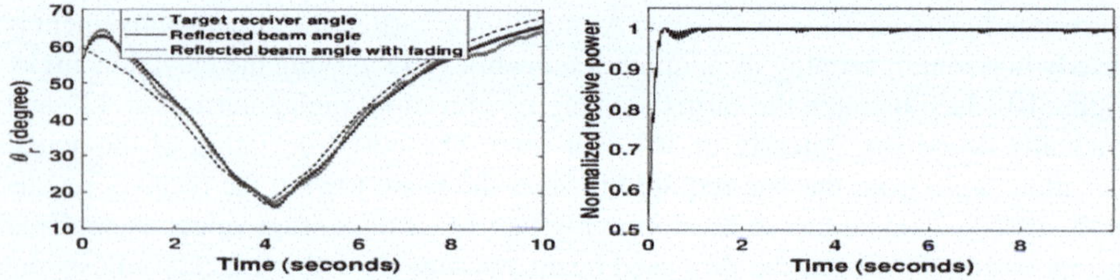

(a) Elevation angle θ_r (b) Received signal power

Fig. 10.14 Mobile receiver with a speed of 72 km/h. (Reproduced from [44] released under a CC BY 4.0 license; https://doi.org/10.1109/TITS.2023.3241214)

that, for this case of 72 km/h, we also consider the effects of shadow fading, which is modeled as a log-normal distribution with a standard deviation of 8.14 dBs, following the design guidelines outlined in [49].

The results of these experiments are presented in Fig. 10.14. One noteworthy observation is that the impact of shadow fading on the output responses is relatively insignificant. This can be attributed to the presence of a LPF incorporated into the ESC scheme, which effectively filters out the high-frequency components induced by fading.

Furthermore, it is evident that the proposed algorithm remains capable of steering the beam toward the target despite the increased speed of the receiver. This underscores the suitability of the proposed method for applications such as V2V and V2I, where mobile targets are encountered. As shown in Fig. 10.14, as the speed increases, the algorithm can still track the beam near the target location. However, there are more fluctuations in the corresponding received signal power, and the algorithm's performance degrades.

The results illustrate that, as expected, an increase in speed has an adverse effect on tracking performance. This raises the question of what speed level is acceptable for the system's performance. To address this, we conducted additional simulations and observed that for a speed of 144 km/h, the algorithm exhibits poor performance in terms of convergence time and received signal power. Similarly, for speeds exceeding 160 km/h, the algorithm fails to converge within a reasonable time frame.

10.8 RIS-Assisted Autonomous Driving

As highlighted in the introduction, RIMs would be useful in cases where NLOS communication significantly impacts the performance of the cooperative driving algorithms. A number of scenarios can be identified and analyzed. In this section, to show the potential of RIS integration for CD, based on our recent work in [42], we indicatively consider the case of a cooperative margin maneuver on an intersection. Figure 10.15 depicts the considered scenario where a T-shaped intersection with a platoon of CAVs coming from the bottom (group C) need to turn right and merge into the main traffic flow.

On the roadside, there are buildings that obstruct visibility and block LoS communications, impairing the use of cameras and of classic vehicular communication

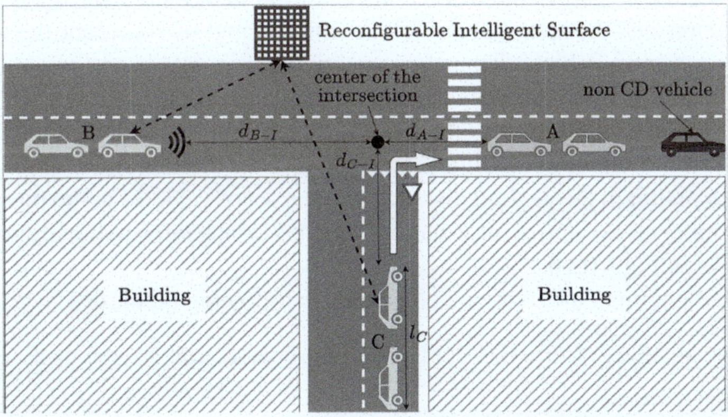

Fig. 10.15 Cooperative driving scenario. (Reproduced from [42] with permission from IFIP)

technologies such as IEEE 802.11p. We assume that, for human-driven vehicles, the traffic coming from the bottom road should yield the right of way to vehicles traveling left to right. However, in a CD scenario, it might be more efficient to coordinate the traffic so that vehicles in group C merge in between groups A and B, if inter-vehicle gaps so allow. This may prevent vehicles in group C to come to a complete stop, and should only very slightly slow down vehicles in group B. From now on, we assume that all groups A, B, and C are in fact vehicle platoons, and that slowing down platoon B to let platoon C merge in is always the most efficient action. We also assume that all platoons are driven by the PATH cooperative adaptive cruise control (CACC) algorithm [50], which maintains a constant inter-vehicle gap between the vehicles. To perform the maneuver, vehicles in platoon C need to know the distance to the last vehicle in platoon A in order to properly regulate their speed and smoothly merge in the intersection. The driving pattern of platoon A might be disturbed by other, non-cooperative vehicles, as depicted in Fig. 10.15. Such information can be shared by the leader of platoon B, which has a clear LoS toward platoon A. To share such information, vehicles could use Cooperative perception (CP). The leader of platoon B could share raw sensor data with the leader of platoon C, instead of data post-processing results (e.g., distance measurements) to build a more complete view of the surrounding environment, especially when the LoS is obstructed [51, 52].

In our example, not only can vehicles in platoon C be aware of the distance to platoon A but also by receiving raw data such as the camera video stream, they can become aware of vulnerable road users (VRUs) behind the corner (see the zebra crossing in Fig. 10.15). In this specific scenario, the use of RISs in the mmWave spectrum can be extremely beneficial because they can overcome the missing LoS problem, as well as providing the large bandwidth required to stream raw sensor data. To formally describe the maneuver, we first introduce the CACC control law employed by the CAV [51]:

$$u_i^{CACC} = \alpha_1 u_{i-1} + \alpha_2 u_0 + \alpha_3(v_i - v_{i-1}) + \alpha_4(v_i - v_0) + \alpha_5(-d_i + d_d). \tag{10.11}$$

In Eq. (10.11), u_i^{CACC} is the desired acceleration of vehicle i (i.e., prior to actuation), v_i represents the speed of vehicle i, while d_i and d_d represent the measured and the desired inter-vehicle distance, respectively. For details on the control gains α_i please refer to [53]. We consider the default parameters defined by PLEXE. Intuitively, (10.11) uses the leading and preceding vehicles' acceleration and speed as well as the distance to the preceding vehicle to compute the acceleration/deceleration command to maintain the desired distance d_d. The set of vehicles in the platoon from which a member takes control information is defined as the information flow topology (IFT), shown in Fig. 10.16.

To implement the maneuver, we consider the following quantities:

- d_{A-I}: the distance from the center of the intersection to the last vehicle in platoon A;
- d_{B-I}: the distance of the leader of platoon B to the center of the intersection;
- d_{C-I}: the distance of the leader of platoon C to the center of the intersection;
- l_C: the length of platoon C.

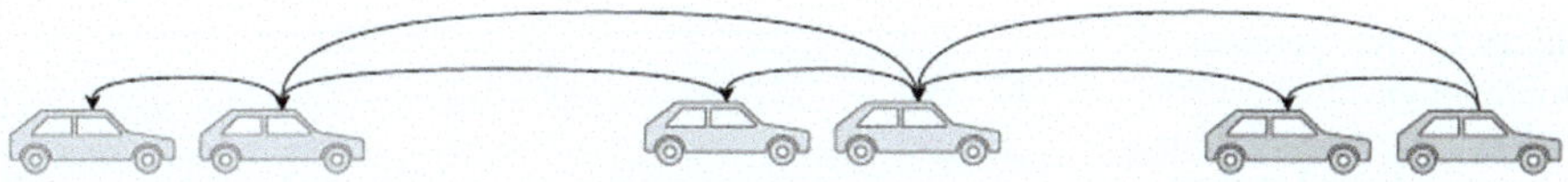

Fig. 10.16 Information flow topology configuration. (Reproduced from [42] with permission from IFIP)

To compute their control actions, the leaders of platoons B and C should feed to the control algorithm a different measured distance d_i. In particular, platoon B should consider the distance $d_i = d_{A-I} + d_{B-I}$ - l_C, i.e., leaving room for platoon C to merge in. Group C, instead, needs to consider $d_i = d_{C-I} + d_{A-I}$, i.e., the distance to the center of the intersection plus the distance to platoon A. By acting on the measured distance, the platoons automatically adjust their distances to match the desired one. However, the above setup is suboptimal, because the vehicles might implement unnecessary control actions. Assume, for example, that the gap between platoons A and B is already large enough to merge platoon C in. As a result, platoon B will accelerate to compensate for the distance error, because $d_d > d_i$. Such unnecessary acceleration wastes fuel. Moreover, the system should only decelerate if there is no room for platoon C. To overcome these problems, we assume the leader to run multiple control algorithms in parallel, coupling the CACC with a standard cruise control (CC) with control function:

$$u_i^{CC} = k_p(v_d - v_i). \tag{10.12}$$

In Eq. (10.12), v_d is the desired speed, while k_p the proportional control gain. For the leaders, the final control law is defined as

$$u_i = min(u_i^{CC}, u_i^{CACC}). \tag{10.13}$$

Equation (10.13) causes the leaders to accelerate only if their current speed is lower than the desired one, but always maintaining a distance larger than d_d, thanks to the CACC. One important detail is how to set the leader and the predecessor for each vehicle. Here, we assume the configuration in Fig. 10.16, i.e., the leaders choose as leader and predecessor the leader and the last vehicle of the platoon preceding them on the road, respectively. This guarantees leaders of platoons B and C to properly follow as if they were a member of the platoon preceding them on the road. The key issue now becomes how to get control data from platoon A to C, given the obstructed LoS. In this case, we assume the leader of platoon B to relay A's data to C through the RIS link. Data from platoon A to B is transmitted through a standard IEEE 802.11p link. In turn, the leader of platoon C can relay data about its members through the same RIS link to platoon B.

We then consider the cooperative merging maneuver by implementing it first in ideal conditions, i.e., perfect communication without the presence of buildings. This allows us to observe the ideal maneuver dynamics. In a second scenario, we use IEEE 802.11p and consider a building close to the intersection, which causes the interruption of the communication between the leaders of platoons B and C

Fig. 10.17 Acceleration profiles of the three platoon leaders. (Reproduced from [42] with permission from IFIP)

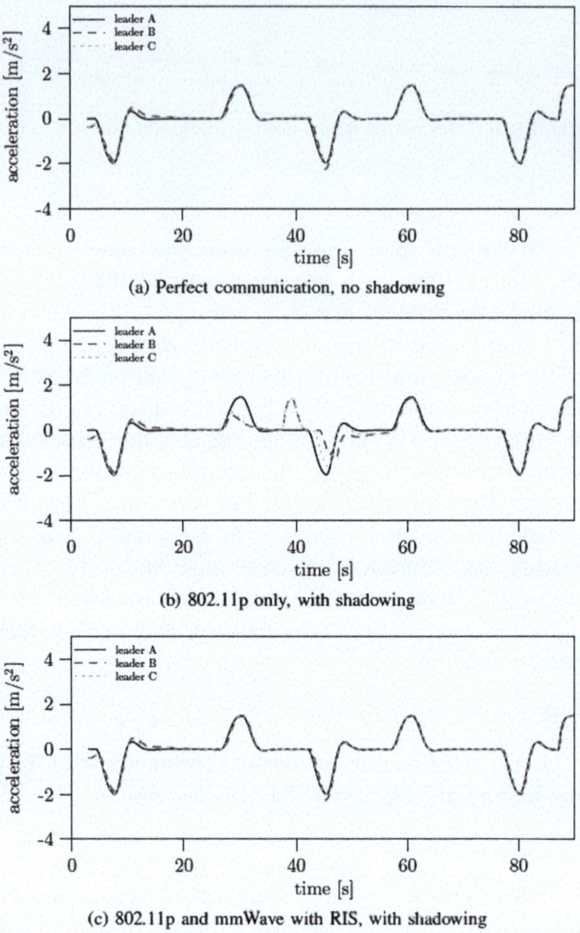

for a certain period of time. Finally, to support the maneuver, we use both IEEE 802.11p and an mmWave link between such leaders to overcome the communication blockage thanks to the RIS. In this simplified scenario, we consider no underlying MAC protocol for mmWave communications. In addition, we assume the traffic infrastructure to be able to track vehicles on the road and automatically reconfigure the RIS.

We start the discussion by first observing the ideal behavior. Figure 10.17a shows the acceleration dynamics for the leaders of platoons A, B, and C, while Fig. 10.18a the virtual distance as computed by the leaders of platoons B and C, i.e., the distance of leader C to the last vehicle in A and the distance of leader B to the last vehicle in C, as depicted in Fig. 10.15. With respect to the acceleration profiles, the changes in speed are due to the human-driven vehicle in front of platoon A which induces traffic perturbations. Thanks to data sharing, all the leaders are capable of adapting

Fig. 10.18 Distances between platoon C leader to platoon A last member and platoon B leader to platoon C last member. (Reproduced from [42] with permission from IFIP)

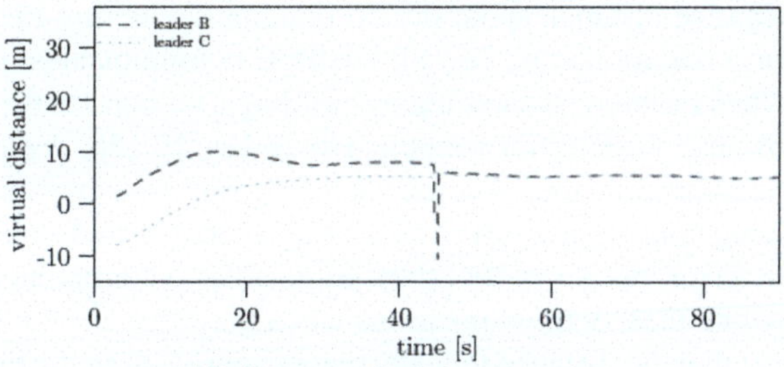

(a) Perfect communication, no shadowing

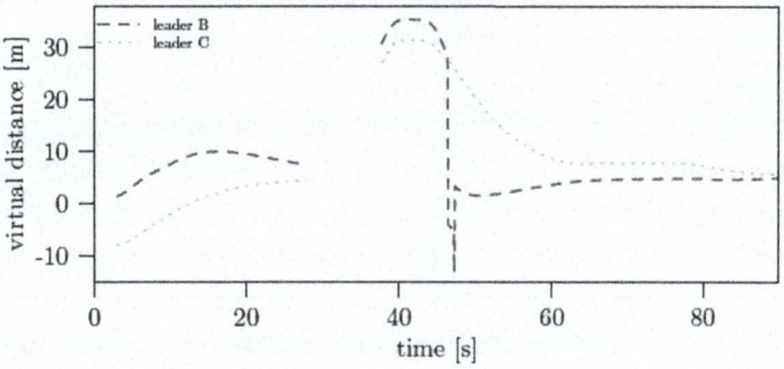

(b) 802.11p only, with shadowing

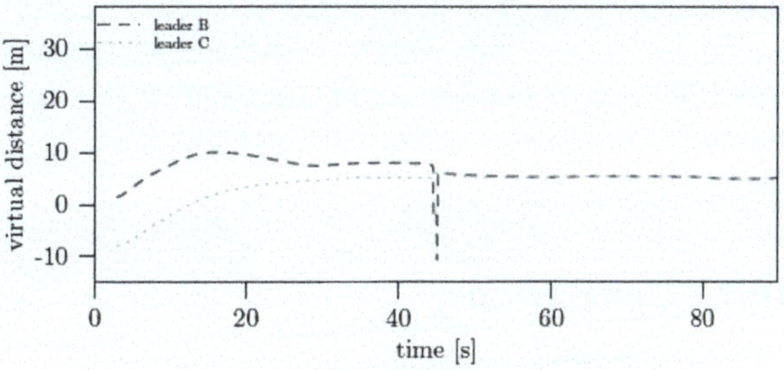

(c) 802.11p and mmWave with RIS, with shadowing

to such changes, especially leader C, which is aware of such perturbations thanks to CP, i.e., leader B forwarding radar information to leader C. With respect to the maneuver, we first need to mention that the minimum target inter-vehicle distance is 5 m, i.e., vehicles will decelerate when closer than such distance to the front vehicle, while they will maintain their actual distance if farther. 5 m is a typical inter-vehicle distance for the PATH CACC [53]. Observing the evolution of the virtual distance in Fig. 10.18a (ideal scenario), we see that both leaders are located too far ahead. Given the initial conditions of this simulation, leader B starts from a virtual distance of 0 m (5 m below target), meaning that it needs to decelerate to leave room for platoon C to merge in, while leader C is 15 m below the target distance, i.e., it needs to decelerate to enter the intersection having the right distance to platoon A. The two platoons adapt their distance according to data received between them while adapting to the disturbances induced by the human-driven vehicle as well. When platoon C merges in the intersection at roughly 45 s, leader B measures a negative distance spike due to platoon C entering the intersection. This just holds for a few moments, i.e., till leader C communicates it has entered the intersection. When enabling shadowing by buildings, we can see the information about the virtual distance missing for roughly 10 s in Fig. 10.18b. As leaders B and C are unable to communicate, they continue driving using data they received in the last message. By being both slightly off position they continue to decelerate trying to compensate for the error, but given that they are unable to communicate they find themselves way behind their optimal positions when they pass the building causing shadowing. The fact that the distance then decreases approaching the target is just due to the fact that the human-driven vehicle is decelerating when platoon C enters the intersection, as shown in Fig. 10.17b. Notice that, in a scenario where vehicles cannot communicate for more than a couple of seconds, the maneuver should most probably be aborted for safety reasons. Here the maneuver continues to show the effects of communication impairments on vehicle dynamics. The final graphs show the acceleration and the virtual distance profiles Figs. 10.17c and 10.18c when adding an mmWave interface supported by a RIS to the vehicles. By comparison with the ideal case, the two scenarios are undistinguishable. This means that, even in the presence of a physical obstacle, vehicles are perfectly capable of communicating thanks to the help of the RIS. This shows the potential of RIS and mmWave applied to CD scenarios and maneuvering. The considered scenario is clearly a proof-of-concept, but the possibility of using mmWave to communicate "around the corner" will enable high-speed Vehicle-to-Vehicle communication (V2V) links, fundamental for future CD systems based on new paradigms such as CP.

10.9 Conclusions

SDMs have shown a great potential as functional artificial skins that can shape and control electromagnetic waves. Several challenges arise in the implementation of SDMs, providing exciting opportunities that can drive research for years to come. These are indicatively: (i) the careful design of the metasurface in order to accommodate the communication channel between the controllers and enable efficient and cost-effective communication while avoiding interference with the metasurface operation, and (ii) the implementation of smart controllers that encompass the actuators along with transceivers for communication and sensors, while respecting the sub-wavelength unit cell dimensioning of metasurfaces that enables their broad functionality. The physical implementation largely depends on the frequency of operation; different tuning mechanisms, design strategies, and limitations become relevant for different frequency regimes. As reconfigurable metasurfaces continue to mature and evolve, we envision that software-defined metasurfaces can have a transformational effect on a broad range of applications including Connected Autonomous Driving. In this chapter, we have considered the Hypersurface as the enabling technology for RIS-supported CAV applications and have outlined design considerations than can significantly affect the overall performance. This includes the workload which affects delays and power consumptions as well as the actuator controller network topology and routing which dictates reconfigurations delays. Such delays are crucial in the stability of feedback-based beam steering techniques when the RIS reflected beam tracks a moving receiver. Finally, we presented simulation results which demonstrate the potential of RIS to support autonomous driving applications. We considered the case of a merging maneuver on an intersection demonstrating the beneficial effects of RIS integration. The area is currently at its infancy and many design challenges remain toward the development of RIS communication adaptive Cooperative Autonomous Driving Algorithms.

Acknowledgements This research has received support from the project BRIDGE2HORIZON /0823A/004 (AUTOMETA), which is implemented under the Cohesion Policy Funds "THALIA 2021-2027" with EU co-funding.

References

1. Karangwa J, Liu J, Zeng Z (2023) Vehicle detection for autonomous driving: a review of algorithms and datasets. IEEE Trans Intell Transp Syst
2. Pavlínek P (2023) Transition of the automotive industry towards electric vehicle production in the east European integrated periphery. Empirica 50(1):35–73
3. Wu X, Cao J, Douma F (2021) The impacts of vehicle automation on transport-disadvantaged people. Transp Res Interdiscip Perspect 11:100447
4. Mordorintelligence (2023) Precise market intelligence and advisory 2023. J.: Effective substances. In: The dictionary of substances and their effects. Royal Society of Chemistry (1999) Available via DIALOG. https://www.mordorintelligence.com/. Accessed 8 Sept 2023
5. Noor-A-Rahim M, Liu Z, Lee H, Khyam MO, He J, Pesch D, Poor HV (2022) 6G for vehicle-to-everything (V2X) communications: enabling technologies, challenges, and opportunities. Proc IEEE 110(6):712–734
6. Liaskos C, Tsioliaridou A, Nie S, Pitsillides A, Ioannidis S, Akyildiz IF (2019) On the network-layer modeling and configuration of programmable wireless environments. IEEE/ACM Trans Netw 27(4):1696–1713
7. Pitilakis A, Tsilipakos O, Liu F, Kossifos KM, Tasolamprou AC, Kwon DH, Tretyakov SA (2020) A multi-functional reconfigurable metasurface: electromagnetic design accounting for fabrication aspects. IEEE Trans Antennas Propag 69(3):1440–1454
8. Segata M, Casari P, Lestas M, Tyrovolas D, Saeed T, Karagiannidis G, Liaskos C (2023) On the feasibility of RIS-enabled cooperative driving. In: 2023 IEEE vehicular networking conference (VNC). IEEE, pp 143–150
9. Zhu Y, Mao B, Kato N (2022) Intelligent reflecting surface in 6G vehicular communications: a survey. IEEE Open J Veh Technol 3:266–277
10. Chen Y, Wang Y, Zhang J, Di Renzo M (2021) QoS-driven spectrum sharing for reconfigurable intelligent surfaces (RISs) aided vehicular networks. IEEE Trans Wirel Commun 20(9):5969–5985
11. Wang J, Zhang W, Bao X, Song T, Pan C (2020) Outage analysis for intelligent reflecting surface assisted vehicular communication networks. In: GLOBECOM 2020-2020 IEEE global communications conference. IEEE, pp 1–6
12. Ai Y, Felipe AP, Kong L, Cheffena M, Chatzinotas S, Ottersten B (2021) Secure vehicular communications through reconfigurable intelligent surfaces. IEEE Trans Veh Technol 70(7):7272–7276

13. Ozcan YU, Ozdemir O, Kurt GK (2021) Reconfigurable intelligent surfaces for the connectivity of autonomous vehicles. IEEE Trans Veh Technol 70(3):2508–2513
14. Saeed T, Aziz W, Pitsillides A, Vassiliou V, Akyildiz I, Taghvaee H, ... Lestas M (2021) On the use of programmable metasurfaces in vehicular networks. In: 2021 IEEE 22nd international workshop on signal processing advances in wireless communications (SPAWC). IEEE, pp 521–525
15. Segata M, Casari P, Lestas M, Tyrovolas D, Saeed T, Karagiannidis G, Liaskos C (2023) On the feasibility of RIS-enabled cooperative driving. In: 2023 IEEE vehicular networking conference (VNC). IEEE, pp 143–150
16. Li A, Singh S, Sievenpiper D (2018) Metasurfaces and their applications. Nanophotonics 7(6):989–1011. https://doi.org/10.1515/nanoph-2017-0120
17. Ali A, Mitra A, Aïssa B (2022) Metamaterials and metasurfaces: a review from the perspectives of materials, mechanisms and advanced metadevices. Nanomaterials 12(6):1027
18. Liaskos C, Pyrialakos GG, Pitilakis A, Tsioliaridou A, Christodoulou M, Kantartzis N, Akyildiz IF (2020) The internet of metamaterial things and their software enablers. Int Telecommun Union J 1(1):55–77
19. Soukoulis CM, Wegener M (2011) Past achievements and future challenges in the development of three-dimensional photonic metamaterials. Nat Photonics 5(9):523–530
20. Tsilipakos O, Tasolamprou AC, Pitilakis A, Liu F, Wang X, Mirmoosa MS, Tretyakov S (2020) Toward intelligent metasurfaces: the progress from globally tunable metasurfaces to software-defined metasurfaces with an embedded network of controllers. Adv Opt Mater 8(17):2000783
21. Xu Z, Li X, Zhao X, Zhang MH, Wang Z (2017) DSRC versus 4G-LTE for connected vehicle applications: a study on field experiments of vehicular communication performance. J Adv Transp
22. Sakaguchi K, Fukatsu R, Yu T, Fukuda E, Mahler K, Heath R, Shimizu T (2021) Towards mmWave V2X in 5G and beyond to support automated driving. IEICE Trans Commun 104(6):587–603
23. Zhao J (2019) A survey of intelligent reflecting surfaces (IRSs): towards 6G wireless communication networks. arXiv:1907.04789
24. Abadal S, Liaskos C, Tsioliaridou A, Ioannidis S, Pitsillides A, Solé-Pareta J, Cabellos-Aparicio A (2017) Computing and communications for the software-defined metamaterial paradigm: a context analysis. IEEE Access 5:6225–6235
25. Liaskos C, Tsioliaridou A, Pitsillides A, Akyildiz IF, Kantartzis NV, Lalas AX, Soukoulis CM (2015) Design and development of software defined metamaterials for nanonetworks. IEEE Circuits Syst Mag 15(4):12–25
26. Di Renzo M, Zappone A, Debbah M, Alouini MS, Yuen C, De Rosny J, Tretyakov S (2020) Smart radio environments empowered by reconfigurable intelligent surfaces: how it works, state of research, and the road ahead. IEEE J Sel Areas Commun 38(11):2450–2525
27. Liaskos C, Nie S, Tsioliaridou A, Pitsillides A, Ioannidis S, Akyildiz I (2018) A new wireless communication paradigm through software-controlled metasurfaces. IEEE Commun Mag 56(9):162–169
28. Saeed T, Abadal S, Liaskos C, Pitsillides A, Taghvaee H, Cabellos-Aparicio A, ... Lestas M (2021) Workload characterization and traffic analysis for reconfigurable intelligent surfaces within 6G wireless systems. IEEE Trans Mobile Comput
29. Saeed T, Skitsas C, Kouzapas D, Lestas M, Soteriou V, Philippou A, ... Pitsillides A (2018) Fault adaptive routing in metasurface controller networks. In: 2018 11th international workshop on network on chip architectures (NoCArc). IEEE, pp 1–6
30. Saeed T, Soteriou V, Liaskos C, Pitsillides A, Lestas M (2020) Toward fault-tolerant deadlock-free routing in hypersurface-embedded controller networks. IEEE Netw Lett 2(3):140–144
31. Duato J (1995) A necessary and sufficient condition for deadlock-free adaptive routing in wormhole networks. IEEE Trans Parallel Distrib Syst 6(10):1055–1067
32. Duato J, Yalamanchili S, Ni L (2003) Interconnection networks. Morgan Kaufmann
33. Taghvaee H, Abadal S, Georgiou J, Cabellos-Aparicio A, Alarcón E (2019). Fault tolerance in programmable metasurfaces: the beam steering case. In: 2019 IEEE international symposium on circuits and systems (ISCAS). IEEE, pp 1–5

34. Wu J (2003) A fault-tolerant and deadlock-free routing protocol in 2D meshes based on odd-even turn model. IEEE Trans Comput 52(9):1154–1169
35. Glass CJ, Ni LM (1992) The turn model for adaptive routing. ACM SIGARCH Comput Archit News 20(2):278–287
36. Taghvaee H, Abadal S, Pitilakis A, Tsilipakos O, Tasolamprou AC, Liaskos C, Alarcón E (2020) Scalability analysis of programmable metasurfaces for beam steering. IEEE Access 8:105320–105334
37. Hosseininejad SE, Rouhi K, Neshat M, Cabellos-Aparicio A, Abadal S, Alarcón E (2019) Digital metasurface based on graphene: an application to beam steering in terahertz plasmonic antennas. IEEE Trans Nanotechnol 18:734–746
38. Yu N, Genevet P, Kats MA, Aieta F, Tetienne JP, Capasso F, Gaburro Z (2011) Light propagation with phase discontinuities: generalized laws of reflection and refraction. Science 334(6054):333–337
39. Cui TJ, Qi MQ, Wan X, Zhao J, Cheng Q (2014) Coding metamaterials, digital metamaterials and programmable metamaterials. Light: Sci & Appl 3(10):e218–e218
40. Kouzapas D, Skitsas C, Saeed T, Soteriou V, Lestas M, Philippou A, Pitsillides A (2020) Towards fault adaptive routing in metasurface controller networks. J Syst Archit 106:101703
41. Zheng P, Ding J, Fei D, Zhang J, Ai B, Li X, Cui T (2023) Field trial measurement and channel modeling for reconfigurable intelligent surface. Digital Commun Netw 9(3):603–612
42. Segata M, Lestas M, Casari P, Saeed T, Tyrovolas D, Karagiannidis G, Liaskos C (2023) Enabling cooperative autonomous driving through mmWave and reconfigurable intelligent surfaces. In: 2023 18th wireless on-demand network systems and services conference (WONS). IEEE, pp 32–39
43. Krstic M, Wang HH (2000) Stability of extremum seeking feedback for general nonlinear dynamic systems. Automatica-Kidlington 36(4):595–602
44. Ashraf N, Saeed T, Taghvaee H, Abadal S, Vassiliou V, Liaskos C, ... Lestas M (2023) Intelligent beam steering for wireless communication using programmable metasurfaces. IEEE Trans Intell Transp Syst
45. Ashraf N, Lestas M, Saeed T, Taghvaee H, Abadal S, Pitsillides A, Liaskos C (2020) Extremum seeking control for beam steering using hypersurfaces. In: 2020 IEEE international conference on communications workshops (ICC Workshops). IEEE, pp 1–6
46. Tan X, Sun Z, Koutsonikolas D, Jornet JM (2018) Enabling indoor mobile millimeter-wave networks based on smart reflect-arrays. In: IEEE INFOCOM 2018-IEEE conference on computer communications. IEEE, pp 270–278
47. Ashraf N, Saeed T, Ansari RI, Lestas M, Liaskos C, Pitsillides A (2019). Feedback based beam steering for intelligent metasurfaces. In: 2019 2nd IEEE Middle East and North Africa COMMunications Conference (MENACOMM). IEEE, pp 1–6
48. Ariyur KB, Krstic M (2003) Real-time optimization by extremum-seeking control. Wiley
49. MacCartney GR, Zhang J, Nie S, Rappaport TS (2013) Path loss models for 5G millimeter wave propagation channels in urban microcells. In: 2013 IEEE global communications conference (GLOBECOM). IEEE, pp 3948–3953
50. Rajamani R, Tan HS, Law BK, Zhang WB (2000) Demonstration of integrated longitudinal and lateral control for the operation of automated vehicles in platoons. IEEE Trans Control Syst Technol 8(4):695–708
51. Kim SW, Qin B, Chong ZJ, Shen X, Liu W, Ang MH, Rus D (2014) Multivehicle cooperative driving using cooperative perception: design and experimental validation. IEEE Trans Intell Transp Syst 16(2):663–680
52. Cigno RL, Segata M (2022) Cooperative driving: a comprehensive perspective, the role of communications, and its potential development. Comput Commun 193:82–93
53. Segata M, Cigno RL, Hardes T, Heinovski J, Schettler M, Bloessl B, ... Dressler F (2022) Multi-technology cooperative driving: an analysis based on PLEXE. IEEE Trans Mobile Comput

Chapter 11
Curb Management for More Efficient Deliveries

Robert B. Binder, Genevieve Giuliano, and Jaehyun Ha

Abstract Curb management is critical for urban functionality, particularly in downtown areas, where various users compete for limited curb space. This chapter presents two Southern California curb management pilot projects funded by the US Department of Energy, focused on promoting zero-emission vehicles (ZEVs). The research employs semi-structured interviews with stakeholders involved in these projects to explore their characteristics, expectations, and limitations. The escalating demand for curb space, driven by ride-hailing services, goods delivery, and alternative transportation modes, is discussed, along with the ensuing challenges like congestion and conflicts between modes. The study highlights the complex relationship between optimizing curb efficiency and promoting ZEVs and the role of automation in enforcement. This chapter offers valuable insights into the intricate trade-offs inherent in managing curb space for a sustainable urban future.

11.1 Introduction

Curb management, the way cities regulate and allocate curb space, has a major impact on the functioning of cities [1]. Especially in downtown areas, there is competition among users at the curb. These users include single-occupancy vehicle parking, loading and unloading of goods from light- to heavy-duty trucks, Transportation Network Companies (TNCs) pickups and drop-offs, public and private transit, and non-motorized modes. The growth of new mobility operators, especially ride-hailing/TNC services and goods delivery, continues, and the demand for access to curb space increases with it [2–4]. Despite the importance of curb activity to the overall success of a business district, management strategies differ as cities adopt different models and strategies of enforcement and pricing [1, 5–7].[1]

R. B. Binder · G. Giuliano (✉) · J. Ha
University of Southern California, Los Angeles, CA, USA
e-mail: giuliano@usc.edu

[1] OpenStreetMap is a trademark of the OpenStreetMap Foundation, and is used with their permission. This product is not endorsed by or affiliated with the OpenStreetMap Foundation.

P. Ioannou and A. A. Malikopoulos (eds.), *Transportation Mobility in Smart Cities*, Springer Tracts on Transportation and Traffic 21,
https://doi.org/10.1007/978-3-031-64769-7_11

This chapter summarizes the history and background of two curb management pilot projects in Southern California and highlights the methodology and results of a series of interviews. These projects are funded by the US Department of Energy and have a primary goal of promoting the use of zero-emission vehicles via curb management policies. The targets are delivery vehicles and TNCs.

Our research objectives were to identify the characteristics, expectations, and limitations of these projects. We conducted semi-structured interviews to understand the pilot projects and learn about the stakeholders involved and their decision-making processes before, during, and after the pilot projects. We also wanted to understand what outcomes each of the municipalities and stakeholders wanted, and how extensively the pilot could impact the future of curb management.

This research is a first step in understanding the potential of curb management to promote use of zero-emission vehicles, and more generally increase the productivity of the curb. We find that interest in promoting zero-emission vehicles is mixed. We find that there are two distinct sets of stakeholders, one that seeks to promote zero-emission vehicles, in-line with the main goals of the project; another that have a greater interest in better managing the curb. The remainder of the chapter is organized as follows. We introduce the curb problem, briefly discuss the relevant literature, present our methodology, and then present our case studies and results. We discuss the key observations and outcomes of the pilot projects and how these outcomes may impact the future of curb management.

11.2 The Curb Problem

Increasing demand for curb space is explained by the growth in ride-hailing/TNC services, goods delivery, as well as bikes, bike share, and scooters [2, 5–8]. This increase in demand is in addition to the needs of traditional public transit modes and commercial parking. Figure 11.1 illustrates the variety of curb demands and how they may be allocated on a city street.

As a result of the growth in e-shopping and omni channel shopping, one of the main drivers of curb demand is the growth in goods deliveries. Along with the growth in e-shopping also comes a demand for more frequent and nearly "instant" deliveries [4]. To achieve desired frequencies and delivery times, there is also a continued diversification of delivery fleets in cities, including parcel trucks, food trucks, car deliveries (Uber Eats, Instacart, GrubHub, etc.), cargo bikes, and personal bicycles. The shift in goods deliveries is shown in Fig. 11.2. The bars are mail volumes, and the lines are parcel volumes for USPS, UPS, FedEx, and Amazon. It can be seen that parcel volumes have grown at an increasing rate since around 2011, reflecting the rise in e-shopping [10, 11]. For instance, the parcel volumes for Amazon were less than 70 billion in 2018 but have jumped to almost 150 billion in 2022 [11]. Relatedly, a projection on worldwide last-mile volumes by value suggests a doubling in market size from just over $100 billion to $200 billion from 2020 to 2027 [12].

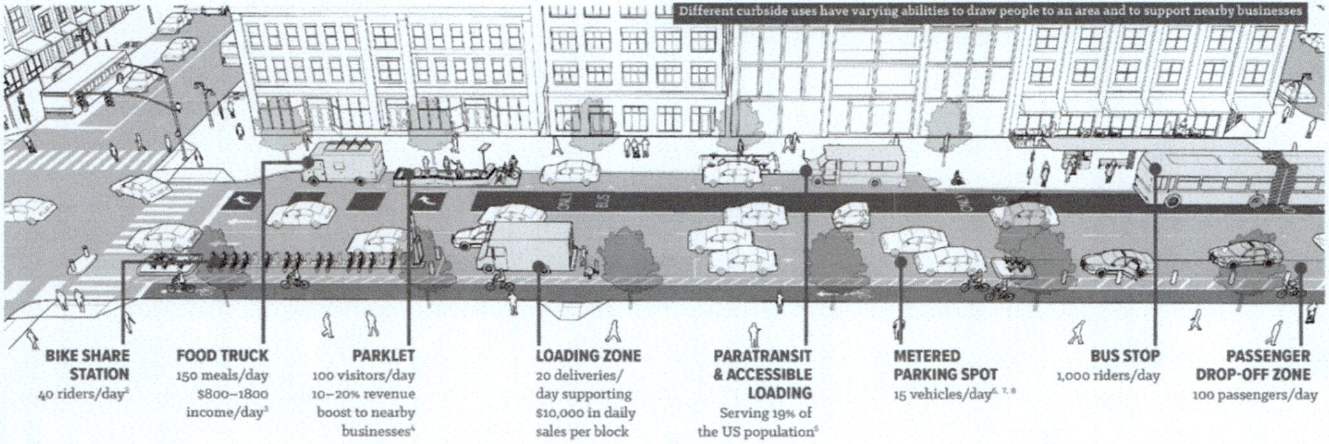

Fig. 11.1 NACTO curb appeal street section diagram. Adopted from [9][2]

[2] Authors received permission from the National Association of City Transportation Officials [9] to adopt the figure for publication with the following requirements (1) NACTO is credited by name somewhere near the image and (2) the image itself isn't edited or modified (November 13, 2023).

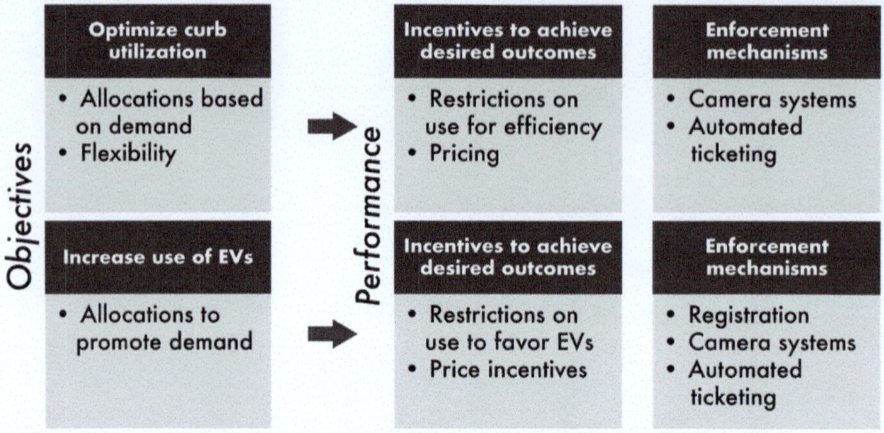

Fig. 11.2 Objective and performance measures of curb demonstration/pilot projects

The challenges that come with the increased demand and pressures on the curb include conflicts between passenger and delivery modes, conflicts between municipal policies and ordinances, double parking and its impacts on traffic flow, pedestrian safety, and the additional congestion and pollution from searching for parking and queueing at the curb. These challenges generate questions of who should have priority at the curb, and under what conditions. How can pricing, citations, and other management tools be implemented? How can policies best be enforced? And finally, should enforcement be automated?

Another dimension of curb management is its relationship to more general GHG reduction goals, including shifting the vehicle fleet to zero-emission technologies. Many countries have extensive zero-emission vehicle policies with targets to phase out gasoline and diesel-powered vehicles within the next few decades. These policies typically include subsidies and other incentives for purchase, priority access to parking and express lanes, or discounted rates for fuel. Curb management could play a role by offering priority to zero-emission vehicles (ZEVs), through free or discounted parking, or restricting curb space only to ZEVs.

The objectives of managing curbs for efficiency and managing curbs to promote electric vehicles (EVs) are not necessarily aligned, as illustrated in Fig. 11.3. To optimize curb productivity, we would manage based primarily on-demand. For example, we could establish a set of prices that would clear the market for each time period by using dynamic pricing. To promote ZEVs, we want to increase demand and therefore provide incentives—free parking or zones restricted only to ZEVs. Switching to a ZEV cannot happen instantaneously, hence these incentives will be inefficient at least in the short run.

No matter the objective, enforcement is a major consideration. Traditional enforcement is conducted by human parking patrols that roam streets and check for parking violations. There is a random dimension to parking enforcement as getting a citation depends on whether the parking patrol sees the violation. The less likely a citation becomes, the more likely the curb rules will be ignored. Consistent enforcement is most efficiently done automatically via cameras. Thus, curb management programs typically also involve some form of automation for enforcement.

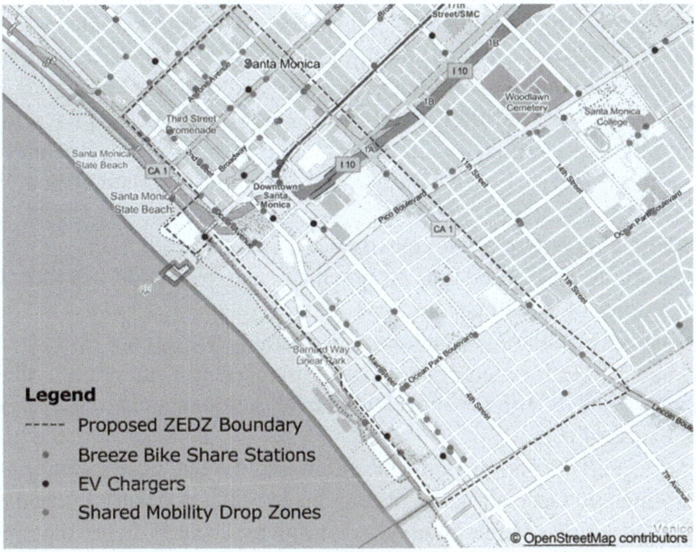

Fig. 11.3 Santa Monica ZEDZ boundary and modalities[3]

Curb management research is aimed at improving productivity while accommodating growing and diverse demand. The established research touches on all curb uses and modes and tends to come in two different forms, either a case study or a set of case studies, of existing policies, or a study on how the implementation of a tool or policy may impact curb management success. Our research includes both forms; only the case studies are presented here.

Using both qualitative and quantitative methods, some researchers consider and examine the implications of parking policies at the curb. In one case, they conceptualize parking policies as a low cost of payment, a low level of quality or investment, but a large share of revenue derived from punishment/fines for municipal governments [13]. Others have examined how some variables determine parking choices when prices increase [5]. Others broaden their scope to consider what US municipalities are doing to adapt to the new pressures on the curb using a semi-structured interview approach [8]. In one case, San Francisco adopted a performance goal of 60–80% occupancy for its metered parking, intended to reduce double parking, reduce cruising for parking, and improve the driver experience. Millard-Ball et al. [14] found that rate changes have helped achieve the City's occupancy goal and reduced cruising by 50%.

The technologies, both in user modes and management strategies, that are impacting curb management policies are also considered. Tirachini [15] found that ride hailing is expected to reduce parking requirements, which shifts the attention of policy-makers toward curb management. Machado-León et al. [7] conducted a

[3] Base map from OpenStreetMap (https://www.openstreetmap.org/copyright); Available under the Open Database License (https://www.opendatacommons.org).

case study of Seattle to examine the dwell time of passenger vehicles while picking up or dropping off passengers. They argue that dwell time is a vital metric for curb management. Finally, researchers developed and evaluated a tool for Arlington County, VA to reconsider the allocation of curb space using six modules (ride-hailing services, commercial loading, on-street parking, transit service, micromobility, and non-transportation uses) to provide the greatest economic and societal benefit to users of the transportation system [16].

To our knowledge, there are no existing examples of curb management policies intended to promote ZEVs, and thus no specific research to build from for this study. There are examples of low emissions zones (LEZs) that restrict access to low or zero-emission vehicles (e.g., Paris) or reduce price of entry for low emission vehicles (e.g., London) [17]. A more focused effort is being planned in Rotterdam, where a zero-emission zone is linked with urban freight consolidation centers to use ZEVs for the last-mile deliveries [18].

The zones in Paris, London, and Rotterdam may be the closest example to ZEV-favoring parking policies, but curb management policies do not currently seek to address ZEV adoption. Municipalities are implementing policies that build from national and federal policies. However when ZEV policies are advocated, the last-mile issue and how larger scale policies can impact it is consistently highlighted [19–21]. In one specific case, following interviews and analysis, Abe et al. [22] revealed that the burden of Zero-Emission Delivery Zones (ZEDZ) on delivery companies is greater than the social benefits such as GHG reductions from large investment in EVs. In this study, however, we observe how considerations of safety, the needs of businesses, and the ability to use new technology to enhance data collection take precedent over larger goals of lowering emissions.

11.3 Methodology

Our curb management research is organized around three questions:

I. What strategies are most effective in improving curb utilization?
II. Is curb management an effective strategy to increase the use of EVs?
III. What are the organizational and institutional factors that may affect curb management policies?

In this chapter, we focus on the third question, the organization and institutional factors that explain curb management policy choices and outcomes. We seek to understand the role and motivations of sponsors and participants and explain observed outcomes. We use a mixed methods approach. First, we conduct historical research to document the origin, development, and outcomes of the pilot projects. The historical research is based on project documents and meetings, as well as interviews with project participants.

Second, we conduct semi-structured interviews to interpret and understand outcomes. We conducted 19 interviews with representatives of different aspects of the projects, from sponsors to delivery businesses to technology developers. Interviewees were selected based on their extent of experience and participation in the pilots. Contacts were obtained from initial lists provided by project sponsors and subsequent referrals. Interviews took place via video conference from October 2022 to December 2022. Interviews were recorded and transcripts were generated from the recordings.

Interview transcripts were analyzed in two ways. First, the authors independently reviewed the transcripts and generated lists of findings. The lists were then discussed and merged. Second, we use text analysis. We were unable to conduct a conventional text analysis using statistical modeling approaches (i.e., Latent Dirichlet Allocation (LDA) topic modeling) due to the limited amount of interview text. Alternatively, we applied the ChatGPT tool to derive summaries, findings, and implications from the interview transcripts. Several recent papers have explored the potential of using ChatGPT to analyze text data especially for the limited size of data and inductive thematic analysis [23, 24]. We asked ChatGPT to provide summaries of the interview responses. Here is one example of the prompts asked to ChatGPT: "The transcripts below were obtained from respondents in an interview on 'What do you see as problems with either the curb or curbside activities?'. Summarize the main points in the responses and prioritize them if they were mentioned multiple times." We combine results from both methods to assess the pilots.

11.3.1 The Case Study Demonstrations

The case study demonstrations are funded by the US Department of Energy (DOE). There are three demonstration sites: the downtown areas of Pittsburgh, PA, Santa Monica, CA, and Los Angeles, CA. The lead is the Los Angeles Cleantech Incubator (LACI), a non-profit established by the City of Los Angeles and the Los Angeles Department of Water and Power. LACI supports and facilitates green energy and technologies. This demonstration is the first major project across multiple municipalities in the US to conduct pilot projects on smart loading zones and how these zones could influence the adoption of zero-emission ride hailing and delivery vehicles. Our research focuses on the Santa Monica and Los Angeles demonstrations.

11.4 History of Santa Monica and Los Angeles Demonstrations and Major Stakeholders

In May 2018, the Transportation Electrification Partnership (TEP) was formed by LACI. The TEP is a multi-year partnership among local, regional, and state stakeholders to make progress toward transportation electrification and zero-emissions goods movement in the Greater LA region in advance of the 2028 Olympic and

Paralympic Games. The TEP published "Roadmaps" to guide the process of electrification. The 2019 Roadmap included the concept of ZEDZs that would be used by ZE delivery trucks and cargo bikes.

In February 2020, LACI issued an expression of interest on "Voluntary Neighborhood Zero-Emissions Delivery Zone." The goals of the voluntary ZEDZ were to (1) reduce air pollution, (2) reduce traffic congestion via smaller footprint vehicles, (3) reduce noise, and (4) become a resource for cities to develop policies on zero-emissions goods deliveries. The objective of the pilot was to prove and demonstrate innovative technologies. The final suite of deployed technologies would be a mix of startup innovations and innovative offerings from established players. More specifically, the voluntary ZEDZs aimed to test technology and operations, community benefits, and model policies. Major stakeholders in addition to DOE and LACI include the participating cities and agencies, the firms or activities to be subject to the ZEDZs, and the technology firm providing the hardware and software required for the demonstration.

11.5 Santa Monica

LACI received three letters of interest. The City of Santa Monica was selected based on it having the highest likelihood for the impact of the pilot project. Its resources include substantial infrastructure to support ZEVs, a variety of sources and uses of micromobility services, a unique mixture and balance of residential and commercial segments, and of the applicants, the highest concentration of deliveries. Santa Monica proposed approximately one square mile of the city that includes key commercial districts such as Main Street and Third Street Promenade, as well as 15,850 residents. The specific boundaries are noted by the red dashed line in Fig. 11.4. Figure 11.4 shows the zone in the context of the downtown area and the related resources within the zone. The diverse representation of residential and commercial makes the zone an appropriate place to test applications for food and parcel delivery using different modalities.

The voluntary ZEDZ was officially launched in February 2021. The stated goals of the pilot are: (1) Advance emissions reduction policies and tools, (2) Pilot curb management solutions, and (3) Demonstrate technology innovations. The ZEDZ includes 8 ZE priority loading zones and 4 control locations. All locations are monitored via cameras. The camera locations (and hence the zone locations) were selected based on (1) power availability, (2) infrastructure availability, and (3) pole/streetlight availability. Figure 11.5 shows the locations of the pilot zones, blue denoting ZEDZs, and red denoting control zones. The priority zones are marked with signs and paint; the control zones are signed as conventional loading zones. There is no special enforcement of the priority zones.

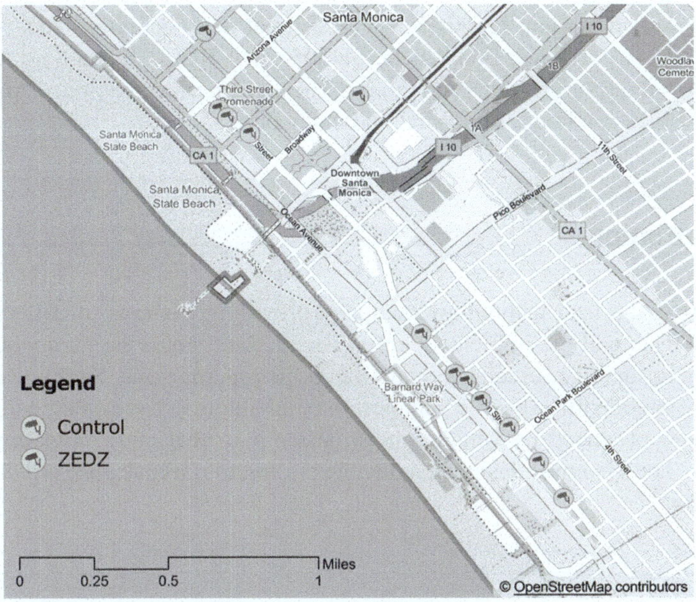

Fig. 11.4 ZEDZ and control zones in Santa Monica[4]

Fig. 11.5 Example of computer vision technology in-use by automotus. Gray colored boxes are overlayed on pedestrian detections due to privacy issues[5]

---

[4] Base map from OpenStreetMap (https://www.openstreetmap.org/copyright); Available under the Open Database License (https://www.opendatacommons.org).

[5] Source: Provided to authors by Automotus, Inc. on Oct. 16, 2023.

The pilot zone brought together over fifteen partners including Ikea, AxleHire, Guayaki, Alsco Uniforms, FoodCycle, Shopify, and REEF Technology to voluntarily deploy and test zero-emission modes. The tech providers included Automotus, Maxwell Vehicles, Circuit, FreeWire, and many others.

11.6 Automotus

Automotus is the lead technology company providing cameras to collect data on usage, and eventually to implement enforcement. Automotus uses computer vision technology to identify vehicles in and around the loading zone. See Fig. 11.6. The technology can identify specific vehicles, making automated enforcement possible. The data are produced by vehicle and timestamp so that curb utilization can be monitored. The system is designed to enable pricing to improve curb utilization and productivity.

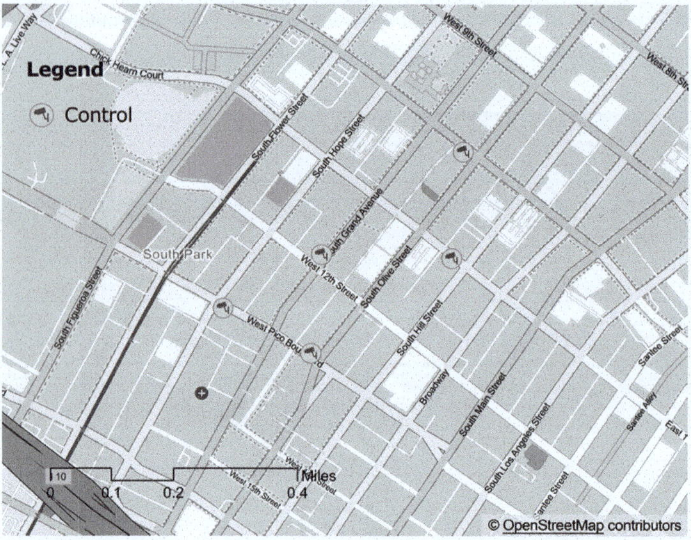

Fig. 11.6 ZEDZs in Los Angeles[6]

[6] Base map from OpenStreetMap (https://www.openstreetmap.org/copyright); Available under the Open Database License (https://www.opendatacommons.org).

11.7 Los Angeles

In November 2019, the City of Los Angeles launched the Urban Movement Labs (UML), to design and deploy transportation technology solutions [25]. One of the UML's flagship programs was the city's first Transportation Technology Innovation Zone, which is a designated area where private sector firms can test transportation technology solutions [26]. Automotus was selected to serve as a private sector partner in this program. The partnership led to the company installing curb management cameras in five targets and two control zones in 2021. The first five ZEDZs were implemented with painting and signage. They are located in downtown Los Angeles. See Fig. 11.7. Two control zones are located outside of the downtown area and are not shown. As in Santa Monica, no enforcement of these zones takes place. Figure 11.8 shows examples of the ZEDZs in Los Angeles. At the time when the picture was taken, the two ZEDZs were occupied by private non-ZE vehicles.

Los Angeles had intended to begin enforcement of the zones, but legal issues within the City of Los Angeles and the State of California related to the use of cameras and Automatic License Plate Recognition (ALPR), which Automotus utilizes in its technology, have not been worked through. As of July 2023, proposed state legislation, AB 1463 is in committee. The bill would regulate the use of ALPR data, require privacy and security of the data, and mandate purging of the data after a given time period [27].

The city continues to collect the camera data. The ability to monitor the curb, especially for understanding when and why double parking occurs, is noted by the City as being an effective use of the cameras currently.

Fig. 11.7 Downtown Los Angeles ZEDZ in use by private non-ZE vehicles

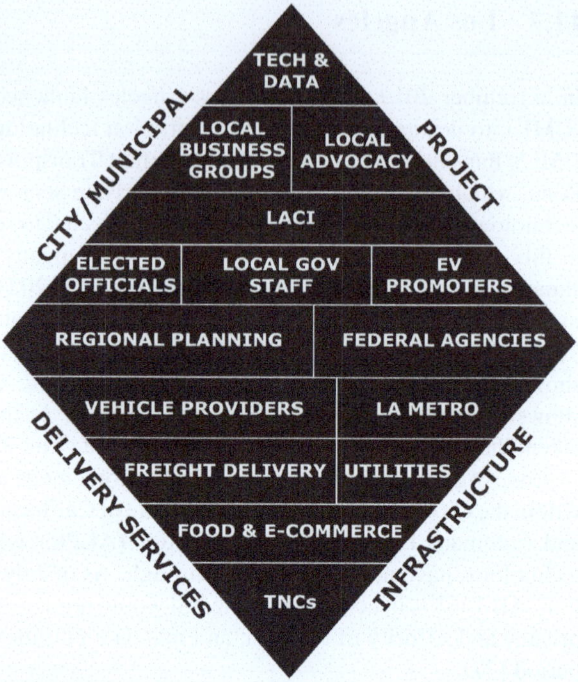

Fig. 11.8 Stakeholder involvement and relationships across categories (City/Municipal; Project; Delivery Services; Infrastructure)

11.7.1 Semi-structured Interviews

The purpose of the interviews was to obtain an in-depth understanding of the pilot projects. The interviews identified key stakeholders, their motivations and decision-making processes, and their perceptions and attitudes regarding the project. Pilots are tests to determine what works and what can be scaled. It is therefore important to understand why cities participated, what their expectations were, and to what extent their expectations have been realized. These experiences can provide insights on the potential for broad implementation as well as guidance on how future projects could be improved.

The pilots have many different dimensions and partners. A conceptualization of these dimensions and the stakeholders is presented in Fig. 11.9. The first dimension is the city. Curbs are public infrastructure owned and managed by cities. Therefore, any type of curb management must involve the city. City stakeholders include both staff and elected officials. City stakeholders also include the communities who influence local decisions: organizations, most often nonprofits, working on advocacy, as well as neighborhood and business improvement district initiatives. The second dimension is the project proponents: sponsors, advocates, and facilitators. In this case there are many. The project has high visibility within the region and complements larger scale efforts to improve curb management. Infrastructure is the third dimension and includes physical infrastructure of the curb, related services and energy sources, and

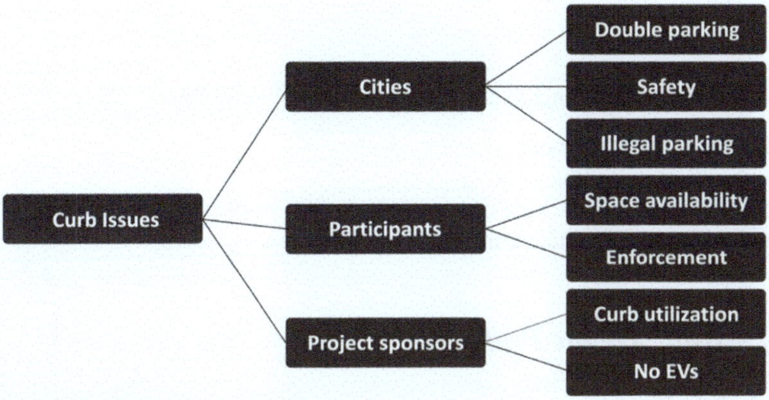

Fig. 11.9 What curb problems exist?

the hardware and software of the system. The final dimension is the target of the pilot—curb users. These include passenger services (e.g., taxi, ride-hailing services) and goods deliveries.

As part of a separate effort, one of our partners on this project, CityFi, was able to interview representatives from four delivery services—AxleHire, Coco, Kiwibot, and Ikea. The notes from these interviews were also included in our textual and thematic analyses.

We identified 40 people representing the four dimensions of stakeholders. Of the 40 requests, 19 interviews were obtained. Of the 19 interviews, four were conducted by project partner CityFi, and the rest were conducted by the authors. Table 11.1 shows the distribution of requests and interviews conducted across the four dimensions. The overall success rate for interviews is satisfactory. As might be expected, it was more difficult to obtain interviews from businesses than from public agencies. Within the cities and local stakeholder group, the interviews are mostly from city staff; we were able to interview just one community group and no local business group.

Table 11.1 Stakeholder interview requests and participation

Stakeholder group	N requested	N received	Percent
Cities and local stakeholders	19	9	47
Project sponsors and advocates	7	4	57
Infrastructure	1	1	100
Delivery services	13	5	38

11.8 Findings

The interviews were organized around the following questions:

1. What curb problems exist?
2. Why participate in this pilot?
3. What outcomes are or were expected, and what would constitute success?
4. What was the experience with the pilot?

11.8.1 What Are the Problems?

We found that different stakeholders had different perspectives on the nature of the curb problem and what should be done about it. Stakeholders tended to be aligned in three groups: cities, other pilot participants, and project sponsors. Figure 11.10 illustrates responses to the first question. Cities were most concerned with parking problems and their associated safety issues, especially double parking. These are ongoing problems for cities that they must deal with daily. One planner stated,

> Double parking can be a real hazard. People parking across a bike lane to park or make their delivery can be hazardous. There are plenty of demands on curb space – parking desires for private vehicles, bus stops, people putting scooters when not riding them, bike share station, and outdoor dining, food and package delivery – making safety important.

Those using the curb were more concerned with parking availability and enforcement. This is understandable since the ZEDZs were not enforced and they were used most frequently by private vehicles. For example, a delivery robot provider stated,

> We never want to be parked blocking anything, but it would be a good thing if there's always available space. The city could, for example, incentivize smaller delivery devices.

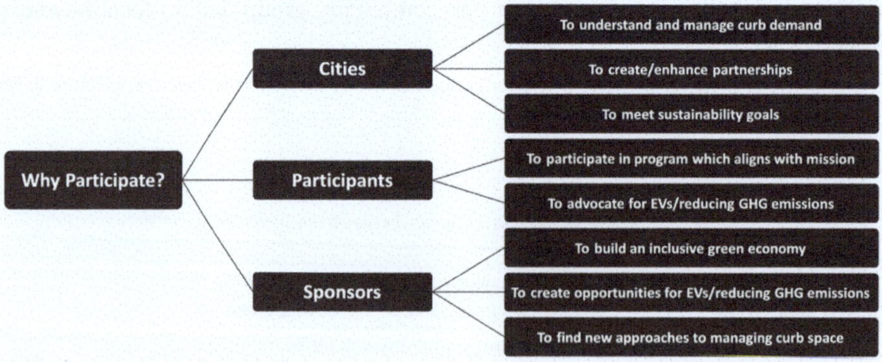

Fig. 11.10 Why participate?

Project sponsors were more concerned with the general problem of curb utilization and productivity and the primary goal of the demonstration, incentivizing the use of EVs. One project sponsor expressed the importance of leveraging curb productivity,

> There are infinite users of the curb, and they all have their different needs. I think having not only the pricing tool, but dynamic utilization would be helpful. The pricing and enforcement pieces are also going to be big challenges.

11.8.2 Why Participate?

Responses to reasons for participating in the pilot also showed differences between these groups (see Fig. 11.11). The most common response for cities was to understand and manage curb demand, followed by creating and enhancing partnerships and meeting sustainability goals. These reasons are consistent with concerns about safety and double parking. One planner from Los Angeles stated,

> By trying to provide incentives for delivery companies to convert their fleet is directly in-line with our sustainability goals. If we could go ahead and make it easier for them to access the curb and reduce double parking, then that is also reducing traffic congestion from vehicles circulating, which are all part of our LA Express Park goals.

Participants were mostly "green" businesses (for example, delivery robot service provider and restaurant food recycler) who were seeking opportunities to fulfill their mission and provide advantages for their business models. The most obvious of these is Automotus, whose entire business is based on automated enforcement for safety and sustainability. For example, a staff of Automotus noted,

> As a curb management company, the pilot project is a great place to play and Automotus has a unique ability to determine what propulsion type it is. And it's also a core tenant of what Automotus is about, trying to improve curb spaces, for humanity at large. But one of those being, how can we help decrease emissions, decrease pollution as well. The pilot project aligned really well with our mission.

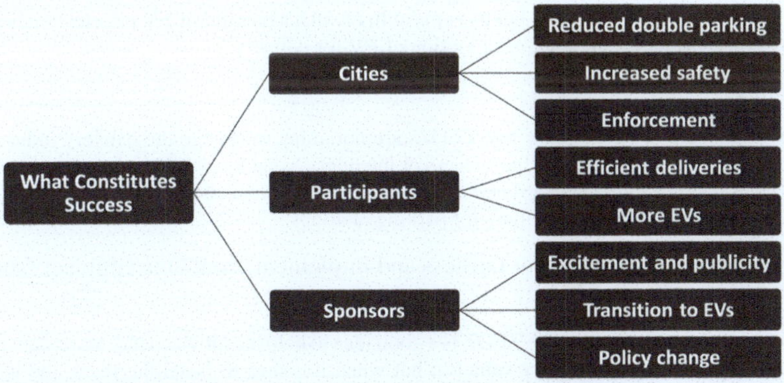

Fig. 11.11 What are your expectations, and what would constitute success?

The primary interest of the sponsors was in creating the green economy, more specifically increasing the use of EVs to reduce GHGs. For them, better curb management was a less important consideration. Moving to clean energy is a major goal for DOE and the main objective of the pilots is to show that EV use can be incentivized by prioritizing them at the curb. In California there are specific GHG reduction goals and a series of regulations that require transportation plans to meet these goals. Local and regional planners are always seeking new strategies to reduce GHGs. As noted above, LACI's mission is to accomplish these goals at the city level. Sponsor responses are very much in-line with organizational goals. For example, one LACI staff mentioned,

> LACI's mission is to create an inclusive green economy. We also look specifically at technologies like Automotus, where we can support startups to be successful and commercialize. The ZEDZ pilot is helpful for us to test the technologies of the startups, where we expect to see market transformation.

11.8.3 Expectations and Success

It is not surprising that with different perceptions of the curb problem and different reasons for participating, expectations for the pilots also differed (see Fig. 11.11). When asked what would constitute the success of the pilots, cities emphasized their current problems: reducing double parking, increasing safety, and effective enforcement. Double parking is a safety hazard and causes more congestion. The safety hazard comes from cars, bicycles, and pedestrians trying to circumvent the double-parked vehicle. It has been observed that major delivery services (e.g., UPS) frequently double park in high-demand parking areas [28, 29]. For delivery firms, time is money. When the cost of queuing or more distant parking exceeds the expected costs of a citation, it is rational to double park and risk the citation. When cities are limited to parking patrol enforcement, the expected cost of citations can be quite low.

City staff were not optimistic about the potential of ZEDZs to impact ZEV adoption. One planner said,

> It is still useful to test this pilot, but it seems unlikely that a handful of EV priority zones are really going to change the regional adoption of EV for delivery.

An engineer stated,

> After realizing that having only five ZEDZs are not going to change the delivery industry, we have very limited hopes for the success of the pilot – what we would hope is that we are able to provide incentives for private companies to make them change to cleaner vehicles and avoid paying millions of dollars for parking citations.

There were also concerns about fairness and implementation. One engineer from the public sector stated,

> I think everybody thinks that the ZEDZ is a good idea to help incentivize vehicles or delivery companies to change to cleaner vehicles; but when it comes to installing them, the local businesses start to question, particularly if they have services or vehicles that don't fall in that category.

Participant expectations are consistent with their reasons for participating in the pilot. They are seeking ways to exhibit their product, expand their businesses, and see curb prioritization as one way of doing so. For example, a delivery robot provider said,

> We wanted to provide a data point for cities to show that our technology can help reduce car trips in dense cities where we can do the delivery much safer in an energy efficient way.

Sponsors have broader and higher-level perceptions of success. First, the demonstrations are seen as a mechanism to gain visibility and support for GHG reduction policies. Second, they are viewed as an important step in the transition to EVs. Sponsors are aware that the EV transition will require strong incentives to purchase and use vehicles that have higher up-front costs and lower performance relative to their conventional ICE counterparts. The ultimate goal is policy change: showing that curb management works and can be scaled. A regional planner noted,

> We want to see that the ZEDZs are being used as intended and frequently occupied by EVs; we also want to see that the delivery companies are not getting many citations because they can't find a space to park.

Another planner observed,

> Regardless of what's really happening on the curbs, the pilot program itself gained a lot of publicity and it helped advance the conversation on what people should be thinking about and working on.

A representative from a non-profit organization identified the biggest benefit of the pilot as,

> it created a space where new approaches to managing the curb and using delivery technologies could be explored.

11.8.4 Experiences and Concerns

Participation in the pilots required a lot of effort from cities and participants. For cities, the pilot needed approval from elected officials, identification of test zones suitable for equipment installation, construction, maintenance, and operation of the ZEDZs, and collaboration with many partners. The last set of questions addressed experiences with the pilot planning and implementation.

Many complexities were revealed in the planning phase. Ideally the ZEDZs would be in high-demand locations and spatially clustered. Location choices were constrained by infrastructure: there had to be a place where the camera could be mounted and would have an adequate view of the zone. These constraints drove the choices of locations. There were additional concerns about zone locations including adequate sidewalk access, conflicts with protected bike lanes, and taking space from other uses (e.g., private parking). With high demand and competition among different users, there were questions of prioritization. Should scarce curb space be reserved exclusively for EVs when there is growing demand from buses, pedestrians, or scooters?

Both cities have created the ZEDZs, installed the signage, and are receiving data from the camera system; however, there is no enforcement. Santa Monica has decided to pause and retain the ZEDZs as voluntary, and Los Angeles remains delayed by legal issues of using ALPR. Most interviewees agreed that enforcement is the key for the success of the pilot, but there appear to be formidable challenges. The first is the logistics of enforcement, which would only be effective if it were consistent. The only practical way is through automation, which generates concerns on data privacy and security.

The second is the politics of enforcement. One city planner commented,

> With the enforcement piece, thinking about who is being punished for utilizing the curb improperly is an interesting thing to consider.

Some interviewees cited equity issues, especially if the ZEDZ policy involved pricing. There were concerns about drivers being unfamiliar with the rules, and these drivers would more likely be lower income people of color. There were also concerns about prioritizing the small number of EVs when curb demand is high. Doing so could add to congestion and double-parking problems. Finally, a city parking manager expressed concerns about losing revenue from parking with the ZEDZs,

> I think that the ZEDZs may help change behavior but may lead to an ongoing revenue loss for the city where we need to figure out how to have that balanced out.

11.9 Discussion

The results provide us with several key observations, which help to explain the outcomes of the pilot projects and how these outcomes may impact the future of curb management.

11.9.1 Goals Mismatch

First, there is a mismatch of project goals between the different stakeholders of the projects. The cities are focused on more modest goals of solving day-to-day problems, while sponsor goals are to use curb management as an incentive to move passenger and goods deliveries to EV fleets. This is a far more ambitious goal that is not yet demonstrated by the outcomes of the pilots to date. Participants have goals that support their business models and give them an advantage in the market.

Modest city goals make sense. Double parking is often a serious problem, as noted, and competition for curb uses is increasing. Cities have an obvious incentive to seek new ways to better address these problems. In addition, in California the use of ALPR for enforcement is not resolved, and the technology has yet to be tested as an effective and issue-free solution for enabling pricing at the curb in the long term. Without automated enforcement, double parking cannot be solved, and any type of priority use of ZEDZs is not practical. Priority use would involve some type of special pass. With automated enforcement it is possible to limit non-compliant vehicles to off-hours or charge a parking fee based on the pass. Enforcement via patrols would either be prohibitively costly or highly variable. There is also a lack of consensus at the local level for automated enforcement. While City of LA is pushing forward with its legislative efforts, City of Santa Monica has no plans to go forward with automated enforcement. Stated reasons include equity concerns (who can afford to purchase EVs, impacts on small vs. large businesses), satisfaction with using the Automotus data for monitoring purposes, and low priority assigned to expanding the number of ZEDZs.

Another possible explanation was not brought up in the interviews. California state law on GHG reduction provides a backstop for local governments. Current California law targets the phaseout of all non-zero-emission new passenger car sales by 2035. A variety of incentives support the target including tax credits and rebates. The Advanced Clean Truck and Advanced Clear Fleet regulations respectively require the purchase and sales of zero-emission trucks starting in 2024, with a goal of a fully transitioned fleet where feasible by 2042 [30, 31]. These regulations are a much stronger incentive for EV adoption than any type of curbside EV priority could provide, although such incentives might support the acceleration of the transition. Furthermore, the state regulatory backstop reduces pressure on local jurisdictions to implement policies that are politically sensitive.

11.9.2 Complexities of Implementation

Implementation turned out to be complicated. As noted above, locating the ZEDZs depended on utilities and infrastructure as well as where a ZEDZ might be attractive. How much demand from other modes and users had to be considered. There were of course the usual problems and delays associated with new technology systems: camera positions that didn't capture images fully, data interruptions, etc. These are an expected part of any pilot. Most technical problems worked themselves out over the course of the pilot.

The bigger challenges were around enforcement and the future of the pilots. What ZEDZ policy could look like was never discussed. In both cities there are signs that say the zones are for zero-emission vehicles only, which means a simple policy of exclusive EV use all day every day. Because there has been no enforcement, we have no information on what would happen if there were. The Automotus data shows that the ZEDZs are used as frequently as the control zones and by the same types of vehicles (mainly private cars). The question is where these users would go if the zones were enforced. There is also the larger political question of whether enforcement would be possible if few EVs use the zones, but many local businesses and residents are inconvenienced.

Interviewees recognized that ZEDZs would have to be deployed at a much larger scale to have an impact on the choice to use or own an EV, yet there seemed to be little appetite for such an expansion on the part of the cities. The pilots were subsidized; actual implementation would need some other source of funding. Without some solid evidence that ZEDZs would work, meaning measurably increase the use of EVs, it is understandable that cities would be hesitant. Cities might be more inclined to pursue automation and enforcement to help solve their more immediate problems of safety and double parking.

11.9.3 Transferability of Findings

Demonstration pilots are voluntary and hence by definition participants are self-selected. Cities must have an interest in solving a problem or enhancing their reputations and must be willing to take the risks involved in testing new policies and technologies. Participants have an interest in selling their products and growing their markets. Sponsors are carrying out policy objectives and seeking ways to push the policy agenda forward. The question is to what extent does this self-selection affect the transferability of findings.

The most obvious self-selection in this case was among the participants: the group is dominated by small to large companies within the small niche market of promoting EVs and sustainability. There were no major delivery companies or TNCs involved in the project. LACI did engage with several delivery companies, but they did not choose to participate. Possible explanations include the smaller scale of the project, the Southern California focus, and where the companies were at in their electrification plans at the beginning of the project. Without some of the most frequent users of curb involved in the demos, it is difficult to predict how ZEDZ policies would work at a larger scale.

One example of the delivery service participant was a growing delivery company aimed at providing "a fully integrated same-day delivery platform that provides e-commerce and traditional businesses with a logistics service that allows them to stay competitive among shifting consumer expectations [32]." The company was already making deliveries in Santa Monica—they just made sure they were doing it with all EVs. The company partnered with another start-up which allowed them to send an electric box truck to their sortation center where everything going out for delivery in Santa Monica was sorted into one separate load and delivered by the EV truck. Making special arrangements for Santa Monica deliveries was worth the trouble for this company given its mission. However, whether additional costs are sustainable in a competitive market remains to be seen.

11.10 Conclusions

Our study was conducted to develop a greater understanding of the implementation, existing policies, and potential policies of curb management in Los Angeles and Santa Monica, California. Through a series of semi-structured interviews, we found that the pilot projects have competing objectives. Stakeholders either align with the goal of ZEDZs as a mechanism for delivery companies to increase ZEV adoption or they are focused more on the ability to gain curb management insight from the implemented technology. The local governments hosting and implementing the pilot projects are generally more interested in the productivity of the curb, and in the double-parking issues that are occurring in and near the zones.

When considering the effectiveness of curb management policies, including those that are in use in the pilot projects in this study, it seems to depend on the magnitude and scale of a ZEDZ program and the enforcement of these zones. The goals of the technology providing the means of enforcement at a larger scale are stymied by the legal problems with ticketing via ALPR. For parking policies to be effective, cities and municipalities must have the legal permission to enforce them through the most efficient means.

We conclude that there are significant challenges to implementing ZEDZs in California. The legal challenges of using ALPR prevent effective enforcement. The comprehensive state policies of regulations and incentives are driving the adoption of EVs, relieving some pressure from cities to layer on additional policies. Proposals to price or restrict curb space are politically difficult because of the general public adversity to pricing what was once free and concerns of equity regarding the incidence of pricing or parking restrictions.

Acknowledgements The research presented in this chapter was funded by the US Department of Energy via a subcontract to the Los Angeles Cleantech Incubator LACI). The authors gratefully acknowledge the interview participants, the leadership of LACI, and the sharing of interview information by CityFi. Maps throughout this chapter were created using ArcGIS® software by Esri. ArcGIS® and ArcMap™ are the intellectual property of Esri and are used herein under license.

Disclaimer The contents of this chapter reflect the views of the authors, and not necessarily those of the project sponsors or partners. All errors and omissions are the responsibility of the authors.

Ethics Statement The study was approved by the University of Southern California Institutional Review Board (IRB) designee (Study ID: UP-22–00,712) on Sep 15, 2022. Informed consent was received from all participants prior to proceeding with individual interviews, with permission given to use the contents of the interviews without direct attribution by name.

References

1. Zalewski A, Buckley S, Weinberger R (2011) Regulating curb space. In: TRB 2012 annual meeting, pp 1–17
2. Shaheen S et al (2020) Mobility on demand planning and implementation: current practices, innovations, and emerging mobility futures. Berkeley
3. Shaheen S, Cohen A (2020) Mobility on demand in the United States. In: Analytics for the sharing economy: mathematics, engineering and business perspectives. Springer International Publishing, Cham, pp 227–254. https://doi.org/10.1007/978-3-030-35032-1_14
4. Giuliano G, Fang J, Binder RB, Ha J, Holmes A (2022) Travel behavior in E-commerce: shopping, purchasing, and receiving, p 7. https://doi.org/10.7922/G2377723
5. Shoup D (2021) Pricing curb parking. Transp Res Part A Policy Pract 154(November):399–412. https://doi.org/10.1016/j.tra.2021.04.012
6. Yu M, Bayram A (2021) Management of the curb space allocation in urban transportation system. Int Trans Oper Res 28(5):2414–2439. https://doi.org/10.1111/itor.12941
7. Machado-León JL, MacKenzie D, Goodchild A (2023) An empirical analysis of passenger vehicle dwell time and curb management strategies for ride-hailing pick-up/drop-off operations. Transportation (Amst). https://doi.org/10.1007/s11116-023-10380-6
8. Butrina P, Le Vine S, Henao A, Sperling J, Young SE (2020) Municipal adaptation to changing curbside demands: exploratory findings from semi-structured interviews with ten U.S. cities. Transp Policy 92(April):1–7. https://doi.org/10.1016/j.tranpol.2020.03.005
9. Roe M, Toocheck C (2017) Curb appeal: curbside management strategies. https://nacto.org/wp-content/uploads/2017/11/NACTO-Curb-Appeal-Curbside-Management.pdf. Accessed 23 July 2023
10. Rodrigue JP (2020) The distribution network of Amazon and the footprint of freight digitalization. J Transp Geogr 88(June):102825. https://doi.org/10.1016/j.jtrangeo.2020.102825
11. Rodrigue JP (2023) Mail Carried by USPS and Parcels Carried by Major Carriers, United States, 2004–2022|The Geography of Transport Systems. https://transportgeography.org/contents/chapter2/information-technologies-and-mobility/mail-parcels-carried-united-states/. Accessed 23 July 2023
12. Placek M (2023) Global Last Mile Delivery Market Size 2020–2027," Statista. https://www.statista.com/statistics/1286612/last-mile-delivery-market-size-worldwide/. Accessed 15 July 2023
13. Manville M, Pinski M (2021) The causes and consequences of curb parking management. Transp Res Part A Policy Pract 152:295–307. https://doi.org/10.1016/j.tra.2021.07.007
14. Millard-Ball A, Weinberger RR, Hampshire RC (2014) Is the curb 80% full or 20% empty? Assessing the impacts of San Francisco's parking pricing experiment. Transp Res Part A Policy Pract 63:76–92. https://doi.org/10.1016/j.tra.2014.02.016
15. Tirachini A (2020) Ride-hailing, travel behaviour and sustainable mobility: an international review. Transp 47(Aug):2011–2047. https://doi.org/10.1007/s11116-019-10070-2
16. Pochowski AL et al (2022) Solving the curb space puzzle through the development of a curb space allocation tool. Transp Res Rec 2676(10):601–621. https://doi.org/10.1177/03611981221090514
17. European Commission (2023) Urban Access Regulations in Europe. https://urbanaccessregulations.eu/countries-mainmenu-147/france/paris#:~:text=ParishasaZFE-m,Parishavethesamestandards
18. de Bok M et al (2021) Simulation of the impacts of a zero-emission zone on freight delivery patterns in rotterdam. Transp Res Rec 2675(10):776–785. https://doi.org/10.1177/03611981211012694
19. Iwan S, Nürnberg M, Jedliński M, Kijewska K (2021) Efficiency of light electric vehicles in last mile deliveries–Szczecin case study. Sustain Cities Soc 74(February). https://doi.org/10.1016/j.scs.2021.103167
20. Jaller A, Ambrose L (2018) Evaluating the use of zero-emission vehicles in last mile deliveries. J Rep (33):2017. https://doi.org/10.7922/G2JM27TW

21. Jaller M, Pineda L, Ambrose H, Kendall A (2021) Empirical analysis of the role of incentives in zero-emission last-mile deliveries in California. J Clean Prod 317(June):128353. https://doi.org/10.1016/j.jclepro.2021.128353
22. Abe R, Forbes M, Marshall E, Navid M, Shepard K (2022) Zero emission delivery zones—an analysis on US implementation. UCLA Institute of Transportation Studies, Los Angeles, CA. https://doi.org/10.17610/T6288G
23. De Paoli S (2024), Thematic analysis with large language models: does it work with languages other than English? A targeted test in Italian, arXiv preprint, arXiv:2404.08488
24. Mostafizer R, Watanobe Y (2023) ChatGPT for education and research: Opportunities, threats, and strategies. App Sci 13(9 may):5783. https://doi.org/10.3390/app13095783
25. Urban Movement Labs (2019) About Urban Movement Labs About. https://urbanmovementlabs.org/about/. Accessed 30 July 2023
26. Carey C (2020) LA Mayor outlines plans for city's first transport innovation zone. Cities Today. [Online]. Available: https://cities-today.com/la-mayor-outlines-plans-for-citys-first-transport-innovation-zone/
27. Lowenthal J, Weiner S (2023) AB-1463 Automated license plate recognition systems: retention and use of information. Sacramento: California Assembly. [Online]. Available: https://leginfo.legislature.ca.gov/faces/billTextClient.xhtml?bill_id=202320240AB1463
28. Schmid J, (Cara) Wang X, Conway A (2018) Commercial vehicle parking duration in New York City and its implications for planning. Transp Res Part A Policy Pract 116(March):580–590. https://doi.org/10.1016/j.tra.2018.06.018
29. Amer A, Chow JYJ (2017) A downtown on-street parking model with urban truck delivery behavior. Transp Res Part A Policy Pract 102:51–67. https://doi.org/10.1016/j.tra.2016.08.013
30. California Air Resources Board, Advanced Clean Cars Program|California Air Resources Board, ww2.arb.ca.gov. https://ww2.arb.ca.gov/our-work/programs/advanced-clean-cars-program. Accessed 30 July 2023
31. California Air Resources Board, Advanced Clean Fleets Regulation Summary|California Air Resources Board, ww2.arb.ca.gov. https://ww2.arb.ca.gov/resources/fact-sheets/advanced-clean-fleets-regulation-summary
32. CB Insights (2023) AxleHire Company Profile. https://www.cbinsights.com/company/axlehire

Chapter 12
Developing a Sustainable Active Mobility Framework Model for Smart Cities

George N. Papageorgiou and Elena Tsappi

Abstract This chapter proposes a smart, active, sustainable mobility framework model to promote accessibility, connectivity, and social inclusion among people in urban areas. In this context, Active Mobility, encompassing activities such as walking and cycling, could emerge as one of the most significant transport modes, sparking higher levels of well-being, quality of life, and sustainability in urban spaces. Historically, city planning authorities have often overlooked Active Mobility, heavily focusing instead on motorized traffic. This unjustified focus has resulted in severe traffic congestion and unbearable air pollution. In contrast, Active Mobility-friendly accessible, and connected urban spaces offer multiple health, economic, and environmental benefits to cities and their inhabitants. With support from the latest advances in Intelligent Transportation Systems (ITS), Information Communication Technology (ICT), Artificial Intelligence, and Augmented Reality, a viable Smart Active Mobility model may be established, representing a paradigm shift in how urban communities use and access transportation. In this envisioned model, active mobility is central and becomes the dominant transport mode for the development of revitalized sustainable smart cities.

12.1 Introduction

As urbanization continues to shape societies globally, cities across the world find themselves navigating the complexities of innovation and sustainability. These expanding urban landscapes, symbols of human advancement, are simultaneously grappling with a multitude of challenges. Foremost among these are the environmental dilemmas stemming from traditional transportation systems that heavily favor

G. N. Papageorgiou (✉) · E. Tsappi
Systema Research Centre, European University Cyprus, 6 Diogenous Street, 2404 Egkomi, Nicosia, Cyprus
e-mail: g.papageorgiou@euc.ac.cy

P. Ioannou and A. A. Malikopoulos (eds.), *Transportation Mobility in Smart Cities*, Springer Tracts on Transportation and Traffic 21,
https://doi.org/10.1007/978-3-031-64769-7_12

motorized vehicles. This entrenched paradigm, notorious for intensifying air pollution, increasing greenhouse gas emissions, and contributing to persistent congestion, profoundly detracts from the quality of urban living. The consequences extend beyond environmental impact, leading to public health issues due to increased sedentariness, economic pressures from excessive road maintenance and healthcare costs, and eroded social connections due to the decline of communal spaces.

In this intricate scenario, active mobility emerges not merely as an alternative but as a pivotal shift. Incorporating non-motorized transport modes such as walking and cycling, active mobility stands as a fundamental element for sustainable urban living. Its benefits go beyond environmental protection, promoting public health, economic efficiency, social interaction, and improved access to urban facilities and services. However, this shift is not solely about adopting different transport options; it's about a comprehensive metamorphosis of urban lifestyles. This profound change requires an integrated strategy that merges sustainable urban planning, innovative policymaking, cultural evolution, technological advancements, and crucially, community involvement.

This chapter urgently advocates for a robust active mobility framework model, one that is multifaceted and all-encompassing. We call for a strategy that synergizes principles of sustainable urban planning, crafts effective policies, encourages cultural adaptability, utilizes technological innovations, and prioritizes community engagement. Such a framework isn't an indulgence but an imperative for safeguarding our urban futures. In this context, we introduce an innovative concept of Smart Active Mobility. The proposed Smart Active Mobility (SAM) framework model combines the latest developments in technology with deep insights from behavioral science, placing active mobility at the forefront of urban planning initiatives. Further, by being flexible in its applicability, it recognizes the unique character of each urban setting and provides a backbone for tailored solutions.

The following section explores the historical backdrop, emphasizing the regrettable neglect of active mobility in past city planning and the diverse negative outcomes of this oversight. We then pivot to a more hopeful narrative, discussing how current technological leaps, particularly in Artificial Intelligence (AI), present a unique opportunity for redemption. This pathway not only seeks to correct past missteps but also aims to transform urban centers into models of sustainability. Our discussion is anchored in empirical research and enriched by practical case studies, demonstrating the realistic integration of active mobility into the urban framework. Through this chapter, we seek to contribute significantly to the academic conversation on sustainable urban mobility, championing a method that is strategic, integrated, and centered around people.

12.2 Background

The twenty-first century confronts a dual challenge: rapid urbanization and escalating environmental crises, thrusting sustainable urban development into the spotlight. Active mobility, with its promise to mitigate traffic congestion, enhance urban livability, and improve public health, is central to this discourse [1–3]. However, a historical bias toward motorized transport in city planning has led to a myriad of issues: increasing traffic, declining air quality, rising greenhouse gas emissions, and a reduction in overall well-being, posing substantial barriers to sustainable urban development [4–6].

Recent advancements in AI highlight the transformative potential of ITS and ICT in reinforcing active mobility and addressing these persistent challenges. Integrating these technologies into the smart city framework could significantly enhance the efficiency and attractiveness of active mobility, encouraging broader adoption among urban populations. Furthermore, contemporary studies emphasize the importance of understanding human behavior and engaging specific demographic segments to promote active mobility [7, 8]. Research illuminates various factors influencing transportation mode choices and the critical role of evolving perceptions toward active mobility, particularly in response to global events such as the COVID-19 pandemic [9].

During the pandemic and in the aftermath of COVID-19, the call for comprehensive active mobility transportation planning has never been more pressing. Such an approach, amalgamating cutting-edge research and technological innovations, addresses the current deficiencies in urban planning. Its implementation would align city transportation frameworks with sustainable development goals, enhancing public health and mitigating the negative environmental impacts of conventional transportation methods. Drawing from a wealth of global studies and practical experiences from various international transportation projects, this chapter aims to develop a framework model that champions accessibility, connectivity, and social inclusion in urban environments. The ultimate goal is to illuminate the potential impacts of such a model as a comprehensive platform designed to invigorate active transportation [10]. This model highlights the shift from traditional motorized transport to sustainable, active transit options, indicating a profound change in urban planning philosophies [1, 11]. It emphasizes the importance of sustainable mobility solutions, infrastructure, and participatory approaches that can drive behavioral change [9, 12, 13].

This innovative transportation framework model could employ System Dynamics (SD) [3] and GIS to assess active mobility routes in urban areas [10]. It would then consider various factors affecting active mobility, including the urban environment, safety measures, weather conditions, and street design [14–16]. Additionally, the model's practicality could be enhanced through the development of a navigation system designed specifically to support active mobility and meet users' needs [7, 8]. This comprehensive approach addresses a significant gap in ITS, which has traditionally focused primarily on motorized vehicles [6, 18, 19]. By providing real-time information and enabling users to make informed decisions about their travel routes, the model is poised to substantially increase the adoption of active mobility, offering wide-ranging benefits for individuals, communities, businesses, governments, economies, and the environment.

12.2.1 Significance of the Proposed Smart Active Mobility Framework Model

The proposed model introduces a new perspective on urban transportation design and management [10], transitioning from a traditional, motorized vehicle-centric approach to an emphasis on active, sustainable mobility [1, 16]. This transition is facilitated by a dedicated Active Mobility Navigation System, specifically tailored to meet the needs of pedestrians and cyclists. Implementing this system requires the integration of data concerning existing infrastructure, public transport connectivity, safety protocols, weather variables, and population density, all aimed at developing a safer, more efficient, and user-friendly infrastructure for walking and cycling [2, 15, 20].

A Smart Active Mobility Navigation System will provide citizens with real-time data, highlighting the most suitable routes for active commuting. Utilizing up-to-date safety and weather data, the system empowers users to make informed decisions, thereby overcoming potential barriers to active mobility [9, 21, 22]. The effectiveness of such navigation tools is supported by previous research on smart pedestrian navigation systems [17, 23–29], and is evident in two European-funded projects in which the authors were involved: “Smart Pedestrian Net (SPN): Smart Cities are Walkable; a Model to Plan a Pedestrian Network and Pedestrian Navigation System” [30–37] and “Active-Ageing, An Intelligent Active Mobility System for the Elderly” [38, 39].

The proposed Smart Active Mobility can be viewed as an evolution of Smart Pedestrian Navigation Systems. These systems have evolved from traditional motorized-based models to ones that are more pedestrian-focused. Instead of primarily catering to vehicles, these systems prioritize pedestrian pathways, safety, walkability, and accessibility. They take into account sidewalks, pedestrian areas, parks, and other pedestrian-friendly spaces when suggesting routes, while also assessing the quality of walking paths, identifying potential obstacles, and considering real-time crowd movements to recommend the most efficient walking routes.

Safety features in SAM leverage current data to alert users about potential hazards, based on recent incidents or areas with limited visibility and offer alternative routes during inclement weather or events causing disruptions. As urban centers become smarter, deeper integration with the city’s infrastructure becomes possible. For example, SAM can provide insights into pedestrian traffic light changes, helping users estimate waiting times or suggesting routes with synchronized pedestrian light sequences. Moreover, Personalized Routing in SAM isn’t just about navigation; it caters to individual preferences, offering everything from wheelchair-friendly routes to paths that avoid allergen-prone areas or feature popular city landmarks. Users can also provide Real-time Feedback, and thanks to Social Integration, the community can contribute updates, ensuring the data remains current. Advanced iterations of SAM might even incorporate Augmented Reality (AR) for an enhanced navigational experience, with devices like smartphones or AR glasses overlaying routes, landmarks, and more on the user's real-world view.

The proposed SAM model, with its emphasis on active transportation modes, embodies these advancements. It acknowledges the unique challenges urban pedestrians face, from navigating congested streets to ensuring safety amidst changing weather conditions, and strives to provide a comprehensive, adaptable, and pedestrian-centric urban transport solution. By harnessing real-time data and integrating System Dynamics (SD) modeling with Geographic Information Systems (GIS), SAM presents itself as a solution for urban planners and policy-makers. Informed by the latest research, the model is attuned to the specific needs and challenges of pedestrians, promising a more sustainable, safe, and efficient urban mobility landscape [6, 24, 40–43].

12.2.2 Structure of the Chapter

This chapter unveils the Smart Active Mobility (SAM) framework model and its potential implications in several sections. It begins by establishing a robust understanding of active mobility and its fundamental principles, exploring its significance within the realm of urban planning, and considering the variety of factors that influence its successful implementation and societal acceptance. This foundation paves the way for an in-depth analysis of ITS and ICT. The potential of these technologies, along with their crucial roles in augmenting active mobility, are dissected, laying the groundwork for the introduction of the SAM model. Following this, the chapter outlines the research methodology, providing insight into the nature of the study, the objectives guiding the investigation, the employed methodological framework, and the anticipated innovative outcomes within the active mobility field. The proposed SAM model forms the main part of this chapter. Here, the conceptual underpinnings of the model, its design, the stages of its implementation, and the potential benefits it holds for urban communities are revealed.

Subsequent sections emphasize the value of international collaboration and standardization for the successful replication in implementing the model. Key global stakeholders and institutions are spotlighted, underlining the synergies and reciprocal benefits achievable through such transnational partnerships. Toward the chapter's end, an analysis of the projected impact of the SAM project on sustainability is presented. Here, the benefits and potential applications of the model, the beneficiaries, and the ways the model fosters a shift toward sustainable mobility are explored. Finally, the chapter culminates in a summary of key findings and their implications for urban planning policy and practice. Future research directions in the field of active mobility are also proposed, laying a foundation for continued exploration and improvement in this critical area of study.

12.3 Understanding Active Mobility: The Core Concepts

Transitioning from the comprehensive insights provided in the previous section, this section expands upon the historical context provided earlier, focusing on the intrinsic characteristics of active mobility and its growing significance in contemporary urban planning. Building on the foundational understanding established in the "Background," we delve into the core concepts of active mobility, underscoring its pivotal role in shaping healthier, more sustainable urban environments. The ensuing segment delves into the complex narrative surrounding the critical role of active mobility in urban planning. It excavates the intricate matrix of relationships between active mobility and various facets of urban life—environmental, social, economic, and health related. Through this exploration, the symbiosis between these spheres and active mobility is brought to the fore. Finally, this section disentangles the myriad of key external and internal factors that modulate active mobility. Presenting a clear framework that elucidates the complex dynamics of these factors, it advocates for their explicit recognition in formulating strategies to enhance active mobility. This detailed overview, while providing critical insights into the elements shaping active mobility, also sets the stage for the proposed Smart Active Mobility (SAM) model.

12.3.1 Definition of Active Mobility

Active mobility encompasses various human-powered modes of transportation, including but not limited to, walking, cycling, wheelchair use, and manually powered scooters, which are essential for daily travel within urban areas. This form of mobility leverages human physical activity as a means of transportation, promoting a healthier, more sustainable way of commuting that contributes to not only physical health but also mental well-being [10, 15, 20]. Beyond the individual benefits, active mobility confers societal advantages such as reduced traffic congestion, decreased carbon emissions, and enhanced urban livability [15, 44]. It is a fundamental component of sustainable urban development, with goals extending to improved accessibility, connectivity, and social inclusion [45]. However, the adoption of active mobility is not without its challenges. It necessitates appropriate infrastructure, supportive policies, public awareness campaigns, and solutions to address safety concerns, diverse weather conditions, and accessibility for individuals with disabilities [13, 15, 44].

The integration of intelligent transportation technologies with active mobility solutions has the potential to establish efficient and sustainable urban mobility networks. Such technological advancements are highlighted in a recent study by Rani and Sharma [45], which delves into AI-enabled autonomy and traffic management systems. Active mobility signifies a transformative approach to urban transportation, prioritizing human-powered modes of travel that benefit personal well-being, environmental sustainability, and overall urban livability. Nevertheless, it demands a multidisciplinary approach, amalgamating elements from individual behavior to infrastructure design and technological innovations, to holistically foster active mobility.

12.3.2 The Importance of Active Mobility in Urban Planning

Building on the foundational concepts previously outlined, this section highlights the burgeoning recognition of active mobility as a linchpin of progressive urban planning. This paradigm shift, driven by active mobility's capacity to tackle pressing environmental and societal challenges, is redefining contemporary urban landscapes [1, 18, 46]. Committing to non-motorized transportation modes, such as walking and cycling, unlocks a plethora of benefits, spanning environmental sustainability, public health, social equity, and economic prosperity [4, 8]. Environmentally, active mobility is instrumental in shrinking a city's carbon footprint. By curtailing dependence on fossil fuel-powered vehicles, cities can drastically reduce greenhouse gas emissions and combat air pollution, thus contributing earnestly to the global climate change initiative [2].

In terms of public health, the perks of active mobility are manifold. Regular engagement in active travel, like walking and cycling, counters sedentary lifestyles, bolstering cardiovascular health, reducing chronic disease risks, and enhancing mental well-being [3, 13]. Additionally, urban spaces that favor walking and cycling can improve public health outcomes and potentially ease strains on healthcare systems [3]. Socially, active mobility fosters more inclusive, lively urban spaces, augmenting a city's overall livability [10, 20]. Infrastructure that is safe and accessible for pedestrians and cyclists encourages social interaction, nurtures community engagement, and cultivates a sense of belonging, thereby enhancing social equity [2, 10].

Economically, the implications of active mobility are substantial. It lessens the need for expensive road infrastructure and parking facilities [7], and can invigorate local economies since pedestrians and cyclists are more apt to frequent businesses on their routes. This boosts local commerce and fosters an economically dynamic urban milieu [4]. For instance, Copenhagen, celebrated for its cycling infrastructure, has reaped these economic advantages, underscoring the link between active mobility and thriving economies. In summary, integrating active mobility into urban planning aligns with the objectives of environmental preservation and sustainability while promoting public health, social equity, and economic vibrancy. Investments in infrastructure, policies, and initiatives championing active mobility can yield vibrant, inclusive, and sustainable urban spaces [9, 18].

12.3.3 External and Internal Factors Influencing Active Mobility

Active mobility, encompassing modes such as walking and cycling, is influenced by a multifaceted array of external and internal factors that shape individual choices and behaviors. A deep understanding of these factors is paramount for efficacious urban planning and the promotion of active mobility [4, 9]. Table 12.1 elucidates the reciprocal relationship between external and internal factors that impact active mobility. It delineates how the physical and infrastructural environment (external factors) interacts with personal attitudes and sociocultural dimensions (internal factors), contributing to an individual's predilection for active modes of transport such as walking and cycling.

Table 12.1 Interrelation of external and internal factors influencing active mobility

Category	Factors	Examples/Details	Interrelation
External	Urban infrastructure and design	Infrastructure like sidewalks, bicycle lanes, and pedestrian crossings encourage active modes of transport [4, 20]	Can influence internal factors like attitudes and personal motivations
External	Land use and density	Compact, mixed-use neighborhoods make walking and cycling more feasible and appealing [2]	Can influence internal factors like attitudes and personal motivations
External	Accessibility to key destinations	Accessibility of places like schools, workplaces, parks, and public transportation hubs influences the choice of active modes of transport [15]	Can influence internal factors like attitudes and personal motivations
External	Safety and security	Safety perceptions like well-lit pathways, low crime, and traffic calming measures can make active mobility more appealing [2, 44]	Can influence internal factors like attitudes and personal motivations
Internal	Attitudes and perceptions	Attitudes, beliefs, and perceptions toward active mobility significantly impact whether individuals engage in walking and cycling [8, 44]	Can influence response to external factors like infrastructure and design
Internal	Personal motivations and preferences	Personal health goals, environmental consciousness, and enjoyment of physical activity can encourage active mobility [8]	Can influence response to external factors like infrastructure and design
Internal	Socio-cultural factors	Social norms, social support, and cultural practices can influence active mobility behaviors [10, 13]	Can influence response to external factors like infrastructure and design

These factors play a crucial role in shaping the proposed Smart Active Mobility (SAM) model, a strategic initiative designed to foster the use of active modes of transport. For instance, the SAM model addresses urban infrastructure and design by prioritizing accessibility and connectivity, and it seeks to shape personal attitudes and perceptions by showcasing the benefits and feasibility of active transport modes. By understanding and incorporating these diverse factors, the model endeavors to affect a significant behavioral shift toward sustainable and active modes of transport. Through improving urban environments and infrastructure, implementing robust safety measures, and accounting for individual motivations, the proposed SAM model through the concepts outlined in this chapter has the potential to revolutionize urban transportation, by championing sustainable and active modes of travel.

12.4 An Overview of ITS- and AI-Based ICT

Building on the insights garnered from the exploration of active mobility's significance, this section delves into the technological engines that can propel this concept into tangible reality: ITS and AI-based ICT. These systems stand at the forefront of revolutionizing how we perceive and navigate our urban spaces, particularly in the context of active mobility. Further, it delineates the essential roles these and other advanced technologies, such as Extended Reality (XR) play in facilitating and promoting active mobility strategies and frameworks. The subsequent discourse provides a meticulous dissection of the potential of ITS and ICT in optimizing active mobility, underscoring their capacity to create innovative solutions to address persistent challenges and augment the overall active mobility experience. In essence, this section illuminates the importance of ITS and ICT as pivotal tools in enhancing active mobility. Further, it provides valuable insights and a strategic framework for urban planners, policy-makers, and stakeholders in encouraging sustainable and health-conscious travel choices and cultivating more intelligent, safe, and habitable urban environments.

12.4.1 Role and Importance of ITS and ICT in Active Mobility

ITS and ICT serve as indispensable tools in the progression of active mobility strategies and frameworks [9, 10, 18, 22]. Their effective integration can notably enhance urban planning, management, and user experiences, paving the way for a more efficient, sustainable, and user-friendly urban environment. ITS, a conglomerate of advanced technological applications, offers innovative services for various modes of transportation and traffic management [5, 46]. By prioritizing active mobility, ITS enhances safety measures for pedestrians and cyclists, delivers real-time updates on routes and traffic conditions, and elevates the overall efficiency of the transport network [9, 22, 45]. For example, ITS can implement alert systems for motorists regarding cyclists' presence and prioritize pedestrians at traffic signals during peak foot-traffic hours [6, 17].

ICT, comprising a wide array of communication devices and applications, plays a significant role in promoting active mobility [19, 47]. It fosters the development of robust platforms for collecting, processing, analyzing, and distributing transportation-related information, thus enabling well-informed travel decisions [18, 21]. For instance, smartphone applications can relay real-time data on efficient and safe cycling routes, considering factors such as traffic conditions, weather, and terrain [9, 18]. In essence, the incorporation of ITS and ICT into active mobility solutions equips citizens, urban planners, and policy-makers with invaluable tools and information to promote sustainable and health-conscious travel decisions. This approach also propels the creation of innovative solutions to urban mobility challenges, fostering smarter, safer, and more loveable cities [4, 13, 14].

12.4.2 Significant Enhancements to ITS and ICT Are also the Advancements in XR, ML, and AI

In recent years, the rapid advancements in Extended Reality (XR), Machine Learning (ML), and Artificial Intelligence (AI) have opened up further possibilities for enhancing ITS/ICT and applications to active mobility. These technologies have the potential to revolutionize how we plan, design, and navigate urban environments, making them more conducive to walking and cycling. Below are some of the significant enhancements these technologies bring to active mobility:

- Extended Reality (XR): This refers to a spectrum of technologies that include Virtual Reality (VR), Augmented Reality (AR), and Mixed Reality (MR) [48, 69]. XR can play a vital role in promoting active mobility by creating immersive and interactive experiences for pedestrians and cyclists. For instance, VR simulations can support urban planners and policy-makers to visualize proposed walking and cycling infrastructure, allowing them to identify potential challenges and optimize designs before implementation [49, 50, 69]. AR applications can provide real-time information to pedestrians and cyclists, such as navigation instructions, nearby points of interest, and safety alerts [41, 51]. Beyond merely providing information, AR technologies can gamify walking and cycling experiences, awarding points or virtual rewards for reaching certain milestones or completing specific routes, thereby encouraging people to engage more in active mobility options. This gamification can be particularly effective in attracting new users and retaining existing ones, making active mobility not just a necessity but an enjoyable experience [69]. XR technologies can also be used to gamify walking and cycling experiences, making them more engaging and enjoyable for users.

- Machine Learning (ML): ML algorithms have the capability to process vast amounts of data and identify patterns and trends. In the context of active mobility, ML can be used to analyze data from various sources, such as GPS data from smartphones and sensors, to understand travel patterns and behavior. This information can be used to optimize pedestrian and cycling routes, predict traffic congestion, and improve safety measures. ML can also enable personalized route recommendations based on individual preferences and habits, encouraging more people to choose active mobility options [52, 53].
- Artificial Intelligence (AI): It complements ML in enhancing active mobility experiences. AI-powered systems can process data in real time and make intelligent decisions to improve traffic flow, optimize signal timings for pedestrians and cyclists, and prioritize active mobility users at intersections. AI algorithms can also be employed for predictive maintenance of cycling infrastructure, ensuring that paths and bike lanes remain in good condition and safe for use. Additionally, AI-powered chatbots and virtual assistants can provide personalized support and information to active mobility users, addressing their queries and concerns [54, 55].

By integrating XR, ML, and AI into active mobility solutions, urban planners and policy-makers can create more efficient, safe, and user-friendly environments for pedestrians and cyclists. These technologies enable data-driven decision-making, enhance user experiences, and contribute to the overall promotion of active mobility in smart sustainable cities. With the current advancements in technological tools, the prospects of the proposed Smart Active Mobility (SAM) Model are promising. This should be accompanied with city planning authorities recognizing the significance of prioritizing active mobility in their urban planning strategies. In this way, the integration of ITS, ICT, XR, ML, and AI is bound to play a pivotal role in transforming urban transportation networks and promoting sustainable and health-conscious travel choices. As technology continues to evolve, so does the potential of the SAM Model. XR technologies, for instance, have the capability to sync effortlessly with wearable devices, delivering real-time navigation details straight into a user's line of sight, thereby elevating safety and user convenience. The refinement in ML and AI algorithms promises enhanced prediction capabilities for travel patterns and traffic flow, resulting in superior route recommendations for active mobility.

12.4.3 Potential of Recent Technological Advancements to Support Active Mobility Optimization

Technological advancements continuously highlight the potential of ITS and ICT in enhancing active mobility. Coupled with advances in artificial intelligence, machine learning, and the Internet of Things (IoT), ITS and ICT introduce innovative solutions to address mobility challenges and enhance the experience for active commuters. Data analytics, supported by ITS and ICT, become instrumental in analyzing travel behaviors, patterns, and preferences [1, 3, 57]. This detailed analysis enables tailored interventions and infrastructure innovations to support active transport modes. In addition, predictive models identify future demands, guiding effective resource allocation [9, 10, 22, 43]. Artificial intelligence plays a pivotal role in dissecting large datasets, offering real time, personalized travel suggestions, flexible route planning, and efficient transport network management [9, 21]. One significant advantage of AI is its ability to predict and mitigate traffic accidents. Using real-time data from various sources, AI systems can forecast potential accident-prone areas, allowing for preventative measures. Analyzing historical accident data, AI identifies common accident patterns and recommends infrastructure changes to prevent their recurrence, thus enhancing safety and building trust among users [42, 57].

On the other hand, IoT devices provide immediate feedback on environmental conditions, air quality, and infrastructure effectiveness. These insights support evidence-based decisions and proactive infrastructure maintenance and upgrades [5, 43, 46]. ITS and ICT further enhance active mobility by supporting demand management and behavior change. Gamification strategies, rewards, and incentives, delivered via mobile applications, promote active transport choices. These initiatives, built on insights from behavior change campaigns, offer educational content and social engagement opportunities to support active mobility [6, 15, 68]. Integrating the capabilities of ITS and ICT, along with the insights from AI and IoT, can pave the way for a more responsive and adaptable urban transport landscape. These technologies not only provide invaluable data but also play a pivotal role in enhancing the overall safety, efficiency, and user experience in active transport systems. As we delve further, we'll explore the implications and potential applications of these advancements within the broader context of urban planning and policy-making.

12.5 SAM, the Proposed Framework Model for Promoting and Developing Active Mobility Networks

As illustrated in the previous section, current technological advancements offer a pivotal opportunity for promoting active mobility. This section delves deeper into the framework of the proposed Smart Active Mobility (SAM) model, specifically designed to evaluate and optimize active mobility networks within urban settings. The SAM model's multidimensional nature is central to its approach, systematically addressing a wide array of factors influencing active mobility.

Inherent in the model's design is the integration of methodologies like System Dynamics (SD), Geographic Information Systems (GIS), and multi-criteria analysis, lending substantial robustness to its structure. A notable aspect of the SAM model is its co-creative design principle, which encourages contributions from a diverse set of stakeholders, including urban planners, policy-makers, researchers, and the general public. This collaborative approach ensures the model is well-rounded and reflective of varied perspectives and needs.

Additionally, this section highlights the potential impacts and benefits the SAM model can usher in, spanning improvements in accessibility and connectivity, to promoting better health outcomes, and fostering environmental sustainability, alongside prospective economic advantages. This version stays closer to your original text, with minimal changes aimed at preserving the integrity of your content while ensuring linguistic clarity and coherence.

12.5.1 The SAM Model: Concept and Design

The Smart Active Mobility (SAM) model is a holistic conceptual framework designed to evaluate and optimize active mobility networks within urban areas. The concept behind SAM is to provide a systematic approach for assessing and improving the accessibility, connectivity, and overall quality of active transportation systems. It encompasses various factors that influence active mobility, including the urban environment, street topology, accessibility, transportation, residential and employment density, land usage, and the built environment.

Incorporating ITS and leveraging methodologies such as System Dynamics (SD) [3, 9, 24, 68], GIS [68], and Time-Quality Assessment (TQA) [17] are essential in the design of SAM. SD enables the modeling of complex urban systems and their dynamic interactions, while GIS aids in the spatial analysis and visualization of mobility networks [10, 17, 68]. Within SAM, a multi-criteria analysis approach evaluates active mobility networks by assigning weights to different criteria based on their relative importance. The TQA method is employed to assess the quality of urban spaces in terms of active mobility [17], considering factors such as travel time, safety, comfort, and aesthetics. The design of SAM emphasizes co-creation, drawing insights from a myriad of stakeholders, including urban planners, policymakers, researchers, and the public, to foster relevancy and effectiveness in diverse contexts.

12.5.2 Implementing the SAM Model: Stages and Methodology

The implementation of the SAM model involves several stages and a well-defined methodology to ensure its successful application in evaluating active mobility networks [9, 22]. The following stages outline the implementation process:

1. Data Collection: The first stage involves gathering relevant data on urban environments, active mobility infrastructure, transportation patterns, and other factors that influence active mobility. This data can be collected through surveys, observations, existing databases, and other sources.

2. Data Analysis: In this stage, the collected data is analyzed using GIS and other analytical tools. Spatial analysis techniques are applied to understand the relationships between different variables, identify patterns, and assess the performance of active mobility networks.
3. Model Development: Based on the analysis of the data, an appropriate SAM system model is developed for a particular city. This involves integrating the identified criteria, weights, and methodologies into a comprehensive framework that can evaluate and optimize active mobility networks.
4. Model Validation: The developed SAM system model is validated using real-world data. This validation process ensures that the model accurately represents the characteristics and dynamics of active mobility networks.
5. Application and Evaluation: The SAM model is then applied to specific urban areas to evaluate their active mobility networks. The model assesses the performance of these networks against the identified criteria, providing insights into areas for improvement and optimization.
6. Feedback and Iteration: Stakeholder feedback and evaluation of the SAM model's application are collected and incorporated into the model. This iterative process allows for continuous improvement and refinement of the model's effectiveness.

The implementation process is also depicted in a diagrammatic form in Fig. 12.1 showing the main interrelationships and feedback loops. The flowchart for implementing the SAM model provides a structured approach to improve active mobility networks in urban areas. The chart is divided into four main stages: Initialization, Model Development, Model Application, and Feedback and Iteration, each with specific steps that guide the process. The Initialization stage is where the process begins, and it includes Data Collection and Data Analysis. During data collection, information about the urban environment, such as existing transportation infrastructure and mobility patterns, is gathered from various sources, like surveys, observations, or existing databases. Following this, the collected data is analyzed using tools like GIS for spatial analysis, helping to understand the geographical distribution and patterns within the data.

The second stage, Model Development, builds upon the insights gained from the data analysis. This phase involves identifying evaluation criteria, assigning them weights based on their importance, and developing an evaluation method that considers these weights. These criteria could be factors like safety, comfort, efficiency, and accessibility, which are crucial in assessing active mobility networks. The Model Application stage is where the developed SAM model is put to use. It can be applied to specific urban areas to assess their current active mobility networks. The application of the model provides insights into the performance of these networks against the established criteria, pointing out areas of strength and opportunities for improvement.

The last stage, Feedback and Iteration, is all about refining the SAM model. Stakeholder feedback is collected and then incorporated into the model to enhance its accuracy and effectiveness. The model is then subjected to iterative refinement and improvement, based on the feedback and results from its application. Finally, the

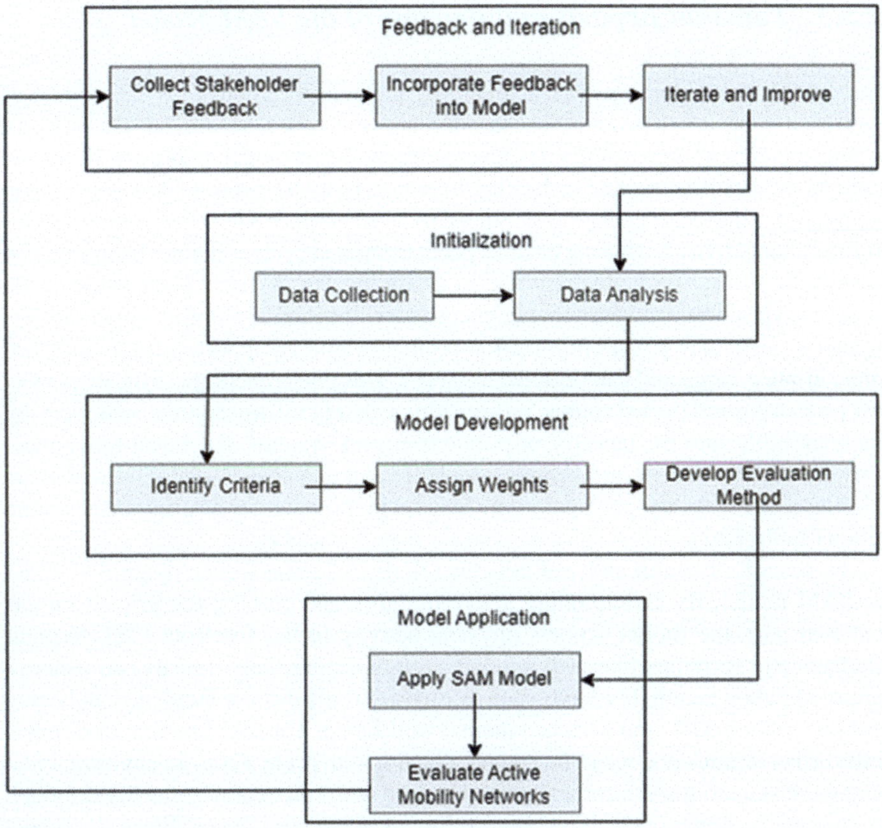

Fig. 12.1 A Flowchart for Implementing the SAM Model

process loops back to the Data Analysis stage. This cyclical process highlights the dynamic and iterative nature of the SAM model. The model is continually updated and improved based on feedback and new data, ensuring its relevance and accuracy in evaluating and optimizing active mobility networks. The flowchart of Fig. 12.1 is essential because it provides a roadmap for systematically implementing the model. Each step is connected, contributing to the progress and development of the next. The structure helps ensure a thorough and holistic approach, integrating input and feedback from various stakeholders to create a comprehensive and robust model. This iterative process is key to achieving a more accurate, relevant, and efficient model that can effectively enhance active mobility networks within urban environments.

12.5.3 Potential Impacts and Benefits of the SAM Model

The SAM model's adoption in urban planning and transportation holds the promise of significant impacts and benefits. The key domain where the SAM model focusses is in its enhancement of accessibility. By leveraging the SAM model, regions with subpar active mobility infrastructure or lacking in connectivity to essential destinations can be pinpointed and remedied [9, 22]. Such a focused approach to infrastructure evolution and connectivity bolstering ensures pedestrians and cyclists benefit from superior accessibility, fostering inclusivity within urban locales. A subsequent merit of the model is its prowess in augmenting connectivity within active mobility networks. Via the model's evaluative and optimization protocols, golden opportunities to enhance connectivity emerge and can be acted upon [9]. This could encompass the genesis of new pedestrian and cyclist routes, refurbishing existing infrastructures, and weaving active mobility into the tapestry of public transport systems. By elevating connectivity, the SAM model unfurls seamless and efficient transit avenues for those that are favoring active mobility, paving the way for a harmonized and cohesive transportation milieu (Table 12.2).

The beacon of health and wellness also finds its luminescence magnified under the SAM model. By championing active mobility and curbing the grip of motorized transport, the model propels physical activity to the forefront [22]. Regular indulgence in activities like walking and cycling during daily transits can usher in enhanced public health dividends such as lower chronic disease incidence, bolstered cardiovascular health, and enriched mental well-being. The SAM model, in essence, crafts urban terrains that wear the badge of health with pride. From an environmental vantage, sustainability is a cornerstone addressed with conviction by the SAM model. By stoking the flames of walking and cycling as green transit modalities, the model nudges urban habitats away from the carbon-thirsty clutches of motorized transport

Table 12.2 Potential impacts and benefits of the SAM model

Impacts	Methods/Actions to achieve impacts	Benefits
Enhanced accessibility	Identification of areas with inadequate infrastructure or poor connectivity	Better accessibility for pedestrians and cyclists
Improved connectivity	Development of new walking and cycling paths	Seamless and efficient travel options
Health and well-being	Encouragement of physical activity and reduced reliance on motorized transport	Improved public health outcomes
Environmental sustainability	Reduction of carbon emissions and environmental impact	Promotion of sustainable and eco-friendly transport
Economic benefits	Stimulation of local businesses and economic vitality	The attraction of investments and economic growth

[12]. This shift trims down carbon emissions and curtails the overarching environmental footprint, translating to a fresher air quality and a more muted carbon silhouette. Through its endorsement of green transit avenues, the SAM model paints urban canvases with bold strokes of environmental conscientiousness.

The SAM model's adoption in urban planning and transportation holds the promise of significant impacts and benefits. The key domain where the SAM model focusses is in its enhancement of accessibility. By leveraging the SAM model, regions with subpar active mobility infrastructure or lacking in connectivity to essential destinations can be pinpointed and remedied [9, 22]. Such a focused approach to infrastructure evolution and connectivity bolstering ensures pedestrians and cyclists benefit from superior accessibility, fostering inclusivity within urban locales. A subsequent merit of the model is its prowess in augmenting connectivity within active mobility networks. Via the model's evaluative and optimization protocols, golden opportunities to enhance connectivity emerge and can be acted upon [9]. This could encompass the genesis of new pedestrian and cyclist routes, refurbishing existing infrastructures, and weaving active mobility into the tapestry of public transport systems. By elevating connectivity, the SAM model unfurls seamless and efficient transit avenues for those that are favoring active mobility, paving the way for a harmonized and cohesive transportation milieu.

12.6 Implications of Implementing SAM

The implementation of sustainable mobility proposals as the Smart Active Mobility (SAM) framework model promises a myriad of transformative effects spanning urban life, environmental conservation, and public health. It's essential, however, to quantify these impacts meticulously to ascertain their efficacy and offer direction for subsequent enhancements [1]. Subsequently, this information should be passed to people engaging in Active Mobility. Among the methods to gauge sustainable mobility outcomes, the Time-Quality Assessment (TQA) stands out, offering robust evaluations of urban spaces crucial for formulating pedestrian and cycling routes [17]. Notably, this methodology has demonstrated its value, as witnessed in its application to children's school commutes by Easton and Ferrari [14]. Further, examining the technological spectrum, the potential of incorporating Internet of Things (IoT) solutions in conjunction with ITS is tremendous [5, 45, 46]. Furthermore, the application of System Dynamics and GIS allows for the assessment of the benefits of active mobility into the near- and long-term future [9].

The use of Extended Reality (XR) technology could introduce realistic simulations and immersive platforms to appraise SAM-based solutions prior to deployment, thus guiding policy-makers and other stakeholders [48]. By sifting through extensive transport datasets, Machine Learning (ML) algorithms can spotlight efficacy, utilization, and potential enhancement zones for sustainable mobility [52, 53]. Artificial Intelligence (AI) amplifies real-time evaluations and feedback mechanisms, capacitating immediate policy decisions rooted in prevailing transport scenarios [54, 55].

The emergence of intelligent multimedia sensors refines the potential of IoT-enhanced autonomous transit systems [23]. Infrastructure compatibility's influence on pedestrian and cyclist preferences, as assessed via Susilo's techniques, can't be overlooked [15, 23]. A nuanced understanding of travel contentment across diverse transit modes is integral to sustainable mobility projects' success. Public perception of active mobility and sustainable transit remains pivotal [18, 55], and this sentiment can profoundly shape the acceptance and triumph of such initiatives. Evaluating sustainable mobility projects is paramount. By harnessing modern tools and techniques, notably XR, ML, and AI, stakeholders can validate goal achievement, anchoring sustainability in transport [65]. Given the unique objectives and contexts of different initiatives, the selection and continuous refinement of methodologies is crucial.

Lastly, the fruitful implementation and propagation of urban transport frameworks, specifically SAM, relies heavily on international collaboration and standardization. Valuing the rich tapestry of insights and expertise from a global pool, SAM's implementation strategy underscores collaborative engagement. The ensuing discourse underscores international collaboration's role in the model's success, highlighting the profuse advantages of mutual learning, policy evolution, and shared resources. Envisioning a global nexus of shared knowledge, SAM seeks to elevate its impact, inspiring comparable global projects. This joint venture aims to sculpt a refined model to champion active mobility, charting the path for sustainable, intelligent, and habitable global cities.

12.6.1 Importance of International Stakeholder Cooperation and Standardization

The realm of urban transportation could rapidly evolve toward sustainability with the proposed model positioned at the forefront. On the other hand, for its successful implementation, international standards and cooperation are indispensable. Urban transportation challenges are inherently complex, transcending local concerns, which underlines the need for collective, cross-border solutions. International standard-setting entities, such as the American National Standards Institute (ANSI), the European Standardization Organization (CEN), and the more general International Standards Organization (ISO), are central to this collaborative approach [57, 58]. While ANSI and CEN as well as other national standardization committees predominantly craft local standards, ISO assumes a broader mandate, encompassing global concerns. Together, their support could enhance the stature of active mobility and sustainable transportation. The authors experience in working with CEN's ITS group revealed the need for establishing a new standard for promoting active mobility and protecting Vulnerable Road Users (VRUs). CEN's operation is underscored by its dedication to voluntary standards. These standards, designed to bolster global compatibility and interoperability, align with the vision of the Single European Market. By ensuring consumer protection and fostering innovation, CEN echoes the evolving societal needs and public aspirations.

ISO, functioning as a non-governmental organization, brings together global thought leaders, striving to formulate international standards that resonate with contemporary global challenges while promoting innovative solutions. Delving deeper into their impact reveals that the guidelines set by ISO are foundational to the trajectory of sustainable transportation and active mobility. Their meticulously crafted standards serve as a beacon for transportation initiatives worldwide. These benchmarks recognized universally, ensure that diverse urban mobility endeavors adhere to a consistent and unified framework, facilitating efficient assessment, comparison, and subsequent optimization. The relevance of active mobility standards amplifies when viewed through the lens of international integration. By adhering to these guidelines, cities and nations lay the groundwork for transportation infrastructures that epitomize quality and safety. This alignment becomes even more pivotal when considering international dynamics such as cross-border mobility or the intent to attract global businesses and tourists. Embracing standardized active mobility infrastructures ensures smooth transitions and interactions, minimizing potential disruptions and ensuring a seamless user experience.

Moreover, the synergy between active mobility standardization and international partnerships accelerates the transformation toward sustainable urban frameworks. Collaborations empower cities by equipping them with the insights to adopt successful strategies, incorporate proven best practices, and preemptively address potential challenges. Over time, this harmonized approach not only streamlines resources but also fosters a global shift toward environmentally conscious, sustainable urban environments. Institutions such as CEN and ISO have undoubtedly etched their significance in the annals of sustainable urban development. Their profound influence extends beyond mere standardization, acting as a catalyst in the global transition toward sustainable future models.

Standardization would further enhance the collaboration between different cities and nations is paramount in magnifying the benefits of active mobility. Harnessing international cooperation results in a rich tapestry of shared knowledge and diversified expertise. Such partnerships enable countries to benefit from technological advancements in domains such as ITS and ICT. Conversely, nations with distinct urban blueprints contribute their unique solutions, enhancing the collective understanding of active mobility. Collaboration is equally transformative in the sphere of policy-making. Countries, united in their mission, can collaboratively sculpt cohesive policy frameworks, design robust infrastructural blueprints, and devise effective funding strategies in alignment with the ethos of active mobility. Through a collaborative lens, countries can extract valuable lessons from shared global experiences, refining their policies and tailoring them to their specific contexts. Combining resources further amplifies research opportunities, paves the way for infrastructural advancements, and promotes the seamless integration of cutting-edge technologies.

This collective endeavor ensures resource efficiency and fast-tracks the tangible realization of active mobility objectives [59–62]. Collaboration and standardization are expected to yield several synergies and benefits:

- Knowledge Exchange: International cooperation allows for the exchange of knowledge, best practices, and research findings between countries [63]. This knowledge-sharing fosters mutual learning and enables the integration of global expertise into the SAM model, making it more relevant and effective across diverse urban contexts.
- Technology Transfer: Advanced countries can share their technological advancements and expertise in ITS and ICT with other participating nations. This technology transfer enables the adoption of innovative solutions and tools to optimize active mobility networks [40]. Countries with limited resources can benefit from access to advanced technologies, enhancing their capacity to implement smart and sustainable transportation systems.
- Policy Development: Collaborative efforts among countries can facilitate the development of supportive policy frameworks and guidelines for promoting active mobility. Drawing on the experiences of various nations, the SAM model can contribute to the formulation of effective policy approaches that align with sustainable urban development goals [64].
- International Collaboration: The SAM model could serve as a platform for international collaboration in addressing common urban challenges related to active mobility [63]. This collaboration fosters a global network of knowledge-sharing and cooperation, inspiring other countries and cities to engage in similar initiatives for sustainable urban transportation.

By fostering international cooperation and standardization, the proposed model can harness the collective wisdom and resources of multiple countries to create a more robust and inclusive approach to promoting active mobility. In this way, collaboration and standardization strengthen innovation, knowledge exchange, and policy development, ultimately advancing the cause of sustainable urban transportation on a global scale.

12.7 Conclusions

This chapter delved deeply into the capabilities of the Smart Active Mobility (SAM) model in enhancing sustainability and refining urban transportation systems. The discussion penetrated the core aspects of active mobility, demonstrating how, when integrated with Intelligent Transportation Systems (ITS) and Information and Communication Technology (ICT), SAM can guide urban transportation toward heightened efficiency, inclusivity, and environmental harmony.

The SAM model embodies a systematic approach that commences with the meticulous collection and analysis of data. This crucial foundation paves the way for the subsequent stages of model development, which are bolstered by ITS, ICT, and emerging technologies such as Extended Reality (XR), Machine Learning (ML), and Artificial Intelligence (AI). These technologies establish the parameters for System Dynamics (SD) modeling and Time-Quality Assessment (TQA). When applied to real-world urban settings, SAM emerges as a powerful instrument for evaluating and enhancing active mobility networks, ensuring they remain adaptable to changing urban needs and technological advancements.

Active mobility's essential role in leading sustainable urban transformation is highlighted by its wide-ranging benefits across environmental, health, and socio-economic domains. Furthermore, the fusion of SAM with technologies like ITS, ICT, XR, ML, and AI significantly magnifies its effectiveness. This convergence presents a clear directive for policy-makers: to prioritize active mobility as a foundational element of sustainable urban development, which involves substantial investment in robust infrastructure, a focus on seamless connectivity, and the deployment of supportive policies and mechanisms.

Peering into the future, in a technological environment that is perpetually evolving, anticipating new trends is crucial. This might involve deeper integration with the Internet of Things (IoT) for real-time data gathering, or the development of AI-driven analytics to provide customized mobility solutions. As we advance, ensuring user privacy and data protection within these technologically sophisticated models is paramount. In conclusion, the potential of SAM to transform urban mobility is limitless. As it intertwines with cutting-edge technologies, its agility and efficiency are set to soar. The future research landscape in this field is brimming with opportunities, and each advancement holds the promise of reshaping urban mobility toward greater sustainability, well-being, and quality of life.

References

1. Lange P, Driessen PP, Sauer A, Bornemann B, Burger P (2013) Governing towards sustainability—conceptualizing modes of governance. J Environ Planning Policy Manag 15(3):403–425
2. Bereitschaft B (2017) Equity in microscale urban design and walkability: a photographic survey of six Pittsburgh streetscapes. Sustainability 9(7):1–20
3. Currie DJ, Smith C, Jagals P (2018) The application of system dynamics modelling to environmental health decision-making and policy—a scoping review. BMC Public Health 18(1):1–11 [Online]. Available: https://doi.org/10.1186/s12889-018-5763-9
4. Methorst R, Monterde i Bort H, Risser R, Sauter D, Tight M, Walker J (eds) (2010) Pedestrians' quality needs. Final Report of the COST project 358
5. Brincat AA, Pacifici F, Martinaglia S, Mazzola F (2019) The Internet of Things for intelligent transportation systems in real smart cities scenarios. In: 2019 IEEE 5th world forum on Internet of Things (WF-IoT), pp 128–132 [Online]. Available: https://doi.org/10.1109/WF-IoT.2019.8767324
6. Papageorgiou G, Maimaris A (2017) Towards the development of intelligent pedestrian mobility systems (ipms). In: 2017 International Conference on Electrical Engineering and Informatics (ICELTICs), pp 251–256 [Online]. Available: https://doi.org/10.1109/ICELTICs.2017.8276505

7. Papageorgiou G, Maimaris A, Demetriou D, Balamou E (2018) Market research and concept study for a smart pedestrian network application. In: Proceedings of the 11th PErvasive technologies related to assistive environments conference, pp 522–528 [Online]. Available: https://doi.org/10.1145/3197768.3201541
8. Scorrano M, Danielis R (2021) Active mobility in an Italian city: Mode choice determinants and attitudes before and during the Covid-19 emergency. Res Transp Econ 86:101031 [Online]. Available: https://doi.org/10.1016/j.retrec.2020.101031
9. Pejic Bach M, Tustanovski E, Ip AW, Yung KL, Roblek V (2020) System dynamics models for the simulation of sustainable urban development: a review and analysis and the stakeholder perspective. Kybernetes 49(2):460–504
10. Marušić D, Goličnik Marušić B (2016) Practical value of user-centred spatial statistics for responsive urban planning. In: Applications of spatial statistics: earth and planetary sciences, pp 127–146
11. Angelidou M, Psaltoglou A, Komninos N, Kakderi C, Tsarchopoulos P, Panori A (2018) Enhancing sustainable urban development through smart city applications. J Sci Technol Policy Manag 9(2):146–169
12. Sallis JF, Bull F, Burdett R, Frank LD, Griffiths P, Giles-Corti B, Stevenson M (2016) Use of science to guide city planning policy and practice: how to achieve healthy and sustainable future cities. Lancet 388(10062):2936–2947
13. Toan TD (2022) Smart planning and management solutions for smart mobility. In: CIGOS 2021, emerging technologies and applications for green infrastructure: proceedings of the 6th international conference on geotechnics, civil engineering and structures. Springer, Singapore, pp 1397–1405
14. Easton S, Ferrari E (2015) Children's travel to school—the interaction of individual, neighbourhood and school factors. Transp Policy 44:9–18 [Online]. Available: https://doi.org/10.1016/j.tranpol.2015.06.007
15. Susilo YO, Cats O (2014) Exploring key determinants of travel satisfaction for multi-modal trips by different traveler groups. Transp Res Part A: Policy Practice 67:366–380 [Online]. Available: https://doi.org/10.1016/j.tra.2014.07.010
16. Naumann S, Graser A, Straub M (2016) Converting OSM data into a routable graph for pedestrians. In: 28th European conference on operational research. Poznan University of Technology, Poznań, Poland
17. Marušić D, Goličnik Marušić B (2017) Time quality-measure for quality of place. In: Certoma C et al (eds) Citizen empowerment and innovation in the Data-Rich City. Springer, Cham
18. Markvica K, Millonig A, Haufe N, Leodolter M (2020) Promoting active mobility behavior by addressing information target groups: the case of Austria. J Transp Geogr 83:102664 [Online]. Available: https://doi.org/10.1016/j.jtrangeo.2020.102664
19. Bıyık C, Abareshi A, Paz A, Arce Ruiz R, Battarra R, Rogers CDF, Lizarraga C (2021) Smart mobility adoption: a review of the literature. J Open Innovat Technol Market Complex 7(2):146
20. Koh PP, Wong YD (2023) Influence of infrastructural compatibility factors on walking and cycling route choices. J Environ Psychol 36:202–213 [Online]. Available: https://doi.org/10.1016/j.jenvp.2013.07.006Bharadiya; JP (2023) Artificial intelligence in transportation systems: a critical review. Am J Comput Eng 6(1):35–45
21. Papageorgiou G, Demetriou G (2020) Investigating learning and diffusion strategies for sustainable mobility. Smart Sustain Built Environ 9(1):1–16 [Online]. Available: https://doi.org/10.1108/SASBE-02-2019-0020. Supported by the Republic of Cyprus and the European Regional Development Fund as part of ERA–NET Cofund Smart Urban Futures (ENSUF) Joint Programming Initiative (JPI) Urban Europe, protocol no. KOINA/ΠΚΠ URBAN EUROPE/1215/11
22. Rehman A, Saba T, Haseeb K, Jeon G, Alam T (2023) Modeling and optimizing IoT-driven autonomous vehicle transportation systems using intelligent multimedia sensors. In: Multimedia tools and applications. Springer, pp 1–15
23. Papageorgiou G, Demetriou G (2019) Developing a system dynamics model for creating a learning sustainable mobility culture. Inl 2019 international conference on control, artificial

intelligence, robotics and optimization (ICCAIRO), pp 189–194 [Online]. Available: https://doi.org/10.1109/ICCAIRO47923.2019.00037
24. Fonseca F, Conticelli E, Papageorgiou G, Ribeiro P, Jabbari M, Tondelli S, Ramos R (2021) Use and perceptions of pedestrian navigation apps: Findings from Bologna and Porto. ISPRS Int J Geo Inf 10(7):446
25. Fonseca F, Papageorgiou G, Tondelli S, Ribeiro P, Conticelli E, Jabbari M, Ramos R (2022) Perceived walkability and respective urban determinants: insights from Bologna and Porto. Sustainability 14(15):9089
26. Fonseca F, Ribeiro P, Jabbari M, Petrova E, Papageorgiou G, Conticelli E et al (2020) Smart pedestrian network: An integrated conceptual model for improving walkability [Online]. Available: https://doi.org/10.1007/978-3-030-45293-3_10
27. Fonseca F, Ribeiro P, Jabbari M, Petrova E, Papageorgiou G, Conticelli E et al (2019) Smart Pedestrian network: an integrated conceptual model for improving walkability. In: International conference on society with future: smart and liveable cities. Springer International Publishing, Cham, pp 125–142
28. Fonseca F, Ribeiro P, Jabbari M, Petrova E, Papageorgiou G, Conticelli E, Tondelli S, Ramos R (2020) Smart pedestrian network: an integrated conceptual model for improving walkability. In: Society with future: smart and liveable cities, lecture notes of the institute for computer sciences, social informatics and telecommunications engineering, vol 318, pp 125–142
29. Papageorgiou G, Balamou E, Maimaris A (2020) Developing a business model for a smart pedestrian network application. In: Fourth international congress on information and communication technology: ICICT 2019, London, vol 2. Springer Singapore, pp 375–381
30. Jabbari M, Ahmadi Z, Ramos R (2022) Defining a digital system for the pedestrian network as a conceptual implementation framework. Sustainability 14(5), art. no. 2528
31. Papageorgiou G, Hadjigeorgiou K, Ness AN (2019) A big data approach to developing a smart pedestrian network (SPN) system. WSEAS Trans Environ Dev 15:371–378
32. Papageorgiou G, Hadjigeorgiou K, Ness AN (2020) An innovative way to promote walking via a smartphone pedestrian navigation application. In: 2020 European navigation conference, ENC 2020 [Online]. Available: https://doi.org/10.23919/ENC48637.2020.9317478
33. Papageorgiou G, Hadjigeorgiou K, Ness AN (2020) Exploring the prospects of developing a smartphone application for pedestrians. In: 2020 19th international symposium INFOTEH-JAHORINA (INFOTEH). IEEE, pp 1–5
34. Papageorgiou G, Ioannou A, Maimaris A, Ness A (2021) A strategic approach for implementing a smart pedestrian network (SPN) system. Open Comput Sci 11(1):208–217
35. Papageorgiou G, Ioannou A, Maimaris A, Ness AN (2019) Managing the implementation of a smart pedestrian network (SPN) system. In: INFORMATICS 2019—IEEE 15th international scientific conference on informatics, proceedings, pp 409–414 [Online]. Available: https://doi.org/10.1109/Informatics47936.2019.9119334.
36. Papageorgiou G, Ioannou A, Maimaris A, Ness AN (2019) A framework for evaluating the benefits of implementing a smart pedestrian network system. INFORMATICS 2019—IEEE 15th international scientific conference on informatics, proceedings, pp 497–502 [Online]. Available: https://doi.org/10.1109/Informatics47936.2019.9119261
37. Papageorgiou G, Ioannou A, Ness AN, Maimaris A (2019) Developing a smart pedestrian network (SPN) system—smartphone app. In: 27th telecommunications Forum, TELFOR 2019 [Online]. Available: https://doi.org/10.1109/TELFOR48224.2019.8971054
38. Papageorgiou G, Prodromou M, Christou G, Maimaris A (2020) Exploring the potential for developing a smart pedestrian navigation system for the elderly market. In: 2020 6th IEEE international conference on information management, ICIM 2020, pp 194–199 [Online]. https://doi.org/10.1109/ICIM49319.2020.245371.
39. Papageorgiou G (2020) Intelligent transportation systems in cyprus–past, present and future; prospects for a Smart Pedestrian Network (SPN). In: Workshop on "Intelligent Transport Systems in Emerging Markets–Enhancing National Policies through Standards," European University Cyprus. European Standardization Organization (CEN)-EUC/CYS

40. Papageorgiou G, Ioannou A, Maimaris A, Ness A (2021) Evaluation of the benefits of implementing a smart pedestrian network system. Open Comput Sci 11(1):224–231
41. Oduor M, Perälä T (2021) Interactive urban play to encourage active mobility: usability study of a web-based augmented reality application. Front Comput Sci 3, art. no. 706162
42. Papageorgiou G, Demetriou G (2019) Examining the impact of learning strategies on developing sustainable mobility mindsets: a systems approach. WSEAS Trans Environ Dev 15:533–543
43. Papageorgiou G, Petrakis C, Georgiades, P, Ioannou N (2019) Investigating the potential of a smartphone application for tourists to promote sustainable mobility. In: 2019 third world conference on smart trends in systems security and sustainability (WorldS4), pp 1–6. IEEE
44. López MCR, Wong YD (2017) Attitudes towards active mobility in Singapore: a qualitative study. Case Stud Transp Policy 5(4):662–670 [Online]. https://doi.org/10.1016/j.cstp.2017.09.005
45. Rani P, Sharma R (2022) Intelligent transportation system for internet of vehicles based vehicular networks for smart cities. Comput Electr Eng 105, art. no. 108543 [Online]. https://doi.org/10.1016/j.compeleceng.2022.108543
46. Gohar A, Nencioni G (2021) The role of 5G technologies in a smart city: the case for intelligent transportation system. Sustainability 13(9), art. no. 5188 [Online]. https://doi.org/10.3390/su13095188.
47. Conticelli E, Maimaris A, Papageorgiou G, Tondelli S (2018) Planning and designing walkable cities: a smart approach. In: Papa R, Fistola R, Gargiul C (eds) Smart planning: sustainability and mobility in the age of change. Springer, Cham, pp 251–269
48. Rauschnabel PA, Felix R, Hinsch C, Shahab H, Alt F (2022) What is XR? Towards a framework for augmented and virtual reality. Comput Hum Behav 133, art. no. 107289
49. Ullmann D, Kreimeier J, Kipke H (2022) Pedaling through a virtually redesigned city: evaluation of traffic planning and urban design factors influencing bicycle traffic. J Urban Mobility 2, art. no. 100032
50. Erath A, Maheshwari T, Joos M, Kupferschmid J, van Eggermond MA (2017) Visualizing transport futures: the potential of integrating procedural 3D modeling and traffic micro-simulation in virtual reality applications. Arbeitsberichte Verkehrs-und Raumplanung 1185
51. Odenigbo IP, Reen JK, Eneze C, Friday A, Orji R (2022) The journey: an AR Gamified mobile application for promoting physical activity in young adults. In: Adjunct proceedings of the 30th ACM conference on user modeling, adaptation, and personalization, pp. 342–353
52. Pisoni E, Christidis P, Cawood EN (2022) Active mobility versus motorized transport? User choices and benefits for society. Sci Total Environ 806, art. no. 150627
53. Ullah Z, Al-Turjman F, Mostarda L, Gagliardi R (2020) Applications of artificial intelligence and machine learning in smart cities. Comput Commun 154:313–323
54. Nikitas A, Michalakopoulou K, Njoya ET, Karampatzakis D (2020) Artificial intelligence, transport, and the smart city: Definitions and dimensions of a new mobility era. Sustainability 12(7), art. no. 2789
55. Lehtiö A, Hartikainen M, Ala-Luopa S, Olsson T, Väänänen K (2023) Understanding citizen perceptions of AI in the smart city. AI Soc 38(3):1123–1134
56. Papageorgiou G, Petrakis C, Ioannou N, Zagarelou D (2019) Effective business planning for sustainable urban development: the case of active mobility. In: ECIE 2019 14th European conference on innovation and entrepreneurship, vol 2, p 759. Academic Conference Publishing Ltd.
57. Caird SP, Hallett SH (2019) Towards evaluation design for smart city development. J Urban Des 24(2):188–209 [Online]. https://doi.org/10.1080/13574809.2018.1469402. Published online: 23 May 2018
58. Huovila A, Bosch P, Airaksinen M (2019) Comparative analysis of standardized indicators for Smart sustainable cities: what indicators and standards to use and when? Cities 89:141–153. VTT Technical Research Centre of Finland Ltd., Espoo, Finland; The Netherlands Organisation for Applied Scientific Research TNO, Utrecht, the Netherlands; Finnish Association of Civil Engineers RIL, Helsinki, Finland [Online]. https://doi.org/10.1016/j.cities.2019.01.029

59. Chumakova O (2017) Features of 'smart city' concept in urban paradigm of globalization. In: MATEC web of conferences, vol 106, p 01030. EDP Sciences [Online]. https://doi.org/10.1051/matecconf/201710601030. Published online: 23 May 2017
60. Saborido R, Alba E (2020) Software systems from smart city vendors. Cities 101, art. no. 102690 [Online]. https://doi.org/10.1016/j.cities.2020.102690. ITIS Software, Universidad de Málaga, Spain. Published: June 2020
61. Nowacki G (2012) Development and standardization of intelligent transport systems. Int J Mar Navig Saf Sea Transp 6(3):403–411. Book reference: Transport systems and processes, 1st edn. CRC Press, 2011 [eBook ISBN: 9780429217364]
62. Karale A (2021) The challenges of IoT addressing security, ethics, privacy, and laws. Internet Things 15, art. no. 100420. Department of Computer Engineering, University of Mumbai, Maharashtra, India. [Online]. https://doi.org/10.1016/j.iot.2021.100420. Received: 3 January 2021, Revised: 18 May 2021, Accepted: 8 June 2021, Published online: 16 June 2021, Version of Record: 27 June 2021
63. Neun MG et al (2020) 'Prioritizing active mobility': a report and recommendations arising from ESOF 2020 addressing the EU Commission on the European Strategy for a Sustainable and Smart Mobility
64. Parkes R (2009) EU mobility partnerships: a model of policy coordination? Euro J Migr Law 11(4):327–345
65. Lopez-Arboleda E, Sarmiento AT, Cardenas LM (2019) Systematic review of integrated sustainable transportation models for electric passenger vehicle diffusion. Sustainability 11, art. no. 2513 [CrossRef]
66. Petticrew M, Roberts H (2006) Systematic reviews in the social sciences: a practical guide. Blackwell Pub, Hoboken, NJ, USA. ISBN 9781405121118
67. Papageorgiou G, Maimaris A, Ness AN (2019) A value creation analysis of implementing an intelligent active mobility system for the elderly market. In: 2019 3rd European conference on Electrical Engineering and Computer Science (EECS), pp 1–5 [Online]. https://doi.org/10.1109/EECS49779.2019.00013
68. Mazzoleni S, Giannino F, Colandrea M, Nicolazzo M, Massheder J (2003) Integration of system dynamics models and geographic information systems. Simile 600(5):304–306
69. Papageorgiou G, Shuying SY (2023) Augmented Reality (AR) a new technology for advertising and marketing digital Communications. In: 2023 zooming innovation in consumer technologies conference (ZINC), pp 56–61

Chapter 13
Centrally Coordinated Routing of Freight in Smart Cities

Pengfei Chen and Petros Ioannou

Abstract The efficient routing and distribution of goods are vital to the survival and sustainability of any smart city. Currently, different truck companies route their trucks on a road network that is often congested and unbalanced in time and space with respect to traffic load distribution. The lack of coordination among truck operators often leads to long waiting times at pick-up and drop-off points and/or along popular routes that trucks usually follow due to overlaps. Due to connectivity and the Internet of Things new opportunities open for centrally coordinated solutions in order to achieve better load balancing in space and time for trucks across the road network. In this chapter, we present a centrally coordinated approach for routing freight in urban environments where truck traffic loads are balanced in time and space in an effort to improve mobility and reduce cost. We assume that freight is moved by trucks using the road network and truck fleets consist of a mix of diesel and electric trucks. The electric trucks add the constraint of charging time, charging station locations and driving range that is less than that of diesel. The routing problem is formulated as an optimization problem with several constraints. A co-simulation load-balancing approach is used to generate routes for trucks that reduce the overall cost. The co-simulation approach employs a traffic simulator in a feedback loop that replaces simple mathematical models often used to predict the states of the network. The simulation model captures the complexity and dynamics of the network and generates predicted states that are used to generate optimum routes and achieve load balancing. Numerical experiments are performed using a traffic simulator of the Los Angeles/Long Beach Metropolitan road traffic network that includes two major ports. The simulation results demonstrate strong potential for the proposed centralized truck routing system to reduce the impact of trucks on traffic and make their routing more efficient.

P. Chen · P. Ioannou (✉)
University of Southern California, Los Angeles, CA, USA
e-mail: ioannou@usc.edu

P. Chen
e-mail: pengfeic@usc.edu

P. Ioannou and A. A. Malikopoulos (eds.), *Transportation Mobility in Smart Cities*,
Springer Tracts on Transportation and Traffic 21,
https://doi.org/10.1007/978-3-031-64769-7_13

13.1 Introduction

In today's world, the efficient movement of goods is a critical factor for the sustainability and well-being of society especially in urban areas with highly concentrated populations. Due to the increasing volume of goods to be processed, deteriorating traffic conditions, and increasing pollution, the sustainability of efficient goods distribution in big cities is in danger. Traffic congestion results in enormous costs to shippers, carriers, and the economy. According to [1], in 2016, around 1.2 billion hours of delay were wasted across the US national highway system, which is equal to approximately $74.5 billion. Trucks account for a significant share of air toxics: about one-third of all nitrous oxides (NO_x) and nearly 30% of all particulate matter of 10 microns or less (PM10) [2]. Trucks also account for a significant share of greenhouse gas (GHG) emissions. The US port sector accounts for 28% of all GHG emissions. Medium- and heavy-duty trucks account for 27% of the transport share [2]. The above statistics motivate the need for an intelligent routing system for trucks. Numerous research efforts have been addressed in this area. Jourquin and Beuthe presented a multi-modal freight model based on a digitized geographical network [3]. Southworth and Peterson developed a multi-layer intermodal shipment model in [4]. The intermodal freight transport model between rail and road has been described in [5–7]. Moreover, Russ et al. and Yamada et al. showed the application of routing and scheduling techniques in multi-modal freight network design [8, 9].

With the recent advances in electric vehicles which can help cut down emissions, the management and operation of electric vehicles in a freight fleet adds more challenges to routing. According to [10], the penetration of electric trucks into the global medium- and heavy-duty market is projected to be 9.4% by 2030. The use of electric trucks adds several constraints which include the limited range of batteries compared to diesel trucks, long charging times, and limited number of locations of charging stations. Any routing system should take into account these constraints. Research efforts for this problem under various constraints can be found in [11–19]. In most of these efforts, simple mathematical models are used to estimate travel times by often ignoring the complexity of the dynamics of the traffic network which is shared with passenger vehicles. Given the size and slower dynamics of trucks their impact on traffic speed is much higher and cannot be ignored.

The impact of trucks on traffic is another point that needs consideration. Currently, truck users route their trucks without any consideration of what other users do which leads to heavy congestion at popular pick-up and drop-off hubs and/or congestion along popular truck routes. As a result, the road network is unbalanced in space and time with respect to truck traffic. To address this issue, the dynamics of the transportation system are analyzed by formulating the problem as a variant of the traffic assignment problem (TAP), which was introduced by Beckmann et al. [20]. It was used to generate an optimum route distribution by minimizing the total cost for transportation planning purposes and develop meaningful operational strategies for the goods movement industry, such as the assignment of multi-modal freight. For example, Guelat and Florian proposed a linear approximation algorithm to solve a multi-modal and multiproduct freight TAP [21]. Castelli et al. used a Lagrangian-based heuristic procedure to solve the freight scheduling problem [22]. Ham, Kim, and Boyce showed the application of Wilson's iterative balancing method in inter-regional multi-modal shipment planning [23]. Zografos et al. developed a dynamic programming-based algorithm for multi-modal scheduling [24]. Moccia et al. solved a multi-modal routing problem with timetables and time windows by integrating a heuristic methodology with a column generation algorithm [25]. Crainic et al. proposed meta-heuristic methods for freight demand distribution in congested urban areas [26, 27]. The main difficulty is that these approaches use mathematical models that may not be able to find a closed-form solution when applied to complex networks such as large-scale multi-modal freight networks that exhibit nonlinear interactions between the routing decisions and traffic network states.

With electric trucks entering the market routing with TAP adds additional constraints due to battery life, location of charging stations, and charging time. Nan et al. presented a mathematical programming model and solution method for the path-constrained traffic assignment problem for electric vehicles in congested networks [28]. Bahrami et al. proposed a complementarity equilibrium model for electric vehicles without violating driving range constraints [29]. Based on the assumption of large adoption of electric vehicles, Faridimehr et al. [30] proposed a two-stage stochastic programming model to determine the optimal network of charging stations for a community as well as the charging decision for each electric vehicle.

Despite the amount of research in TAP, there are many issues that need to be addressed. First, the complexity of modeling the nonlinear behavior at the traffic flow level that is caused by the non-homogeneous dynamics of different vehicle classes at the vehicle level is immense. The complexity of the real system cannot be possibly captured by TAP mathematical models, especially with the integration of electric trucks. Second, the development of accurate mathematical models to describe traffic characteristics has always been a challenge and is becoming more of a challenge if electric trucks are included in the model. The constraints introduced by electric trucks dealing with charging and battery depletion behavior, etc. make the mathematical modeling even more complicated. The complexity of traffic dynamics, the size and dynamics of trucks further complicated by the new constraints, and dynamics of electric trucks make it impossible to use simple mathematical models to accurately estimate traffic states that are used to generate optimum routes. The traffic flows and states can be better predicted using simulation models that are more complex and can capture phenomena that cannot be formulated with simple models. Some researchers tried to solve the centralized multi-modal routing and scheduling problem from the aspect of dynamic traffic assignment based on user equilibrium using simulation-based methods as in the unimodal road routing and scheduling problem [31–37]. The idea of user equilibrium-based traffic assignment is to adjust the traffic flows iteratively until the trip costs for the same demands are equal and the costs of the used routes are less than the costs of the other unselected routes. Peeta and Mahmassani formulated and developed a simulation model-based method for the dynamic user equilibrium problem given fixed demand for the road transport mode [31]. The proposed method was also applied for multi-modal transport [32–

34]. The user equilibrium between supply and demand has been also studied when trip costs from all shippers to the same destination are in equilibrium [35, 36]. Other solution algorithms such as cross entropy-based [37] and gap-based [34, 38] methods were proposed to give better performance than the method of successive average (MSA) that is widely used for user equilibrium problems. The system optimal routing problem can also be solved by the above algorithms by using the marginal costs of the routes in the equilibrium solution instead of the trip costs [32]. Due to the fact that the number of routes increases exponentially with the size of the transportation network, it is computationally prohibitive to evaluate the marginal costs of all possible routes by using simulations. The marginal costs of the routes can be estimated approximately from the link states of the simulation model [32]. The co-simulation type of methods utilizes traffic simulation models which can run in real time to provide the information and predicted states of the traffic network to choose routes that are more likely to be close to optimality than those based on simplified mathematical models. The challenge is how these simulation models can be integrated with optimization tools to generate more realistic outcomes.

The contribution of this chapter is the development and evaluation of a centrally coordinated load-balancing approach for a mixed fleet of trucks that is computationally feasible. Specifically, the contributions are:

- A centrally coordinated mixed freight truck dynamic routing system is formulated and constructed. The system achieves a minimum system cost by generating optimum routes that balance the truck loads on the road network.
- A co-simulation optimization method is used to solve the mixed freight routing problem. The method achieves load balancing across the road network.
- The system is tested under realistic scenarios assuming different percentages of penetrations of electric trucks and background traffic conditions.

The remainder of this chapter is organized as follows. Section 13.2 deals with the problem formulation and solution methodology. Section 13.3 presents the numerical results of the proposed system. Finally, conclusions are presented in Sect. 13.4.

13.2 Problem and Methodology

In this section, the formulation of the mixed freight truck fleet routing problem is presented and a centrally coordinated routing system with a co-simulation optimization method is proposed to address this problem. Specifically, this routing problem deals with drayage trucks transshipping containers and demand is expressed as containers to be transported from origin to destination. The aim of the routing problem is to provide each truck in the fleet with a route with a minimal total system cost which includes energy and time cost.

13.2.1 Formulation

We consider the road network to be a directed graph $G(E, V)$, where E is the set of all links and V is the set of all nodes. Among all the nodes, a subset of them are origin nodes, denoted as O, i.e., $O \subset V$. Another subset of nodes are destination nodes, denoted as D, i.e., $D \subset V$. For a certain pair of origin and destination nodes $(i, j), i \in O, j \in D$, the demand volume is $q_{i,j}$. All the truck types are included in a set U. To represent the distribution of trucks, we use m_i^u as the number of available trucks of type u at node i. To cope with the temporal dimension, we discretize the time horizon into $|K|$ time intervals and use K as the set of all the time intervals. The following notation is used in the formulation to follow:

- $R_{i,j}^u$: The set of routes for trucks of type u from i to $j, i \in O, j \in D$.
- $X_{i,j,r,k}^u$: The number of trucks of type u from i to $j, i \in O, j \in D$, using route r in route set $R_{i,j}^u$ with a departure time k. The collection of $X_{i,j,r,k}^u$ is denoted as X.
- $S_{i,j,r,k}^u(X)$: The average service cost per container fulfilled by a truck of type u from i to $j, i \in O, j \in D$, using route r in route set $R_{i,j}^u$ with a departure time k.

Given the above notation, we formulate the problem as follows:

$$\min_X \sum_{k\in K}\sum_{i\in O}\sum_{j\in D}\sum_{u\in U}\sum_{r\in R_{i,j}^u} S_{i,j,r,k}^u(X) X_{i,j,r,k}^u \tag{13.1}$$

$$\sum_{k\in K}\sum_{u\in U}\sum_{r\in R_{i,j}^u} X_{i,j,r,k}^u = q_{i,j}, \ \forall i \in O, j \in D \tag{13.2}$$

$$\sum_{k\in K}\sum_{j\in D}\sum_{r\in R_{i,j}^u} X_{i,j,r,k}^u \le m_i^u, \ \forall i \in O, u \in U \tag{13.3}$$

$$X_{i,j,r,k}^u \ge 0. \tag{13.4}$$

Equation (13.1) is the objective function, which aims to minimize the sum of the service cost of all the freight loads that are assumed to be containers. $S_{i,j,r,k}^u(X)$ is the unit service cost of transporting a container with a truck of type u using route r from i to j at time k given X. The cost $S_{i,j,r,k}^u(X)$ is given by

$$S_{i,j,r,k}^u(X) = C_{i,j,r,k}^u(X) + \eta T_{i,j,r,k}^u(X), \tag{13.5}$$

where $C_{i,j,r,k}^u(X)$ is the cost of the consumed energy, $T_{i,j,r,k}^u(X)$ is the travel time, and η is the value of time. The energy and travel time costs depend on the dynamics of the traffic network. The dynamics of the traffic network can be expressed as nonlinear dynamic functions of all decision variables, denoted as X, and will be discussed in the following sections. Note here, the energy cost is calculated based on the dynamics of the traffic network [39]. More specifically, the energy cost coefficient of each truck type is formulated as a polynomial function of the speed of the road link, where the parameters of the function are estimated using regression over a set of testing data. Here we assume one truck can only load one container, so the total number of trucks for an O/D pair is equal to the demand of the O/D pair, as shown in Eq. (13.2). Equation (13.3) represents the constraints on the availability of a certain type of truck at each node.

The dynamics of a traffic network are highly nonlinear and exhibit the following temporal-spatial relations: traffic flow dynamics in a link and between links. The dynamics in a link describe how the traffic flow moves from the upstream end of a link to the downstream end, while the dynamics between links describe how the traffic flow propagates across the traffic network. In most of the literature on vehicle routing, the complex dynamics of the traffic network are overly simplified and the dynamics between links are ignored. The optimum routes of these methods may not be optimum in practice as they are generated by ignoring important traffic dynamics. In our approach, we connect the traffic flow on each route with the number of trucks of a certain type on a road edge as follows: First, a subset of nodes in the traffic network including all O/D nodes as well as the nodes necessary for the routing of freight vehicles are chosen. These nodes are called service nodes. Then a set of virtual service links connecting the service nodes are constructed. To differentiate the link in the road network, the virtual service links are called segments. Let $x^u_{l,k}$ be the number of trucks of type u passing through segment l at time k. Then the relations between routes and service segments can be shown as follows:

$$\sum_{i\in O}\sum_{j\in D}\sum_{u\in U}\sum_{r\in R^u_{i,j}}\sum_{\tau\leq k} X^u_{i,j,r,k}\delta^u_{l,r,\tau,k} = x^u_{l,k}, \tag{13.6}$$

where $l \in L, k \in K$ and $\delta^u_{l,r,\tau,k} = 1$ when the truck of type u uses route r with departure time τ passing through segment l at time k, otherwise, $\delta^u_{l,r,\tau,k} = 0$. As for the relations between the service segment and traffic network links, we denote as $t^p_{l,k}$ the travel time on path p if a truck departs from the origin of segment l at time k. Assume links constituting path p to be $e_{p,1}, e_{p,2}, \ldots, e_{p,N_p}$, where N_p is the total number of links on path p. We define $\xi_{e,k}$ as the entering time at link e of a truck with a departure time k from the origin of that path. With $w_{e,k}$ to be the travel time of link e at time k, we now write the travel time of a path as follows:

$$t^p_{l,k} = \sum_{n_p=1}^{N_p} w_{e_{p,np},\xi_{k,e_{p,np}}} \tag{13.7}$$

$$\xi_{k,e_{p,1}} = 1 \tag{13.8}$$

$$\xi_{k,e_{p,np+1}} = \xi_{k,e_{p,np}} + w_{e_{p,np},\xi_{k,e_{p,np}}}, \tag{13.9}$$

where $n_p = 1, \ldots, N_p - 1$. To make the notation simpler, we let $\hat{w}_{p,n_p,k} \equiv w_{e_{p,np},\xi_{k,e_{p,np}}}$ to denote the travel time of link e_{p,n_p} on path p with the path departure time being $\xi_{k,e_{p,np}}$. Then the mixed freight routing problem can be formulated in the form of a service-road two-layer structure:

$$\min_{y^u} TC = \sum_{k\in K}\sum_{l\in L}\sum_{p\in P_l}(c_{l,k}^{p,u} + \eta t_{l,k}^{p,u})y_{l,k}^{p,u}, \tag{13.10}$$

where TC stands for the total cost of the assignment with mixed freight vehicles, which is a combined value of energy consumption cost and travel time cost. $c_{l,k}^{p,u}$ is the energy consumption coefficient for trucks of type u passing through path p of segment l at time k, $t_{l,k}^{p,u}$ is the travel time of path p in segment l that departs at time k, $y_{l,k}^{p,u}$ is the number of trucks of type u assigned to pass through path p of segment l at time k, and η is the value of time. This objective function is constrained by Eqs. (13.6)–(13.9) as well as the following equations:

$$\sum_{p\in P_l} y_{l,k}^{p,u} = x_{l,k}^{u},\ \forall l \in L, k \in K \tag{13.11}$$

$$y_{l,k}^{p,u} \geq 0,\ \forall l \in L, p \in P_l, k \in K, \tag{13.12}$$

which represents the relation between variables from the service network and the simulated traffic network. In summary, we achieve the following changes to make routing decisions closer to a theoretical optimum:

- We use a traffic simulation model to capture the complex dynamics and interactions of background traffic with the loaded trucks instead of using simplified mathematical models. The simulation model provides a far more accurate description of the traffic dynamical characteristics to be used by the optimum route generator.
- We construct a service network layer to efficiently utilize the simulation model and apply the optimization algorithms.
- In the algorithm, the direction and step size are properly chosen based on the knowledge of the marginal cost.

In the next subsection, we present a co-simulation optimization method for solving the multi-layer routing problem.

13.2.2 Solution Methodology

Figure 13.1 gives a general overview of the method. The inputs to the routing system are demands, vehicle-related characteristics, and other predetermined parameters. Demands represent the number of containers to be transferred from origin to destination nodes. The truck characteristics include the physical (weight, length, frontal area, etc.) and dynamic (max speed, acceleration, etc.). The energy consumption (the amount of energy consumed based on the dynamic states) characteristics for each type of truck are included in the predetermined parameters. Based on the energy consumption characteristics of diesel/electric trucks, the cost coefficients on each segment of both types of trucks are calculated under different traffic conditions. We

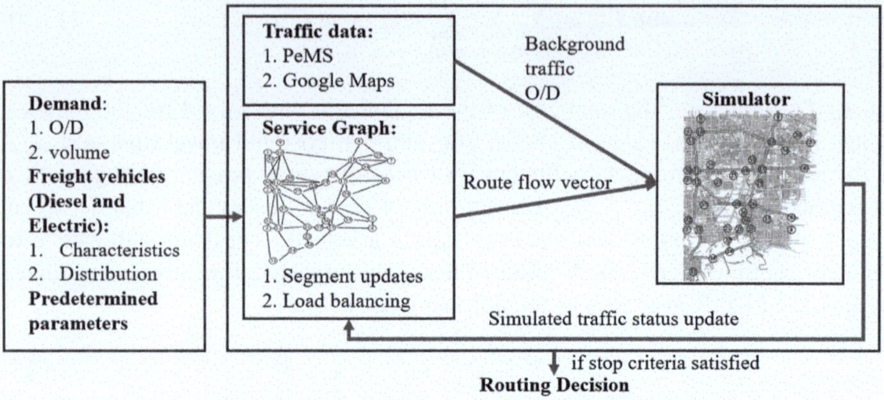

Fig. 13.1 Framework of proposed mixed freight dynamic routing system

use an emission model from National Renewal Energy Laboratory (NREL) to calculate the emissions. In the routing system, two major components, the service graph and traffic simulator, form a feedback iterative loop where iterations stop when certain criteria are satisfied. The service graph acts as a central coordinator in practice whose aim is to assign trucks to fulfill demands at minimal system cost. A real-time traffic simulator is used to capture the dynamical characteristics of traffic and provide traffic status such as travel times along the links and routes as well as estimates of the energy cost of diesel and electric trucks depending on the simulated traffic flow. The information from the simulator is used by the service graph optimization component to update the status, such as the marginal cost of each service segment. Then the route cost is updated based on the cost of the service segment. Based on the simulated route cost, the route collection for each O/D pair is updated as well. The updates of route collection start by trying to find if there is a new minimum-cost route. If there is, add the new route to the route collection; otherwise, do nothing. Then given the updated route collection, the assignment of diesel/electric trucks for each O/D pair is updated by solving an integer combinatorial programming problem using a properly selected efficient step size. The new assignment is then generated in the form of a route flow vector and passed to the traffic simulator, which starts another iteration. Other than the route flow vector from the service graph, the background traffic data accounts for the other part of the road network. The background traffic flow is obtained from various sources, such as PeMS [42] and Google Maps [43]. The assignment traffic flow is generated by the optimizer in the service graph. The co-simulation optimization procedure iterates in a feedback loop that involves the traffic simulator and service graph optimization. Through this procedure, the states of the route flow vector and road network feedback are sequentially updated until both states converge. The difficulty in this procedure is to calculate the marginal cost of each route, which represents the change in the total cost as a result of adding one unit of demand on that route. Since the total cost TC of Eq. (13.10) is complex, the marginal cost with respect to a route cannot be calculated directly. One way to calculate the marginal cost is to use Monte Carlo to simulate the impact of one unit of demand on each route at each time. However, it is impractical to enumerate all routes due to the fact that the number of possible routes grows exponentially with respect to the service network size. Our proposed approach bypasses this issue and works as follows:

Step 1: Initialize cost coefficients based on the physical features such as speed limit for each segment l and iteration number $n = 0$. Initialize the diesel/electric route collections for each O/D pair based on the segment cost calculated with the cost coefficients. Establish the initial route flow vector $X^{(0)}$ by assigning the portion of demands in the origin node to electric trucks with the portion of demand to be equal to the portion of electric trucks in the mixed fleet.

Step 2: If $n > 1$, check if the objective function value of the current iteration converges, i.e., $|TC(X^{(n)}) - TC(X^{(n-1)})| < \varepsilon$; ε is set to be a small number. If it converges, then stop the procedure and return with route flow vector; otherwise, continue to the next step.

Step 3: Input the route flow vector $X^{(n)}$ into the traffic simulator and obtain the marginal cost of each segment.

Step 4: Update the marginal cost of each segment as well as diesel/electric routes for each O/D pair and check whether there is a new minimal marginal cost route. If there is, then add it to the route collection.

Step 5: Solve the following optimization problem for each origin node o to obtain a feasible route flow vector $\hat{X}^n$.

$$\min_X \sum_{u\in U}\sum_{k\in K}\sum_{j\in D}\sum_{r\in R^u_{o,j}} MC^u_{o,j,r,k} X^u_{o,j,r,k} \tag{13.13}$$

$$\sum_{u\in U}\sum_{k\in K}\sum_{r\in R^u_{o,j,k}} X^u_{o,j,r,k} = q_{o,j},\ \forall j \in D \tag{13.14}$$

$$\sum_{k\in K}\sum_{j\in D}\sum_{r\in R^u_{o,j,k}} X^u_{o,j,r,k} \le m^u_o,\ \forall u \in U, \tag{13.15}$$

where $MC^u_{o,j,r,k}$ is the marginal cost of route r from o to j with a truck of type u departing at time k. The marginal cost of a route is the sum of the marginal costs of the segments along it. The computation of the marginal cost of a segment will be addressed in the next subsection.

Step 6: Set the route flow vector for the next iteration as $X^{(n+1)} = X^{(n)} + \lambda^{(n)} \cdot (\hat{X}^n - X^{(n)})$, where $\lambda^{(n)}$ is the step size at the nth iteration, and go back to step 2. The step size $\lambda^{(n)}$ at the nth iteration is selected as in [40].

$$
\begin{aligned}
X^{(n+1)} &= X^{(n)} + \lambda^{(n)} \cdot (\hat{X}^n - X^{(n)}) \\
\lambda_{i,j}^n &= \min\{\lambda_{max}, \\
&\sum_{i \in O} \sum_{j \in D} \lambda_{i,j}^{n-1} \cdot \frac{\sigma(q_{i,j})}{\sum_{i \in O} \sum_{j \in D} \sigma(q_{i,j})}\},
\end{aligned}
\tag{13.16}
$$

where $\sigma(q_{i,j})$ is the standard deviation of the marginal cost of all the routes by demand $q_{i,j}$ and λ_{max} is the upper bound of the step size.

Different from the load-balancing cases with single-type vehicles, we address the problem with two types of trucks: diesel and electric trucks. Due to that, the type of steepest descent direction used in the work of [40] may not be feasible for the mixed freight case. The update of each iteration should consider not only the marginal cost of a certain route but also the availability of each type of truck at a certain node. In step 5, a linear programming subproblem is formulated by explicitly imposing the availability constraints at each node for each type of truck as shown in Eqs. (13.13)–(13.15).

The marginal cost of a service segment represents the change in the total cost if one unit of demand/container is changed on the segment, which can be calculated through the marginal cost of a path in the segment based on the traffic status from the traffic simulator. Let $MCP_{l,k}^{p,u}$ be the marginal cost of path p of segment l with departure time k for the trucks of type u. It can be formulated as follows:

$$
\begin{aligned}
MCP_{l',k'}^{p',u'} &= \frac{\partial TC}{\partial y_{l',k'}^{p',u'}} \\
&= \frac{\partial \sum_{k \in K} \sum_{l \in L} \sum_{p \in P_l} (c_{l,k}^{p,u} + \eta t_{l,k}^{p,u}) y_{l,k}^{p,u}}{\partial y_{l',k'}^{p',u'}} \\
&= c_{l',k'}^{p',u'} + \eta t_{l',k'}^{p',u'} + \\
&\quad \eta \sum_{k \in K} \sum_{l \in L} \sum_{p \in P_l} y_{l,k}^{p,u} \frac{\partial t_{l,k}^{p,u}}{\partial y_{l',k'}^{p',u'}} + \\
&\quad \sum_{k \in K} \sum_{l \in L} \sum_{p \in P_l} y_{l,k}^{p,u} \frac{\partial c_{l,k}^{p,u}}{\partial y_{l',k'}^{p',u'}},
\end{aligned}
\tag{13.17}
$$

where the first two terms are the cost of the path and the third term describes the travel time cost change due to the impact on the link travel time based on the dynamics of the traffic system. The fourth term accounts for the change of energy cost associated with the changes in link volume and can be calculated approximately using the traffic network states from the simulator. A detailed derivation procedure is presented in the appendix. The final form of marginal cost is

$$
\begin{aligned}
MCP_{l',k'}^{p',u'} &\approx c_{l',k'}^{p',u'} + \eta t_{l',k'}^{p',u'} \\
&+ \sum_{k\in K}\sum_{l\in L}\sum_{p\in P_l}\sum_{n_p=1}^{N_p} \mathbb{1}_{e'_{p,np},\xi_{k'},e'_{p,np}}(e_{p,n_p},\xi_{k,e_{p,np}}) \\
&\cdot y_{l,k}^{p,u}\frac{1}{v_{p,n_p,k}\Delta t}(\eta + \frac{\partial h^u(v_{p,n_p,k})}{\partial \hat{w}_{p,n_p,k}} d_{p,n_p}) B_{p,n_p,k}\hat{z}_{p,n_p,k}^{\beta_{e_{p,np}}-1} \\
&\forall l' \in L, p' \in P_{l'}, k' \in K, u' \in U,
\end{aligned}
\tag{13.18}
$$

where $\frac{\partial h^u(v_{p,np,k})}{\partial \hat{w}_{p,np,k}}$ is obtained from Eq. (13.30). Since the first and second terms are decomposable with respect to the links, the marginal costs of the paths belonging to the same segment will be placed in equilibrium by running a dynamic assignment algorithm. Then the marginal cost for a segment $MC_{l',k'}^{u'}$ is approximated by its marginal cost of path $MCP_{l',k'}^{p',u'}$,

$$
MC_{l',k'}^{u'} \approx MCP_{l',k'}^{p',u'}, \forall l' \in L, k' \in K, u \in U. \tag{13.19}
$$

From Eq. (13.18), we can see that calculating the marginal cost of a segment requires the following information from the traffic simulator:

- The state of trucks $\mathbb{1}_{e'_{p,np},\xi_{k'},e'_{p,np}}(e_{p,n_p},\xi_{k,e_{p,np}})$.
- The basic traffic network status $(\hat{w}_{p,n_p,k}, \hat{z}_{p,n_p,k}, v_{p,n_p,k}, h^u(v_{p,n_p,k}))$.
- The aggregated segment-level information $(c_{l,k}^{p,u}, t_{l,k}^{p,u}, y_{l,k}^{p,u})$.

With the marginal cost of each segment updated, route collections are updated by checking whether there are new lower marginal cost routes. Then the route flow vector X is updated to move along the descent direction with the step size described in Eq. (13.16) with the knowledge of the updated marginal cost. The algorithm stops when no more improvement on the total cost can be gained.

13.3 Numerical Evaluation

This section presents the evaluation of the proposed approach using a regional transportation network. The road network covers the Los Angeles/Long Beach terminal port area from the south to I-105 freeway in the north. Lane characteristics such as length, capacity, and speed limit are incorporated in the traffic simulator for this network. The freight vehicles from and to the terminal port area account for a large amount of traffic around the area and have a great impact on the environment. The road traffic network and its corresponding service network are shown in Fig. 13.2. The traffic simulator used is a macroscopic traffic simulator VISUM [41].

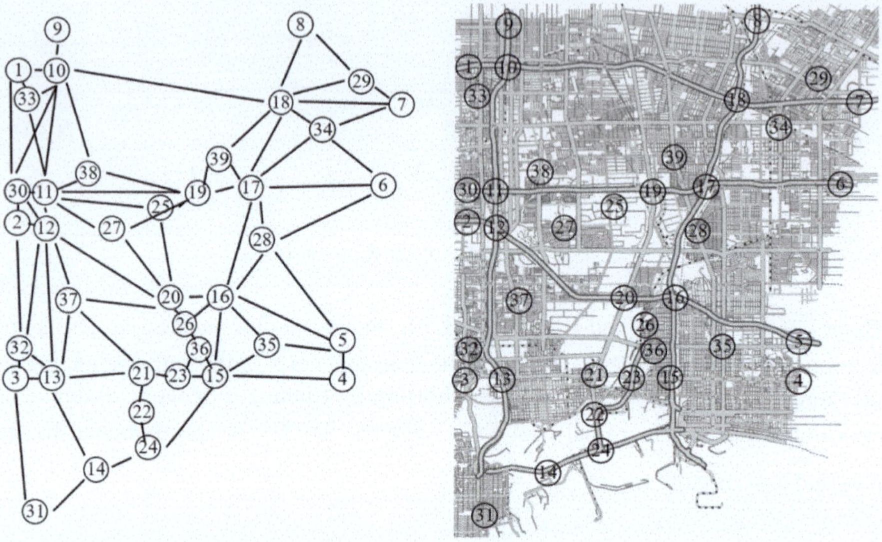

(a) Service network (b) Associated road network

Fig. 13.2 Service network and road network overview

We use freeway traffic flow data from PeMS [42] and arterial way data from Google Maps [43]. Traffic data in three time periods are extracted from the raw data: from 2 am to 6 am representing a light background traffic condition, from 12 pm to 4 pm representing a medium background traffic condition, and from 7 am to 11 am representing a heavy background traffic condition. The extracted traffic data are then processed (formatted/truncated/aggregated) to fit the format of the traffic simulator. We assume the total number of all demands to be 1715 containers. The locations of service network nodes are presented in numbers within a circle in Fig. 13.2. The service network nodes are composed of O/D nodes as well as intersections of freeways and major arterial ways. The service nodes also play the role of charging stations. To make sure the routes of electric trucks are feasible, we assume every charging station has enough capacity for charging and electric trucks always get charged the amount of electricity they consumed on the previous segment along the route. The interval used for the numerical experiment is 30 min.

The routing system is tested under different scenarios of various percentages of electric vehicles. The experimental scenarios are constructed in the following manner: three background traffic conditions (light, medium, and heavy) are constructed and under each background traffic condition, the percentage of electric vehicles in the fleet is varied from 0 to 100% in increments of 10%. Let v be the speed of the truck, according to [39], the energy cost coefficient function for diesel trucks is

$$h^d(v) = 11.327 - 0.649v + 0.045v^2 - 0.001v^3. \tag{13.20}$$

For electric trucks, it is

$$h^e(v) = 66.552 - 7.670v + 0.291v^2 - 0.004v^3. \quad (13.21)$$

The hourly time value is set to be \$60 in terms of the year 2020 US Dollar value [44]. We assume the charging cost to be the same as the time cost at \$60 per hour. In our simulation, we assume that it takes 4 hours for a full charge which is equivalent to \$240. The experiments are run on a desktop computer with a 4.2 GHz CPU and 16G memory. The results include total costs in unit of dollar of the assignment (with and without charging time cost), the weight in unit of gram of several emissions (CO, NO_x, CO_2, $PM_{2.5}$) as well as fuel consumed in unit of kg. The emissions are calculated by the modified EPA model MOVES [45] with speed as input and emissions in units of g/mile as output. The results under light, medium, and heavy traffic conditions are shown in Figs. 13.3, 13.4, 13.5, 13.6, 13.7, 13.8, 13.9, 13.10, and 13.11. We can see that by increasing the percentage of electric trucks, all types of emissions are decreasing as expected and as shown in Figs. 13.3, 13.4 for light traffic and Figs. 13.6, 13.7 for medium traffic and Figs. 13.9, 13.10 for heavy traffic. Another observation is that by increasing the percentage of electric trucks the overall cost does not decrease. The cost of charging time becomes more and more dominant as the percentage of electric trucks increases. This is shown in Figs. 13.5, 13.8, and 13.11 for light, medium, and heavy traffic, respectively.

In summary, the following observations can be made from the experiments:

- The total cost without including charging cost decreases as the percentage of electric vehicles increases.
- The total cost that also includes the charging cost tends to increase in general with increasing percentage of electric trucks in the fleet.

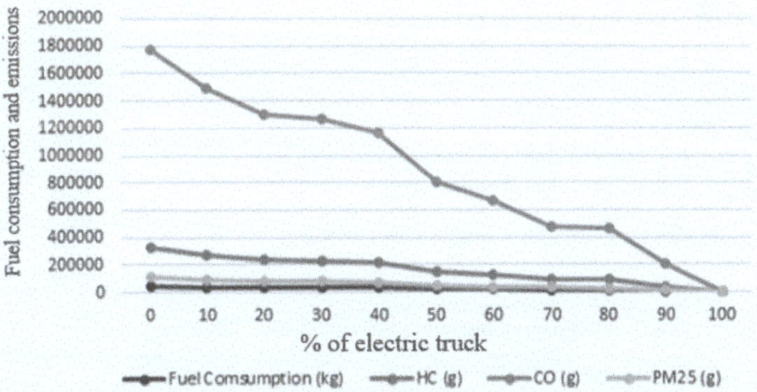

Fig. 13.3 Fuel consumption, HC, CO, $PM_{2.5}$ emission versus percentage of electric trucks under light traffic condition. The fuel consumption, and emissions of HC, CO, $PM_{2.5}$ decrease as the penetration of electric trucks in the fleet increases

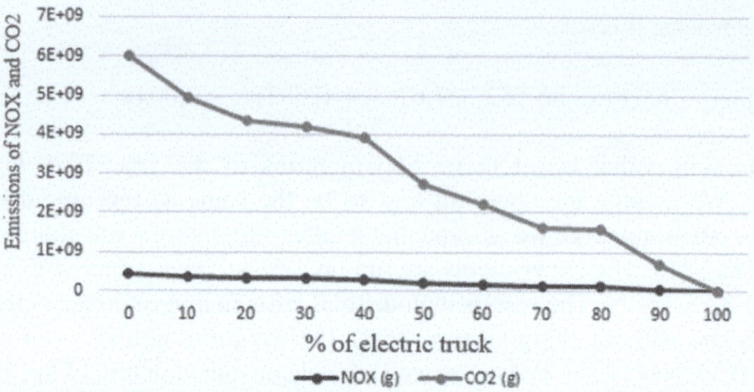

Fig. 13.4 NO_X, CO_2 emission versus percentage of electric trucks under light traffic condition. The emissions of NO_X and CO_2 decrease as the penetration of electric trucks in the fleet increases

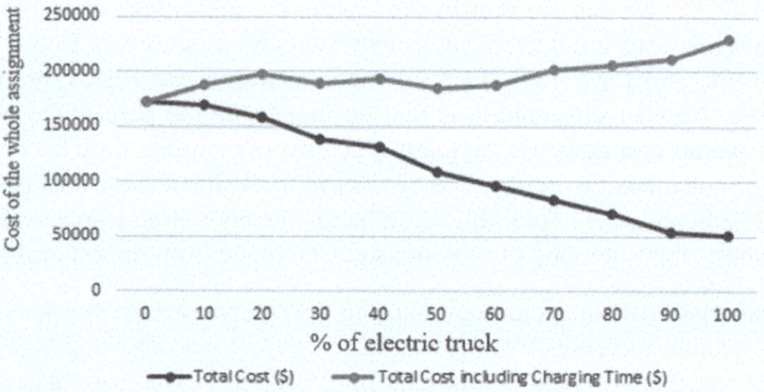

Fig. 13.5 Total cost with/without charging time cost versus percentage of electric trucks under light traffic condition. The total cost without charging time cost decreases as the penetration of electric trucks in the fleet increases, but with the inclusion of charging time the total cost increases as the penetration of electric trucks increases

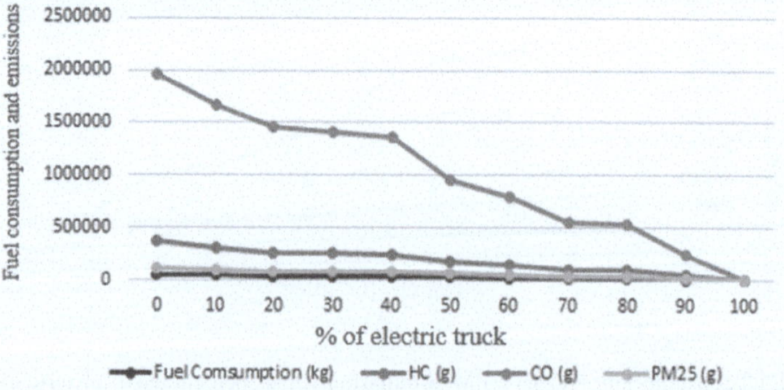

Fig. 13.6 Fuel consumption, HC, CO, $PM_{2.5}$ emission versus percentage of electric trucks under medium traffic condition. The fuel consumption, and emissions of HC, CO, $PM_{2.5}$ decrease as the penetration of electric trucks in the fleet increases

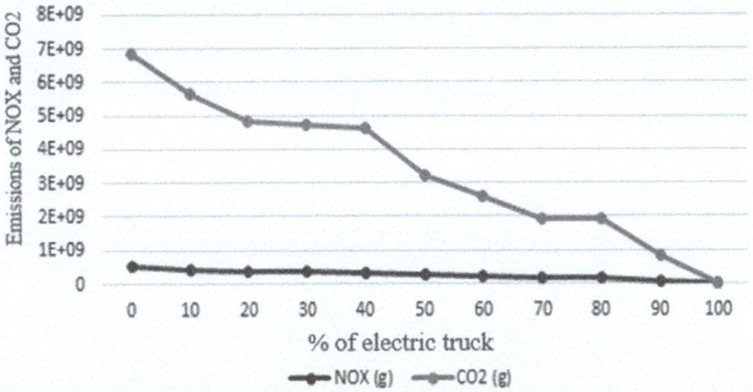

Fig. 13.7 NO_x, CO_2 emission versus percentage of electric trucks under medium traffic condition. The emissions of NO_x and CO_2 decrease as the penetration of electric trucks in the fleet increases

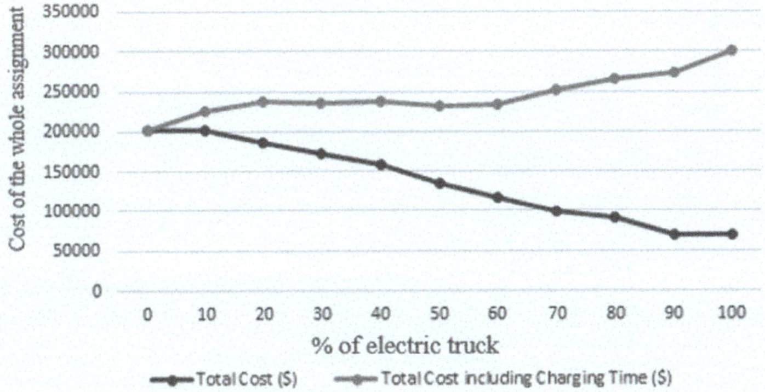

Fig. 13.8 Total cost with/without charging time cost versus percentage of electric trucks under medium traffic condition. A similar pattern is shown in the setting of medium traffic condition. The total cost without charging time cost decreases as the penetration of electric trucks in the fleet increases, but with the inclusion of charging time the total cost slightly increases as the penetration of electric trucks increases

- The difference between the total cost including and excluding charging time is the charging time, which we assume that the time of the driver waiting for charging is included. Due to that, we can observe that if charging is done off-duty the total cost can be reduced considerably.
- As expected the emissions go down as the number of electric vehicles increases in the fleet.

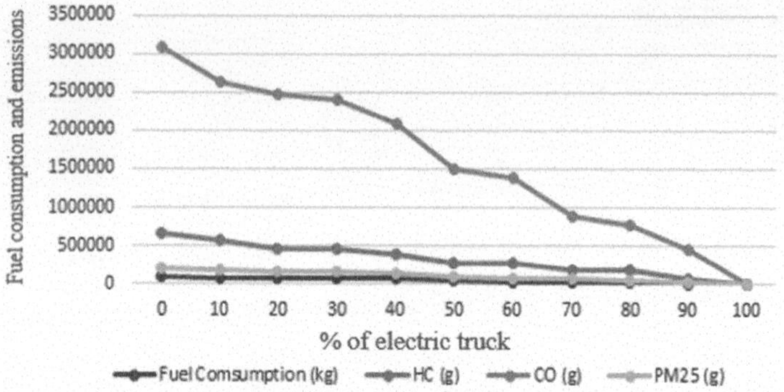

Fig. 13.9 Fuel consumption, HC, CO, $PM_{2.5}$ emission versus percentage of electric trucks under heavy traffic condition. The fuel consumption, and emissions of HC, CO, $PM_{2.5}$ decrease as the penetration of electric trucks in the fleet increases

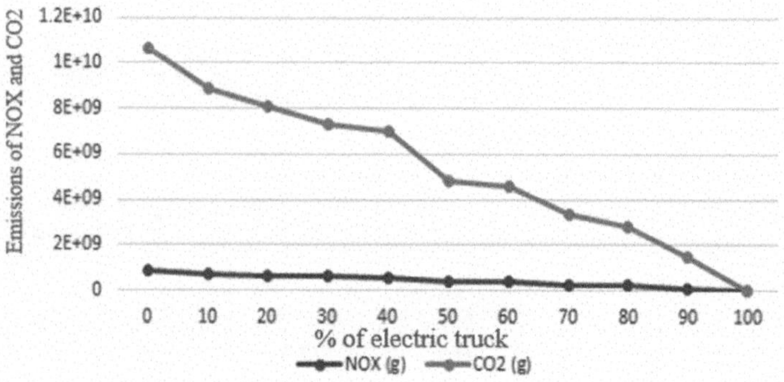

Fig. 13.10 NO_x, CO_2 emission versus percentage of electric trucks under heavy traffic condition. The emissions of NO_x and CO_2 decrease as the penetration of electric trucks in the fleet increases

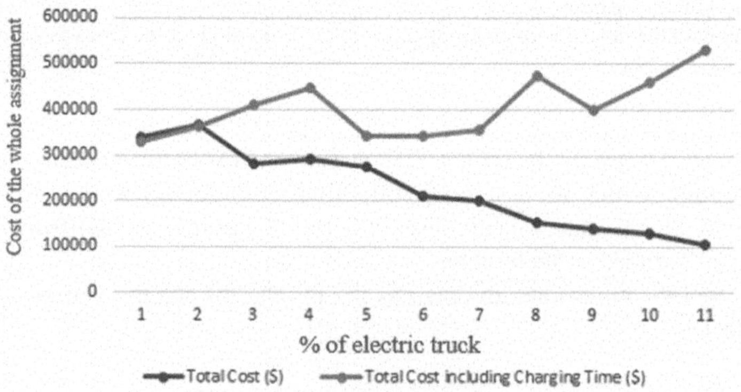

Fig. 13.11 Total cost with/without charging time cost versus percentage of electric trucks under heavy traffic condition. The total cost without charging time cost decreases as the penetration of electric trucks in the fleet increases, but with the inclusion of charging time the total cost slightly increases as the penetration of electric trucks increases

13.4 Conclusion

In this chapter, we proposed a centrally coordinated truck routing system based on a co-simulation load-balancing approach to generate optimum truck routes that minimize the overall system cost. The method considers the inclusion of electric trucks with their penetration varying from 0 to 100% which takes into account battery life, charging times, range, and location of charging stations. The multi-layer co-simulation optimization method used consists of one layer for the traffic simulator and another layer for the service network. The traffic simulator is used to accurately predict the states of the transportation system and the service network to generate the optimum routes. The numerical experiments have been performed on a realistic traffic network in the Los Angeles/Long Beach area that includes the two ports. The results show that the use of electric trucks can notably reduce emissions and total cost without including the charging time cost. Since the charging is done during working hours and includes the driver cost, it makes the operational cost of electric trucks comparable to or higher than diesel trucks. From this observation, we suggest scheduling charging during driver-off hours or during times that the driver is idle for job purposes to make the operational cost of electric trucks lower. Another important aspect of the proposed approach is that routes are generated to minimize a system cost. This does not imply that all individual user costs are also minimized as in some cases some users will benefit and some will lose when compared with the user optimal route decisions. This phenomenon known as the price of anarchy is well known [46]. Introducing pricing and incentives for participation so that the individual cost is always less or equal to that generated by the coordinator is an important topic of interest and some preliminary results are presented in [47].

Acknowledgements This work has been supported by the National Science Foundation under grant CNS-1932615 and in part by the South Coast Air Quality Management District (AQMD).

Appendix: Derivation of Marginal Cost

The marginal cost represents the change in the total cost if one unit of demand/container is changed on the path and is formulated as follows:

$$\begin{aligned}
MCP_{l',k'}^{p',u'} &= \frac{\partial TC}{\partial y_{l',k'}^{p',u'}} \\
&= \frac{\partial \sum_{k\in K}\sum_{l\in L}\sum_{p\in P_l}(c_{l,k}^{p,u}+\eta t_{l,k}^{p,u})y_{l,k}^{p,u}}{\partial y_{l',k'}^{p',u'}} \\
&= c_{l',k'}^{p',u'} + \eta t_{l',k'}^{p',u'} + \eta \sum_{k\in K}\sum_{l\in L}\sum_{p\in P_l} y_{l,k}^{p,u}\frac{\partial t_{l,k}^{p,u}}{\partial y_{l',k'}^{p',u'}} + \\
&\sum_{k\in K}\sum_{l\in L}\sum_{p\in P_l} y_{l,k}^{p,u}\frac{\partial c_{l,k}^{p,u}}{\partial y_{l',k'}^{p',u'}},
\end{aligned} \tag{13.22}$$

where the first two terms are the cost of the path and the third term describes the travel time cost change due to the impact on the link travel time based on the dynamics of the traffic system. The fourth term accounts for the change of energy cost associated with the changes in link volume and can be calculated approximately using the traffic network states from the simulator. According to the derivative chain rule and Eq. (13.7),

$$\frac{\partial t_{l,k}^{p,u}}{\partial y_{l',k'}^{p',u'}} = \sum_{n_p=1}^{N_p} \frac{\partial \hat{w}_{p,n_p,k}}{\partial y_{l',k'}^{p',u'}} = \sum_{n_p=1}^{N_p} \frac{\partial \hat{w}_{p,n_p,k}}{\partial \hat{z}_{p,n_p,k}}\frac{\partial \hat{z}_{p,n_p,k}}{\partial y_{l',k'}^{p',u'}}, \tag{13.23}$$

where $\hat{z}_{p,n_p,k}$ is the traffic volume of the link e_{p,n_p} on path p with the path departure time being $\xi_{k,e_{p,np}}$. The term $\frac{\partial \hat{w}_{p,np,k}}{\partial \hat{z}_{p,np,k}}$ represents the travel time change in link e_{p,n_p} at time $\xi_{k,e_{p,np}}$ caused by changing the link volume by one unit. One of the most commonly used relationships between link volume and travel time is the Bureau of Public Roads (BPR) function [48]:

$$w_e = t_f\left(1+\alpha_e\left(\frac{z_e}{\text{cap}_e}\right)^{\beta_e}\right), \tag{13.24}$$

where w_e is the link travel time, t_f is the link free-flow travel time, z_e is the vehicle volume on link e, and cap_e is the road link capacity. α_e and β_e are parameters for the model and can be estimated through historical traffic data. Then the link travel time derivative $\frac{\partial \hat{w}_{p,np,k}}{\partial \hat{z}_{p,np,k}}$ based on Eq. (13.24) can be written as follows:

$$\frac{\partial \hat{w}_{p,n_p,k}}{\partial \hat{z}_{p,n_p,k}} = \frac{\alpha_{e_{p,np}}\beta_{e_{p,np}}t_f\hat{z}_{p,n_p,k}^{\beta_{e_{p,np}}-1}}{\text{cap}_{e_{p,np}}} \equiv B_{p,n_p,k}\hat{z}_{p,n_p,k}^{\beta_{e_{p,np}}-1}. \tag{13.25}$$

The calculation of the term $\frac{\partial \hat{z}_{p,np,k}}{\partial y_{l',k'}^{p',u'}}$, in the view of the service network level, can be estimated in an aggregated way. We introduce $\mathbb{1}_{e',t'}(e,t)$ be the indicator equal to 1 when $e = e^{'}$ and $t = t^{'}$, and 0 otherwise. Using $v_{p,n_p,k}$ to represent the speed of the n_p-th link in path p at time $\xi_{k,e_{p,n_p}}$, we get

$$\frac{\partial \hat{z}_{p,n_p,k}}{\partial y^{p^{'},u^{'}}_{l^{'},k^{'}}} \approx \mathbb{1}_{e^{'}_{p,n_p},\xi_{k^{'},e^{'}_{p,n_p}}}(e_{p,n_p},\xi_{k,e_{p,n_p}})\frac{1}{v_{p,n_p,k}\Delta t}, \tag{13.26}$$

where $e^{'}_{p,n_p} \equiv e_{p^{'},n_{p^{'}}}$. With Eqs. (13.23)–(13.26), we get

$$\frac{\partial t^{p,u}_{l,k}}{\partial y^{p^{'},u^{'}}_{l^{'},k^{'}}} \approx \sum_{n_p=1}^{N_p} \frac{\mathbb{1}_{e^{'}_{p,n_p},\xi_{k^{'},e^{'}_{p,n_p}}}(e_{p,n_p},\xi_{k,e_{p,n_p}})}{v_{p,n_p,k}\Delta t}\frac{\partial \hat{w}_{p,n_p,k}}{\partial \hat{z}_{p,n_p,k}}. \tag{13.27}$$

The term $\frac{\partial c^{p,u}_{l,k}}{\partial y^{p^{'},u^{'}}_{l^{'},k^{'}}}$ is obtained by applying the chain rule as

$$\frac{\partial c^{p,u}_{l,k}}{\partial y^{p^{'},u^{'}}_{l^{'},k^{'}}} = \sum_{n_p=1}^{N_p} \frac{\partial h^u(v_{p,n_p,k})}{\partial v_{p,n_p,k}} \cdot \frac{\partial v_{p,n_p,k}}{\partial \hat{w}_{p,n_p,k}}\frac{\partial \hat{w}_{p,n_p,k}}{\partial y^{p^{'},u^{'}}_{l^{'},k^{'}}} d_{p,n_p}, \tag{13.28}$$

where $h^u(v)$ is the energy consumption coefficient function for the trucks of type u and d_{p,n_p} is the length of link e_{p,n_p}. In this chapter, we assume that the energy consumption coefficient $h^u(v)$ satisfies a cubic polynomial function of the speed v:

$$h^u(v) = b^u_0 + b^u_1 v + b^u_2 v^2 + b^u_3 v^3, \tag{13.29}$$

where the parameters of the polynomial are estimated for both electric and diesel trucks using offline experiments demonstrated in the numerical evaluation section. The relationship between speed v and its travel time w is simply $v = \frac{d}{w}$, where d is the length of a road link. Then we obtain

$$\begin{aligned}\frac{\partial h^u(v_{p,n_p,k})}{\partial \hat{w}_{p,n_p,k}} &= \frac{\partial h^u(v_{p,n_p,k})}{\partial v_{p,n_p,k}}\frac{\partial v_{p,n_p,k}}{\partial \hat{w}_{p,n_p,k}} \\ &= -\frac{(b^u_1 + 2b^u_2 v_{p,n_p,k} + 3b^u_3 v^2_{p,n_p,k})d_{p,n_p}}{\hat{w}^2_{p,n_p,k}}.\end{aligned} \tag{13.30}$$

By combining Eqs. (13.22)–(13.30), the marginal cost of a path in a segment can be approximately calculated by

$$
\begin{aligned}
MCP_{l',k'}^{p',u'} &\approx c_{l',k'}^{p',u'} + \eta t_{l',k'}^{p',u'} \\
&+ \sum_{k\in K}\sum_{l\in L}\sum_{p\in P_l}\sum_{n_p=1}^{N_p} \mathbb{1}_{e'_{p,np},\xi_{k'},e'_{p,np}}(e_{p,n_p},\xi_{k,e_{p,np}}) \\
&\cdot y_{l,k}^{p,u}\frac{1}{v_{p,n_p,k}\Delta t}\left(\eta + \frac{\partial h^u(v_{p,n_p,k})}{\partial \hat{w}_{p,n_p,k}}d_{p,n_p}\right)B_{p,n_p,k}\hat{z}_{p,n_p,k}^{\beta_{e_{p,np}}-1} \\
&\forall l' \in L, p' \in P_{l'}, k' \in K, u' \in U.
\end{aligned}
\tag{13.31}
$$

References

1. Hooper A (2018) Cost of congestion to the trucking industry: 2018 update. American Transportation Research Institute, Atlanta, GA
2. Environmental Protection Agency (EPA) (2020) U.S. transportation sector greenhouse gas emissions, 1990–2018. U.S. Office of Transportation and Air Quality. https://nepis.epa.gov/Exe/ZyPDF.cgi?Dockey=P100ZK4P.pdf
3. Jourquin B, Beuthe M (1996) Transportation policy analysis with a geographic information system: the virtual network of freight transportation in europe. Transp Res Part C: Emerg Technol 4(6):359–371
4. Southworth F, Peterson BE (2000) Intermodal and international freight network modeling. Transp Res Part C: Emerg Technol 8(1–6):147–166
5. Nijkamp P, Reggianib A, Tsanga WF (1999) Comparative modelling of interregional transport flows. Res Memorandum 2
6. Arnold P, Peeters D, Thomas I (2004) Modelling a rail/road intermodal transportation system. Transp Res Part E: Log Transp Rev 40(3):255–270
7. Macharis C, Bontekoning YM (2004) Opportunities for or in intermodal freight transport research: a review. Eur J Oper Res 153(2):400–416
8. Russ BF, Yamada T, Castro JT, Yasukawa H (2005) Optimising the design of multimodal freight transport network in Indonesia. J East Asia Soc Transp Stud 6:2894–2907
9. Yamada T, Russ BF, Castro J, Taniguchi E (2009) Designing multimodal freight transport networks: a heuristic approach and applications. Transp Sci 43(2):129–143
10. Wyatt D (2020) Electric trucks 2020-2030: analysis of the medium and heavy-duty electric truck markets, with granular market forecasts for battery. https://www.idtechex.com/en/research-report/electric-trucks-2020-2030/710#:~:text=IDTechEx%20forecast%20the%20penetration%20of,to%20be%209.4%25%20by%202030
11. Gendreau M, Potvin J-Y, Bräumlaysy O, Hasle G, Løkketangen A (2008) Metaheuristics for the vehicle routing problem and its extensions: a categorized bibliography. In: The vehicle routing problem: latest advances and new challenges. Springer, pp 143–169
12. Ropke S, Cordeau J-F (2009) Branch and cut and price for the pickup and delivery problem with time windows. Transp Sci 43(3):267–286
13. Baldacci R, Mingozzi A (2009) A unified exact method for solving different classes of vehicle routing problems. Math Program 120(2):347–380
14. Ceselli A, Righini G, Salani M (2009) A column generation algorithm for a rich vehicle-routing problem. Transp Sci 43(1):56–69
15. Baldacci R, Bartolini E, Mingozzi A, Roberti R (2010) An exact solution framework for a broad class of vehicle routing problems. CMS 7(3):229–268
16. Vidal T, Crainic TG, Gendreau M, Prins C (2013) Heuristics for multi-attribute vehicle routing problems: a survey and synthesis. Eur J Oper Res 231(1):1–21

17. Toth P, Vigo D (2014) Vehicle routing: problems, methods, and applications. SIAM
18. Bettinelli A, Ceselli A, Righini G (2014) A branch-and-price algorithm for the multi-depot heterogeneous-fleet pickup and delivery problem with soft time windows. Math Program Comput 6(2):171–197
19. De Giovanni L, Gastaldon N, Sottovia F (2019) A two-level local search heuristic for pickup and delivery problems in express freight trucking. Networks 74(4):333–350
20. Beckmann MJ, McGuire CB, Winsten CB (1955) Studies in the economics of transportation
21. Guelat J, Florian M, Crainic TG (1990) A multimode multiproduct network assignment model for strategic planning of freight flows. Transp Sci 24(1):25–39
22. Castelli L, Pesenti R, Ukovich W (2004) Scheduling multimodal transportation systems. Eur J Oper Res 155(3):603–615
23. Ham H, Kim TJ, Boyce D (2005) Implementation and estimation of a combined model of interregional, multimodal commodity shipments and transportation network flows. Transp Res Part B: Methodol 39(1):65–79
24. Zografos KG, Androutsopoulos KN (2008) Algorithms for itinerary planning in multimodal transportation networks. IEEE Trans Intell Transp Syst 9(1):175–184
25. Moccia L, Cordeau J-F, Laporte G, Ropke S, Valentini MP (2011) Modeling and solving a multimodal transportation problem with flexible-time and scheduled services. Networks 57(1):53–68
26. Crainic TG, Gobbato L, Perboli G, Rei W (2016) Logistics capacity planning: a stochastic bin packing formulation and a progressive hedging meta-heuristic. Eur J Oper Res 253(2):404–417
27. Crainic TG, Ricciardi N, Storchi G (2004) Advanced freight transportation systems for congested urban areas. Transp Res Part C: Emerg Technol 12(2):119–137
28. Jiang N, Xie C, Waller S (2012) Path-constrained traffic assignment: model and algorithm. Transp Res Record: J Transp Res Board 2283:25–33
29. Bahrami S, Aashtiani HZ, Nourinejad M, Roorda MJ (2017) A complementarity equilibrium model for electric vehicles with charging. Int J Transp Sci Technol 6(4):255–271
30. Faridimehr S, Venkatachalam S, Chinnam RB (2018) A stochastic programming approach for electric vehicle charging network design. IEEE Trans Intell Transp Syst 99:1–13
31. Peeta S, Mahmassani HS (1995) System optimal and user equilibrium time-dependent traffic assignment in congested networks. Ann Oper Res 60(1):81–113
32. Mahmassani HS, Zhang K, Dong J, Lu C-C, Arcot VC, Miller-Hooks E (2007) Dynamic network simulation-assignment platform for multiproduct intermodal freight transportation analysis. Transp Res Rec 2032(1):9–16
33. Zhou X, Mahmassani HS, Zhang K (2008) Dynamic micro-assignment modeling approach for integrated multimodal urban corridor management. Transp Res Part C: Emerg Technol 16(2):167–186
34. Lu C-C, Mahmassani HS, Zhou X (2009) Equivalent gap function-based reformulation and solution algorithm for the dynamic user equilibrium problem. Transp Res Part B: Methodol 43(3):345–364
35. de Cea Ch J, Soto A et al (2003) A multi-modal supply–demand equilibrium model for predicting intercity freight flows. Transp Res Part B: Methodol 37(7):615–640
36. Li Z-C, Lam WH, Wong S, Sumalee A (2010) An activity-based approach for scheduling multimodal transit services. Transportation 37:751–774
37. Ma T-Y, Lebacque J-P (2013) Dynamic system optimal routing in multimodal transit network. Transp Res Rec 2351(1):76–84
38. Verbas Ö, Mahmassani HS, Hyland MF (2016) Gap-based transit assignment algorithm with vehicle capacity constraints: simulation-based implementation and large-scale application. Transp Res Part B: Methodol 93:1–16
39. Chen P, Ioannou P, Dessouky M (2021) Mixed freight dynamic routing using a co-simulation optimization approach. IEEE Trans Intell Transp Syst
40. Zhao Y, Ioannou PA, Dessouky MM (2018) Dynamic multimodal Freight Routing using a co-simulation optimization approach. IEEE Trans Intell Transp Syst

41. PTV Group (2010) VISUM 11.5. https://www.ptvgroup.com/en/solutions/products/ptv-visum/
42. Chen C (2003) Freeway performance measurement system (PeMS). California Partners for Advanced Transportation Technology, UC Berkeley
43. Google (2020) Google Maps. https://www.google.com/maps
44. Kawamura K (2000) Perceived value of time for truck operators. Transp Res Rec 1725(1):31–36
45. Zhang Y, Ioannou PA (2016) Environmental impact of combined variable speed limit and lane change control: a comparison of moves and cmem model. IFAC-PapersOnLine 49(3):323–328
46. Koutsoupias E, Papadimitriou C (2009) Worst-case equilibria. Comput Sci Rev 3(2):65–69
47. Papadopoulos A-A, Kordonis I, Dessouky M, Ioannou P (2018) Coordinated freight routing with individual incentives for participation. IEEE Trans Intell Transp Syst 20(9):3397–3408
48. U. B. of Public Roads (1964) Office of Planning. Urban Planning Division. Traffic assignment manual for application with a large, high speed computer. US Department of Commerce

Chapter 14
A Control Framework for Socially Optimal Emerging Mobility Systems

Andreas A. Malikopoulos

Abstract Connected and automated vehicles (CAVs) provide the most intriguing opportunity for enabling users to significantly improve safety and transportation efficiency by monitoring network conditions and making better operating decisions. CAVs, however, could alter the tendency to travel, eventually leading to a high traffic demand and causing rebound effects (e.g., increasing vehicle miles traveled). This chapter provides a control framework to distribute travel demand in a given transportation network, resulting in a socially optimal mobility system that travelers would be willing to accept. A "socially optimal mobility system" implies a mobility system that (1) is efficient (in terms of energy consumption and travel time), (2) mitigates rebound effects, and (3) ensures equity in transportation.

14.1 Introduction

Emerging mobility systems, such as connected and automated vehicles (CAVs) and shared mobility, provide significant opportunities to improve safety and reduce pollution, energy consumption, and travel delays [22, 79]. CAVs are typical cyber-physical systems where the cyber component (e.g., data and shared information through vehicle-to-vehicle and vehicle-to-infrastructure communication) can aim at optimally controlling the physical entities (e.g., CAVs and non-CAVs). The cyber-physical nature of such emerging mobility systems requires large quantities of shared information through vehicle-to-vehicle and vehicle-to-infrastructure communication, which, in turn, are associated with significant technical challenges. This has given rise to a new level of complexity [45] in modeling and control [29]. It is expected that CAVs will gradually penetrate the market and interact with human-driven vehicles in ways that will improve safety and transportation efficiency over the next several years [64, 77, 80]. However, different levels of vehicle automation in

A. A. Malikopoulos (✉)
Cornell University, 527 College Avenue Ithaca, Ithaca, NY 14853-3501, US
e-mail: amaliko@cornell.edu

P. Ioannou and A. A. Malikopoulos (eds.), *Transportation Mobility in Smart Cities*, Springer Tracts on Transportation and Traffic 21,
https://doi.org/10.1007/978-3-031-64769-7_14

the transportation network can have differing impacts on different transportation efficiency metrics [76] ranging from an improvement of 45% to a deterioration of 60% as compared to the baseline of no automation. Moreover, we anticipate that efficient transportation and travel cost reduction might alter human travel behavior, causing rebound effects. For example, one impact of improving efficiency is a decrease in travel cost and a subsequent increase in the willingness to travel. The latter would increase overall vehicle miles traveled, which, in turn, might negate the benefits in terms of energy and travel time.

While several studies have shown the benefits of emerging mobility systems to reduce energy and alleviate traffic congestion in specific transportation scenarios, one key question that still remains unanswered is "How can we develop a mobility system that can enhance accessibility, safety, and equity in transportation without causing rebound effects, while also gaining the travelers' acceptance?" This chapter aims to address this question using a control framework with two attributes:

1. Distribute travel demand in a given transportation network, resulting in a socially optimal mobility system.
2. All travelers will willingly accept this distribution of travel demand.

In our exposition, the notion of a "socially optimal mobility system" implies that the mobility system is efficient (in terms of energy consumption and travel time), mitigates rebound effects, and ensures equity in transportation. Specifically, the control framework involves the following two research steps:

Step 1: Aggregate the preferences of the travelers into a collective, system-wide set of recommendations, e.g., routing choices, modes of transportation, while the private information of the travelers is not publicly known; and

Step 2: Develop control technologies allowing CAVs to navigate automatically and to co-exist with human-driven vehicles (HDVs) safely and efficiently in a mixed-traffic environment.

Step 1 will identify the new congestion patterns of an optimized mobility system. In contrast, Step 2 will improve transportation efficiency under safety guarantees in response to the new levels of imposed travel demand. More specifically, the two research steps address the following challenges: In Step 1, we formulate and solve an optimization problem in which the decision variables are the optimal routes and the selection of a transportation mode for all travelers so as to maximize a social utility function. A suitable approach to address this problem is considering a decentralized traveler decomposition using mechanism design theory [35]. In mechanism design, we are concerned with implementing system-wide optimal solutions to problems involving multiple agents—in this case, travelers—each with private information about preferences, e.g., individual tolerance to traffic delay, the value of time and money, and preferred travel time. In this context, a social planner faces the problem of aggregating the travelers' preferences into a collective, system-wide decision when the travelers' private information is not publicly known. Thus, mechanism design entails the social planner solving an optimization problem with incomplete information.

In Step 2, we need to have an optimal coordination framework which, in conjunction with large amounts of data from vehicles and the infrastructure, will improve safety and efficiency in a mixed traffic environment consisting of CAVs and human-driven vehicles. In particular, the coordination framework will be used by CAVs to navigate optimally (in the sense of minimizing travel times and energy consumption) while also guaranteeing safety in different traffic settings, e.g., crossing a signal-free intersection without stopping, merging at roadways or a roundabout, and executing automated passing maneuvers. This step will ensure that CAVs co-exist and interact safely with other human-driven vehicles and pedestrians.

The one feature that sharply distinguishes the proposed framework from previous approaches reported in the literature to date is that it considers simultaneous optimization of the travel demand (Step 1) and transportation network efficiency (Step 2).

14.1.1 A Brief Review of Mechanism Design

Mechanism design theory is concerned with implementing system-wide optimal solutions to problems involving multiple agents, each with private information about preferences. It can be viewed as the art of designing the rules of a game to achieve a specific desired outcome. A popular example of mechanism design, taken from [2], is a cake-cutting problem involving a mother with two kids. The mother must design a mechanism to make her kids share a cake equally. The centralized solution of this problem is for the mother to slice the cake equally and give a slice to each kid. The decentralized mechanism design solution is to (1) allow one kid to slice the cake into two pieces and (2) allow the other kid to determine who gets which piece. This mechanism achieves the desired outcome of the kids sharing the cake equally without the mother's intervention. In this example, the mother is called the "social planner." In the context of mechanism design, the social planner typically faces the problem of aggregating the preferences of multiple agents into a collective, system-wide decision when the agents' private information might not be publicly known. Thus, mechanism design entails the social planner to solve an optimization problem with incomplete information. The underlying structure used is to induce a game among the agents so that the desired system-wide solution is implemented in an equilibrium of the induced game.

Numerous economists and mathematicians have rigorously studied the mechanism design theory since the 1950s to provide insights and solutions to different economic topics. The theory started with the seminal contributions of Leonid Hurwicz and Jacob Marschak, who were interested in resource allocation problems and controlling (through incentives) individual agents. In parallel, Kenneth Arrow, Gérard Debreu, and Herbert A. Simon also worked on problems with incomplete information and how to bound rationality. In 1961, William Vickrey's seminal work on auctions was published, paving the way for Hurwicz's theoretical framework to be applied to design incentives based on the agents' information for a simple yet formidable problem of an auction. Arrow and Debreu's work was instrumental in establishing the interconnected relation of information and decision-making and its role in influencing behavior in the efficient allocation of limited resources. Much later in the 1970s and the 1980s, Peter Diamond, Oliver Hart, Jean-Jacques Laffont, Eric Maskin, James Mirrlees, and Sherwin Rosen worked independently on principal-agent problems focusing on how one can design a contract between a principal (e.g., institution corporation) and a rational agent efficiently. As a continuation of Vickrey's work, Ronald Coase, Jerry R. Green, Theodore Groves, and John Ledyard made significant contributions to the design of incentives for public goods problems. Furthermore, Roger Myerson, Paul Milgrom, and Robert Wilson expanded the Vickrey auction to address more complicated scenarios and complex problems. The theory of mechanism design represents the confluence of microeconomics [4, 19, 31–34, 41, 53, 55–57, 61], and social choice theory [1, 2], while it equally draws from auction [74], optimization [5, 6], and game theory [21, 23, 54, 63, 65, 72].

14.2 Outline of the Control Framework

It is expected that CAVs will gradually penetrate the market and disturb travelers' behaviors along with their mobility tendencies, resulting in unintended consequences (rebound effects). These consequences may be additional energy use and greenhouse gas emissions, challenging equity in transportation, and significant alterations in the density of urban areas [3, 28, 42]. Consequently, we are compelled to reassess the relationship between mobility and social life [10, 67] and provide solutions that consider the decision-making processes of travelers.

Our aim is to develop a holistic and rigorous framework to capture the societal impact of emerging mobility systems [16] and provide solutions that mitigate any potential rebound effects, e.g., increased vehicle miles traveled, increased travel demand, and empty vehicle trips, while enhancing accessibility, safety, and equity in transportation. In our approach, we consider a finite set of travelers who seek to travel in a given transportation network of a city (or a big metropolitan area) where a central authority (the social planner in our exposition) seeks to ensure the efficient allocation and operation of the different modes of transportation available in the transportation network of the city. We call these different modes "mobility services." A few examples are privately owned CAVs and human-driven vehicles, shared mobility vehicles (e.g., Uber, Lyft), bicycles, and public transit (e.g., train, bus, and light rail). The travelers make a request via a smartphone app (Fig. 14.1) to use a service to satisfy their mobility needs, i.e., desired origin–destination. The social planner, which can be visualized as a central computer, compiles all travelers' origin–destination requests and other information, e.g., preferred travel time, value of time, to provide a travel recommendation to each traveler with the goal of achieving a socially optimal mobility system. In this context, the social planner formulates and solves an optimization problem in which the decision variables are the mobility services that maximize the travelers' social welfare (utilities). A suitable approach

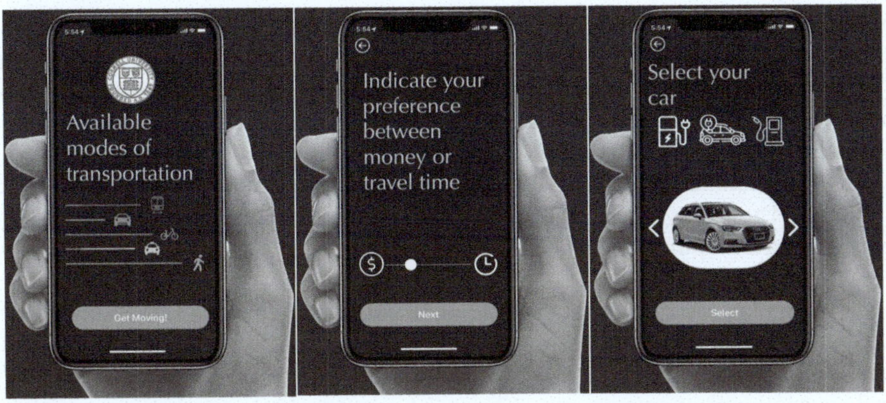

Fig. 14.1 The app of the proposed framework

to address this problem is to consider a decentralized decomposition of the travelers using mechanism design theory [11, 35]. In mechanism design, we are concerned with how to implement system-wide optimal solutions to problems involving multiple agents—in this case, travelers—each with private information. Thus, mechanism design entails the social planner solving an optimization problem with incomplete information.

The social planner's collective recommendations must achieve three objectives: (1) respect and satisfy the travelers' preferences, (2) guarantee equity in travel recommendations, and (3) ensure that no mobility system will become congested. We assume the city supports connected and automated mobility technologies on public transit infrastructure and a transportation network. Consequently, the social planner is fully aware of the system's capabilities and the network's capacity. In other words, the social planner is fully capable of computing the maximum capacity of each mobility service and the associated costs aimed at providing travel recommendations to all travelers. The social planner's objective is to design appropriate monetary incentives, e.g., tolls, fares, subsidies, to guarantee the realization of the desired outcome, i.e., maximize the social welfare of all travelers. The travelers will, in turn, accept or reject the social planner's recommendations.

14.3 Socially Optimal Management of Travel Demand

As we move to increasingly complex emerging mobility systems with an expanded feature space, fundamentally new approaches are needed to understand the impact on system behavior [44]. The approaches reported in the literature to date have considered emerging mobility systems without deliberating on human decision-making and perception. To develop and operate a socially optimal mobility system, technological and information management innovations need to be integrated with the social dimensions to ensure adoption by the drivers, travelers, and the public.

The standard approach to alleviate congestion in transportation has been the management of travel demand. Some approaches have considered congestion pricing/tolling [75], while others have considered the application of mechanism design to provide a solution to individual route selection under different congestion traffic scenarios. The theory of mechanism design was developed for the implementation of system-wide optimal solutions to problems involving multiple rational agents, each with private information and conflicting interests [52]. It can be viewed as the art of designing the rules of a game to achieve a specific desired outcome. Mechanism design has broad applications spanning different fields, including microeconomics, social choice theory, and control engineering. Applications in engineering include communication networks [62], social networks [24], transportation routing [9], online advertising [38], smart grid [66], multi-agent systems [68], and resource allocation problems [81]. Due to the economic nature of congestion in transportation, auctioning has also been proposed [36] to create a market of tolls in a network of roads. Auctions are processes for allocating goods among bidders, so the challenge of auction design can only be understood by studying the demands of the participants [58]. Auction design has been the focus of significant results on multi-object auctions and matching market problems [27, 30]. On the one hand, auctions have been proposed to design pricing schemes with tolls in a network of roads, leading to a spark of studies in auctioning techniques [36, 59, 60, 71, 73]. On the other hand, this approach has significant limitations: (1) the implementability of auction-based tolling on highways is not straightforward due to the dynamic and fast-changing nature of transportation systems; (2) it is also uncertain how the public, e.g., travelers, passengers, and drivers, will respond concerning toll roads in an auction setting. Understanding the travelers' interests, i.e., willingness-to-pay, the value of time, and the impacts on different sociodemographic groups, becomes imperative for a socially efficient design of an emerging mobility system.

In our approach, we seek to "design" a socially optimal mobility system that assigns mobility services to a finite group of travelers by considering their personal travel preferences. By a socially optimal mobility system, we mean a mobility system that is (1) efficient (in terms of energy consumption and travel time), (2) mitigates rebound effects, and (3) ensures equity in transportation. We want to ensure that the emerging mobility system is incentive compatible (travelers always report their personal travel preferences truthfully), individually rational (travelers always benefit from voluntarily participating in the system), and weakly budget balanced (the system always generates revenue from each traveler that can be used for maintenance of the road network, etc.).

Our mobility system is managed by a social planner who aims to allocate $m \in \mathbb{N}$ mobility services to $n \in \mathbb{N}$ travelers, where $n \geq m$. We denote the nonempty set of travelers by $\mathcal{I} = \{1, 2, \dots, n\}$ and the nonempty set of mobility services by $\mathcal{J} = \{1, 2, \dots, m\}$. For example, each mobility service $j \in \mathcal{J}$ can either represent privately owned CAVs and HDVs, shared mobility vehicles (e.g., Uber, Lyft), bicycles, and public transit (e.g., train, bus, light rail). For our purposes, we can think of

$\mathcal{G} = (\mathcal{V}, \mathcal{E})$ representing a smart city network with a road and public transit infrastructure. Next, a traveler $i \in \mathcal{I}$ seeks to travel using these mobility services in a transportation network from their current location $o_i \in \mathcal{V}$ to their desired destination $d_i \in \mathcal{V}$. So, each traveler $i \in \mathcal{I}$ is associated with the origin–destination pair (o_i, d_i) represented by an undirected multigraph $\mathcal{G} = (\mathcal{V}, \mathcal{E})$, where each node in $\mathcal{V}$ represents a different neighborhood, and each link $e \in \mathcal{E}$ represents a sequence of city roads and a public transit connection. On the other hand, each mobility service $j \in \mathcal{J}$ is associated with several different links. Therefore, we do not have to limit the number of mobility services that connect any origin o_i to any destination d_i. Furthermore, each traveler $i \in \mathcal{I}$ seeks to travel in the network $\mathcal{G}$ with only one mobility service $j \in \mathcal{J}$ that satisfies their origin–destination pair (o_i, d_i), while each service $j \in \mathcal{J}$ can be used by multiple travelers.

The traveler–service assignment is a vector $\mathbf{a} = (a_{11}, \ldots, a_{ij}, \ldots, a_{nm}) = (a_{ij})_{i \in \mathcal{I}, j \in \mathcal{J}}$, where a_{ij} is a binary variable of the form: $a_{ij} = 1$, if $i \in \mathcal{I}$ is assigned to $j \in \mathcal{J}$, and $a_{ij} = 0$, otherwise. The set of travelers with the same origin–destination pair is

$$\mathcal{I}_k = \{i \in \mathcal{I} \mid (o_i, d_i) = (o_k, d_k)\}, \; k = 1, 2, \ldots, N, \tag{14.1}$$

where $N \in \mathbb{N}$ is the number of sub-classes over the complete set of travelers, i.e., $\mathcal{I} = \bigcup_{k=1}^{N} \mathcal{I}_k$. The justification of the introduction of $\mathcal{I}_k$ is that in an emerging mobility system, we can acquire verifiable location data of travelers either by using a global positioning system or estimating the average number of travelers using public transit [20]. By partitioning the set of travelers in $N \in \mathbb{N}$ sub-classes, the traveler–service assignment of sub-class $\mathcal{I}_k$ is given by $\mathbf{a}_k = (a_{ij})_{i \in \mathcal{I}_k, j \in \mathcal{J}}$. Since each mobility service $j \in \mathcal{J}$ can be utilized by multiple travelers simultaneously, we introduce a metric of "co-travelers." Hence, for any traveler, $i \in \mathcal{I}_k$, $k = 1, \ldots, N$, the number of co-travelers having the same origin–destination pair and using the same service $j \in \mathcal{J}$ is computed by

$$\psi_i(\mathbf{a}_k) = \sum_{\ell \in \mathcal{I}_k} a_{\ell j} - a_{ij}. \tag{14.2}$$

A traveler $i \in \mathcal{I}_k$, $k = 1, \ldots, N$, is characterized by a tuple of personal travel preferences, denoted by π_i, and given by $\pi_i = (\theta_i, \eta_{ij}, \delta_{ij})$, where $\theta_i \in \mathbb{R}_{\geq 0}$ is the preferred travel time, $\eta_{ij} \in \mathbb{N}$ is the maximum preferred number of co-travelers with the mobility service j, and $\delta_{ij} \in [0, 1]$ is a monetary value that traveler i is willing to pay (or accept as a compensation) when using mobility service j. Naturally, a traveler i's preferred travel time θ_i is a non-negative real number and represents how fast traveler $i \in \mathcal{I}_k$ wishes to reach their destination. Similarly, traveler i's preferred number of co-travelers $\eta_{ij} \in \mathbb{N}$ represents the maximum tolerable number of other travelers using mobility service j, and δ_{ij} represents what potentially drives a traveler's behavior, i.e., the value of time. In particular, δ_{ij} represents a monetary value that traveler $i \in \mathcal{I}_k$ is willing to pay to save time (or accept as compensation) for service $j \in \mathcal{J}$. For each traveler $i \in \mathcal{I}_k$, π_i is considered private information, known only to traveler $i \in \mathcal{I}_k$. Hence, $(\pi_i)_{i \in \mathcal{I}_k}$, $k = 1, \ldots, N$ are unknown information to the social planner. This is one of the key challenges in the proposed mobility system, i.e., "how do we incentivize the travelers to be truthful and elicit the private information needed to provide a socially optimal solution for the whole system?" The answer to this question is given in Sect. 14.3.1.

Next, we introduce an "inconvenience" metric for any traveler $i \in \mathcal{I}_k$ using any mobility service $j \in \mathcal{J}$. Quantitatively, the inconvenience metric can represent the extra monetary value of travel dis-utility from any costs, travel delays, or violation of personal preferences caused by using the mobility service $j \in \mathcal{J}$. The mobility inconvenience metric for traveler $i \in \mathcal{I}_k$, $k = 1, \ldots, N$, assigned to service $j \in \mathcal{J}$ is a continuous function

$$\phi_i(\pi_i, \tilde{\theta}_i(\mathbf{a}_k), \psi_i(\mathbf{a}_k)) \in \mathbb{R}_{\geq 0}, \tag{14.3}$$

where π_i is the tuple of the personal travel preferences of $i \in \mathcal{I}_k$, $\tilde{\theta}_i(\mathbf{a}_k) \in \mathbb{R}_{\geq 0}$ is the experienced travel time, and $\psi_i(\mathbf{a}_k)$ is the number of co-travelers. Note that the mobility inconvenience metric ϕ_i strictly increases when $\tilde{\theta}_i(\mathbf{a}_k)$ and/or $\psi_i(\mathbf{a}_k)$ increases. That is because, from a modeling perspective, traveling with time delays or during peak times can cause significant inconveniences to any traveler $i \in \mathcal{I}_k$.

Next, a traveler's satisfaction is captured by a valuation function $v_i(\mathbf{a}_k)$, which can reflect the traveler's willingness-to-pay for their travel, i.e.,

$$v_i(\mathbf{a}_k) = \bar{v}_i - \phi_i\left(\pi_i, \tilde{\theta}_i(\mathbf{a}_k), \psi_i(\mathbf{a}_k)\right), \tag{14.4}$$

where $\bar{v}_i \in \mathbb{R}_{>0}$ is the value gained by traveler $i \in \mathcal{I}_k$ when their origin–destination pair (o_i, d_i) is satisfied using mobility service $j \in \mathcal{J}$ without any travel delays or travel inconveniences, i.e., $\tilde{\theta}_i(\mathbf{a}_k) = \theta_i$ and $\psi_i(\mathbf{a}_k) = 0$. We call $\bar{v}_i$ the *maximum willingness-to-pay*. Naturally, for any traveler $i \in \mathcal{I}_k$ and any service $j \in \mathcal{J}$, we have $v_i(\mathbf{a}_k) \in [0, \bar{v}_i]$, where $v_i(\mathbf{a}_k) = 0$ means that traveler $i \in \mathcal{I}_k$ is unwilling to utilize mobility service $j \in \mathcal{J}$. Although our analysis will treat the satisfaction function $v_i(\mathbf{a}_k)$ in its most general form, given by (14.4), one can explicitly define it as follows:

$$v_i(a_{ij}) = \begin{cases} \bar{v}_i, & \text{if } \phi_i = 0, \\ \lambda_i \cdot \bar{v}_i, & \text{if } \phi_i = (1 - \lambda_i) \cdot \bar{v}_i, \\ 0, & \text{if } \phi_i = \bar{v}_i, \end{cases} \tag{14.5}$$

where $\lambda_i \in (0, 1)$ is a discount rate. When $\phi_i = 0$, we say that traveler i's personal travel requirements π_i are satisfied using the CAV mobility service without any co-travelers. When $\phi_i = \bar{v}_i$, then we say that traveler i's personal travel requirements are not satisfied. Lastly, when $\phi_i = (1 - \lambda_i) \cdot \bar{v}_i$, we say that traveler i's personal travel requirements π_i are satisfied using any mobility service.

The total utility $u_i(\mathbf{a}_k)$ of traveler $i \in \mathcal{I}_k$, $k = 1, \ldots, N$, is given by

$$u_i(\mathbf{a}_k) = v_i(\mathbf{a}_k) - p_i(\mathbf{a}_k), \tag{14.6}$$

where $v_i(\mathbf{a}_k)$ is the maximum willingness-to-pay and $p_i(\mathbf{a}_k)$ is the mobility payment traveler $i \in \mathcal{I}_k$ is required to make to use service $j \in \mathcal{J}$ (e.g., pay road tolls or buy a public transit ticket). The operating cost of service $j \in \mathcal{J}$, denoted by $r_j \in \mathbb{R}_{>0}$, is given by

$$r_j(\mathbf{a}_k) = \sum_{i \in \mathcal{I}_k} r_{ij}(a_{ij}), \tag{14.7}$$

where $r_{ij}(a_{ij}) \in \mathbb{R}_{>0}$ is traveler i's corresponding share of the operating cost of vehicle $j \in \mathcal{J}$. Intuitively, the operating cost r_{ij} captures traveler i's fair share of the costs of mobility service $j \in \mathcal{J}$. These costs can be associated with fuel/energy consumption, drivers' labor reimbursement, and environmental impact. Moreover, the operating cost r_{ij} can be thought as the minimum acceptable payment, i.e., $p_i \geq r_{ij}$, by traveler $i \in \mathcal{I}_k$ using service $j \in \mathcal{J}$.

In the modeling framework described above, we impose the following assumptions:

Assumption 14.1 For all sub-classes $\mathcal{I}_k$, $k = 1, \ldots, N$, $N \in \mathbb{R}$, any traveler $i \in \mathcal{I}_k$ is modeled as a selfish decision-maker with private information $\pi_i = (\theta_i, \eta_{ij}, \delta_{ij})$. Traveler i's objective is to maximize their total utility

$$u_i(\mathbf{a}_k) = v_i(\mathbf{a}_k) - p_i(\mathbf{a}_k). \tag{14.8}$$

Assumption 14.2 For any sub-class $\mathcal{I}_k$, $k = 1, \ldots, N$, and for any traveler $i \in \mathcal{I}_k$, the maximum willingness-to-pay realized from using any mobility service $j \in \mathcal{J}$ must outweigh the operating costs, i.e., for all $i \in \mathcal{I}_k$, we have $\bar{v}_i > r_{ij}(a_{ij})$, where $r_{ij}(a_{ij})$ is the specific operating cost imposed by traveler $i \in \mathcal{I}_k$ using mobility service $j \in \mathcal{J}$.

The first assumption essentially indicates that each traveler is selfish in the sense that they are only interested in their own well-being. In economics, such behavior is called "strategic" since agents attempt to misreport or lie about their private information to the social planner if that means higher individual benefits. The second assumption implies it is always beneficial to travel when it can be guaranteed that there will be no travel delays or inconveniences.

In the proposed framework, travelers request (via a smartphone app) in advance a travel recommendation from the social planner that satisfies their origin–destination. Given the travelers' origin–destination pairs, the social planner distributes all travelers to different sub-classes. Thus, travelers from the same neighborhood have the same origin. Similarly, travelers going to the same neighborhood have the same destination. The social planner's task is to satisfy all travel requests and provide recommendations to the travelers, e.g., which mobility service to use. Hence, we are interested in minimizing the travel inconvenience of all travelers and the operating costs, which is equivalent to maximizing the utility of each traveler. Thus, the social planner formulates the following optimization problem for each sub-class $\mathcal{I}_k$, $k = 1, \ldots, N$,

$$J_1(\mathbf{a}_k) = \min_{\mathbf{a}_k}[\omega_1 \sum_{i \in \mathcal{I}_k} \phi_i(\pi_i, \tilde{\theta}_i(\mathbf{a}_k), \psi_i(\mathbf{a}_k)) + \omega_2 \sum_{j \in \mathcal{J}} r_j(\mathbf{a}_k)], \tag{14.9}$$

where ω_1 and ω_2 are factors that normalize the terms. The problem is subject to the following constraints: (1) each traveler $i \in \mathcal{I}_k$ is assigned at most one mobility service, (2) service j must not exceed its maximum usage capacity, and (3) the traveler's assignment impose equity in transportation [18] using the mobility equity metric [7].

14.3.1 Desired Properties of the Mobility System

The problem in (14.9) is a mixed-integer programming model, and standard algorithmic approaches exist to find its global optimal solutions or, in worst-case scenarios, their approximations. Note, though, that these approaches assume complete information of all parameters and variables in the model. Such an assumption is unreasonable to expect from strategic decision-makers. Thus, in our framework, travelers are not expected to report their private information truthfully. This turns our problem into a preference elicitation problem. Next, we discuss how we can elicit the necessary private information of all travelers using monetary incentives in the form of mobility payments (e.g., tolls, fares, fees).

Since $(\pi_i)_{i \in \mathcal{I}_k}$, $k = 1, \ldots, N$, are unknown information for the social planner, the question becomes "How do we incentivize the travelers to be truthful and elicit the private information needed to provide a socially optimal solution for the whole system?" We employ the mobility payments $p_i(\mathbf{a}_k)$ reported in [17] for each traveler $i \in \mathcal{I}_k$ to look similar to the celebrated Vickrey–Clarke–Groves (VCG) mechanism [19, 32, 74]. Our payments are significantly different within our context as we have considered the travelers' personal travel preferences and introduced the capacity constraints for each mobility service along with the equity in transportation constraint. In contrast, the VCG mechanism does not have any constraints, and thus, the proposed mobility system is considerably different from the VCG mechanism.

Therefore, a direction for future research should show that the mobility system satisfies (1) incentive compatibility and (2) individual rationality despite our departure from VCG. Incentive compatibility means that all travelers are incentivized to report their personal travel preferences truthfully regardless of what other travelers report. Individual rationality implies that all travelers voluntarily participate in the mobility system in the most potent form. Informally, we compare the utility of a traveler $i \in \mathcal{I}_k$ under two possible scenarios: traveler $i \in \mathcal{I}_k$ participates in the mobility market, and traveler $i \in \mathcal{I}_k$ rejects any travel recommendations from the social planner and simply uses their self-owning vehicle (CAV or conventional vehicle). The mobility system is individually rational if for any traveler $i \in \mathcal{I}_k$, we have $u_i(\mathbf{a}_k) \geq u_i(\widehat{\mathbf{a}}_k)$, where $\widehat{\mathbf{a}}_k$ denotes the traveler assignment in which traveler $i \in \mathcal{I}_k$ rejects social planner's travel recommendations and instead uses their self-owning vehicle. Finally, future research should show that the proposed mobility system is guaranteed to generate revenue from each traveler. This revenue can be used to maintain and update the infrastructure of the city's transportation network over the years.

One particular limitation of the proposed framework is that we consider travelers' preferences to be static. This implies that if any preferences of the travelers change, then the social planner would have to recompute the solution of the optimization problem in the mobility system to get an updated traveler–service assignment.

14.4 Optimal Coordination of Connected and Automated Vehicle with Human-Driven Vehicles

This section addresses the optimal coordination of CAVs with the neighbor human-driven vehicles (HDVs). We consider a team consisting of CAVs and HDVs that is about to encounter a given traffic scenario (e.g., crossing a signal-free intersection, merging at roadways or a roundabout, cruising in congested traffic, passing through a speed reduction zone, and lane-merging or passing maneuvers) with the common objective to coordinate in this scenario and avoid stop-and-go driving [14, 15, 43, 49, 78]. The implications of the latter are that the vehicles do not have to come to a full stop, thereby conserving momentum and fuel while also improving travel time. For example, consider a signal-free intersection (Fig. 14.2) with a team of CAVs and HDVs. The region at the center of the intersection, called *merging zone*, is the area of potential lateral collision of the vehicles. The intersection has a *control zone* inside of which the CAVs can communicate with each other. The objective of the team of CAVs and HDVs is to cross the intersection without the use of traffic lights, without creating congestion, and under the hard safety constraint of collision avoidance. We should emphasize that the proposed framework can be applied to any traffic scenario. We use an intersection here just as a reference for our exposition.

We model the communication between the team members with the word-of-mouth communication structure that we have previously developed [25, 26]. In a word-of-mouth communication structure, every member of the team communicates with her neighbors with delays in communication. This is a non-classical information structure [47] where the topological and temporal restrictions in communication mean that information propagates slowly through the team members.

In our modeling framework, we consider a number of $K \in \mathbb{N}$ members in the team of CAVs and HDVs with a decentralized information structure. At time $t = 0, 1, \ldots, T$, $T \in \mathbb{N}$, the state of the team X_t takes values in a finite set $\mathcal{X}_t$ and the decision U_t^k associated with the team member $k \in \mathcal{K}$, $\mathcal{K} = \{1, \ldots, K\}$, takes

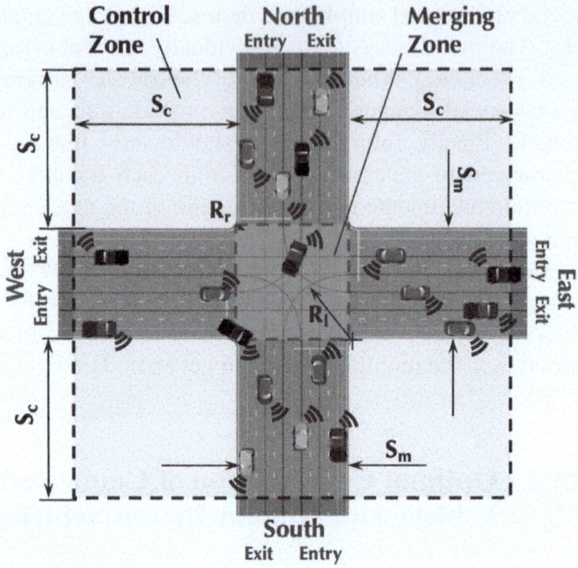

Fig. 14.2 A signal-free intersection with human-driven and connected automated vehicles

values in a finite set $\mathcal{U}_t^k$. Let $U_t^{1:K} = (U_t^1, \ldots, U_t^K)$ be the team's decision at time t. Starting at the initial state X_0, the evolution of the team is described by the state equation $X_{t+1} = f_t\left(X_t, U_t^{1:K}, W_t\right)$, where W_t is a random variable corresponding to the external, uncontrollable disturbance to the team's mission that takes values in $\mathcal{W}$. The sequence $\{W_t : t = 0, \ldots, T\}$ is a sequence of independent random variables which is also independent of the initial state X_0. At time $t = 0, 1, \ldots, T$, every team member $k \in \mathcal{K}$ makes an observation Y_t^k according to the equation $Y_t^k = h_t^k(X_t, V_t^k)$, where V_t^k is a random variable corresponding to the noise of the observation and takes values in the finite set $\mathcal{V}^k$.

To capture the delay typically encountered between vehicle-to-vehicle and vehicle-to-infrastructure communication, we consider that the team has n-step delayed information sharing. Namely, at time t, team member k observes Y_t^k, and the n steps past observations $Y_{t-n}^{1:K}$ and decisions $U_{t-n}^{1:K}$ of all team members. Thus, at time t, the information available at the team member k is (Δ_t, Λ_t^k), where $\Delta_t := (Y_{t-n}^{1:K}, U_{t-n-1}^{1:K})$, $Y_{t-n}^{1:K} = \{Y_{t-n}^1, \ldots, Y_{t-n}^K\}$, $U_{t-n-1}^{1:K} = \{U_{t-n-1}^1, \ldots, U_{t-n-1}^K\}$, is the information known to all team members and $\Lambda_t^k := (Y_{t-n+1:t}^k, U_{t-n+1:t-1}^k)$ is the additional information known at the team member $k \in \mathcal{K}$ only. Note that the n-step delayed information sharing can also be asymmetric, i.e., for each subsystem $k \in \mathcal{K}$, $Y_{t-n_i}^k$, $U_{t-n_i}^k$, where n_i, $i \in \mathcal{K}$, are constant but not necessarily the same for each k. The collection $\{(\Delta_t, \Lambda_t^k); t = 0, \ldots, T\}$, is the information structure of the team and captures who knows what about the team and when. Note that Λ_t^k of each robotic vehicle $k \in \mathcal{K}$ is "richer" than the information Λ_t^j of each human-driven vehicle $j \in \mathcal{K}$, since the observation Y_t^k of k also includes information from other CAVs, whereas HDVs can make only local observations, e.g., distance from the preceding vehicle, etc.

Let $\mathcal{D}_t$ be the space of all possible realizations of Δ_t, and $\mathcal{L}^k$ be the space of all possible realizations of Λ_t^k. The team member k makes a decision according to a control law g_t^k, i.e., $U_t^k = g_t^k(\Delta_t, \Lambda_t^k)$. The problem for each team member is to derive its optimal control law g_t^k such that the collective optimal strategy $g = \{g_t^k; k \in \mathcal{K}; t = 1, \ldots, T\}$ will enable the team to pass through the traffic scenario without stop-and-go driving. The latter can be modeled as the minimization of an expected total cost

$$J(g) = \mathbb{E}^g\{\sum_{t=1}^{T} c_t(X_t, U_t^{1:K})\}, \tag{14.10}$$

where $c_t(X_t, U_t^{1:K})$ corresponds to travel delay of the team and the expectation is with respect to the joint probability of the random variables designated by the choice of g. Deriving the solution of (14.10) has the following conceptual difficulties: (1) the functional optimization problem of selecting a sequence of strategies is not trivial, as the set of the class of strategies is infinitely large and (2) the domain of the control strategies given by the information $\{(\Delta_t, \Lambda_t^{1:K}); t = 0, \ldots, T\}$ increases with time, causing significant implications on storage requirements and real-time implementation. These difficulties can be circumvented by using the conditional probability of the state given the data available as a sufficient statistic, i.e., $\Pi_t := \mathbb{P}(X_{t-n}|\Delta_t)$. The conditional probability is called information state, Π_t, and takes values in a time-invariant space. Using the information state can help us restrict our attention to control strategies in a time-invariant domain. Such results, where data that are increasing with time are "compressed" to a sufficient statistic taking values in a time-invariant space, are called structural results. The structural results are related to the concept of separation, namely, the information state does not depend on the control strategy. This has been called a one-way separation between estimation and control. An essential consequence of this separation is that for any given choice of control strategies until time t and a given realization of the system variables till time t, the information states at future times do not depend on the choice of the control strategy at time t but only on the realization of control action at time t. Thus, the future information states are separated from the choice of the current control strategy. This fact is crucial for deriving the optimal control strategy where, at each step, the optimization problem is to find the best control action for a given realization of the information state.

The team's information structure and structural results can provide the framework to derive optimal control prescription functions that will yield the optimal decisions of CAVs. Since the structural results can help us restrict our attention to control strategies in a time-invariant domain. Thus, the optimal planning strategy of the CAVs can be derived a priori even before they start evolving in the field encountering a specific traffic scenario. Then, while each robotic vehicle in the team $k \in \mathcal{K}$ operates, its control prescription function $\Gamma_t^k : \mathcal{L}^k \to \mathcal{U}^k$ maps the k's information Λ_t^k at time t to her decision, e.g., $U_t^k = \Gamma_t^k(\Lambda_t^k)$. Note, the prescription functions are derived by the information structure of the team through Δ_t, i.e., $\Gamma_t^{1:K} = \psi_t(\Pi_t)$.

Thus, the problem for each team member $k \in \mathcal{K}$ is reformulated as to derive its optimal planning strategy $\psi_t^{k^*}$ so that the team's planning strategy $\boldsymbol{\psi}_t^* = \{\psi_t^{k^*}; t = 1, \ldots, T; k \in \mathcal{K}\}$ minimizes the expected total cost

$$\bar{J}(\boldsymbol{\psi}) = \mathbb{E}^{\psi}\{\sum_{t=1}^{T} c_t(\Pi_t, \Gamma_t^{1:K})\}, \tag{14.11}$$

where $c_t(\Pi_t, \Gamma_t^{1:K})$ corresponds to travel delay of the team, and the expectation is with respect to the joint probability of the information state Π_t and prescription functions $\Gamma_t^{1:K}$ variables designated by the choice of $\boldsymbol{\psi}_t$. The solution of (14.11) can be derived using standard techniques for partially observed Markov decision processes [40]. If the observation space of the system is finite, then (14.11) has a finite-dimensional characterization. In particular, the explicit solution to (14.11) is a piecewise linear concave function of the information state [69].

The optimal planning strategy $\boldsymbol{\psi}_t^*$ yields the optimal control prescription function $\Gamma_t^{k^*}$ with respect to the information state Π_t, i.e., $\Gamma_t^{k^*} = \psi_t^{k^*}(\Pi_t)$. Since $U_t^k = \Gamma_t^k(\Lambda_t^k)$ and $\Gamma_t^k(\cdot) = \psi_t^{k^*}(\mathbb{P}(X_{t-n}|\Delta_t))$, the optimization problem is equivalent to the original problem (14.10), and hence, $g_t^k(\Delta_t, \cdot) = \Gamma_t^k(\cdot) = \psi_t^{k^*}(\Pi_t)$. $\Delta_t \in \mathcal{D}_t$ is the information known by all team members at time t, thus the prescription function $\Gamma_t^k(\cdot)$ can be derived by each robotic vehicle $k \in \mathcal{K}$ on her own without any centralized intervention. This is a key property of the proposed framework that allows each robotic vehicle k to adapt its decision and thus improve the efficiency of the team in the field based on what each team member can learn from the information $\Lambda_t^k \in \mathcal{L}_t^k$ of her environment as discussed next.

Since we can separate the information state from the prescription function [48], we can develop a mechanism to learn the statistic $\mathbb{P}(X_{t-n}|\Delta_t)$ using standard machine learning (ML) techniques that will aim CAVs at adapting their decisions designated by the control prescription functions in situations where they encounter different behaviors from what they already know about human driving. ML has been used extensively for entitling autonomy features in automotive systems [46]. The optimal control prescription functions designate how each team member will coordinate and collaborate over variable time scales (recall that their domain is time-invariant) and without any centralized (human) intervention while there exist environmental disturbances $\{W_t; t = 0, \ldots, T\}$. At time t, all team members know the shared information Δ_t, the optimal planning strategy $\psi_t^*(\cdot)$, and the optimal prescription functions $\Gamma_t^{1:K}(\cdot)$. Each CAV $k \in \mathcal{K}$ will update her optimal planing strategy according to $\hat{\psi}_t^{k^*}(\cdot) = \Psi_t^k(\Delta_t, \psi_t^{k^*}(\cdot), \Lambda_t^k)$, where Ψ_t^k is a learning function that will use local information and observations Λ_t^k to update the control prescription functions $\Gamma_t^{1:K}(\cdot)$. Ψ_t^k will adapt the planning strategy CAV k in a way to enhance what k knows about drivers' behavior. In addition, the learning function Ψ_t^k will enhance the robustness of each CAV k in the presence of a contested communication environment, as described by the sequence of independent random variable $\{V_t^k : t = 0, \ldots, T; k = 1, \ldots, K\}$ corresponding to the noise of each team's member observation. The control prescription function of k will then be derived according to $\hat{\Gamma}_t^{k^*}(\cdot) = \hat{\psi}_t^{k^*}(\mathbb{P}(X_{t-n}|\Delta_t) = \hat{\psi}_t^{k^*}(\Pi_t)$. This will aim team member k at feeding the realization of her local information Λ_t^k into her updated control prescription function to derive the optimal decision, i.e., $U_t^{k^*} = \hat{\Gamma}_t^{k^*}(\Lambda_t^k)$.

14.4.1 Evaluation and Experiments in a Scaled Smart City

Ongoing research includes implementing and validating the proposed framework in the Information and Decision Science Lab's (IDS^{3}C) (1:25) testbed (Fig. 14.3) [12]. This testbed occupies a 20 by 20 feet area and includes 50 robotic cars (both CAVs and human-driven), which can replicate real-world traffic scenarios in a small and controlled environment.

IDS^{3}C has six driver emulation stations (Fig. 14.4) interfaced directly with the robotic cars which allow us to explore human driving behavior. IDS^{3}C can help us prove concepts beyond the simulation level and understand the implications of errors/delays in the vehicle-to-vehicle and vehicle-to-infrastructure communication as well as their impact on energy usage. In several recent efforts, we have used IDS^{3}C to implement and validate control algorithms for coordinating CAVs at traffic scenarios, such as merging roadways [39, 70], roundabouts [13], intersections [50, 51], and corridors [8, 12, 78]. We have also used IDS^{3}C to transfer policies derived in simulation using reinforcement learning techniques [14, 37].

We can apply the proposed framework over various real-world driving scenarios deemed characteristic of typical commutes. A typical vehicle commute includes

Fig. 14.3 A view of the IDS Lab's scaled smart city (IDS^{3}C)

Fig. 14.4 A driver emulation station interfaced directly with the robotics vehicles

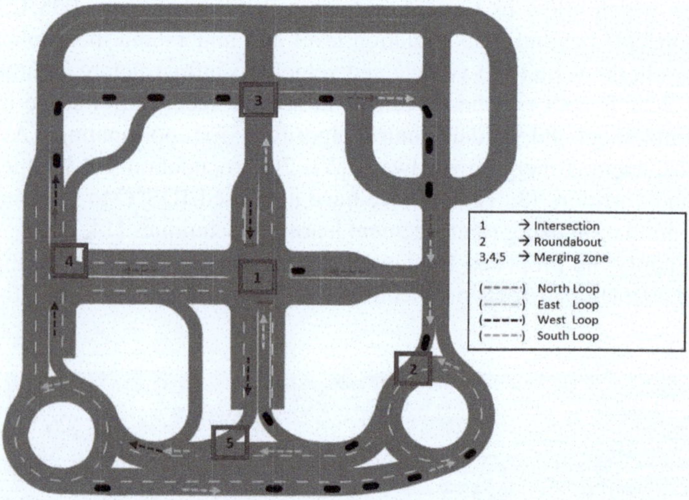

Fig. 14.5 Traffic scenarios of at the scaled smart city

merging at roadways, crossing signalized intersections, cruising in congested traffic, and passing through speed reduction zones. The robotic cars will be able to cross intersections in locations 1, 3, and 4 (Fig. 14.5), merge at a roundabout in location 2 (Fig. 14.5), and merge at roadways in location 5 (Fig. 14.5) by optimizing transportation efficiency.

14.5 Concluding Remarks and Discussion

It is expected that CAVs will gradually penetrate the market and interact with HDVs in a way to improve safety and transportation efficiency [64, 77, 80] over the next several years. However, we anticipate that efficient transportation and travel cost reduction might alter human travel behavior, causing rebound effects, e.g., by improving efficiency, travel cost is decreased, hence willingness-to-travel is increased. The latter would increase overall vehicle miles traveled, which in turn might negate the benefits in terms of energy and travel time.

We expect the proposed framework presented in this chapter to enhance our understanding of the rebound effects, travel demand and capacity changes, human reception, adoption, and use of emerging mobility systems as it addresses a complex multidimensional research problem focusing on societal needs. The framework captures the societal impact of CAVs and provides solutions that mitigate any potential rebound effects, e.g., increased vehicle miles traveled, increased travel demand, and empty vehicle trips while enhancing accessibility, safety, and equity in transportation.

Acknowledgements This research was supported by NSF under Grants CNS-2149520 and CMMI-2219761.

References

1. Acemoglu D, Golosov M, Tsyvinski A (2008) Political economy of mechanisms. Econometrica 76(3):619–641
2. Acemoglu D, Golosov M, Tsyvinski A (2010) Dynamic mirrlees taxation under political economy constraints. Rev Econ Stud 77(3):841–881
3. Anderson JM, Nidhi K, Stanley KD, Sorensen P, Samaras C, Oluwatola OA (2014) Autonomous vehicle technology: a guide for policymakers. Rand Corporation
4. Armstrong M (1996) Multiproduct nonlinear pricing. Econometrica 64(1):51
5. Bandi C, Bertsimas D (2014) Optimal design for multi-item auctions: a robust optimization approach. Math Oper Res 39(4):1012–1038
6. Bandi C, Bertsimas D (2012) Tractable stochastic analysis in high dimensions via robust optimization. Math Program 134(1):23–70
7. Bang H, Dave A, Malikopoulos AA (2023) Routing in mixed transportation systems for mobility equity. (In review, arXiv:2309.03981)
8. Beaver LE, Chalaki B, Mahbub AM, Zhao L, Zayas R, Malikopoulos AA (2020) Demonstration of a time-efficient mobility system using a scaled smart city. Veh Syst Dyn 58(5):787–804
9. Bian Z, Liu X (2019) Mechanism design for first-mile ridesharing based on personalized requirements part i: theoretical analysis in generalized scenarios. Transp Res Part B: Methodol 120:147–171
10. Bissell D, Birtchnell T, Elliott A, Hsu EL (2020) Autonomous automobilities: the social impacts of driverless vehicles. Curr Sociol 68(1):116–134
11. Borgers T, Krahmer D, Strausz R (2015) An introduction to the theory of mechanism design, 1st edn. Oxford University Press
12. Chalaki B, Beaver LE, Mahbub AMI, Bang H, Malikopoulos AA (2022) A research and educational robotic testbed for real-time control of emerging mobility systems: from theory to scaled experiments. IEEE Control Syst 42(6):20–34

13. Chalaki B, Beaver LE, Malikopoulos AA (2020) Experimental validation of a real-time optimal controller for coordination of cavs in a multi-lane roundabout. In: 31st IEEE intelligent vehicles symposium (IV), pp 504–509
14. Chalaki B, Beaver LE, Remer B, Jang K, Vinitsky E, Bayen A, Malikopoulos AA (2020) Zero-shot autonomous vehicle policy transfer: from simulation to real-world via adversarial learning. In: IEEE 16th international conference on control & automation (ICCA), pp 35–40
15. Chalaki B, Malikopoulos AA (2021) Time-optimal coordination for connected and automated vehicles at adjacent intersections. IEEE Trans Intell Transp Syst
16. Chremos IV, Malikopoulos AA (2022) An analytical study of a two-sided mobility game. In: 2022 American control conference (ACC), pp 1254–1259
17. Chremos IV, Malikopoulos AA (2023) Socioeconomic impact of emerging mobility markets and implementation strategies. In: Kolmanovsky I, Murphey Y, Watta P (eds), AI-enabled technologies for autonomous and connected vehicles. Springer
18. Chremos IV, Malikopoulos AA (2023) Mobility equity and economic sustainability using game theory. In: 2023 American control conference (ACC) (to appear)
19. Clarke EH (1971) Multipart pricing of public goods. Public Choice 11(1):17–33
20. Coleri S, Cheung SY, Varaiya P (2004) Sensor networks for monitoring traffic. In: Allerton conference on communication, control and computing, pp 32–40
21. Condorelli D (2012) What money can't buy: efficient mechanism design with costly signals. Games Econom Behav 75(2):613–624
22. Connor WD, Wang Y, Malikopoulos AA, Advani SG, Prasad AK (2020) Impact of connectivity on energy consumption and battery life for electric vehicles. IEEE Trans Intell Veh 6(1):14–23
23. Dasgupta P, Hammond P, Maskin E (1979) The implementation of social choice rules: some general results on incentive compatibility. Rev Econ Stud 46(2):185–216
24. Dave A IV, Chremos AA Malikopoulos (2022) Social media and misleading information in a democracy: a mechanism design approach. IEEE Trans Autom Control 67(5):2633–2639
25. Dave A, Malikopoulos AA (2020) Structural results for decentralized stochastic control with a word-of-mouth communication. In: 2020 American control conference (ACC). IEEE, pp 2796–2801
26. Dave A, Malikopoulos AA (2021) The prescription approach for decentralized stochastic control with word-of-mouth communication. arXiv:1907.12125
27. Demange G, Gale D, Sotomayor M (1986) Multi-item auctions. J Polit Econ 94(4):863–872
28. Fagnant DJ, Kockelman K (2015) Preparing a nation for autonomous vehicles: opportunities, barriers and policy recommendations. Transp Res Part A: Policy Pract 77:167–181
29. Ferrara A, Sacone S, Siri S (2018) Freeway traffic modeling and control. Springer
30. Gale D, Shapley LS (1962) College admissions and the stability of marriage. Am Math Mon 69(1):9–15
31. Green J, Laffont JJ (1977) Characterization of satisfactory mechanisms for the revelation of preferences for public goods. Econometrica 45(2):427
32. T. Groves. Incentives in teams. *Journal of the Econometric Society*, pages 617–631, 1973
33. Groves T, Ledyard J (1977) Optimal allocation of public goods: a solution to the free rider problem. Econometrica 45(4):783
34. Groves T, Loeb M (1975) Incentives and public inputs. J Public Econ 4(3):211–226
35. Hurwicz L, Reiter S (2006) Designing economic mechanisms. Cambridge University Press
36. Iwanowski S, Spering W, Coughlin WJ (2003) Road traffic coordination by electronic trading. Transp Res Part C: Emerg Technol 11(5):405–422
37. Jang K, Vinitsky E, Chalaki B, Remer B, Beaver L, Malikopoulos AA, Bayen A (2019) Simulation to scaled city: zero-shot policy transfer for traffic control via autonomous vehicles. In: Proceedings of the 10th ACM/IEEE international conference on cyber-physical systems, pp 291–300
38. Kakade SM, Lobel I, Nazerzadeh H (2013) Optimal dynamic mechanism design and the virtual-pivot mechanism. Oper Res 61(4):837–854
39. Nakka SKS, Chalaki B, Malikopoulos AA (2022) A multi-agent deep reinforcement learning coordination framework for connected and automated vehicles at merging roadways. In: 2022 American control conference (ACC), pp 3297–3302

40. Kumar PR, Varaiya P (1986) No title. Estimation, identification and adaptive control, stochastic systems
41. Laffont JJ, Robert J (1996) Optimal auction with financially constrained buyers. Econ Lett 52(2):181–186
42. Litman T (2020) Autonomous vehicle implementation predictions: implications for transport planning. Victoria Transport Policy Institute, Victoria, BC
43. Mahbub AMI, Malikopoulos AA (2021) Conditions to provable system-wide optimal coordination of connected and automated vehicles. Automatica 131(109751)
44. Malikopoulos AA (2016) A multiobjective optimization framework for online stochastic optimal control in hybrid electric vehicles. IEEE Trans Control Syst Technol 24(2):440–450
45. Malikopoulos AA (2016) A duality framework for stochastic optimal control of complex systems. IEEE Trans Autom Control 61(10):2756–2765
46. Malikopoulos AA (2011) Real-time, self-learning identification and stochastic optimal control of advanced powertrain systems. ProQuest
47. Malikopoulos AA (2023) On team decision problems with nonclassical information structures. IEEE Trans Autom Control 68(7):3915–3930
48. Malikopoulos AA (2023) Separation of learning and control for cyber-physical systems. Automatica 151(110912)
49. Malikopoulos AA, Beaver LE IV, Chremos. (2021) Optimal time trajectory and coordination for connected and automated vehicles. Automatica 125(109469)
50. Malikopoulos AA, Zhao L (2019) A closed-form analytical solution for optimal coordination of connected and automated vehicles. In: 2019 American control conference (ACC). IEEE, pp 3599–3604
51. Malikopoulos AA, Zhao L (2019) Optimal path planning for connected and automated vehicles at urban intersections. In: Proceedings of the 58th IEEE conference on decision and control, 2019. IEEE, pp 1261–1266
52. Mas-Colell A, Whinston MD, Green JR (1995) Microeconomic theory. Oxford University Press
53. Maskin E (1999) Nash equilibrium and welfare optimality. Rev Econ Stud 66(1):23–38
54. Maskin E, Tomas S (2002) Implementation theory. In: Arrow KJ, Sen AK, Suzumura K (eds) Handbook of social choice and welfare, vol 1, Chap 5. Elsevier, pp 237–288
55. Maskin ES (2000) Auctions, development, and privatization: efficient auctions with liquidity-constrained buyers. Eur Econ Rev 15
56. Maskin ES (2008) Mechanism design: how to implement social goals. Amer Econ Rev 98(3):567–576
57. Mathevet L (2010) Supermodular mechanism design. Theor Econ 5(3):403–443
58. Mishra D, Parkes DC (2009) Multi-item Vickrey-Dutch auctions. Games Econom Behav 66(1):326–347
59. Olarte R, Haghani A (2018) Introducing and testing a game-theoretic model for a lottery-based metering system in Minneapolis. Transp Policy 62:63–78
60. Raphael J, Maskell S, Sklar E (2015) From goods to traffic: first steps toward an auction-based traffic signal controller. In: International conference on practical applications of agents and multi-agent systems, pp 187–198
61. Reichelstein S, Reiter S (1988) Game forms with minimal message spaces. Econometrica 56(3):661–692
62. Renou L, Tomala T (2012) Mechanism design and communication networks. Theor Econ 7(3):489–533
63. Richter M (2019) Mechanism design with budget constraints and a population of agents. Games Econom Behav 115:30–47
64. Rios-Torres J, Malikopoulos AA (2018) Impact of partial penetrations of connected and automated vehicles on fuel consumption and traffic flow. IEEE Trans Intell Veh 3(4):453–462
65. Roughgarden T, Talgam-Cohen I (2019) Approximately optimal mechanism design. Ann Rev Econ 11:355–381

66. Samadi P, Mohsenian-Rad H, Schober R, Wong VW (2012) Advanced demand side management for the future smart grid using mechanism design. IEEE Trans Smart Grid 3(3):1170–1180
67. Sheller M, Urry J (2000) The city and the car. Int J Urban Reg Res 24(4):737–757
68. Shoham Y, Leyton-Brown K (2008) Multiagent systems: algorithmic, game-theoretic, and logical foundations. Cambridge University Press, Cambridge
69. Sondik EJ (1971) The optimal control of partially observed Markov processes. PhD thesis, Stanford University
70. Stager A, Bhan L, Malikopoulos AA, Zhao L (2018) A scaled smart city for experimental validation of connected and automated vehicles. In: 15th IFAC symposium on control in transportation systems, pp 130–135
71. Teodorovic D, Triantis K, Edara P, Zhao Y, Mladenovic S (2008) Auction-based congestion pricing. Transp Plan Technol 31(4):399–416
72. Thomson W (1996) Concepts of implementation. Jpn Econ Rev 47(2):133–143
73. Vasirani M, Ossowski S (2011) A computational market for distributed control of urban road traffic systems. IEEE Trans Intell Transp Syst 12(2):313–321
74. Vickrey W (1961) Counterspeculation, auctions, and competitive sealed tenders. J Financ 16(1):8–37
75. Vickrey WS (1969) Congestion theory and transport investment. Am Econ Rev 59(2):251–260
76. Wadud Z, MacKenzie D, Leiby P (2016) Help or hindrance? the travel, energy and carbon impacts of highly automated vehicles. Transp Res Part A: Policy Pract 86:1–18
77. Zhang Y, Cassandras CG (2019) An impact study of integrating connected automated vehicles with conventional traffic. Ann Rev Control 48:347–356
78. Zhao L, Malikopoulos AA (2018) Decentralized optimal control of connected and automated vehicles in a corridor. In: 2018 21st international conference on intelligent transportation systems (ITSC), pp 1252–1257
79. Zhao L, Malikopoulos AA (2022) Enhanced mobility with connectivity and automation: a review of shared autonomous vehicle systems. IEEE Intell Transp Syst Mag 14(1):87–102
80. Zhao L, Malikopoulos AA, Rios-Torres J (2018) Optimal control of connected and automated vehicles at roundabouts: an investigation in a mixed-traffic environment. In: 15th IFAC symposium on control in transportation systems, pp 73–78
81. Zou B, Kafle N, Wolfson O, Lin JJ (2015) A mechanism design based approach to solving parking slot assignment in the information era. Transp Res Part B: Methodol 81:631–653

Chapter 15
Cost-Sharing Mechanisms in Transportation

Tanvir Ibna Kaisar and Maged Dessouky

Abstract The efficient use of transportation infrastructure can greatly reduce travel times, fuel consumption, and environmental pollution. It is well known that a rational driver makes selfish routing decisions to lower their transportation cost when there is no coordination. These selfish route choices lead to the user equilibrium solution which can be far away from the actual social optimum solution. On the contrary, achieving the social optimum reduces the collective travel time, fuel cost, and pollution to the highest possible extent without requiring any additional infrastructure. However, this cannot be accomplished without a fair cost-sharing mechanism because, otherwise, some drivers would benefit at the expense of others. Therefore, the presence of a cost-sharing mechanism that distributes the overall benefits among the stakeholders in a fair manner is vital to establish a stable coordination. We review the working principles and the properties of the core cost-sharing mechanisms that are commonly used in transportation and logistics, as well as some of the most recent advancements in this field. Although there is a plethora of possible collaboration opportunities in transportation and logistics, we focus on vehicle sharing, ridesharing, shipping, and supply chain (SC) collaboration. We also identify possible future research directions in this area. To do so, we have surveyed 46 articles, most of which were published between 2016 and to date.

15.1 Introduction

The idea of the sharing economy is not new. Throughout mankind's history, sharing has been a common social practice among family members and friends [1]. However, in modern society, the idea of sharing has been reshaped and sharing is now commonplace among strangers (Peer-to-Peer or P2P), Business-to-Business (B2B), and

---

T. I. Kaisar · M. Dessouky (✉)
University of Southern California, Los Angeles, USA
e-mail: maged@usc.edu

T. I. Kaisar
e-mail: kaisar@usc.edu

P. Ioannou and A. A. Malikopoulos (eds.), *Transportation Mobility in Smart Cities*, Springer Tracts on Transportation and Traffic 21,
https://doi.org/10.1007/978-3-031-64769-7_15

Business-to-Consumer (B2C) entities [2]. This is mostly because of technological advancements and the sprawling use of the internet which facilitates sharing among a wide spectrum of individual and organizational forms [3, 4]. Although the sharing economy has many different definitions in the literature [5], at its heart, the idea of the sharing economy is to share underutilized resources among multiple agents motivated by a social or individual benefit. The sharing economy has a tremendous economic impact. In 2015, PwC predicted the cumulative global revenue generated by the sharing economy to grow to $335 billion dollars by 2025 from a mere $15 billion dollars in 2014 [6].

In the transportation sector, the sharing economy has been growing exponentially, thanks to the rapid developments in information and communications technology and the use of mobile or computer applications. Uber, a popular ride-hailing service, has completed more than 23 million trips worldwide per day on average in 2022 and is currently operating in over 70 countries and 10,000 cities [7]. As this trend continues, the global ride-hailing industry is projected to experience significant expansion, with its total market value estimated to reach $285 billion by 2030 [8].

Under the broader umbrella of the transportation sharing economy, there exists a specific model known as the *cost-sharing economy*, and it is important to distinguish between the two. The sharing economy in transportation refers to a model where individuals or organizations share their vehicles or transportation resources with others on a short-term basis. It generally involves peer-to-peer sharing platforms that allow people to rent or borrow vehicles from individuals or companies. Examples of sharing economy services in transportation include ride-hailing services such as Uber and Lyft. Cost-sharing economy in transportation, on the other hand, focuses specifically on the collaborative sharing of transportation costs among participants. In this model, individuals or entities pool their resources and share the costs of transportation services to make them more affordable for everyone involved. The primary goal is to distribute the expenses across multiple users, leading to cost savings for each participant. It has emerged in the form of vehicle-sharing [9], ridesharing [10], facility or capacity sharing [11], etc. The motivation for sharing in these settings can be social, environmental, financial, or a combination of these.

The cost-sharing mechanism design plays a vital role in driving the transportation cost-sharing economy. This is because, in the absence of a mechanism to incentivize the participants to collaborate, rational individuals make selfish decisions that result in the user equilibrium (UE) solution. However, the UE solution is generally inferior from a societal perspective. Hence, the lack of coordination results in resource wastage, more congestion, and increased pollution. Coordination between selfish/competing agents in the transportation settings has tremendous potential of making the system more efficient without needing any major infrastructural enhancement.

Two major questions that researchers have been trying to answer in the cost-sharing economy context are (1) how to form a group of agents (i.e., a coalition) in such a way that maximizes the system efficiency, and (2) how to allocate costs (or benefits) to the members of such a group in a fair and equitable fashion so that every member is satisfied and the group remains stable [12]. The success of collaboration depends on whether the costs (or benefits) realized from the collaboration are divided among the members of the group in an equitable manner or not [13]. We investigate the following research questions in the transportation cost-sharing economy context: how the cost–benefit sharing mechanisms have evolved in recent years, what are the most commonly used cost allocation methods, in what contexts the cost mechanisms have typically been used, what are the properties and how good is the performance of these cost allocation methods, and what are the future research implications.

The rest of the article is arranged as follows. Section 15.2 provides a description and motivation of the problem settings that we have considered. In Sect. 15.3 and Sect. 15.4, the performance measures and cost-sharing methods that are typically employed in a transportation setting are briefly discussed and some summary statistics are also provided. Table 15.6 at the end of Sect. 15.4 provides a brief description of the problem setting, the cost-sharing methods that are employed, performance measures that are used, the desirable properties that are achieved, and whether the problem setting is static or dynamic in the context of the 46 articles that we have considered in this study. The last section presents the summary of this study and outlines potential research gaps in the transportation cost-sharing economy.

Out of the 46 articles, the majority of the articles (43) were published on or after 2016. We chose this time period since there is a review article that was published in 2016 [14]. Additionally, we reviewed three more articles published in 2015 which are highly cited in the literature. The articles selected encompass a wide variety of journals and conference proceedings spanning operations research, transportation, computer science, and other research fields, as indicated in Table 15.1.

15.2 Problem Setting

We have classified all the articles into four categories namely, vehicle sharing, ridesharing, shipping, and supply chain (SC) collaboration. In this section, we point out the motivation for sharing in each problem setting and the specific problem scenario that is considered.

Vehicle Sharing: The most common forms of vehicle sharing are carsharing and bike sharing. The popularity of vehicle sharing has been increasing since the last decade. By October 2014, 33 countries were operating carsharing systems possessing over 104,000 vehicles in more than 1531 cities, and serving around 4.8 million members [15]. In the US, over 1.4 million members shared 15,224 vehicles under 21 operators, as of January 1, 2018 [16]. There are two major modes of service in the vehicle sharing business: (1) the round-trip service, where the vehicle pickup and drop-off points are the same, and (2) the one-way service, where the pickup and drop-off points are located at two different designated stations. The one-way vehicle sharing is more convenient for the users; however, it also poses the challenge of supply–demand imbalance. One way to rebalance the supply and demand volume is to use staff members to directly relocate the vehicles from low-demand

Table 15.1 Number of articles per journal

Journal/Conference proceeding	Count
Transportation Research Part B: Methodological	6
IEEE Transactions on Intelligent Transportation Systems	5
Transportation Research Part C: Emerging Technologies	3
Transportation Science	3
European Journal of Operational Research	2
Journal of the Operational Research Society	2
AAAI Conference on Artificial Intelligence	1
Applied Intelligence	1
Asia-Pacific Journal of Operational Research	1
City Logistics 1: New Opportunities and Challenges	1
Computers and Industrial Engineering	1
Expert Systems with Applications	1
International Conference on Advances in Geographic Information Systems	1
International Conference on Wireless Algorithms, Systems, and Applications	1
Joint Rail Conference	1
Knowledge-Based Systems	1
Manufacturing and Service Operations Management	1
Mathematical Problems in Engineering	1
Omega	1
Soft Computing	1
Theoretical Computer Science	1
Transportation	1
Transportation Research Interdisciplinary Perspectives	1
Transportation Research Procedia	1
Transportmetrica B: Transport Dynamics	1

areas to high-demand areas. However, this approach typically results in significant operational expenses. Therefore, a lot of research endeavors have been dedicated to addressing this issue by providing incentives to the users to facilitate cross off the vehicle relocation [17–24]. Users agreeing to certain relocation tasks proposed by the provider receive payments in return for their service or the inconvenience they may face.

Ridesharing: As we can see from Table 15.2, ridesharing receives the most attention in the transportation cost-sharing economy research due to the possibility of reducing the cost of transportation, improving the vehicle utilization rate, reducing traffic, and reducing pollution [25]. Most of the ridesharing articles are either concerned with reducing the total transportation costs [10, 12, 23, 26–29] or vehicle miles [30, 31], Within the domain of ridesharing, researchers have considered different contexts. These include first-mile and last-mile ridesharing [10, 30, 32], pri-

Table 15.2 Problem setting

No.	Problem setting	Articles	Count
1	Vehicle sharing	[17–24]	8
2	Ridesharing	[10, 12, 26–34, 46–53]	19
3	Shipping	[35–40, 54–59]	12
4	SC collaboration	[11, 41–45, 60]	7
Total			**46**

oritized ridesharing [32], gender-preferenced ridesharing [28], riders with common destination [33], door-to-door ridesharing [34], etc.

Shipping: In collaborative shipping or freight problems, cost savings for the logistic service providers under improved capacity utilization is the main goal. Reducing congestion [35, 36] and decreasing carbon emissions [37, 38] have also motivated shipping collaboration. Shipping through a common consolidation center are considered in [39, 40]. Tinoco et al. also considered the cost of inventory associated with collaboration in shipping [38].

SC Collaboration: In this setting, multiple firms on the same level (horizontal SC) or different levels (vertical SC) collaborate together to improve the utilization of facilities or distribution centers and achieve better responsiveness. Collaboration in horizontal SC and vertical SC are investigated in [11, 41–45], respectively.

Table 15.2 below groups the articles in the four problem settings we have considered.

15.3 Performance Measures: Desirable Properties and Performance Metrics

In this section, we provide a detailed description of commonly used performance measures for cost-sharing mechanisms from the perspective of a transportation cost-sharing economy. To measure the performance of a certain cost-sharing mechanism, researchers resort to several desirable properties that the mechanism is supposed to satisfy or approximately satisfy. The first few of these, namely, Budget-Balanced (BB), Individual Rationality (IR), and Incentive Compatibility (IC) are the most widely used performance measures. Some of the lesser common desirable properties are Economic Efficiency (EE), player monotonicity, and symmetry.

In addition to the desirable properties mentioned above, several performance metrics such as Price of Anarchy (PoA), Price of Stability (PoS), percentage cost reduction, computational efficiency, matching rate, etc., are also used to judge the solution quality of a cost-sharing mechanism.

Let us introduce some notations first. Let $N = \{1, 2, 3, ...n\}$ define the set of members in a coalition and $v : 2^N \rightarrow \mathbb{R}$ be a *characteristic function*. N is called the grand coalition. Let P be the set of all the subsets of N. S is defined as the subset (sub-coalition or simply coalition) of N such that $S \subseteq N$. Therefore, $v(S)$ denotes the characteristics value function for every subset $S \subset N$ and $v\{\emptyset\} = 0$. In our cost-sharing mechanism context, the value function represents the cost. The payoff vector $x = (x_1, x_2, x_3, ...x_n)$ represents the cost (or profit) allocated to each individual $j \in N$, where $x_j \in \mathbb{R}$.

Budget-Balanced (BB) Property

The total payment from all the players belonging to a coalition should at least cover the cost of the coalition. In a ridesharing context, this would mean that total payment by the riders of a group offsets the cost to the drivers and possibly some profit. Mathematically:

$$\sum_{j \in S} x_j = v(S) \quad \forall S \in P. \tag{15.1}$$

If the mechanism is not budget-balanced, then a subsidy is required to keep the mechanism afloat. For example, the government may provide subsidies to cover the deficit if the mechanism is not budget-balanced. Oftentimes, it is difficult to obtain a perfect budget-balanced. That is why researchers often instead focus on obtaining an approximate budget-balance instead [61]. This is called $\beta-$budget balance, which is defined as

$$v(S)/\beta \leq \sum_{j \in S} x_j \leq v(S) \quad \forall S \in P, \ \beta \geq 1. \tag{15.2}$$

Individual Rationality (IR)

A mechanism satisfies the individual rationality property if no player pays more than their willingness-to-pay. Coalitional rationality is also defined similarly where a coalition never pays more than its willingness-to-pay.

Incentive Compatibility (IC)

This means providing truthful information is in the best interest of all the players. In other words, it is a dominant strategy for all the players to report truthful information or bid truthfully. Given that v_i is the true private valuation or willingness-to-pay for a player to receive a service, the IC property satisfies the inequality:

$$v_i - p_i' \leq v_i - p_i, \tag{15.3}$$

where p_i and p_i' denote the payments of i for his truthful bid of v_i and misreported bid of v_i', respectively. A cost-sharing mechanism is called *strategyproof* if it satisfies the IC property.

Economic Efficiency (EE)
Economy efficiency is satisfied when a cost-sharing mechanism maximizes the overall social welfare or system efficiency [62]. For example, in a freight-consolidation context, this may refer to the objective of minimizing total travel-miles across the system. Similar to the concept of $\beta-$budget balance property, α-approximate economic efficiency is defined as the cost allocation solution producing social cost at most a factor α times the optimum social cost [12, 63].

It is important to distinguish between *economic efficiency* and just *efficiency*. The former refers to maximizing the social welfare (or minimizing the social cost), while the latter allocates the benefits fully among the members of the coalition [35, 54].

Symmetry
If two players have the same standalone costs, then their payment should be the same when they are in a coalition. Let c_i denote the standalone cost of player i, i.e., the player would incur c_i if it chooses to operate alone instead of joining a coalition. Then the symmetry property states

$$p_i = p_j \quad \text{if} \quad c_i = c_j, \tag{15.4}$$

where p_i denotes the payment for player i when player i operates in a coalition.

Player Monotonicity
Player with higher standalone cost should pay more while operating in a coalition. That is:

$$p_i \geq p_j \quad \text{if} \quad c_i \geq c_j. \tag{15.5}$$

Price of Anarchy (PoA) and Price of Stability (PoS)
Price of Anarchy (PoA) and Price of Stability (PoS) measure how a stable cost allocation solution outcome fares against the system optimum. PoA is defined as the ratio of the *worst* stable outcome and the optimal outcome, whereas PoS is defined as the ratio of the *best* stable outcome and the optimal outcome [64]. A good cost-sharing mechanism should have a small PoA and PoS [65].

Although these two metrics are originally defined in non-cooperative Nash Equilibrium settings, they are equally applicable in any equilibrium [66, 67]. References [46, 68] have used the notion of PoA and PoS to evaluate the solution quality provided by the stable cost allocation (imputation). Also, [26] used a similar metric called *social optimality ratio*, which is the ratio between *a* stable allocation and the social optimum.

Percentage Cost Reduction Relative to Standalone Cost
This refers to the decrease in cost due to the agents cooperating in a group compared to the scenario where each agent operates individually.

Computational Efficiency

Despite having desirable properties, many cost-sharing mechanisms suffer from computational complexity. As the number of players increases, so does the possible combinations of coalitions. This makes some cost mechanisms computationally intractable. For instance, for the Shapley value and nucleolus, which are two of the most commonly used cost-sharing methods in cooperative game theory, computational effort increases exponentially as the number of players increases [69].

Other

Detour and matching/participation rate are two commonly used metrics in a ridesharing context. The detour ratio measures the fraction of vehicle miles traveled to accommodate a new rider compared to the current riders' (or only the driver's) direct vehicle miles. This metric is used in [46, 47]. Matching or participation rate refers to what fraction of all the potential customers currently present in the system actually accept a ride. This metric is used in [26, 28].

Table 15.3 provides an overview of which performance measures and desirable properties are used in the 46 articles that we have considered.

15.4 Cost Allocation Methods

In this section, we describe the common cost allocation methods that are employed in the transportation cost-sharing economy. Many of the methods are from the cooperative game theoretic concepts.

The Core

As mentioned earlier in Sect. 15.3, the cost (or profit) allocated to each individual $j \in N$ is represented by the payoff vector $x = (x_1, x_2, x_3, ...x_n)$. An instance of a payoff vector is also known as an *imputation*. An imputation or cost-allocation vector is said to lie in the *core* if it is budget-balanced (i.e., satisfies Eq. (15.1)) and *stable* [13]. The stability property (individual or group rationality) states that no subset of players have incentive to collude from the coalition to achieve a higher expected payoff (less cost). This would happen if a subset of players could be better off by defecting from the current coalition and forming a new coalition. Such a subset of coalition is called a *blocking coalition* [65]. Mathematically, stability can be expressed as below:

$$\sum_{j \in S} x_j \leq v(S) \; \forall S \in P. \tag{15.6}$$

Table 15.3 Performance measures and desirable properties

	PoS	% cost reduction	Comp. eff.	BB	IR	IC	EE	Efficiency	Sym.	Other
[24]				✓[1]	✓	✓[a 2]	✓			
[29]		✓		✓	✓					
[30]					✓	✓				Price nonnegative
[23]					✓			✓		
[31]				✓	✓			✓		Jain's fariness
[32]			✓			✓		✓	✓	Additivity
[46]	✓			✓	✓	✓[a]				Detour ratio
[47]			✓		✓	✓	✓			Detour ratio
[27]			✓		✓	✓[a]				
[48]				✓[a]	✓	✓				
[10]				✓			✓			
[28]		✓				✓				Matching rate
[12]				✓	✓	✓	✓[a]			
[26]				✓	✓	✓				Matching rate
[33]				✓	✓	✓				
[49]					✓	✓	✓			
[34]				✓	✓	✓				Online fairness, immediate response
[50]				✓	✓	✓				Online fairness, immediate response, reduced burden of first passenger, fairness in driver's cost
[51]					✓	✓				
[52]				✓	✓	✓				
[53]					✓	✓		✓	✓	Player monotonicity
[36]				✓	✓	✓				

(continued)

Table 15.3 (continued)

	PoS	% cost reduction	Comp. eff.	BB	IR	IC	EE	Efficiency	Sym.	Other
[37]		✓		✓	✓	✓				Superadditivity
[38]				✓	✓	✓				Subadditivity, Stability
[54]		✓		✓	✓			✓		
[35]				✓[a]	✓	✓[a]		✓		
[39]				✓[a]	✓	✓	✓			
[40]				✓[a]	✓	✓[a]	✓			
[55]				✓	✓	✓				
[57]		✓		✓[a]		✓	✓			Percentage travel time reduction
[58]		✓		✓[a]		✓	✓			
[59]		✓	✓	✓[a]	✓		✓			
[56]		✓	✓	✓[a]	✓	✓	✓			
[44]				✓[a]	✓	✓[a]				
[41]				✓	✓	✓		✓		
[45]				✓	✓					
[42]		✓		✓	✓	✓				
[11]		✓			✓					
[43]				✓[a]	✓	✓				Online fairness, immediate response
[60]				✓	✓	✓				
[19]					✓					Service rate, vehicle utilization
[18]						✓	✓			
[20]				✓		✓				
[21]										Bike usage, revenue
[22]					✓	✓	✓			
[17]										Profit earned by the operator, cost incurred by the passengers, and percentage of carsharing trips, etc.

✓ means exactly satisfied, ✓[a] means approximately satisfied.

The core can be mathematically defined as

$$Core(v) = x_j \in \mathbb{R}^n : \sum_{j \in N} x_j = v(N), \sum_{j \in S} x_j \leq v(S) \quad \forall S \in P. \tag{15.7}$$

The core allocation is what holds the coalition together. In this sense, the concept of the core resembles a Nash equilibrium in non-cooperative game theory. However, the core may not always exist (i.e., budget-balance and stable cost allocation may not always exist). The core is said to be empty in this case. Sometimes, the stability constraint is relaxed to force the core to be nonempty [70]. This is called *least core* or $\epsilon - core$ (Eq. 15.8).

$$\epsilon - core(v) = x_j \in \mathbb{R}^n : \sum_{j \in N} x_j = v(N), \sum_{j \in S} x_j \leq v(S) + \epsilon \quad \forall S \in P. \tag{15.8}$$

Although stability is an important criterion to make a coalition sustainable, [13] note that slightly relaxing the stability constraint as in Eq. (15.6) is acceptable since, in reality, sub-coalitions are hard to form. The reason is that forming a new sub-coalition also comes with the difficulty in managing and coordinating the newly formed sub-coalition, possible membership fees, etc.

The Shapley Value Method

The Shapely value method is one of the most commonly applied methods for cost-sharing mechanism design in a collaborative environment. The Shapley value to each player j is defined by

$$Shapley(x_j) = \sum_{S \subseteq N : j \in S} \frac{(n - |S|!) \cdot (|S| - 1)!}{n!} \cdot [v(S) - v(S \setminus j)] \; \forall S \in P. \tag{15.9}$$

Intuitively, the Shapley value assigns costs proportional to the average marginal contributions of each player. It has some desirable properties, namely BB, IR, group-rationality, and symmetry. Also, it can provide values even if the core is empty. However, it does not always belong to the core, even if the core is nonempty. Nevertheless, as explained before, this issue may be circumvented by utilizing $\epsilon-$ core or least-core concept. Another disadvantage of using the Shapley value is that as the number of players grow, the computation of the Shapley value quickly becomes intractable and inefficient. Researchers often resort to the approximate computation of the Shapley value to reduce the computational effort [71].

The Vickrey–Clarke–Groves Mchanism (VCG Mchanism)

Player x_i's payment can be calculated as follows, given that v^*_{-i} is the collaborative value of S to all players except player x_i, $(v_{-i})^*$ is the value of the S generated when x_i is excluded from the system:

$$VCG(x_j) = v^*_{-i} - (v_{-i})^*. \tag{15.10}$$

This is the payment that the VCG mechanism charges for player i.

The VCG mechanism is generally IC, EE, and IR [12, 72]. That is, the mechanism is strategyproof (truthtelling). However, there are some disadvantages of the VCG mechanism. It is computationally expensive to calculate, and it is not BB.

The Nucleolus

The idea of the nucleolus was first introduced by [73]. The nucleolus attempts to maximize the minimum satisfaction in a coalition game. To define the nucleolus, we need to define the excess of the game. The excess of the game measures the satisfaction in a coalition game. Higher excess results in higher satisfaction. Mathematically, excess is defined by $\epsilon(x, v, S) = v(S) - \sum_{j \in S} x_j$. The excess vector e at x can be defined as non-decreasingly ordered values of excess for all $S \in P$ such that $e = \epsilon(x, v, S_1), \epsilon(x, v, S_2)......\epsilon(x, v, S_p)$, where $p = 2^N - 1$.

$$Nucleolus = \{x | x > y \ \forall \text{ payoff vectors } y \in \mathbb{R}^p\}. \tag{15.11}$$

The nucleolus always exists and lies in the core provided that the core is nonempty. The nucleolus provides a unique and stabilizing solution to the cost-allocation problem. However, the calculation of the nucleolus can be highly demanding as the number of players grows.

The Moulin Mechanism

The Moulin mechanism was first proposed by [74, 75]. It works in the following way. The iteration begins with each player placing a bid. A cost-sharing function is used to offer shared costs to each player simultaneously. A random player having his bid less than the offered shared cost is removed from the group and the iteration continues. The Moulin mechanism induces truthfulness (IC) conditional on the fact that the cost-sharing method is cross-monotonic. This means the cost share of a player can only increase if any player(s) are removed, or equivalently, cost shares decrease if more players are added to the group.

The Acyclic Mechanism

This is similar to the Moulin mechanism. The major difference is that prices are offered sequentially in the acyclic mechanism, whereas the prices are offered simultaneously in the Moulin mechanism. The sharing method does not need to be cross-monotonic in this case.

The Proportional Cost Sharing

The proportional cost allocation methods distribute cost among the players based on their usage or demand in a coalition, depending on the context. For example, in a ridesharing context, this could be based on the direct vehicle miles that would be traveled by a rider if they were traveling alone. In freight transportation, this might be the amount of goods that are needed to be transported by a single supplier.

The proportional cost-sharing method is simple to implement, easy to calculate, computationally inexpensive, and usually BB and IR. However, in many cases, this method does not satisfy the IC property [23, 29].

The Equal Profit Method (EPM)
The concept of EPM was first introduced by [76]. The EPM attempts to minimize the maximum difference in cost savings between all pairs of players. This minimization problem is formulated as a linear programming problem. The EPM is generally IR, BB, and stable [42] but it needs similar computational effort to the Shapley method [77].

The Equal Cost-Sharing Method
As the name suggests, costs are divided equally among the players in the equal cost-sharing method.

The Egalitarian Sharing Method
This is similar to equal cost sharing. However, in this case, costs are allocated in such a way that every player receives the same utility. The desirable properties may or may not be satisfied depending on the context. However, this method is not truthful in general [29].

Other Cost Allocation Methods
Although not very common, there are other cost-sharing mechanisms used sporadically in the literature such as alternative cost avoided method (ACAM) [42] and constrained equal gains mechanism (CEG) [53]. We also found an article that proposed a cost-sharing mechanism by forming a Stackelberg game model [45]. Although the idea of the Stackelberg model belongs to non-cooperative game, the researchers tailored it to the cooperative game setting in their case. Some researchers resort to auction theory for pricing. These include double-auction [48], reverse auction [20], and second-price Vickrey auction scheme [51]. Another approach in cost allocation problems is to first divide the cost of a player into two portions- *separable cost* and *non-separable cost*. Separable cost refers to the cost that can be directly attributed to the player. That is, it is the increase in the coalition cost if a player joins the coalition [78]. In most cases, the sum of these separable costs for each player is less than the coalitional cost. The remaining cost is called non-separable cost. Many researchers employ different cost allocation methods to divide up these non-separable costs in a fair manner. Such methods include ACAM [42], Cost Gap Allocation [44, 78], weighted methods [79], etc.

Ad hoc Methods
Apart from the methods described above, there is a wide range of literature that employs ad hoc methods tailored to a particular scenario. Such methods are mostly based on the principles of mathematical programming (linear and nonlinear) techniques. In most of these cases, the desirable properties are enforced by modeling them explicitly as constraints.

Table 15.4 Cost-sharing mechanisms and references

	Cost-sharing mechanisms	Articles	Count
1	Shapley method	[32, 35–37, 42, 53]	6
2	Nucleolus	[27, 36, 60]	3
3	Vickrey–Clarke–Groves (VCG)	[12, 18, 22, 24, 32]	5
4	Core	[33]	1
5	Moulin mechanism	[39]	1
6	Acyclic mechanism	[40]	1
7	Proportional	[11, 21, 23, 26, 28, 31, 34, 35, 38, 43, 50, 53, 54, 56, 57, 57]	16
8	Ad hoc (most are LP based)	[10, 17, 19, 20, 30, 34, 41, 43, 47, 49, 50, 55, 59]	13
9	Equal cost sharing	[26, 29, 46, 52, 53, 57]	6
10	Egalitarian (equal utility)	[11, 26]	2
11	Equal Profit Method (EPM)	[36, 42]	2
12	Alternative Cost Avoided Method (ACAM)	[42]	1
13	Auction-based	[20, 48, 51]	3
14	Constrained Equal Gains Mechanism (CEG)	[53]	1
15	Cost gap method	[44]	1

Table 15.5 Cost-sharing mechanisms and static versus dynamic case in different problem settings

Problem setting	Cost-sharing mechanisms														S/D[3]	
	Shapley	Nucleolus	VCG	Core	Moulin	Acyclic	Proportional	Ad hoc	Equal	Egalitarian	EPM	ACAM	Auction	CEG	Static	Dynamic
Vehicle sharing			3				2	3							5	3
Ridesharing	2	1	2	1			4	7	5	1			2	1	12	8
Shipping	3	1			1	1	6	2	1		1				13	0
SC collaboration	1	1					1	3		1	1	1			6	1

Tables 15.4 and 15.5 provide some statistics about different cost-sharing mechanisms that are used in the articles we have reviewed in this study.

Finally,[1] Table 15.6 provides a brief description on the problem setting, whether the problem setting is static or dynamic, how the cost-sharing model is set up in the problem context, the performance measures that are used, and the pricing methods that are employed.

15.5 Summary and Research Gaps

Designing cost-sharing mechanisms in vehicle sharing and ridesharing has received a lot of focus over the years. However, there are still some variations that have not been explored much. For example, round-trip ridesharing and carsharing in the long term. In the former, the rider is guaranteed to have a ride back to the origin after some period, whereas the latter involves sharing unused cars of others over an extended period of time (days, weeks, or months). Many transportation services (especially carsharing and ridesharing services) require a real-time balance of supply and demand, with each request to be fulfilled quickly. This is still an active research area.

In SC collaboration, researchers have mainly focused on reducing logistics costs. However, there are still a plenty of opportunities for sharing in the rest of the supply chain. For example, at the operations level, firms can work together to share the excess production or labor capacity. This would provide a safety net to sudden demand surges and thus, result in better demand responsiveness. Originated in Germany, *maschinenring* is an idea that brings together small family-owned farms to share costly agricultural equipments that would otherwise be infeasible to own for individual use [80]. The same idea can be applied in industrial settings where expensive equipment can be shared among multiple small firms [81]. This idea of machine sharing has not been explored in the literature and thus provides an ample opportunity to model, analyze, and construct proper cost-sharing mechanisms to encourage small firms in forming such coalitions. In recent years, [41, 82] have analyzed the revenue sharing mechanisms in the airline industry and medical-service providers, respectively. Aside from collaborating among competing airlines, there is a cooperation opportunity between the airlines and its customers as well. Oftentimes, people have to cancel their travel plans and expect a refund on their air tickets. In most of these cases, the airline charges a penalty for doing so. Instead, the airline can match the ticket to another person requesting the same journey and distribute the price of the ticket and cost of rescheduling to the persons involved, which will eventually result in higher customer satisfaction.

In many cases, a central authority or aggregator enforces the price or allocates the benefits. This is common in vehicle sharing [23] and ridesharing [47]. The service provider is also present in SC collaboration [44], and urban freight transportation [36] studies. Investigating the role of a central authority in the collaboration can be an interesting research effort. However, whether the presence of a central authority

[1] S/D means Static/Dynamic.

Table 15.6 Brief description of the problem setting, cost-sharing mechanism, performance measures, static/dynamic (S/D) nature of each article reviewed

References	Brief description of the problem setting	Cost-sharing mechanism	Performance Measures	S/D
[24]	Carsharing	VCG with single-side reward (SSR) pricing policy	BB, EE, partial IC, IR	S
[23]	Carsharing	Proportional	IR, EE	S
[17]	One-way carsharing	Ad hoc (bi-level optimization, nonlinear program)	Profit earned by the operator, cost incurred by the passengers, and percentage of carsharing trips, etc.	S
[19]	Pricing and relocation strategy in one-way carsharing	Ad hoc (bi-level optimization, nonlinear program)	Stability of the mechanism, service rate, vehicle utilization	S
[18]	Demand (cars) rebalancing in one-way carsharing	VCG auction	IC, EE	S
[20]	Demand (bikes) rebalancing in bike sharing systems	Ad hoc (bipartite matching between a set of tasks and a set of users)	IC, BB	D
[21]	Bike rebalancing in bike sharing systems	Proportional	Bike usage, revenue	D
[22]	One-way carsharing	VCG	EE, IC, individual and coalition rationality	D
[29]	Dynamic taxi and ridesharing problem	Equal cost sharing	BB, Fair, IR, passengers pay around 30% less on average compared to private rides	D
[30]	First-mile ridesharing problem where passengers are picked up from different locations and dropped off at a common transit hub (e.g., a train station)	Ad hoc (based on LP)	IR, IC, and price non-negativity (generates revenue for the service provider)	S
[31]	Ridesharing problem where passengers and drivers are matched dynamically in real time	Proportional	Fair (guaranteed by Jain's fairness index), IR, efficient, BB	D

(continued)

Table 15.6 (continued)

References	Brief description of the problem setting	Cost-sharing mechanism	Performance Measures	S/D
[32]	Prioritized last-mile ridesharing	Shapley, VCG	The VCG mechanism is efficient and truthful. Efficiency, symmetry, null player property, polynomial time computation, and additivity for Shapley value	S
[46]	Dynamic ridesharing matching problem between a single rider and single driver with the objective of minimizing system-wide vehicle-miles	Equal cost-sharing method	IR, nearly stable, BB, PoS, Detour distance	D
[47]	Ridesharing matching problem where drivers are assumed to be commuters and they share the ride with at most one passenger	Ad hoc	Efficient, stable, IR, BB. Solution can be obtained in polynomial time. Also calculated detour cost as [46]	S and D
[27]	Ridesharing	Nucleolus	IR, approximately stable, computational efficiency	S
[48]	Dynamic ridesharing pricing that has tunable parameters to ensure high matching rate	Ad hoc (based on double auction)	IR, IC, and weakly BB	D
[10]	First mile ridesharing problem where users take ridesharing services to reach the rail transit hub	Ad hoc (based on LP)	BB, EE	S
[28]	Ridesharing problem where commuters can submit their roles as driver or passenger and are matched based on their gender preference	Proportional	Stable, Percentage of distance savings and cost savings	S

(continued)

Table 15.6 (continued)

References	Brief description of the problem setting	Cost-sharing mechanism	Performance Measures	S/D
[12]	Ridesharing	VCG	Approximate EE, IC, IR, BB	S
[26]	Ridesharing	1. Equal sharing, 2. Proportional sharing, 3. Egalitarian Cost Sharing, and 4. Segment-based cost sharing	Stable, BB, IR	S
[33]	Analyzed a ridesharing mechanism where multiple passengers have a common destination	Core	IR, stable, and BB	S
[49]	Ridesharing mechanism where each commuter reports a probability of partaking a trip along with other information such as locations and times.	Ad hoc	IR, IC, EE	S
[34]	Ridesharing problem where passengers request door-to-door rides without any fixed route or schedule	Ad hoc (based on proportional and marginal)	Online fairness, immediate response, IR, BB, and IC (ex-post)	D
[50]	Ridesharing problem where drivers act as commuters along with the other commuters who are looking for a shared ride	Ad hoc (based on proportional)	Driver-out-of-Coalition mechanism satisfies all properties except the reduced burden property. Driver-in-Coalition mechanism satisfies all properties upto a reasonable degree. Passengers Predicting mechanism satisfies all properties except the BB property	D
[51]	Ridesharing	Second price Vickrey auction	IC, IR	D

(continued)

Table 15.6 (continued)

References	Brief description of the problem setting	Cost-sharing mechanism	Performance Measures	S/D
[52]	Cooperative game-theoretic model to price the rapid-transit rail network system users in an equitable manner	Equal cost-sharing method	The core is nonempty, the cost allocation always lies in the core. Core properties, coalitional rationality	S
[53]	Ridesharing	1. Equal cost sharing, 2. proportional sharing, 3. Shapley method, 4. constrained equal gains mechanism, or CEG	CEG is the only mechanism that ensures IC, IR, efficiency, symmetric, player monotone (up to 4 players)	S
[36]	Collaboration between municipality and freight forwarders to transport goods from a city distribution center to various locations of the city	Nucleolus, Shapley value, and Equal Profit Method (modified, EPMLP)	Usual properties of the Shapley method, EPM, and nucleolus. EPM: IR, belongs to the nonempty core	S
[37]	Cooperative truck scheduling problem (CTP), where the truckers form a coalition to fulfill orders cooperatively to increase their profit and decrease emission	Shapley (iterative method to approximate Shapley value solutions)	Superaddtivity of the game, coalitional rationality, and BB	S
[38]	Collaborative shipping problem where two shippers aggregate their goods to share vehicles	Proportional	Subadditivity, Stability, IR, IC, BB	S
[54]	Cooperation opportunity among urban freight stakeholders, where multiple carriers and receivers collaborate and partake in after-hour deliveries	Proportional	Can be unfair, efficient, IR, BB, Cost savings	S

(continued)

Table 15.6 (continued)

References	Brief description of the problem setting	Cost-sharing mechanism	Performance Measures	S/D
[35]	Urban freight consolidation	Proportional, Shapley	Efficiency by both proportional and Shapley methods. Stability decreases as the number of players increases for both methods. Shapley method performed best in terms of satisfying IR and stability	S
[39]	Freight consolidation problem where shipments go to a common destination through a consolidation center	Moulin	Coalitional IC, approximate budget-balanced (α-BB), and EE	S
[40]	Freight consolidation problem where shipments go to a common destination through a consolidation center	Acyclic	Approximate budget-balanced (2-BB), weakly group strategyproof, economic efficiency	S
[55]	Collaborative transportation where there is a leading company that acts in forming the alliance and coordinates the collaboration	Ad hoc (LP)	Stable, IR, and efficient	S
[57]	Value of time-based personalized freight routing	Proportional, equal cost sharing	BB (on average), IC, economic efficiency, percentage cost and travel time reduction	S
[58]	Freight routing	Proportional	Approximately BB, IC, economic efficiency, percentage cost reduction	S
[59]	Freight routing considering the individual routing preferences of the truckers'	Ad hoc	Approximately BB, IR, economic efficiency, percentage cost reduction, runtime	S

(continued)

Table 15.6 (continued)

References	Brief description of the problem setting	Cost-sharing mechanism	Performance Measures	S/D
[56]	Cooperation among truck drivers in freight routing	Proportional	BB (on average), IR, IC, economic efficiency, percentage cost reduction, runtime	S
[44]	Vehicle routing optimization with pickup and delivery problem where multiple distribution centers and pickup centers in the network collaborate to achieve greater benefit	Ad hoc	IR, stability (measured as the distance of a solution from the core center)	S
[41]	Service collaboration among multiple firms having multiple demand sources and facilities in different geographic locations	Ad hoc	BB, stable, efficient	S
[45]	Collaboration between one manufacturer and one retailer where the manufacturer sells new and remanufactured products through the retailer	Stackelberg game (miscellaneous)	IR, BB	S
[42]	Cooperative carrier facility location problem for horizontal logistic service providers	Shapley value, Alternative Cost Avoided Method (ACAM), and Equal Profit Method (EPM)	Group rationality, IR satisfied by Shapley and EPM; Stability is satisfied only by EPM	S
[11]	Cooperative game involving multiple firms to serve customer zones from geographically dispersed distribution centers	Egalitarian, Proportional	In general, Shapley cost allocation method generates higher cost savings than the other two methods. However, proportional and egalitarian methods generate more coalitions compared to Shapley method	S

(continued)

Table 15.6 (continued)

References	Brief description of the problem setting	Cost-sharing mechanism	Performance Measures	S/D
[43]	Horizontal cooperation among multiple firms	Ad hoc (based on proportional and POCS)	Online fairness, budget balance (approx. in HPOCSD), immediate response, individual rationality, and ex-post incentive compatibility (violated by HPOCSrO)	D
[60]	Investigated the collaboration among airlines, tourist attractions, and travel agencies in the tourism supply chain	Nucleolus	Stable, IR, group-rationality	S

is better and crucial to facilitate collaboration needs to be rigorously investigated. Specially, the cost and benefit of having the presence of a for-profit central authority who requires payment to coordinate the collaboration should be thoroughly studied.

From Tables 15.4 and 15.5, we observe that the cooperative game theory (CGT)-based cost-sharing mechanisms are widely used. This may be because CGT mechanisms satisfy many of the desirable properties exactly or it is relatively easy to modify them to satisfy the properties approximately. However, CGT-based methods are computationally difficult to calculate which can make them unrealistic for use in practical applications where simpler rules such as proportional, equal, egalitarian methods are easy to calculate. In this regard, making CGT-based methods run in polynomial time by relaxing some of its properties or modification of simple cost mechanisms to approximately satisfy the desirable properties can be a potential research direction. Among non-CGT methods, equal, proportional, and mathematical program based cost-sharing mechanisms are frequently used because of their simplicity, computational efficiency, and ability to take advantage of the special structure of the problem under consideration.

From our review, we have found that few articles have tackled the problem of cooperation in dynamic settings under shipping and SC collaboration (Tables 15.5 and 15.6). Constructing the pricing mechanism in a dynamic setting would result in a firm's improved dynamic capabilities [83].

We observe from Table 15.3 that BB, IR, and IC are the most researched performance metrics for cost mechanisms, whereas less attention is given to important metrics such as PoA, percentage cost reduction, and computational efficiency.

For the cost-sharing economy in transportation, we have investigated the most recent advancements in cost allocating strategies under four problem settings. We have answered the research questions that we posed earlier and discussed the research gaps that exist in the literature. We also point out some emerging scenarios where the cost mechanism models are still under-researched. We hope this study will stimulate further research, innovation, and policy discussions that foster the growth and sustainability of various transportation sharing platforms.

Acknowledgements Funding for this work has been partially supported by the National Science Foundation Award CNS-1932615.

References

1. Belk R (2010) Sharing. J Consum Res 36(5):715–734
2. Schor J et al (2016) Debating the sharing economy. J Self-govern Manag Econ 4(3):7–22
3. Ganapati S, Reddick CG (2018) Prospects and challenges of sharing economy for the public sector. Gov Inf Q 35(1):77–87
4. Belk R (2014) You are what you can access: sharing and collaborative consumption online. J Bus Res 67(8):1595–1600
5. Acquier A, Daudigeos T, Pinkse J (2017) Promises and paradoxes of the sharing economy: an organizing framework. Technol Forecast Soc Chang 125:1–10
6. PricewaterhouseCoopers L (2015) The sharing economy: consumer intelligence series. PricewaterhouseCoopers LLP, New York, NY, USA
7. Uber Technologies I (2023) Uber press release. https://investor.uber.com/news-events/news/press-release-details/2023/Uber-Announces-Results-for-Fourth-Quarter-and-Full-Year-2022/default.aspx
8. Huston C (2017) Ride-hailing industry expected to grow eightfold to $285 billion by 2030. https://www.marketwatch.com/story/ride-hailing-industry-expected-to-grow-eightfold-to-285-billion-by-2030-2017-05-24
9. Tafreshian A, Masoud N (2022) A traveler incentive program for promoting community-based ridesharing. To appear in transportation science
10. Bian Z, Liu X (2018) A detour-based pricing mechanism for first-mile ridesharing in connection with rail public transit. In: ASME/IEEE joint rail conference, vol 50978, p V001T04A003. American Society of Mechanical Engineers
11. Jouida SB, Guajardo M, Klibi W, Krichen S (2021) Profit maximizing coalitions with shared capacities in distribution networks. Eur J Oper Res 288(2):480–495
12. Liu Y, Zhang C, Zheng Z, Chen G (2021) Cost-sharing mechanism design for social cost minimization in ridesharing systems. In: International conference on wireless algorithms, systems, and applications. Springer, pp 277–289
13. Özener OÖ, Ergun Ö, Savelsbergh M (2013) Allocating cost of service to customers in inventory routing. Oper Res 61(1):112–125
14. Guajardo M, Rönnqvist M (2016) A review on cost allocation methods in collaborative transportation. Int Trans Oper Res 23(3):371–392
15. GH d. Huang AC, Kai, An K (2020) Solving the station-based one-way carsharing network planning problem with relocations and non-linear demand. Transp Res Part C: Emerg Technol 90:1–17
16. Shaheen S, Cohen A (2020) Innovative mobility: carsharing outlook carsharing market overview, analysis, and trends
17. Lu R, G. H. d. Correia A, Zhao X, Liang X, Lv Y (2021) Performance of one-way carsharing systems under combined strategy of pricing and relocations. Transp B: Transport Dyn 9(11):134–152

18. Hara Y, Hato E (2017) A car sharing auction with temporal-spatial od connection conditions. Transp Res Proc 23:22–40
19. Yang S, Wu J, Sun H, Qu Y, Wang DZ (2022) Integrated optimization of pricing and relocation in the competitive carsharing market: a multi-leader-follower game model. Transp Res Part C: Emerg Technol 138:103613
20. Lv H, Zhang C, Zheng Z, Luo T, Wu F, Chen G (2020) Mechanism design with predicted task revenue for bike sharing systems. Proc AAAI Conf Artif Intell 34:2144–2151
21. Zhang J, Meng M, David Z (2019) A dynamic pricing scheme with negative prices in dockless bike sharing systems. Transp Res Part B: Methodol 127:201–224
22. Sasaki S, Yamamoto E, Namerikawa T (2022) Construction of one-way car-sharing system considering coalition rationality based on mechanism design. IEEE Trans Control Netw Syst 9(4):1655–1665
23. Hsieh F-S, Zhan F-M, Guo Y-H (2019) A solution methodology for carpooling systems based on double auctions and cooperative coevolutionary particle swarms. Appl Intell 49(2):741–763
24. Li R, Nie Y, Liu X (2020) Pricing carpool rides based on schedule displacement. Transp Sci 54(4):1134–1152
25. Agatz N, Erera AL, Savelsbergh MW, Wang X (2011) Dynamic ride-sharing: a simulation study in metro Atlanta. Proc Soc Behav Sci 17:532–550
26. Chau SC-K, Shen S, Zhou Y (2022) Decentralized ride-sharing and vehicle-pooling based on fair cost-sharing mechanisms. IEEE Trans Intell Transp Syst 23(3):1936–1946
27. Lu W, Quadrifoglio L (2019) Fair cost allocation for ridesharing services-modeling, mathematical programming and an algorithm to find the nucleolus. Transp Res Part B: Methodol 121:41–55
28. Peng Z, Shan W, Jia P, Yu B, Jiang Y, Yao B (2020) Stable ride-sharing matching for the commuters with payment design. Transportation 47(1):1–21
29. Santos DO, Xavier EC (2015) Taxi and ride sharing: a dynamic dial-a-ride problem with money as an incentive. Expert Syst Appl 42(19):6728–6737
30. Bian Z, Liu X (2019) Mechanism design for first-mile ridesharing based on personalized requirements part iÃƒÂ¢Ã¢â‚¬Â¬Ã‚Â¯: theoretical analysis in generalized scenarios. Transp Res Part B: Methodol 120:147–171
31. Zhang C, Xie J, Wu F, Gao X, Chen G (2020) Pricing and allocation algorithm designs in dynamic ridesharing system. Theoret Comput Sci 803:94–104
32. Levinger C, Hazon N, Azaria A (2015) Cost allocation for prioritized ride-sharing. Unsubmitted. https://www.ariel.ac.il/wp/noam-hazon/wp-content/uploads/sites/317/2021/02/19.famas_.ridesharing_cost_allocation.pdf
33. Krishnaprasad S, Tripathi RR (2022) A pricing mechanism to improve capacity utilisation in ridesharing. J Oper Res Soc 73(4):741–753
34. Furuhata M, Daniel K, Koenig S, Ordonez F, Dessouky M, Brunet M-E, Cohen L, Wang X (2014) Online cost-sharing mechanism design for demand-responsive transport systems. IEEE Trans Intell Transp Syst 16(2):692–707
35. Janjevic M, Al Farisi A, Nsamzinshuti A, Ndiaye A (2018) Investigating the benefits of shipper-driven collaboration in urban freight transport and the effects of various gain-sharing methods. City logistics 1: new opportunities and challenges, pp 105–124
36. Dahlberg J, Engevall S, Göthe-Lundgren M (2018) Consolidation in urban freight transportation–cost allocation models. Asia-Pac J Oper Res 35(04):1850023
37. Schulte F, Lalla-Ruiz E, Schwarze S, González-Ramírez R, Voß S (2019) Scalable core and shapley value allocation methods for collaborative transportation, tech. rep., Working paper
38. Tinoco SVP, Creemers S, Boute RN (2017) Collaborative shipping under different cost-sharing agreements. Eur J Oper Res 263(3):827–837
39. Zhang W, Uhan NA, Dessouky M, Toriello A (2018) Moulin mechanism design for freight consolidation. Transp Res Part B: Methodol 116:141–162
40. Zhang W, Uhan NA, Dessouky M, Toriello A (2021) Acyclic mechanism design for freight consolidation. To appear in transportation science

41. Liu H, Yu Y, Benjaafar S, Wang H (2022) Price-directed cost sharing and demand allocation among service providers with multiple demand sources and multiple facilities. Manuf Ser Oper Manag 24(1):647–663
42. Verdonck L, Beullens P, Caris A, Ramaekers K, Janssens GK (2016) Analysis of collaborative savings and cost allocation techniques for the cooperative carrier facility location problem. J Oper Res Soc 67(6):853–871
43. Zou H, Dessouky MM, Hu S (2021) An online cost allocation model for horizontal supply chains. Transp Res Part C: Emerg Technol 122:102888
44. Wang Y, Zhang J, Assogba K, Liu Y, Xu M, Wang Y (2018) Collaboration and transportation resource sharing in multiple centers vehicle routing optimization with delivery and pickup. Knowl-Based Syst 160:296–310
45. Chen H, Dong Z, Li G, He K (2021) Remanufacturing process innovation in closed-loop supply chain under cost-sharing mechanism and different power structures. Comput Ind Eng 162:107743
46. Wang X, Agatz N, Erera A (2018) Stable matching for dynamic ride-sharing systems. Transp Sci 52(4):850–867
47. Yan P, Lee C-Y, Chu C, Chen C, Luo Z (2021) Matching and pricing in ride-sharing: optimality, stability, and financial sustainability. Omega 102:102351
48. Zhang J, Wen D, Zeng S (2015) A discounted trade reduction mechanism for dynamic ridesharing pricing. IEEE Trans Intell Transp Syst 17(6):1586–1595
49. Zhao D, Ramchurn SD, Jennings NR (2015) Incentive design for ridesharing with uncertainty. arXiv:1505.01617
50. Hu S, Dessouky MM, Uhan NA, Vayanos P (2021) Cost-sharing mechanism design for ride-sharing. Transp Res Part B: Methodol 150:410–434
51. Asghari M, Shahabi C (2017) An on-line truthful and individually rational pricing mechanism for ride-sharing. In: Proceedings of the 25th ACM SIGSPATIAL international conference on advances in geographic information systems, pp 1–10
52. Rosenthal EC (2017) A cooperative game approach to cost allocation in a rapid-transit network. Transp Res Part B: Methodol 97:64–77
53. Chen E, Lev O, Procaccia AD, Wu X, Zick Y Getting your fare share. Unsubmitted. https://agt2017.net.technion.ac.il/files/2017/07/AGTIJCAI2017_paper_8_ChenLPWZ.pdf
54. Bean WL, Joubert JW (2021) An agent-based implementation of freight receiver and carrier collaboration with cost sharing. Transp Res Interdiscip Perspect 11:100416
55. Liu N, Cheng Y (2020) Allocating cost to freight carriers in horizontal logistic collaborative transportation planning on leading company perspective. Math Prob Eng 2020
56. Kordonis I, Dessouky MM, Ioannou PA (2019) Mechanisms for cooperative freight routing: incentivizing individual participation. IEEE Trans Intell Transp Syst 21(5):2155–2166
57. Papadopoulos A-A, Kordonis I, Dessouky MM, Ioannou PA (2021) Personalized pareto-improving pricing-and-routing schemes for near-optimum freight routing: An alternative approach to congestion pricing. Transp Res Part C: Emerg Technol 125:103004
58. Papadopoulos A-A, Kordonis I, Dessouky M, Ioannou P (2018) Coordinated freight routing with individual incentives for participation. IEEE Trans Intell Transp Syst 20(9):3397–3408
59. Papadopoulos A-A, Kordonis I, Dessouky MM, Ioannou PA Personalized freight route recommendations with system optimality considerations: a utility learning approach. Working paper
60. Bai J, Chen Y (2021) An analysis of the triadic relationship between airline, destination, and travel agency: applying nucleolus theory to assess power loss and the profit allocation approach. Soft Comput 25(8):5881–5893
61. Brenner J, Schäfer G (2007) Cost sharing methods for makespan and completion time scheduling. In: Annual symposium on theoretical aspects of computer science. Springer, pp 670–681
62. Roughgarden T, Sundararajan M (2009) Quantifying inefficiency in cost-sharing mechanisms. J ACM (JACM) 56(4):1–33
63. Mehta A, Roughgarden T, Sundararajan M (2009) Beyond moulin mechanisms. Games Econom Behav 67(1):125–155

64. Donahue K, Kleinberg (2021) Optimality and stability in federated learning: a game-theoretic approach. In: Advances in neural information processing systems, vol 34
65. Chau SC-K, Xu J, Bow W, Elbassioni K (2019) Peer-to-peer energy sharing: effective cost-sharing mechanisms and social efficiency. In: Proceedings of the tenth ACM international conference on future energy systems, pp 215–225
66. Fielbaum A, Kucharski R, Cats O, Alonso-Mora J (2022) How to split the costs and charge the travellers sharing a ride? aligning system's optimum with users' equilibrium. Eur J Oper Res 301(3):956–973
67. Amar HM, Basir OA (2018) A game theoretic solution for the territory sharing problem in social taxi networks. IEEE Trans Intell Transp Syst 19(7):2114–2124
68. Gollapudi S, Kollias K, Panigrahi D (2019) You get what you share: incentives for a sharing economy. Proc AAAI Conf Artif Intell 33:2004–2011
69. Kimms A, Kozeletskyi I (2016) Shapley value-based cost allocation in the cooperative traveling salesman problem under rolling horizon planning. EURO J Transp Logist 5:371–392
70. Kimms A, Kozeletskyi I (2016) Core-based cost allocation in the cooperative traveling salesman problem. Eur J Oper Res 248:910–916 2 2016
71. Castro J, Gómez D, Tejada J (2009) Polynomial calculation of the shapley value based on sampling. Comput Oper Res 36(5):1726–1730
72. Kamar E, Horvitz E (2009) Collaboration and shared plans in the open world: studies of ridesharing. In: Twenty-first international joint conference on artificial intelligence, Citeseer
73. Schmeidler D (1969) The nucleolus of a characteristic function game. SIAM J Appl Math 17(6):1163–1170
74. Moulin H (1999) Incremental cost sharing: Characterization by coalition strategy-proofness. Soc Choice Welfare 16(2):279–320
75. Moulin H, Shenker S (2001) Strategyproof sharing of submodular costs: budget balance versus efficiency. Econ Theor 18(3):511–533
76. Frisk M, Göthe-Lundgren M, Jörnsten K, Rönnqvist M (2010) Cost allocation in collaborative forest transportation. Eur J Oper Res 205(2):448–458
77. Rodríguez-Pereira J, Balcik B, -È. Rancourt M, Laporte G (2021) A cost-sharing mechanism for multi-country partnerships in disaster preparedness. Prod Oper Manag
78. Tijs SH, Driessen TS (1986) Game theory and cost allocation problems. Manage Sci 32(8):1015–1028
79. Cruijssen F, ArgusI B (2012) Collaboration concepts for comodality (co^3). ArgusI BV
80. Brandes W (1991) Maschinenring-their role for farmers' income and impacts on structural change in west Germany. Jpn J Farm Manag 29(3):78–85
81. Matzler K, Veider V, Kathan W (2015) Adapting to the sharing economy. MIT Cambridge, MA, USA
82. Olgun MO (2021) Collaborative airline revenue sharing game with grey demand data," *Central European Journal of Operations Research*, pp 1–22, 2021
83. Kang S, Moon T (2016) Supply chain integration and collaboration for improving supply chain performance: A dynamic capability theory perspective. In: 49th Hawaii international conference on system sciences (HICSS). IEEE, pp 307–316

Chapter 16
Autonomous Driving and People with Disabilities in Greece: Challenges and Promises in Perspective

Elia Vardaki, Maria-Despoina Psaromiligkou, and Vassiliki Petousi

Abstract One of the promises of self-driving vehicles is equity to all social groups. Physically impaired people, people with disabilities, currently face unequal conditions in transportation that restrict their freedom of movement; unfavorably impact their medical, social, and economic life; and even accentuate conditions of social exclusion. Autonomous driving carries the potential to deliver accessibility to transportation and create more favorable living conditions for people with mobility impairments. However, there is still limited research, compared to the general population, addressing the challenges people with mobility disabilities face; the prospects autonomous driving can provide to them; and, more importantly, the attitudes and the perspectives of people with disabilities toward autonomous vehicles and autonomous driving. Nevertheless, such research is of importance since it can inform relevant research and policies and the design of transportation systems advancing social justice and equity. The current article, based on a recent survey, aims at filling the abovementioned gaps concentrating its focus on the Greek context from a user's perspective.

E. Vardaki (✉)
Dynamic Systems and Simulation Laboratory, School of Production Engineering and Management, Technical University of Crete. Associate Researcher in the Gender Lab, Department of Sociology, University of Crete, Rethymno, Greece
e-mail: evardaki@tuc.gr

M.-D. Psaromiligkou
Post-graduate student in Bioethics, Department of Sociology, University of Crete, Rethymno, Greece
e-mail: bioethics277@fks.uoc.gr

V. Petousi
Director of the Gender Lab, Department of Sociology, University of Crete, Rethymno, Greece
e-mail: petousiv@uoc.gr

P. Ioannou and A. A. Malikopoulos (eds.), *Transportation Mobility in Smart Cities*, Springer Tracts on Transportation and Traffic 21,
https://doi.org/10.1007/978-3-031-64769-7_16

16.1 Introduction

Transportation eases our daily lives, creating a pattern of routes that help us sustain our social and economic life. Cities reflect this pattern of our daily movements to the point that transportation services are the sole means of our mobile selves. Traditional forms of transportation services, however, have proven to create unequal relations to various social groups. Specifically, People with Disabilities (PwD) are significantly affected by the unequal access to transportation services. This results in low mobility and difficulties accessing places and spaces, creating the conditions for, what the research calls "transport disadvantage" [7–9]. Such a disadvantage has social and psychological effects on PwD leading to social, economic, and cultural exclusion [11]. For PwD, transportation to access basic everyday needs, such as medical care, education, employment, personal, and recreational activities, is a struggle. Therefore, transportation services are more than commuting; transportation creates relations and accessibility to basic social and economic structures, molds relations of equity, inclusiveness, and trust among different communities.

Autonomous Vehicles (AVs) and its related technology promise to increase accessibility to various marginalized groups like seniors, children, and PwD [15, 21, 23, 25]. AVs can change the lives of PwD by offering flexibility, accessibility, and releasing them from the burden to rely on others or on public services for their transportation. Although the AVs have the potential to help PwD accessing and participating in major aspects of social, economic, and medical structures, there is still a need for more analyses and information pertaining to the factors and conditions of usage and accessibility [6, 20, 26].

Within this broader context our research on user's perspective was designed to map 1) certain characteristics of the current mode of transportation of PwD in Greece that limits their mobility as well as some of the related problems they face and 2) explore the perceptions and attitudes of this vulnerable groups toward AV, the usage and acceptance of AVs as well as the factors that may relate to their perceptions and attitudes.

The first section of this article briefly reviews research on PwD, the second section presents the methodology and the data collected. The third section analyzes the data of the main findings by focusing on the factors that influence acceptance rates, while the fourth and final section discusses the results of the findings shedding light on the factors of the deployment of Avs by PwD and possible directions for further future research on AVs and PwD in Greece.

16.2 Disability and Automobility

According to the World Health Organization (WHO), PwD comprises 15% of the total population worldwide. Eurostat and the European Commission estimate that on average 24% of the EU population has some kind of disability [13, 14]. On the other hand, specific groups may be affected differentially from physical and other disabilities. In the EU, in particular, women with disabilities account for 26.1% of the total female population and men with disabilities for 21.8% of the total male population. Moreover, while PwD between 16 and 65 years old comprise 17.9% of the total PwD population, people aged 65 and over account for 48,5% of the same population. In the next few decades, these numbers are going to increase, because of the aging of the population, environmental reasons such as car accidents, conflicts, dietary issues, either undernourishment or obesity, and the globally increasing chronic health conditions associated with cardiovascular diseases and diabetes [24]. These observations and estimates, however, are of global significance not only in number, but mainly because of the similar social problems and challenges that PwD face worldwide: social exclusion, lack of proper education, unemployment, inability to access healthcare systems, prejudice, low self-esteem, and higher rates of depression. The above can be significantly intensified by transportation conditions [4, 12].

Research shows that PwD are challenged by conventional transportation modes. In public transportation PwD must struggle with malfunctions or lack of ramps and/or lifts, absence or blockage of lower curbs, the lack of audible equipment, and the delayed or canceled services [8, 11]. The use of paratransit, that is, a point-to-point system can be free or subsidized by local administrations, and more flexible compared to public transportation [10]. However, the availability of paratransit can be low, costly to local governments, and offers less flexibility than private mode [1, 7]. Private transportation on the other hand accords flexibility but driving can be challenging for some individuals, who may thus have to rely on others for their transportation, decreasing their self-esteem and independence consequently [26]. Finally, the cost of private transportation tends to be high and can be afforded by very few. As a result, the "transport disadvantage" PwD come up against when they use mobility services is maintained [9].

AVs, on the other hand, could potentially remedy such a disruptor in PwD lives by creating the circumstances for accessibility. However, to diffuse this technology into the communities of PwD, it is important to understand their disposition toward AVs, the needs that should be accommodated, and their acceptability rate of autonomous transportation. There is an increasing interest in the research community in the relationship between PwD and AVs technology [3, 6, 20]. Research findings, however, produce conflicting and at times contradictory results in terms of acceptance and use of AV. Specifically, one group of authors found that PwD hold a positive attitude toward AV and its use [5, 6, 17]. Similarly, Chng and Cheah (2020) [5] and Cordts et al. (2021) [6] concluded that people with physical disabilities had an overall positive attitude toward AV. Correspondingly, Hwang et al. (2021) [17] report that their sample, mostly comprised of people with physical disabilities, held positive attitudes toward AV, and willing to use if it was cheaper than conventional vehicles. In contrast, other groups of researchers found that PwD are not enthusiastic about AV [3, 18]. Specifically, Bennett et al. (2019) [3] report that around 64% of PwD in their sample had negative or neutral views of AV. Moreover, Kassens-Noor et al. (2020) [18] found that half of the sample of PwD have lower odds to ride an AV compared to others. Despite the different findings, safety, reliability, accessibility, and cost appear to be the four most important factors on the willingness to drive an AV. Although these factors were found to be significant for the general population as well [20, 22, 23], given the special conditions present in PwD and the associated vulnerabilities it is necessary to investigate how these factors specifically affect PwD and interrelate to their transport mobility and their overall welfare [6, 12].

In Greece, the country in focus is estimated that approximately 23,1% have some type of impairment. Comprehensive and analytical official national data are scant. The National Association of People with Disabilities (NAPwD) questions the reliability of the available data and warns of the consequences this may have for funding, policies, quality of healthcare provisions, insurance, and the overall well-being of PwD.

Reports of agencies such as the Observatory of Issues of Disability (OID) of the NAPwD, however, point to high levels of prejudice and social exclusion among PwD in Greece that in turn result in unemployment, poverty, poor health care, low education attainment, and even domestic violence. According to the above reports, only 15.8% of PwD complete higher education in Greece compared to the 32.4% of the total population. Moreover, employment is significantly low for PwD. Although highly variable depending on the area of the country, the Employment Index of PwD drops as little as 9.5% for disabled persons aged 20–64 years [13].

Issues of accessibility and transportation are particularly challenging for PwD in Greece in almost every available report on disability. A dedicated OID report details transportation hurdles for PwD [13]. Although the emphasis is on Athens and the Region of Attica analogous are the conditions in most parts of the country. According to this report 33.6% of PwD face severe problems of accessibility in their area of residence related to sidewalks, public transport stops, shops, or services. The abovementioned report points to the unreliability of public transport, the lack of access for PwD such as ramps, or stripes, and other facilities and equipment that restrict their mobility and augment their challenges. Consequently, PwD must rely on caretakers and family members to meet their everyday needs of mobility and transportation, significantly increasing their dependence, and, in the long-run, social exclusion and isolation.

Finally, the report calls for Disability Action Plan that will ensure (a) equal opportunities for all citizens, including PwD; (b) equal participation in the local society and economy; and (c) highlight the importance of their integration for local sustainable economic development. There is no prior research on AV acceptability among PwD for Greece. Given the significance of mobility and transportation for PwD's social inclusion and independence, the potential of AVs to reduce or even minimize such problems along with the lack of relevant data for PwD overall, in Greece even more so, the importance of studying PwD's attitudes toward AVs is evident.

Thus, we designed an exploratory study that investigates the factors that relate to PwD's willingness to use AVs. We understand [6, 12] that different types of disability may result in differential attitudes toward AVs. In our study, we emphasize exclusively on physical mobility disabilities and other impairments (e.g., deafness) that may restrict a person's driving ability and thus their mobility. Additionally, we investigate PwD's knowledge and understanding of AV technology and we report on certain characteristics of the current mobility status of people in our sample. As an intermediate step toward our goal, we investigate our sample's knowledge and understanding of AV technology.

Based on the above, the goal of our research is twofold: (1) to describe certain characteristics of the current mode of transportation of people in our sample and their related problems and (2) to explore the factors that may relate to our sample's willingness to use AVs when these will become available.

16.3 Methodology

To meet our goals, a web survey was designed including a total of 109 questionnaires divided into 4 sections. The first section included questions on sociodemographic characteristics and the second addressed characteristics of our sample's current mode of transportation. The third section of the survey inquired into our respondents' knowledge of AVs and asked participants to rate several statements related to factors such as trust, safety, risk, and intention to use. In the last section, participants were asked to give their opinion on the validity and the direction of research on AV and PwD.

Most questions were close-ended and measured on a 6-point Likert scale. The survey was available through a web page https://autovehbio.weebly.com created specifically for the purposes of the research. The web page provided information on the research, made available the information sheet and informed consent forms for participation to the research and two questionnaires: one for disabled persons and one for caregivers. In this presentation, we focus on findings from our sample of disabled persons. More importantly, the web page provided information on AVs including a video explaining AV technology at the most progressed level SAE 5.

Our research is based on a convenience sample. Our sampling was based on the use of personal contacts and facilitators (people holding key positions in organizations of disabled persons who agreed to forward recruitment information to members of their organizations) as well as implementation of the snowball technique.

Data for this study are anonymous. No potentially identifying information of participants was collected. Our research was approved by the Ethics Committee of the University of Crete.

Data was analyzed using SPSS Version 29 with license from the University of Crete.

16.4 Results

16.4.1 Demographics

A total of 109 people replied to our questionnaire. This is a relatively small sample compared to the overall population of PwD in the country. However, it allows analysis of the data for the exploratory research that we set up to do.

Women made up over half of our sample (55%). [VP1] The majority of our participants fall within 35 and 55 years of age (n = 39, % = 35, 8) followed by the age group 56–75 and over (n = 39, % = 34.9) and the age group 18–35 (n = 31, % = 29.4) (Table 16.1).

Participants live in 12 different cities and towns in Greece. Approximately 45.9% of our sample resides in Athens, 13,8% in Komotini, a small town in the North of Greece, and 11% in Patra, a rather large city in Peloponnese. Overall, and despite its small size, our sample covers a large geographical area of the country.

Table 16.1 Sample description: Demographics

Demographic characteristic	n	%
Gender		
Male	49	45.00
Female	60	55.00
Age		
18–25	11	10.10
26–35	21	19.30
36–45	30	27.50
46–55	9	8.30
56–65	24	22.00
66–75	10	9.20
66+	14	12.80
Place of residence		
Metropolitan area	59	54.10
Cities over 100,000	19	17.40
Cities under 100,000	31	28.40
Type of disability		
Moving disability	89	81.70
Hearing disability	20	18.30
Education level		
(Up to and) Secondary education	20	18.30
Tertiary education	46	42.20
Post-graduate education	43	39.40

For analytical purposes and keeping in mind that moving and traffic conditions would differ depending on the size of a town/city, place of residence was grouped in three categories based on size of population: Metropolitan areas including Athens and Thessaloniki, cities over 100,000 people (Patra, Volos, Chania) and cities under 100,000 population (Komotini, Drama, Rethymno, Sparti, Livadia, Pyrgos, Serres). Based on this regrouping, most of our sample lives in metropolitan areas (54.10%) followed by people living in cities with population less than 100,000 people (28.40%). However, the percentage of people living in metropolitan areas and large cities comprise a very high proportion of our sample (approximately 72%) (Table 16.1). Consequently, we can deduce that most of our participants face significant traffic and parking problems usual in metropolitan areas and large cities.

The focus is on people with moving disabilities and/or disabilities that restrict/impair driving. Based on their responses, participants' disabilities included among others paraplegia, amputation, and multiple sclerosis. These represent 87.7% of participants' impairments. However, approximately 18% of our sample reported hearing impairments and other impairments that impact upon driving and thus their moving ability and consequently raise issue of accessibility. For descriptive and analytical purposes, impairments were grouped into two categories: moving impairments (e.g., paraplegia, amputation, etc.) and hearing impairments (e.g., deafness, etc.).

Our sample tends to be highly educated. Specifically, 42.2% of our sample has completed tertiary education, 39.40% has a post-graduate degree (mostly master's degree) while 5.5% of our sample hold a doctoral degree (Table 16.1).

16.4.2 Current Conditions of Transportation

Based on their responses most of our participants utilize private transportation means (89%) and correspondingly prefer to use private transportation means (94.50%). Over 70% of our sample (71.60%) do not drive themselves because they are unable to drive (60.60%), face traffic and parking problems (27.50%), or find driving difficult or tiring (11.90%) (Table 16.2).

Most of them (89%) have not been involved in traffic accidents neither as drivers nor as passengers while those who have been involved in traffic accident have experienced a small number of traffic accidents (Table 16.2).

Participants were asked to rate the importance of their main reason of transportation. Specifically, they were asked: "Think of your regular weekly schedule and the transportation required. Rate the importance of each reason of transportation indicated below by marking 1 = not important to 5 very important."

As shown in the table (Table 16.3) medical reasons account for the most important transportation reason for most of our sample (93.6%) which correspond with the results of similar research in PwD [6, 12, 26]. Personal reasons ranked second in importance (70.60%) followed by social (55.60%) while transportation for work ranked 4th in importance (45.90%). Other reasons for transportation ranked fifth in importance. Approximately 42% of our sample reported as other reasons for

Table 16.2 Sample description: Current conditions of transportation

Transportation	n	%
Main means of transportation		
Private	97	89.00
Public (+taxi)	12	11.00
Preference of means of transportation		
Private means	103	94.50
No preference	6	5.50
Driving		
No	78	71.60
Yes	31	28.40
Main transportation problem		
Unable to drive	66	60.60
Traffic/parking	30	27.50
Driving is tiring/difficult	13	11.90
Involvement in traffic accident (driver)		
No	89	81.70
Yes	20	18.30
Involvement in traffic accident (other driver)		
No	98	89.99
Yes	11	10.10
Number in traffic accidents involved		
0	79	72.50
1	21	19.30
2+	9	8.20

Table 16.3 Sample description: Current conditions of transportation

Reason	Importance		
	Important	Neutral	Not-important
Medical	93.60	2.80	3.70
Personal	70.60	16.50	12.80
Social	56.90	23.90	19.30
Work	45.90	16.50	37.60
Other	42.20	31.20	26.60

transportation sports, walking dogs, emergency trips, leisure/tourism, and groceries. Our data shows the importance of transportation for people with disabilities not only because, as expected, of medical reasons but also because of personal and social reasons. From the above, the importance of transportation for social inclusion and connectivity becomes evident.

16.4.3 Knowledge of AV Technology

Research has shown that knowledge of AV technology similar to knowledge of other types of innovative technologies has an impact on people's acceptance of the specific technology. Thus, of interest to our research was participants' knowledge of AV technology. As shown by their responses, most of our participants do not have adequate knowledge about AVs. Specifically, approximately 59% (59,70%) of our sample stated that they heard of AVs for the first time on our survey while 21.10% of our respondents stated that they had some knowledge of AVs. A small percentage of respondents (8.30%) stated that they had adequate knowledge of AVs.

16.4.4 Willingness to Travel on an AV

Asked specifically about their willingness to travel on an AV approximately 72% (71.6%) of our sample responded positively while a 16.50% declared that they could not decide and 12% stated a negative attitude toward traveling on an AV (Table 16.4).

To investigate the factors that are likely to relate to people's willingness to travel on an AV we performed a series of chi-square tests. Participants' willingness did not differ in statistically significant ways based on gender, place of residence, type of disability, and knowledge about AV. On the contrary, sample participants' willingness to travel with an AV differed in statistically significant ways based on age and education level (Table 16.4). Specifically, more willing to travel on an AV are people in the age group 36–55 followed by people in the age group 18–35. Least likely to want to travel on an AV are people in the age group 56–75 and over. Similarly, people in the same age group are the ones most likely not to be able to decide (Table 16.4).

People who have complete tertiary education are the ones most likely to want to travel on an AV while people who have completed secondary education are the ones

Table 16.4 Willingness to travel on an AV by age and education level

Statement: *I want to travel with an AV*	*n*	%	Age			Educational level		
			18–35	36–55	56–75+	(Up to) Secondary education	Tertiary education	Post-graduate education
YES	78	71.6	37.2	39.7	23.1	11.5	50.0	38.5
Can't decide	18	16.5	0.0	27.8	72.2	22.2	22.2	55.6
NO	19	11.9	23.1	23.1	53.8	53.8	23.1	23.1
Total	109	100	Chi-square 19,993 $p < 0.001$			Chi-square 17,086 $p < 0.001$		

most likely to not want to travel on an AV. Interestingly, the people most likely to report that they cannot decide whether they would want to travel on an AV are people with post-graduate education (Table 16.4).

16.4.5 Willingness to Buy and Use of AV

To examine the conditions under which PwD are willing to buy or use AV, we asked to state to what extent they agree or disagree with several statements on a scale from 1 to 5 (1 = fully disagree to 5 = fully agree and 6 = I don't know). Responses were recorded into four new categories: 1 = fully disagree and disagree, 2 = neither agree nor disagree, 3 = agree and fully agree, and 4 = I don't know. Safety, reliability, and usefulness rank higher among the conditions that need to be met for someone to be willing to buy or use and AV. Most of our participants would agree to use or buy an AV if they were convinced for its safety (81.7%), reliability (81.7%), and usefulness (74.3%) (Table 16.5).

Chi-square tests for safety, reliability, and usefulness and variables such as education, type of disability, place of residence, and knowledge about AV did not produce reliable results. A difference although not statistically significant was noted for gender. Women (52.5%) more so than men stated that they would agree to buy or use an AV when it would become affordable. Chi-square tests for affordability and age produced statistically significant results at the 0.001 level (Table 16.5). Of those agreeing with the statement that they would buy or use an AV if it became affordable 41% belonged to the 18–35 age group followed by the 36–55 age group (39.3%). Of those disagreeing with the statement on affordability, 50% belong to the age group 56–75 and over.

Table 16.5 Conditions for buying and using an AV

Statements (ranked by frequency/agree): *I would buy or use an AV*	Agree (%)	Neither agree nor disagree (%)	Disagree (%)	Don't know (%)
If I were convinced for their safety	81.7	4.6	11.0	2.8
If I were convinced for their reliability	81.7	4.6	11.0	2.8
If I were convinced for their usefulness	74.3	11.0	11.0	3.7
When it becomes affordable	56.0	14.7	12.8	16.5
When it becomes possible	45.9	19.3	13.8	21.1
If I saw others use it	25.7	23.9	45.0	5.5
Chi-square (affordable$*$ age) 35.071 $p < 0.001$				

16.4.6 Conditions of Use of AV Compared to Use of Current Transportation Means

Among the issues we wanted to address in our research was a comparison between our sample's appreciation of their current mode of transportation and the potential future use of AV. Our participants were asked to state to what extent they agree or disagree on a scale from 1 to 6 (1 = fully disagree to 5 = fully agree and 6 = I don't know) with several statements. Responses were recorded into four new categories: 1= fully disagree and disagree, 2 = neither agree nor disagree, 3 = agree and fully agree, and 4 = I don't know.

As shown in Table 16.6, 44% of our sample disagrees with the statement that their current means of transportation meets their needs. On the other hand, approximately 34% (33.9%) of our sample state that they would prefer to use an AV because they believe it would meet their needs. An almost equally high proportion (33.0%) of our sample however neither agrees nor disagrees with this statement while another 22.9% of the sample states that they don't know whether an AV would meet their needs (Table 16.6).

Table 16.6 Preference of current means of transportation vs. AV and conditions of use of AV

Statements	$\bar{x}$	Median	Agree (%)	Neither agree nor disagree (%)	Disagree (%)	Don't know (%)
For my main purpose of transportation						
I prefer my current one because my needs are met	1.76	2.00	20.2	35.8	44.0	0.0
I would prefer to use an AV because I believe it will meet my needs	2.69	3.00	33.9	33.0	10.1	22.9
If for my main purpose of transportation had to choose between my current mode of transportation or an AV, I could choose the one that						
Allows better use of transportation time	2.89	3.00	90.8	5.5	2.8	0.9
Is better tested in the market 2.86	3.00	89.0	5.5	4.6	0.9	
Is cheaper 2.66	3.00	74.3	17.4	8.3	0.0	
Allows the driver full control in case of an accident	2.44	3.00	46.5	22.9	20.2	7.3
Is friendlier to the environment	2.44	3.00	36.7	54.4	11.0	0.9

Chi-square tests were estimated for each of the statements in Table 16.6 with gender, age, education, place of residence, type of disability, and knowledge about AV.

Preference of the current means of transportation differs in statistically significant ways between ages (chi-square 9.852 $p < 0.050$) and type of disability (chi-square 9.633 $p < 0.050$). Specifically, people in the 18–35 age group (43.8%) are more likely to disagree that their current means of transportation meets their needs, people in the age group 36–55 (46.2%) are most likely to neither agree nor agree with the statement, while people in the age group 56–75+ are more likely (45,5%) to agree that their current mode of transportation meets their needs.

Moreover, people with moving disabilities are more likely (85.4%) than people with hearing disabilities to disagree that their current condition of transportation meets their needs. Similarly, they are more likely (89.7%) to hold a neutral position.

Age, type of disability, and knowledge about AV differentiate preferences for an AV in statistically significant ways (chi-square 12.927 $p < 0.05$, chi-square 12.917 $p < 0.050$, and chi-square 33.778 $p < 0.050$, respectively). Specifically, participants in the age groups 18–35 and 36–55 are more likely to (40,5% for both groups) agree that an AV would better meet their transportation needs. People in the age group 56–75+ are more likely (45.5%) to disagree that an AV would meet their transportation needs. The same group of people report the highest percentage of I don't know (60%). Moreover, people who heard about AV for the first time in the context of this research stated (84%) that they do not know whether an AV would meet their needs, were neutral (75%) or disagreed (45.5%). This indicates that knowledge about AV is important for the implementation of this transportation mode.

Chi-square tests for the rest of the statements concerning conditions of acceptance of AV compared to the current mode of" and gender (chi-square 14.780 $p < 0.001$). Specifically, women more so than men (65.4%) agreed that if they had to choose between their current mode of transportation and an AV, they would prefer to use the cheaper one.

16.4.7 Safety

Attitudes toward safety were measured with two statements: "I would feel safe with AV's circulation if I were a pedestrian" and "I would feel safe with AV circulation if I were a passenger in one of them." Overall, participants appear to be rather uncertain about safety issues as indicated by the mean of their responses to the above statements (mean of 1st statement = 1.8 and for 2nd statement = 2.3) (Table 16.7). There is, however, some indication that they may tend to report a higher sense of security as passengers compared to their feeling of security as pedestrians. However, the median of their responses is 2 indicating a tendency toward a neutral position (2 = neither agree nor disagree) or else an uncertainty about whether they would be safe as pedestrians and/or passengers.

Chi-square tests did not produce any statistically significant differences in relation to gender, age, education, etc. However, additional exploratory analyses (not presented here) provide indications that gender and age are variables that need to be further studied in relation to attitudes toward all aspects of AV.

16.4.8 Privacy

Participants were asked to state their agreement to the following two statements: "In order to improve my transportation I would feel comfortable sharing data on destination with other passengers" and "In order to improve my transportation I would feel comfortable sharing my phone data with certified private companies." Based on the mean of their responses, participants appear to feel somewhat more comfortable sharing information about their destination with other passengers than sharing phone data with private companies (mean 2.11 and 1.9, respectively). However, the median of their responses is 2 indicating a neutral position (2 = neither agree nor disagree) (Table 16.7).

Table 16.7 Perceptions of safety, privacy, and use of time

Statements	$\bar{x}$	Median	Agree (%)	Neither agree nor disagree (%)	Disagree (%)	Don't know (%)
Safety						
I would feel safe with AV's circulation if I						
Were a pedestrian	1.8	2.0	28.4	34.9	21.1	15.6
Were a passenger in one of them	2.3	2.0	11.9	33.0	46.8	8.3
Privacy						
In order to improve my transportation, I would feel comfortable sharing data on destination						
With other passengers	2.1	2.0	34.9	25.7	33.9	5.5
With certified private companies	1.9	2.0	25.7	21.1	46.8	6.4
Use of time[a]						
If I were traveling on an AV, I would rather						
Interact with other passengers	2.30	3.00	59.6	11.0	29.4	0.0
Observe driving behavior	2.09	2.00	35.8	37.6	26.6	0.0
Do something fun	2.11	2.00	33.0	22.0	33.0	0.0
Take a nap 1.81	1.00	32.1	17.4	50.5	0.0	
Make phone calls	1.77	1.00	27.5	22.0	50.5	0.0
Do something else	1.47	1.00	2.8	17.4	71.6	0.0

[a] Statements ranked by frequency/agree

Chi-square tests showed that participants' attitudes toward sharing information on destination with other passengers differs in statistically significant ways in relation to age (Chi-square 35.3 $p < 0.001$). As it may be expected given the differential experiences familiarity, and notions of privacy differ between older and younger generations. In our sample, 60.5% of the people who stated they would feel comfortable sharing data on their destination with other passengers belongs to the 18–35 age group. On the other hand, 54.1% of those stating disagreement belong to the 56–75+ age group. The same group accounts for the highest percentage (66.7%) who stated that they do not know.

Similarly, comfort sharing phone data with certified private companies differs in statistically significant ways in relation to age (chi-square 30.848 $p < 0.001$) Thus, 60.7% of those stating they would feel comfortable sharing phone data with certified private companies belong to the 18–35 age group while 51.06% of those stating disagreement with the statement belong to the 56–75+ age group. Noted, however, that for this statement 34.8% of the 18–35 age group and the 56–75+ age group stated a neutral position (neither agree nor disagree). For the age group 36–55, the equivalent percentage was 30.4%. This may indicate a more "reserved" or uncertain position when private companies are involved.

16.4.9 Use of Time

As our study shows (Table 16.7) making better use of transportation time is an important consideration in the participants' choice between their current means of transportation and an AV. With a number of related statements, we looked into how they would make better use of their transportation time. Based on our findings almost 60% (59.6%) of our sample they would rather interact with other passengers when riding an AV followed by 35.8% of our sample that would observe the driving behavior of the vehicle.

Chi-square tests showed statistically significant differences for three types of activities: "interact with other passengers," "observe driving behavior," and "do something fun." Specifically, the preference to interact with other passengers differed in statistically significant ways based on type of disability (chi-square 9.44 $p < 0.05$). People with moving disabilities are more likely (90.8%) compared to people with hearing impairments to report they would rather interact with other people while riding on an AV. Although this may be expected (hearing impairments by function make casual interaction more challenging), it does point to the necessity that while designing AVs type of disability should be considered.

Observance of driving behavior differed in statistically significant ways on the basis of gender (chi-square 18.939 $p < 0.001$). Men (71.8%) more so than women were likely to report while riding an AV they would rather observe the driving behavior of the vehicle.

"Doing something fun" while riding in an AV differed in statistically significant ways based on gender (chi-square 6.75 $p < 0.05$), age (chi-square 26.047 $p < 0.001$), type of disability (chi-square 8.221 $p < 0.05$), and knowledge about AV (chi-square 16.550 $p < 0.051$). Men (51%) more so than women (49%) are likely to what to do something fun while riding an AV. Similarly, younger people in the 18–35 age group more (49,9%) than other age groups are likely to report that they would rather do something fun. Likewise, people with moving impairments (87,8%) more so than people with hearing impairments would like to do something fun while riding in an AV. Finally, people who have no knowledge about AV (heard here for the first time) are much more likely than people with some or much knowledge about AV to disagree (77.8%) about doing something fun while riding on an AV.

16.5 Discussion

"Transportation disadvantage" is a severely affects PwD. They need to struggle with in their everyday lives that creates unequal conditions in transportation, restricting their freedom of movement; unfavorably impact their medical, social, and economic life; and even accentuating conditions of social exclusion. Autonomous driving carries the potential to deliver accessibility to transportation and create more favorable living conditions for people with mobility impairments. After all, one of the promises of self-driving vehicles is equity to all groups. However, for AVs to eventually deliver their promise, the social groups to which they are addressed to (e.g., PwD) need to favorably accept their use.

In recent years, research has paid increasing attention to the conditions that need to be present for PwD to accept the use of AV technology [6, 9, 12, 20, 26]. In our research, we addressed this issue by focusing on Greece for which relevant research is missing and we mapped the current conditions of transportation for PwD and the factors that may relate to their acceptance and use of AVs. Our aim was to understand from a user's perspective and context specific, the attitudes and perception of PwD for AVs in Greece.

Our research shows that people in our sample tend to use (Table 16.1) and prefer to use private (Table 16.2) means of transportation. This is easily explained considering, as mentioned above, the lack of infrastructure and dependable public services catering to the needs of PwD that are nonetheless state subsidized. Even so, a large proportion of our sample (44%) report that their current means of transportation does not meet their needs (Table 16.6). Although our data does not provide conclusive evidence, taking into consideration at least one element of the "transportation disadvantage" dependency on others—we could point as one of the potential explanations the very large proportion of our sample, almost 72% (71.60%) who do not drive themselves (Table 16.2). This points to conditions of dependency for their mobility one element that AVs could potentially minimize. Consequently, it is easily understandable why approximately over one-third of our sample (33.9%) would prefer an AV over their current mode of transportation (Table 16.6). Nevertheless, another third of our sample (33.0%) are uncertain about whether they would prefer an AV over their current mode of transportation (Table 16.6).

Our analysis further showed that preference of an AV over current mode of transportation differs in statistically significant ways, among others, based on knowledge about AV. The more people in our sample claim they know about AV, the more likely it is to consider an AV that would better meet their needs of transportation. This points to the importance of informing people, potential users, that is, of the AV technology and the impact it may have on their everyday lives [21]. This need was well revealed in our research. The larger proportion of our sample had a low level of knowledge about AV. Many of them heard about this technology for the first time in the contexts of the current research. Our analysis also pointed to the importance of age as a differentiating factor, since younger people were more likely to hold a positive attitude toward the use of AVs versus their current use of transportation, and the type of disability [6].

This over positive attitude toward AVs compared to their current mode of transportation was further attested in our sample's willingness to travel on an AV. Almost 72% (71.6%) of our sample would be willing to travel on an AV (Table 16.4). Nevertheless, age and education level were found to be an influential factor since young people and those who had completed tertiary education were more likely to do so (Table 16.4). The effect of the age factor, however, is mixed in the general population with the Becker and Axhausen 2017 [2] stating that the younger generation is more likely to be informed and willing to use an AV than older generation, while Nordhoff et al. [23] did not find it of any significance in the willingness to use AVs.

International research has found that PwD perceptions about safety, reliability, and usefulness of AV technology relates to their level of AV technology acceptance and consequently their willingness to purchase and use AVs [16, 22, 27]. Our research shows that similar finding holds true for our sample. Safety ranked first among the conditions upon which our respondents would base their decision to buy or use an AV followed by reliability and usefulness (Table 16.5). Affordability, a condition of particular importance in studies in the general population [22], ranked fifth in our sample. Nevertheless, chi-squared tests showed that affordability differed in statistically significant ways in relation to age (Table 16.5). Specifically, young people (aged 18–35) were more likely than other age groups to take affordability into account when deciding about whether to buy or use an AV. Although our data cannot lead to definite conclusions, we could infer that young people who are most likely to have lower income compared to other age groups are more likely to consider this factor.

Another important finding in our research is the differentiation based on age, gender, and type of disability in the use of time while riding and AV. Younger people and people with mobility disabilities (compared to people with physical disabilities) are more likely to use their time interacting with other passengers or doing "fun" things. Men, on the other hand, more so than women are likely to observe the vehicle's driving behavior. What is important here to note, once again, is that PwD should not be considered as a homogeneous social group. Sociodemographic factors such as gender, age, education, potentially income, and type of disability should be correlated, analyzed, and considered in the development of technology and most definitely in policies.

The above are of particular importance in Greece, and in other countries or places where infrastructure, facilities, policies, and social attitudes toward PwD are lacking or are inadequate. Meeting the needs of PwD for independence, mobility, and social participation of crucial for their socio-political participation. Especially since AVs technology promises equity and social inclusivity to deliver, the particularities of this group everyday lives need to be seriously considered. Nevertheless, it is the voices of this group that need to be heard through a context-specific and targeted research.

In addition to all other factors, in Greece, special attention must be given to the diffusion of information about AV technology and its potential among PwD. Noteworthy, at least in our sample, is the low level of prior knowledge of AVs which is due to the low integration of such technology to the Greek transportation landscape. This lack of knowledge is a matter for further investigation since it shows that all the research done in Greece on autonomous vehicle remains hidden on the ivory towers of academia and research hence not diffused to the public or influence the designing of future transportation policies. For PwD, specifically, innovative technology can further improve the overall quality of lives by maximizing their accessibility to all social, economic, and cultural resources.

The present research is an initial step toward understanding the current needs and the perceptions of the emerging AV technology among a relatively small sample of PwD. Given, however, the lack of similar research in the country, our research can be used as a steppingstone first for future research on larger samples more inclusive of additional forms of disabilities and a wider range of socio-economic and education backgrounds of participants. Future research should also employ more engaging methodologies that will allow a deeper inquiry into conditions and perceptions. Secondly and final point, our research can inform policy-makers to account all the parameters of accessibility and acceptance as well as to carefully consider the intertwining of variables in designing and implementing equal and inclusive transportation policies that can remedy the "transportation disadvantage" of people with disabilities.

Ethics Statement The study was approved by the Research Ethics Committee of the University of Crete (Ref. No 102/16.06.2022). Informed consent was obtained from study participants before the completion of the questionnaire. No personal identifiers were collected in the study and participants cannot be identified. Data collection and anonymization was reviewed and approved by the Research Ethics Committee of the University of Crete.

References

1. Bazaras D, a. Verseckiene, (2013) Analyzing the aspect of organizing paratransit service in Vilnius. Transport 28(1):60–68
2. Becker F, Axhausen KW (2017) Predicting the use of Automated Vehicles. First results from the pilot survey, ETH Zurich, ETH Zurich Research Collection

3. Bennett R, Vijaypogal R, Kottasz R (2019) Attitudes towards autonomous vehicles among people with physical disabilities. Transp Res Part A: Policy Pract 127:1–17
4. Bezyak J, Sabella S, Gattis R (2017) Public transportation: investigation of barriers for people with disabilities. J Disability Policy Stud 28(1):52–60
5. Chng S, Chea L (2020) Understanding autonomous road public transport acceptance: a study of Singapore. Sustainability 12(12)
6. Cordts P, Cotten S, Qu T, Bush TR (2021) Mobility challenges and perceptions of autonomous vehicles for individuals with physical disabilities. Disability Health J (14)
7. Currie G, Delbosc A (2011) transport disadvantage: a review. new perspectives and methods in transport and social exclusion research. G. Currie. Bingley, Emerald Group Publishing Limited, pp 12–25
8. Currie G, Stanley J (2007) No way to go: Transport and social disadvantage in Australian communities
9. Darcy S, Francis BP (2018) On the road again: The barriers and benefits of automobility for people with disabilities. Transp Res Part A(107) 229–245
10. Deka V, Gonzales E (2014) The generators of paratransit trips by persons with disabilities. Transp Res Part A: Policy Pract 70:181–193
11. Delbosc A, Currie G (2011) Transport problems that matter-Social and psychological links to transport disadvantage. J Transp Geogr 19(1):170–178
12. Dicianno B, Sivakathan S, Sundaram AS, Satpute S, Kulich H, Powers E, Deepak N, Russell R, Cooper R, Cooper RA (2021) Systematic review: automated vehicles and services for people with disabilities. Neurosci Lett 761:136103
13. Disability Issues Observatory (2022). In Attica, the highest percentage of people with severe disabilites, who face problems relating to accessibility: Proposal of the National Confederation of People with Disabilites for a regional Action Plan in Attica. https://www.paratiritirioanapirias.gr/el/news/newsletters/82/sthn-attikh-ypshlotero-pososto-atomwn-me-sobarh-anaphria-sth-xwra-poy-antimetwpizoyn-problhmata-prosbasimothtas-protash-ths-esamea-gia-ena-perifereiako-sxedio-drashs-sthn-attikh
14. European Council (2022) Disability in the EU: facts and figures, Council of the European Union General Secretariat. https://www.consilium.europa.eu/en/infographics/disability-eu-facts-figures/
15. Etzioni S, Hamadneh J, Elvarsson AB, Esztergar-Kiss D, Djukanovic M, Neophytou SN, Sodnik J, Polydoropoulou A, Tsouros I, Pronello C, Thomopoulos N (2020) Modeling cross-national differences in automated vehicle acceptance. Sustainability 12
16. Fagnant DJ, Kockelman K (2015) Preparing a nation for autonomous vehicles: opportunities, barrieres and policy recommendations. Transp Res Part A 77:167181
17. Hwang J, Li W, Staugh L, Lee C, Turnbul K (2020) A focus group study on the potential of autonomous vehicles as a viable transportation option: Perspective from people with disabilities and public transport agencies. Transp Res Part F: Traffic Psychol Behav 70:260–274
18. Kassens-Noor E, Karamchandani ZK, Kai M (2020) Willingness to ride and perceptions of autonomous public transit. Transp Res Part A: Policy Pract 138:92–104
19. Kyriakidis M, Happee R, d. W. J.C.F (2015) Public opinion on automated driving: results of an international questionnaire among 5000 respondents. Transp Res Part F. 322:126–140
20. Kyriakidis M, Sodnik J, Stojmenova K, Elvarsson A, Pronello C, Thomopoulos N (2020) The role of human operators in safety perception of AV deployment-Insights from a large European survey. Sustainability 9166(12)
21. Milakis D, Mueller S (2021) The societal dimension of the automated vehicles transition: Towards a research agenda. Cities(113)
22. Milakis D, van Arem B, van Wee B (2017) Policy and society related implications of automated driving: a review of literature and directions for future research. J Intell Transp Syst 21(4):324–348
23. Nordhoff S, Winter JD, Kyriakidis M, Arem BV, Happee R (2018) Acceptance of driverless vehicles: results from a large cross-national questionnaire study. J Adv Transp 1–22

24. World Health Organization. (2021) Disability and Health-some key facts. Retrieved https://www.who.int/news-room/fact-sheets/detail/disability-and-health, 2023
25. Othman K (2021) Public acceptance and perception of autonomous vehicles: a comprehensive review. AI and Ethics(1): 355–387
26. Petrovic D, Mijailovic R, Pesic D (2022) Persons with physical disabilities and autonomous driving: the perspective of the driving status. Transp Res Part A 164:98–110
27. Zhang T, Zeng W, Yanxuan Z, Tao D, Li G, Qu X (2021) What drives people to use automated vehicles? A meta-analytical review. Acc Anal Prevent 159

Chapter 17
Intelligent Transportation System Solutions in Disadvantaged Communities

Matthew J. Barth, Kanok Boriboonsomsin, Guoyuan Wu, and Peng Hao

Abstract Many cities across the world are looking to use technology and innovation to improve the overall efficiency and safety for their residents. At the heart of these smart-city plans, a variety of intelligent transportation system technologies can be used to improve safety, enhance mobility measures (e.g., traffic flow), and minimize environmental impacts of a city's mobility ecosystem. Early implementations of these ITS technologies often take place in affluent cities, where there are many funding opportunities and suitable areas for deployment. However, it is critical that we also develop smart city solutions that are focused on improving conditions of disadvantaged and environmental justice communities, whose residents have suffered the most from unmitigated urban sprawl and its environmental and health impacts. As a leading example, Inland Southern California has grown to be one of the largest hubs of goods movement in the world. Numerous logistics facilities such as warehouses, rail facilities, and truck depots have rapidly spread throughout these communities, with the local residents bearing a disproportionate burden of truck traffic, poor air quality, and adverse health effects. Further, the majority of residents have lower-wage jobs and very few mobility options, other than low-end personal car ownership. To improve this situation, UC Riverside researchers have focused their smart city research on these impacted communities, finding innovative solutions to eco-friendly traffic management, developing better-shared (electric) mobility solutions for the community, improving freight movements, and enhancing the transition to vehicle electrification. Numerous research and development projects are currently underway in Inland Southern California, spanning advanced smart city modeling and impact analysis, community outreach events, and real-world technology demonstrations. This chapter describes several of these ITS solutions and their potential for improving many cities around the world.

M. J. Barth (✉) · K. Boriboonsomsin · G. Wu · P. Hao
Bourns College of Engineering-Center for Environmental Research and Technology, University of California-Riverside, Riverside, CA, USA
e-mail: matthew.barth@ucr.edu

P. Ioannou and A. A. Malikopoulos (eds.), *Transportation Mobility in Smart Cities*, Springer Tracts on Transportation and Traffic 21,
https://doi.org/10.1007/978-3-031-64769-7_17

17.1 Introduction

One of the key goals of any "Smart City" is to improve a city's operational efficiency and promote economic growth while also improving the quality of life for its citizens. This is typically accomplished by using a variety of smart technologies and data analyses. Smart city characteristics include having effective and highly functional public transportation, progressive city planning, better infrastructure based around technology, and a variety of environmental initiatives that are focused on the public health. This leads to local citizens being able to live and work within the city, effectively uses a number of city resources.[1]

Every city that has goals of becoming "smarter" will have different characteristics based on their current conditions and future plans. Some cities will have clear pathways for improvement, while other cities will have greater challenges due to inherent burdens such as poor air quality, chronic public health problems, lack of affordable housing, and increased levels of poverty. These "disadvantaged" communities desperately want to improve their conditions; one of the key questions is how to take advantage of smart-city technology, infrastructure, and techniques to make these improvements happen.

In this chapter, we examine a variety of Intelligent Transportation System (ITS) solutions that are focused on disadvantaged communities. The general goal of ITS technology is to improve safety, enhance mobility measures (e.g., traffic flow), and minimize environmental impacts of a city's mobility ecosystem. To date, most of the early implementations of these ITS technologies often take place in affluent cities, where there are many funding opportunities and suitable areas for deployment. However, it is critical that we also examine smart city solutions that are focused on improving conditions of disadvantaged and environmental justice communities, whose residents have suffered the most from unmitigated urban sprawl and its environmental and health impacts.

As a case study, we focus on Inland Southern California, including projects in the cities of Riverside and San Bernardino, California. Much of Inland Southern California is designated as "disadvantaged" based on California's CalEnviroScreen modeling tool [1]. Our research addresses a number of ITS topics for this region, including improved traffic flow along major arterial roadways, smart intersections, shared mobility solutions, and improved routing of heavy-duty trucks. This chapter draws on a number of recent publications on these topics, listed in the reference section.

17.1.1 Riverside's Smart City Partnership

The City of Riverside, California, and the University of California-Riverside (UCR) have joined together in the last several years to pursue a number of activities focused on making Riverside a "Smart City." Riverside is one of the largest cities in Inland

[1] OpenStreetMap is a trademark of the OpenStreetMap Foundation, and is used with their permission. This product is not endorsed by or affiliated with the OpenStreetMap Foundation.

Southern California and a key component of Southern California's economy. Riverside has a population of approximately 320,000 residents, medium densification of housing, four expanding universities, and steady growth in local and regional industrial development. A significant amount of rail freight passes through Riverside, coming to and from the ports of Los Angeles and Long Beach, which import over 40% of the goods coming in to the United States. Riverside has over 400 signalized intersections and experiences a number of transportation issues on a daily basis including traffic congestion and related air pollution. Riverside is part of Southern California's "Inland Empire," which has had a long history of dealing with air quality issues, suffering from some of the worst air pollution in the country dating back to the 1960s. However, with aggressive air quality regulations primarily aimed at the transportation sector, criteria pollutant emissions have been reduced by as much as 90% over the last half century.

More recently, greenhouse gas (GHG) emissions are now being aggressively targeted in the region to deal with climate change issues. The State of California has already taken steps to restructure environmental guidelines to de-emphasize the addition of roadway capacity as a congestion and environmental mitigation measure—an effort in which Riverside has been an active participant. Local leadership is now re-envisioning Riverside and embracing smart-city technology in order to bring relief to congested roadways, reduce emissions, encourage mode-shifts, improve safety, update parking requirements and zoning, and significantly enhance the quality of life of the City's residents and patrons. At the heart of this effort is the recent launching of Riverside's Transformative Climate Communities Program.

17.1.2 Transformative Climate Communities (TCC) Program

Funded by the State of California's Strategic Growth Council, the City and its partners received $31.2 M in 2020 to pursue a broad community-based effort to empower Riverside's Eastside area, with the goal of creating new economic opportunities, and improving the health and well-being of Riverside residents. The project area centers around the City's 7th and Chicago housing project, known as Entrada, and contains most of the Eastside from the City's downtown Metrolink station all the way to UCR, as shown in Fig. 17.1. There are a number of projects that are being developed with this TCC funding, including providing high-quality multimodal transportation, affordable housing, urban greening, solar energy, and workforce development training in the targeted region. UCR researchers are working closely with the City, project leaders, and community stakeholders to identify and track specific indicators of project quality and assess public health, economic development, greenhouse gas reductions, and other outcomes for the entire program. This TCC Program is progressing well, more details can be found at https://storymaps.arcgis.com/stories/b5fffd6ae7744b8ab82bb8a725c25c36.

Since UCR is part of the Eastside community, UCR has ongoing efforts to improve local economic opportunities and health. This includes efforts at improving mobility

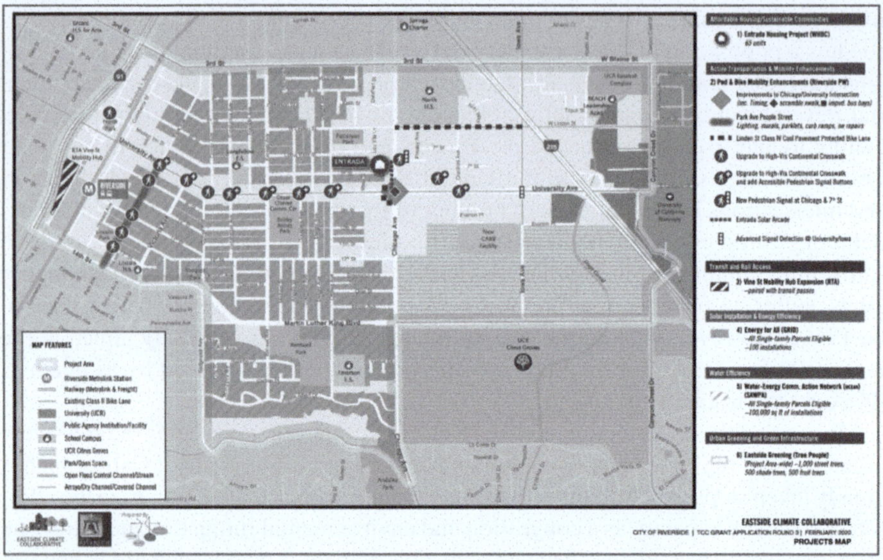

Fig. 17.1 City of Riverside's Eastside community, with various projects being developed as part of the Transformative Climate Communities Program. (permission for reusing this figure granted from the City of Riverside; original source: https://storymaps.arcgis.com/stories/b5fffd6ae7744b8ab82bb8a725c25c36. Base map accessed from ArcGIS Online, see https://www.esri.com/en-us/legal/copyright-proprietary-rights; Map image is the intellectual property of Esri and is used herein under license. Copyright © 2020 Esri and its licensors. All rights reserved)[2]

between the UCR campus and downtown Riverside, centered around a designation of an "Innovation Corridor," described below. In addition, UCR has been researching ways to reduce the impacts of goods movement, warehousing, and supply chain logistics in the region. This includes developing and assessing new technology and techniques to reduce the impacts heavy trucks have on local neighborhoods and working closely with the Center for Community Action and Environmental Justice on community outreach.

17.1.3 Riverside's Innovation Corridor and Smart Intersection

As part of Riverside's Transformative Climate Communities Program, the City and UCR have set out to create an "Innovation Corridor," a 6-mile section of University Avenue between the UCR campus and downtown (see Fig. 17.2 and [2]). This area

[2] In several figures in this document, OpenStreetMap has been utilized. OpenStreetMap is a trademark of the OpenStreetMap Foundation, and is used with their permission. This product is not endorsed by or affiliated with the OpenStreetMap Foundation.

was selected due to its proximity to an expanding transit and alternative transportation network, research institutions associated with UCR, and the ever-expanding entertainment destinations in the downtown region. Along this corridor, the traffic signal controllers were updated to be compatible with SAE connectivity standards. Further, several key intersections along this corridor have been set up with short-range communication roadside units. With this communications capability, Signal Phase and Timing (SPaT) messages from the traffic signal controllers can be directly transmitted to vehicles equipped with similar communication technology, traveling along the corridor. In addition to the SPaT messages, other information can be broadcast, such as positioning correction information and map information on the intersection configuration.

In addition to the communication capability between the traffic signals and the equipped vehicles, the Innovation Corridor also has several air quality monitors located along the roadway.

The overarching goal of this Innovative Corridor is to serve as a key testbed for connected and automated vehicles applications, for improving safety, traffic flow, and reducing pollutant emissions. Since this arterial corridor cuts directly through local neighborhoods, impacts from these applications should bring direct benefits to the community. One of the key connected vehicle applications that has been tested along the Innovation Corridor is described in Sect. 17.2.

Another key part of Riverside's Innovation Corridor, is a smart-intersection located at University Avenue and Iowa Street. This smart-intersection has been equipped with various surveillance systems, including a GridSmart fisheye camera system, as well as stationary LiDAR sensors that provide a detailed 3D view of the intersection and various road users, including cars, trucks, bicyclists, and pedestrians. This intersection is illustrated in Fig. 17.3, showing the location of the sensors at the intersection. This smart-intersection also serves as a key research testbed, where

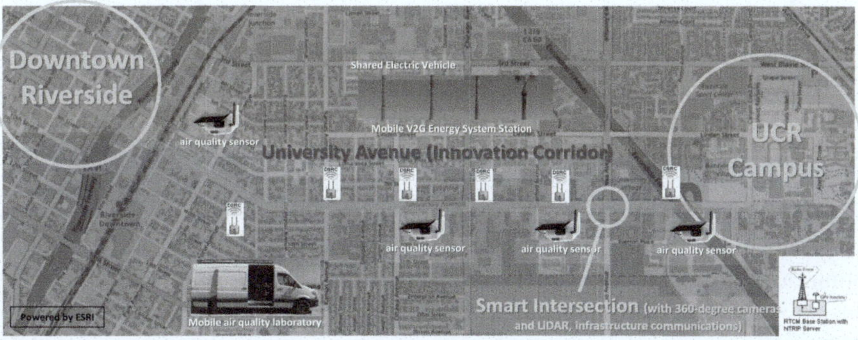

Fig. 17.2 City of Riverside's Innovation Corridor (i.e., a section of University Avenue between UCR and downtown), adapted from [11], permission for reusing this figure granted from UC Davis Institute of Transportation Studies; Base map from OpenStreetMap (https://www.openstreetmap.org/copyright); available under the Open Database License (https://www.opendatacommons.org)

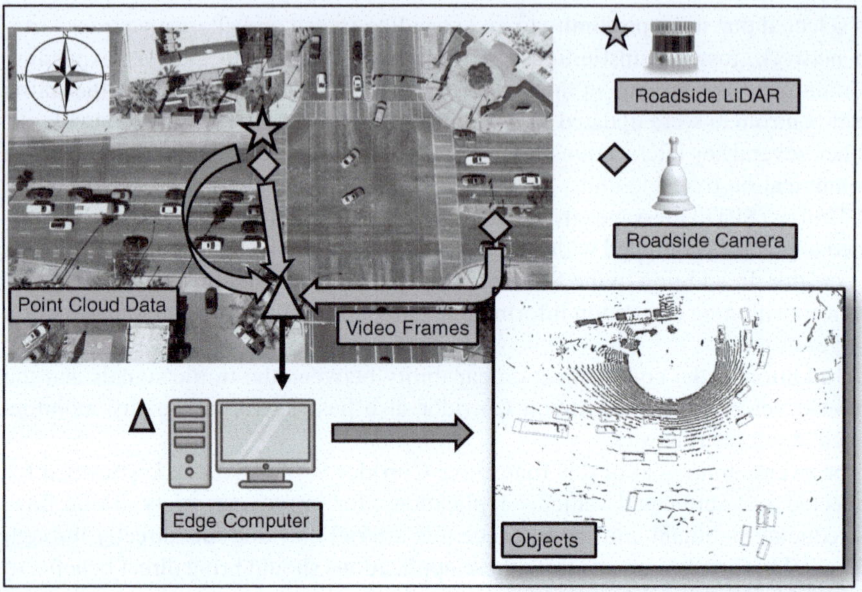

Fig. 17.3 City of Riverside's smart intersection (i.e., intersection of University Avenue and Iowa Street, along the Innovation Corridor)

we can conduct experiments in improving road user safety, mobility, and minimize emissions.

17.2 Eco-Friendly Cooperative Traffic Optimization

It is well known that roadway congestion is detrimental from many perspectives: it negatively impacts our mobility, leading to longer travel times, increased emissions, and increasing the total amount of fuel consumption. There are a number of congestion mitigation measures that can be deployed, many of which take advantage of ITS technology. For example, traffic signal optimization on our arterial roadways has long been recognized as a critical component in improving transportation efficiency and mitigating congestion at signalized intersections [3]. Traffic signal timing optimization aims to determine the ideal allocation of green, yellow, and red signal phases to minimize delays, reduce travel times, enhance traffic flow, and maximize intersection capacity [4]. With advances in sensors, communications, and computing, traffic-adaptive signal controllers can now utilize real-time vehicle detector data to measure traffic flow, which can in turn be used to optimize traffic passing through an intersection. Further, in a connected vehicle environment, vehicles can directly communicate with the infrastructure to provide information on vehicle status (e.g., location, speed, lane position, etc.) to traffic signal controllers, allowing for better signal timing optimization.

17.2.1 Vehicle Trajectory Planning

In addition to optimizing traffic signal timing, it is also possible to have vehicles adjust their speed trajectories based on information from the traffic signals. This trajectory planning optimization can be done in an eco-friendly way, to where energy consumption and emissions are minimized. Over the years, extensive research has been conducted on this type of vehicle trajectory optimization, resulting in the development of various models, algorithms, and optimization techniques. Commonly referred to as "Eco-Approach and Departure at Signalized Intersections" (EAD) in North America and "Green Light Optimized Speed Advisory" (GLOSA) in Europe, various pilot deployments have taken place around the world, including in Riverside California, along the Innovation Corridor.

We have carried out a variety of EAD experiments both in simulation as well as in the real-world, utilizing the enhanced infrastructure along the Innovation Corridor (more details can be found in [5]. The goals of this research are to demonstrate how connected vehicles can improve both mobility and environmental factors along a signalized arterial roadway. Like other CAV applications that involve determining optimal speed profiles for vehicles traveling within an urban transportation network, the EAD application utilizes:

- the SPaT data from the upcoming traffic signals;
- map and route information (e.g., stop-bar location, road grade, road speed limit, turning movement);
- downstream traffic conditions such as queue length; and
- the vehicle's state and powertrain limitations (e.g., vehicle position, instantaneous speed, acceleration/deceleration limit), to determine the optimal recommended speed profile that can minimize the target vehicle's energy consumption and tailpipe emissions when approaching to and departing from signalized intersections.

The EAD application inherently smooths traffic flow, thereby improving mobility, reducing energy consumption, and lowering tailpipe emissions. The advisory speed profile and other relevant information produced from the EAD algorithm can be conveyed to the driver typically through a human–machine interface (HMI or driver–vehicle interface, DVI), or through partial automation (e.g., a form of adaptive cruise control). In our research, we primarily utilize an HMI to provide the drivers with an advisory speed. Figure 17.4 illustrates a generalized system architecture for the EAD application. Technical detail on the application can be found in [2] and related publications.

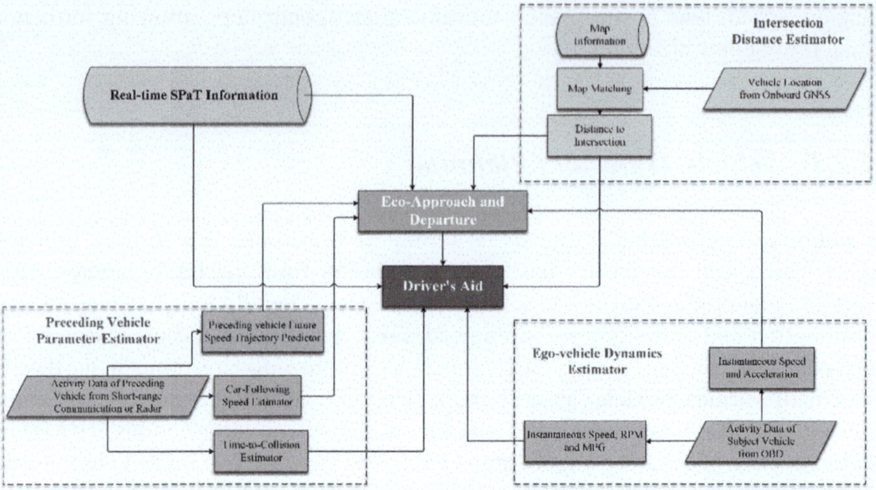

Fig. 17.4 The generalized system architecture of the EAD application that was implemented

We have carried out extensive testing of the EAD connected vehicle algorithm, both in simulation and the real world, with the goal to reduce the idling time at intersections, and avoid unnecessary accelerations, while also allowing for safe driving. The EAD algorithm calculates an optimal velocity to minimize fuel consumption as described in [6]. For our experiments, the signal controllers along the Innovation Corridor were set up to transmit SPaT information, providing a timestamp for the minimum time remaining and maximum time remaining to the connected vehicles in the experiment. Two instrumented vehicles are generally utilized in the experiments, where one test vehicle is utilized that fully implemented the connected vehicle EAD application, while the other vehicle is used as a comparison vehicle, driven normally with traffic without the EAD application. The experiments were conducted at various times throughout a typical weekday (e.g., between 10:00AM and noon, and 1:30PM–3:30PM). During the experiments, the actual fuel consumption from the vehicles were recorded in real time, along with detailed trajectory information (i.e., vehicle speed and position at 1 Hz). Once the vehicle trajectories were collected, they were used as input to well-calibrated vehicle emissions models to also estimate emission reductions.

As an example of our results, Table 17.1 shows the CO_2 and fuel consumption for scenarios carried out both in simulation as well as the real world. In general, the simulation results tended to provide slightly greater reductions compared to the real-world results. But, in general, the EAD application generally provides reductions in the range of 5–15%.

Table 17.1 EAD experimental results

Results		No EAD	EAD	Improvement (%)
Simulation	CO_2 (g/mi)	541.04	479.02	11.46
	Fuel (g/mi)	163.62	144.86	11.5
Real-World	CO_2 (g/mi)	430.7	402.3	6.6
	Fuel (g/mi)	137.63	128.5	6.63

17.2.2 Vehicle Trajectory and Traffic Signal Timing Co-optimization

In recent years, there has been growing interest in developing methods for the co-optimization of traffic signal timing and adjusting vehicle trajectories at signalized intersections [5]. This is motivated by the fact that traditional traffic signal control methods, which typically focus on optimizing traffic flow, can lead to inefficient vehicle movements and increased fuel consumption [5]. By also making adjustments to vehicle speed trajectories, it would be possible to improve both on traffic flow as well as achieving lower energy consumption and emissions. Again, this co-optimization leverages advances in connected and automated vehicle technologies, enabling vehicles to communicate with each other and with the traffic signal system in real time, and to adjust their speed under the prevailing traffic conditions.

As part of our research, we have developed a co-optimization technique referred to as "Eco-friendly Cooperative Traffic Optimization" or ECoTOp, described in detail in [5, 8]. The ECoTOp framework combines vehicle eco-trajectory planning optimization and traffic signal optimization to achieve enhanced transportation performance at individual signalized intersections. The goal of the ECoTOp framework is not to just find a single optimized solution for traffic throughput and lower emissions but to provide a system that can be dynamically adjusted to select the most suitable optimization strategy based on real-time traffic conditions and environmental considerations.

The ECoTOp system is comprised of two major modules: a traffic signal optimization module and the eco-trajectory planning module (from [5]). These modules are designed to interact with each other but often have inherent conflicts when pursuing their own optimization goals. The traffic signal optimization module seeks reliable information from incoming vehicles, including current dynamic state and future movement, as early as possible. It also requires the flexibility to adjust the signal timing plan at any time to adapt to the dynamic traffic conditions. On the other hand, the eco-trajectory planning module requires reliable signal timing information ahead of time to plan trajectories that can reduce energy consumption and emissions. However, vehicles operating under the eco-trajectory plan still require some flexibility in operation to improve safety, mobility, and energy performance in certain situations, such as emergency braking and lane changing.

In general, it is not feasible to create an ideal co-optimized system in an isolated intersection with mixed traffic due to two main reasons. Firstly, unconnected vehicles with human drivers typically have diverse driving styles, which can sometimes be unpredictable. Secondly, the limitations in sensing, communication, and control sometimes make it impossible to predict the time and state when each vehicle enters the system, even for connected and autonomous vehicles. In a mixed traffic environment, each module must make trade-offs to achieve an integrated optimization. This section presents the method for interaction between the two modules.

The ECoTOp system achieves the co-optimization of signal control and vehicle trajectory by following the flowchart depicted in Fig. 17.5. The ECoTOp system works as follows for the co-optimization of signal control and vehicle trajectory [5]:

(1) The number of vehicles is determined for each lane in the network within the communication range.
(2) The number of vehicles is calculated that can be served within the current green time for each lane.
(3) Based on these values, two things are determined: the time required for the delayed vehicles controlled by the phase for that lane to pass through the intersection, and also the time required for maximum throughput.
(4) This information is then used as input to a Sequential Least Squares Programming (SLSQP) optimizer in order to obtain the new phase times.

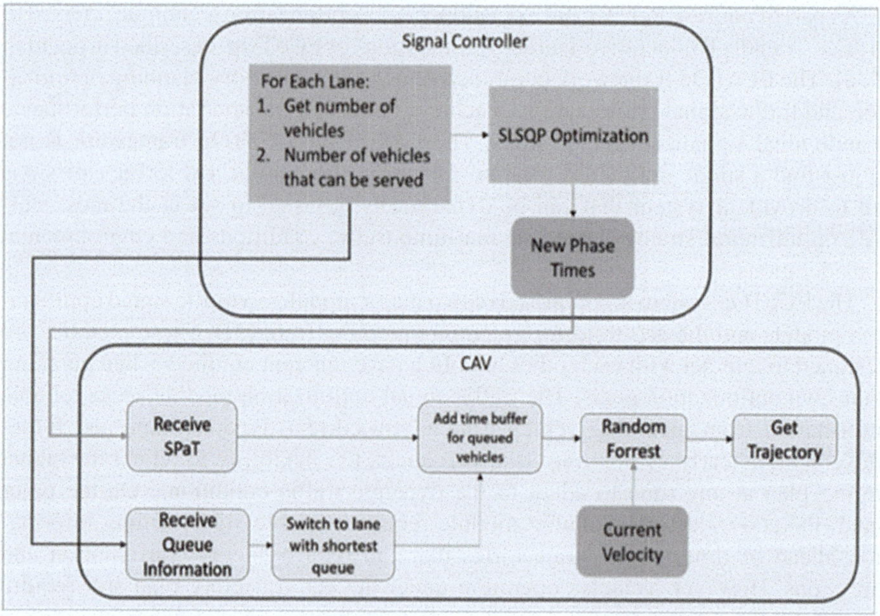

Fig. 17.5 ECoTOp algorithm flowchart (from [5], permission for reusing this figure granted by author David Oswald)

(5) Simultaneously, each connected vehicle receives the Signal Phase and Timing (SPaT) information and queue information.
(6) As in the standard EAD algorithm, the queue information is used to possibly change lanes and to calculate timing for the trajectory planning.
(7) Given this timing and the current vehicle velocity, a Random Forest-based trajectory optimizer is used to obtain each individual vehicle trajectory.
(8) Finally, the signal controller is updated with the optimal phase timing obtained from the SLSQP optimizer.

In order to evaluate the effectiveness of our ECoTOp system, we carried out a high-fidelity simulation of Riverside's smart intersection along the Innovation Corridor (University and Iowa Avenues). This simulation was carefully calibrated with measured traffic volumes and turning movements, along with the current phase and timing information. Different levels of the connected vehicle penetration rate were used and compared to three other scenarios: (1) a baseline scenario where the intersection timing is setup as it is today, with no vehicles performing EAD; (2) an optimized traffic signal timing algorithm (without vehicles performing EAD); and (3) a pure EAD scenario, with current standard signal timing. The results of this are depicted in Fig. 17.6, where we show CO_2 emissions (proportional to energy consumption) and traffic throughput as a measure of mobility. Figure 17.6 thereby consists of four quadrants: the upper left shows the results of the baseline scenario (no EAD, no signal optimization); the upper right shows the results for only optimizing signal timing; the lower left shows EAD trajectory planning only; and the lower right shows the results of the co-optimized ECoTOp system (from [5]). Figure 17.6 illustrates the results for 100% connected vehicle penetration rate and a volume to capacity ratio (V/C) of 0.82.

There are several takeaways from these results: As shown in the lower left quadrant, the EAD trajectory planning algorithm results in a CO_2 savings of approximately 11.7%, as expected; however, the EAD-only scenario has a small negative impact on mobility (i.e., throughput is decreased by 7%). For the upper right quadrant, the signal optimization algorithm improves throughput by approximately 17.5%, with negligible impacts on CO_2 emissions (±1%). When the full ECoTOp system is in place, there is both a CO_2 benefit (~7%) and a throughput increase (~9%). Again, this is a snapshot of a typical result, a more expansive sensitivity analysis is carried out and reported in [5].

It is clear that there are trade-offs between the environmental benefits and mobility (i.e., throughput) in our co-optimization scenarios. Another important feature of the ECoTOp system is that it can be made adaptable based on dynamically changing traffic conditions. Essentially, the parameters of the ECoTOp system can be changed depending on the traffic conditions (e.g., traffic volume), changes in the vehicle mix, as well as the desire of the City to either emphasize environmental improvements or mobility, or some mixture. For example, if a bad air quality day is forecasted, it may make sense to minimize emissions for that day (note that criteria air pollutants track closely with CO_2) at the expense of slightly lower traffic throughput. Conversely, the City may choose to maximize throughput after a major public event when it is in the

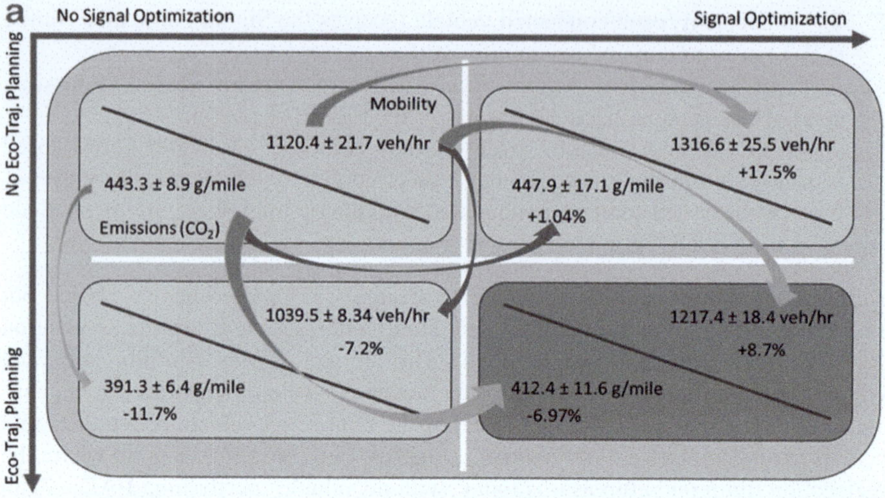

Fig. 17.6 ECoTOp experimental results (lower right), compared to baseline (upper left), EAD-only trajectory planning (lower left), and signal optimization only (upper right) (from [5], permission for reusing this figure granted by author David Oswald)

interest to clear traffic quickly. By simply dialing in the appropriate parameters to the ECoTOp system, the overall system can be made adaptable and versatile, as part of the overall smart city design.

17.3 Riverside Shared Mobility

Like many cities in California, the City of Riverside is moving forward with multiple options and strategies for personal mobility, with the idea of reducing our dependency on a personal-automobile ownership model that is very common today, particularly in Southern California. As part of Riverside's Transformative Climate Communities Program (described in Sect. 17.1), the City is exploring future deployment of shared mobility. Many forms of shared mobility have been discussed, following the general shared mobility definition of shared-use of a vehicle, bicycle, or other mode that enables users to have short-term access to transportation modes on an "as-needed" basis [9]. This included examining carsharing, personal vehicle sharing (PVS, including peer-to-peer (P2P) carsharing and fractional ownership), scooter sharing, bike-sharing, transportation network companies (TNCs, also known as ridesourcing or ridehailing such as Uber and Lyft), ridesharing (i.e., carpooling, vanpooling), microtransit, and courier network services as defined in [9].

17.3.1 Shared Mobility Planning

As part of our research program, we have engaged with the City of Riverside to help develop a shared mobility plan. Our research activities include assisting with community outreach in evaluating what types of shared mobility might work in the city, and conducting extensive modeling of zero-carbon carsharing options within the City. We are also working closely with a local carsharing company called StratosShare [10]. This local company is an on-demand carsharing operator that exclusively rents low-carbon vehicles (e.g., EVs, hydrogen fuel cell electric vehicles) by the hour or day to the public, throughout Inland Southern California. Drivers download the StratosShare app, create accounts, select time of use as well as pickup/drop-off locations, and push to start. StratosShare was founded under the vision to provide a low-cost zero-emission shared-use transportation to Inland Southern California. StratosShare is already up and running, working with the California Energy Commission and Toyota in deploying 15 vehicles in disadvantaged communities throughout Riverside and San Bernardino Counties. These vehicles are strategically located at train stations, universities, airports, and downtown locations to provide a first/last-mile transportation solution. StratosShare is planning to expand their system with additional vehicles and additional locations, so they played a strong role in Riverside's shared mobility planning.

To better understand the mobility needs of the community, we worked with the City of Riverside and StratosShare to develop and carry out a targeted community survey, consisting of a number of questions that were directed at establishing existing travel patterns and needs. In addition, several questions were aimed directly at the potential of a zero-emission carsharing system. The results and analysis of these surveys can be found in [11] and have informed the team on how to best deploy shared mobility, and also served as input to a comprehensive travel demand model, described in further detail below. In summary, the survey indicated that City residents utilize private vehicles for most of their travel, but the community in general was very much in favor of improving other modes such as walking, public transit, biking, and shared mobility. Most survey respondents agreed that it is important to create a sustainable community, by making it easier and safer to use active and public transportation. The majority of respondents were in favor of zero-emission carsharing.

17.3.2 Shared Mobility Modeling

A critical part of our research was to develop extensive shared mobility modeling tools, working with StratosShare and the City of Riverside. Specifically, the focus was on evaluating different scenarios for a carsharing system utilizing a zero-emission fleet. From the literature, it is clear that carsharing systems show great potential in reducing vehicle ownership, achieving VMT and GHG reductions, encouraging alternative transportation modes, and increasing access and mobility of disadvantaged communities (e.g., see [9, 12–15]). Further, it has been shown that vehicle electrification and increased automation have the potential to reduce carsharing GHG emissions, along with reducing private vehicle ownership (see, e.g., [16–18]). For all of these reasons, Riverside wants to increase zero-emission carsharing operations within the community.

It is important to note that many of the existing carsharing studies focus on analyzing the impact of the carsharing program *after* their real-world deployment. It is more challenging to predict the impact of carsharing *before* implementation due to the lack of behavioral data from the participants. To overcome this issue, two methodologies have been applied in existing research. The first method is primarily survey-based and has been applied in many cases (see, e.g., [19–21]). Another mechanism for predicting carsharing success is based on discrete-choice modeling. For example, in a carsharing study for London, a Perceived Activity Set (PAS) model was created to build a conceptual framework of shared mobility, referring to a set of out-of-home activities that encompass their potential travel needs when making decisions that structurally affect their accessibility [22]. This modeling method focused on the long-term impact of carsharing, including the decision to purchase or sell a car or a bike, and the decision to subscribe the transit/carsharing membership. Another study in Rotterdam, Netherland, developed a method in modeling the short-term impact of carsharing using discrete choice model, choosing between five conventional modes: car driver, car passenger, public transport, cycling, and walking. Carsharing was considered as a new mode that was introduced within the mode choice, meaning that a new utility function was required for the carsharing alternative, consisting of variables that are likely to explain carsharing demand.

In our research, the goal was to predict the impact of the potential deployment of zero-emission carsharing in the City of Riverside. A hybrid model was developed with three key components: survey data, discrete-choice model, and agent-based simulation. In our modeling effort, we first derived travel demand data and travelers' activity schedules, and then applied this to the BEAM model (Behavior, Energy, Autonomy, and Mobility), developed by Lawrence Berkeley National Laboratory [23]. BEAM is a mesoscopic simulation model for urban transportation systems with particular support on shared mobility modeling, energy estimation, and computation over large-scale networks. With proper travel demand and travelers' activity schedules, BEAM can evaluate traffic condition, energy consumption and air quality for the entire network, and predict the mode choice and routing decision for each individual agent.

The trips in BEAM are associated with the travelers' demographic information synthesized by *PopGen* and *CEMDAP*, other modeling components of the BEAM framework [23]. A discrete choice model was applied to describe the model choice behavior with existing means of transportation, e.g., car driving, car passenger, public transit, cycling, and walking. The parameters for this model were adopted from literature, and then calibrated using data from our Eastside Climate Collaborative Survey (see Sect. 17.3.1) and other localized data. We then introduced carsharing service into this discrete choice model to study its impact on travel behavior and its benefit on fossil fuel savings and greenhouse gas reduction considering the mode shift from private cars to zero-emission carsharing vehicles.

17.3.2.1 BEAM Model Development and Calibration

In a discrete-choice model, typically there are three types of variables: (1) variables that represent the level-of-service data of a certain mode (e.g., travel time and cost); (2) dummy variables that represent the characteristics of a person (e.g., age, gender, and income) or a household (e.g., number of cars and income); and (3) an alternative specific constant to represent variables that are not present in the utility function but still affect the mode choice. The traveler's daily trip activity data and the corresponding person/household attributes are critical to estimate the mode share for transportation. In our modeling efforts, we introduced a simulation-based data collection method using the BEAM tools which provided calibrated level-of-service data and person/household data to support the discrete choice model, as illustrated in Fig. 17.7.

To acquire personal and household information, we generated raw data using *PopGen*, and calibrated them using latest survey data, including information on population, age, and income for each zone in the City. Next in the calibration process, we utilized data from the Southern California Association of Governments (SCAG) [24]. The SCAG region encompasses 6 counties (Imperial, Los Angeles, Orange, Riverside, San Bernardino, and Ventura) and 191 cities in an area covering more than 38,000 square miles. SCAG develops long-range regional transportation plans including sustainable communities' strategies and growth forecast components, regional transportation improvement programs, regional housing needs allocations and a portion of the South Coast Air Quality management plans [24]. Zone-based

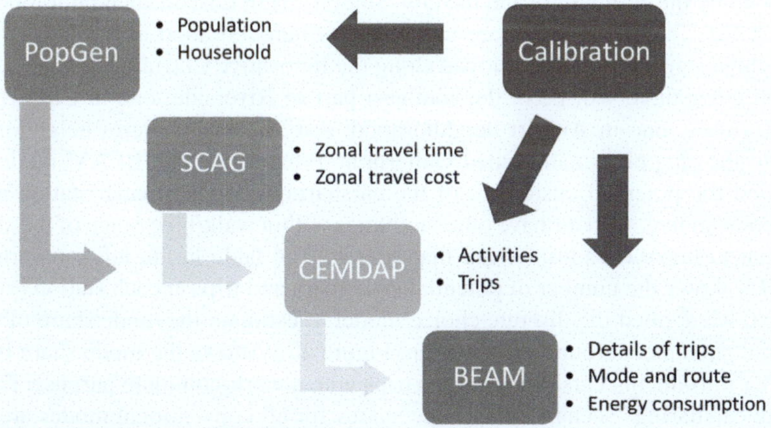

Fig. 17.7 System diagram for data collection (from [11], permission for reusing this figure granted from UC Davis Institute of Transportation Studies)

level-of-service travel data (LOS-data) from SCAG is necessary to analyze carsharing alternatives, identifying travel time, travel cost, along access/egress time for public transit.

Next, we utilized Comprehensive Econometric Micro-simulator for Daily Activity-travel Patterns (*CEMDAP*) software, representing a system of econometric models that represent the decision-making behavior of individuals [25]. It is one of the first systems to comprehensively simulate the activity–travel patterns of workers as well as non-workers in a continuous time domain. Given various land-use, socio-demographic, activity system, and transportation level-of-service attributes as input, the system provides as output the complete daily activity–travel patterns for each individual in the household [25]. With the data from *PopGen* and SCAG, *CEMDAP* creates daily activities for each person in the region of study (i.e., Riverside).

The personal/household data from *PopGen* and trip data from *CEMDAP* were then loaded into BEAM to derive the mode-specific level-of-service data for all the travelers. The BEAM model implementation for the City of Riverside was coded and calibrated using multiple data sources. This Riverside BEAM model was then able to serve as a powerful platform to evaluate the performance of the current shared mobility systems and predict the results of the future deployment, including link-by-link trajectories, mode and routing decisions, and energy consumption and CO_2 production. Further details of the Riverside BEAM model development and calibration are provided in [11].

17.3.2.2 BEAM Modeling Results

To evaluate the potential impact of zero-emission carsharing in the City of Riverside, we first identified potential locations of carsharing stations and the demand around the station, based on community input data. Figure 17.8 shows six potential locations suggested by the community and the city. Station 1 is in Eastside neighborhood near Mission Inn. Stations 2 and 3 are in University neighborhood near UCR campus. These three stations will be deployed along the Innovative Corridor (see Sect. 17.1), and the other three will be in the southern part of Riverside, one in Casa Blanca neighborhood, one in Airport neighborhood, and one in La Sierra neighborhood close to the shopping mall named Galleria at Tyler. Using the BEAM model, we identified the potential customers of the carsharing service at each station as the travelers who live, work or have other activities within walk/bike range of the station and plan to make roundtrips from that station. The table at the left-top corner of Fig. 17.8 shows the number of potential customers and trips at each station.

Next, we applied the discrete choice model to estimate the mode share of those potential customers around each station. Figure 17.9 shows the mode share before and after introducing carsharing for the trips with a work-commute purpose. For the "before carsharing" scenario, the mode shares for all conventional modes are very close to the survey results, showing good performance in coefficient calibration. For the "after carsharing" scenario, the mode share for carsharing for work trips was

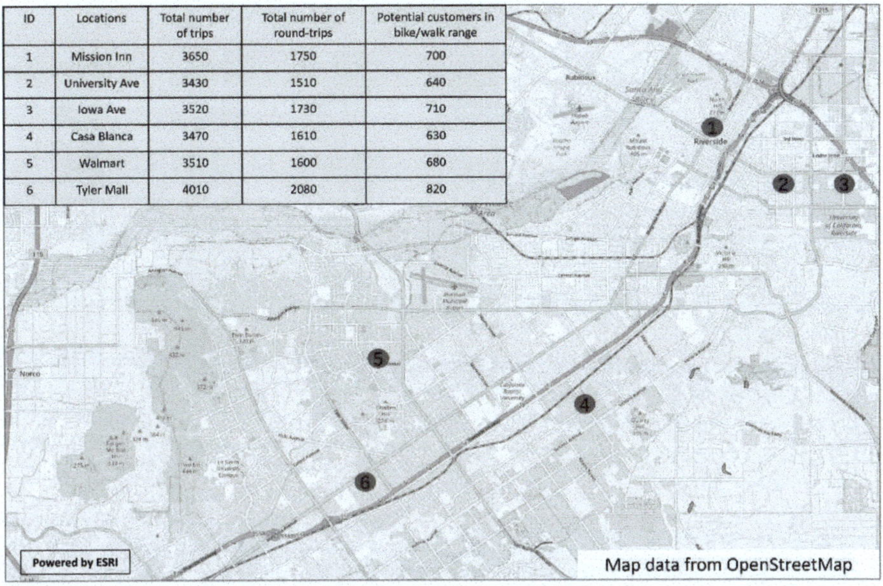

ID	Locations	Total number of trips	Total number of round-trips	Potential customers in bike/walk range
1	Mission Inn	3650	1750	700
2	University Ave	3430	1510	640
3	Iowa Ave	3520	1730	710
4	Casa Blanca	3470	1610	630
5	Walmart	3510	1600	680
6	Tyler Mall	4010	2080	820

Fig. 17.8 Potential locations and travel demand of carsharing stations, from [11], permission for reusing this figure granted from UC Davis Institute of Trans-portation Studies; Base map from OpenStreetMap (https://www.openstreetmap.org/copyright); available under the Open Database License (http://www.opendatacommons.org)

between 8 and 13%. About two thirds of the carsharing trips are shifted from car driving trips, and the rest are shifted from other modes including car passenger, public transit, cycling, and walking. Due to the introduction of carsharing services, it was predicted that trips from single driver cars would be reduced by 8–12%, trips as car passenger (including carpool and taxi) would be reduced by 10–26%, and trips as transit passenger would be reduced by 4–16%. Walking and bicycle trips were less impacted by carsharing, with 6 and 3% reduction, respectively.

For non-work trips, we evaluated trips associated with education, as well as trips "for other purposes" (e.g., shopping, etc.). It was found that for education-based trips, the carsharing mode share shift is less than 1%, having little impact on conventional modes. For trips with other purposes, the carsharing mode share shift was predicted to be between 15 and 23%, as shown in Fig. 17.10. Among all the new carsharing trips, 39% of them are shifted from car driving trips.

It is clear that the potential for carsharing services can enhance the accessibility of the local residents near the stations, especially for the household that do not own a private car. According to the numerical results, for the people without cars in their households, the mode share of carsharing for work trip increases by 17–40%. Considering the high correlation between income and car ownership, the carsharing program would significantly improve the accessibility of the disadvantaged communities.

When considering that the carsharing fleet would be zero-emissions (electric or fuel cell), the mode shift from gasoline private vehicles to electric carsharing vehicles

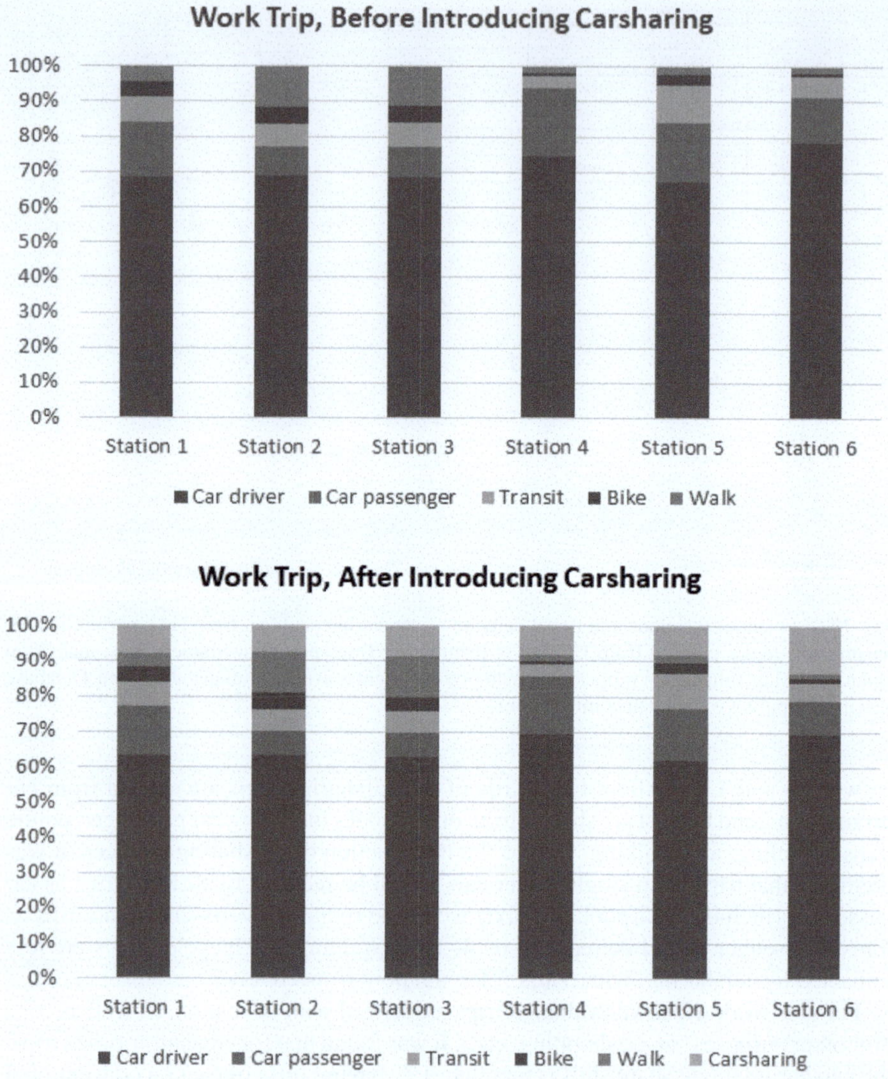

Fig. 17.9 Mode shares for work trips before (top) and after (bottom) introducing carsharing (from [11], permission for reusing this figure granted from UC Davis Institute of Transportation Studies)

would significantly reduce fuel consumption and greenhouse gas emissions. Based on our modeling results, it is expected that greenhouse gas emissions would go down by 10%, simply by reducing the use of the private gasoline-powered cars after introducing carsharing in the neighborhood. Besides the direct mode-shift impact shown in the proposed model, carsharing would also serve as the last-mile solution for public transit and further reduce private car trips from this carsharing-transit synergy. In the long term, the reduction of car ownership will further decrease the

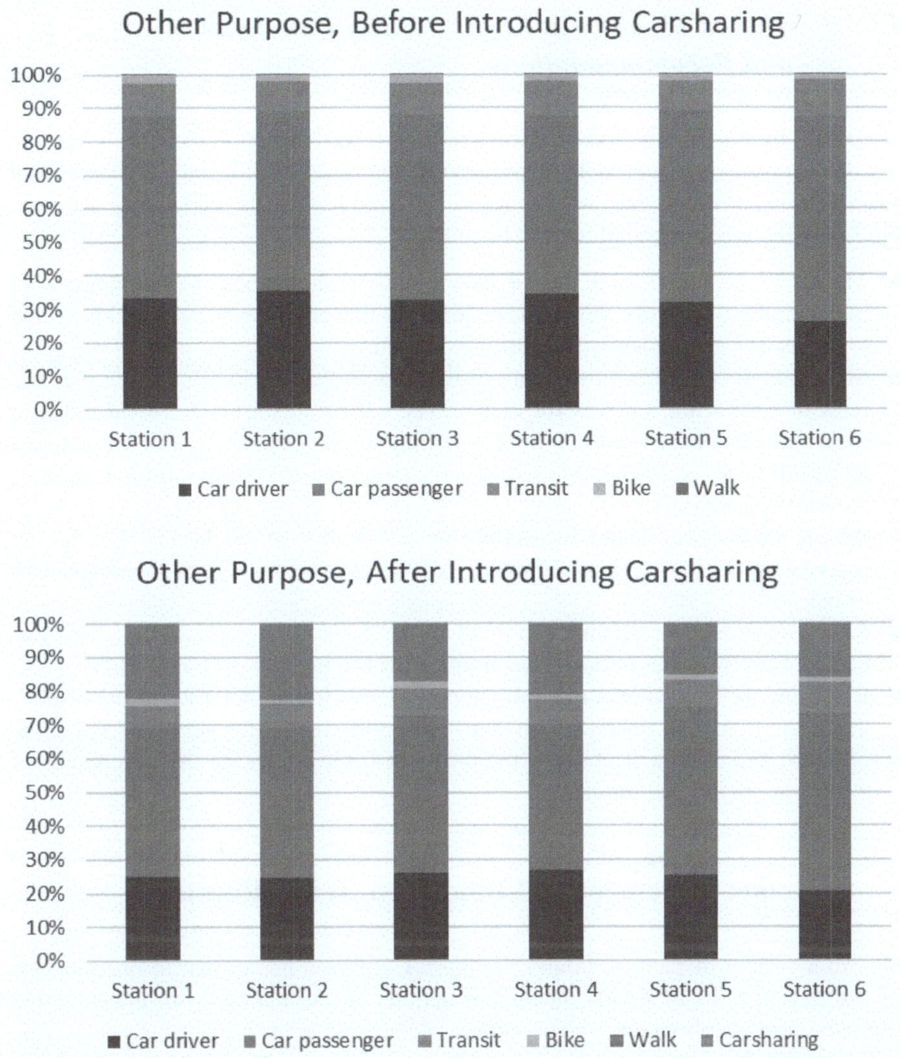

Fig. 17.10 Mode shares for non-work trips before (top) and after (bottom) introducing carsharing (from [11], permission for reusing this figure granted from UC Davis Institute of Transportation Studies)

travel demand and therefore reduce fuel consumption and greenhouse gas emissions. These benefits are being explored in future research.

17.3.3 Shared Mobility in Riverside: Key Conclusions and Recommendations

As shown by our survey and modeling results, a zero-emission carsharing service can be a promising approach to improving the accessibility and environmental sustainability of communities. To successfully implement the carsharing services, the following recommendations can be made:

- The preferable locations of the carsharing stations should be in a community with high population density, where the residents of that community can quickly access to the station conveniently by walking or cycling.
- Since people who don't have cars in the household will likely have higher use rates of carsharing services for their travel, car ownership is a critical index in identifying the best locations for carsharing stations. As such, a disadvantaged area with lower car ownership will receive larger benefits from introducing zero-emission carsharing.
- Age is another key factor impacting the acceptance of carsharing. Carsharing receives higher interest in a community with higher percentage of people aged 45 or below.
- Reducing access/egress time can increase the popularity of carsharing. This can be achieved by easier parking and simple check-out/check-in transactions.
- Based on our Riverside community survey, it was found that 40% of the respondents are not that interested in zero-emission carsharing services. As such, outreach activities will play a critical role to further increase the acceptance of carsharing in the community.

17.4 Goods Movement and Innovative Truck Routing

Due to a variety of factors, Inland Southern California has grown to become one of the largest hubs of goods movement activity in the nation, with considerable infrastructure, employment, and economics connected to the logistics industry. This logistics industry will continue to grow as an important part of the economy, but it is critical that it be managed in a way that the quality of life in the communities is preserved, negative environmental impacts are minimized, and good-paying jobs are prevalent. Imports throughout the nation continue to increase, placing increased stress on the current logistics supply chain. Currently, more than 44% of the nation's goods pass through Inland Southern California on their way to their final destinations—as such, the nation greatly depends on a properly functioning goods movement system in Inland Southern California.

It is clear that goods movement activities in the region have grown substantially over the last few decades and there are many underlying positive elements, including a rich flow of goods and wealth, as well as substantial employment opportunities. But, in addition to these positive elements, there are also a number of negative elements, including:

- Air quality in the region is poor, notably due to the huge flow of diesel-fueled heavy vehicles (both trucks and locomotives), generating nitrogen oxides (NOx), particulate matter, and greenhouse gas emissions; this has adversely affected the public health and quality of life in the region.
- Traffic congestion in the region is severe across all roadway types (freeways, arterial roadways), negatively affecting the quality of life in the region. This leads to not only increased emissions but also an economic loss due to time spent in traffic.
- The employment situation in the region is quite volatile, with many of the logistics related employment opportunities being part-time, temporary, low-paying, and lacking upward career mobility. Employment is also under the threat of replacement by automation, and subject to pandemic-related and other negative health considerations.
- Land use has been negatively affected, where greenery is being replaced by densely packed warehouses, storage facilities, truck and trailer parking, and massive railyards; this has decreased the social and environmental attractiveness, and precluded the implementation of facilities from higher value sectors such as high-tech, manufacturing, and R&D parks.
- Social equity has suffered, where the impacts of air pollution, congestion, and other measures are significantly higher on disadvantaged communities in the region.

It is important to note that the poor air quality associated with goods movement is not due to warehouse operations itself, but instead is mostly due to its associated transportation activities. The internal operation of warehouses typically has a high degree of automation and are highly electrified that can be accomplished cleanly with renewable electricity. Indeed, several warehouses have been awarded a high level of green building certificates. The majority of the poor air quality is associated with the heavy-duty freight transportation sector, which includes diesel trucks and locomotives. This accounts for well over 90% of the emissions associated with warehouse operations. We currently rely on heavy-duty vehicles in high volume that travel significant distances in the region. This flow of heavy-duty vehicles to/from logistics facilities in Inland Southern California is also a major cause of congestion and road infrastructure deterioration on roadways all over the region. While the majority of these heavy vehicles are diesel-powered, there is currently a big push to transition these vehicles to zero emissions, based on battery or fuel cell technology.

The proliferation of freight facilities and heavy-duty vehicles is also caused by the fact that most of the companies operating in the Inland Southern California logistic ecosystem are in many ways operating in silos. Warehouses are typically built in areas that have low land prices; this usually drives where warehouses are located. It is often the case that commercial zoning is adjacent to disadvantaged communities and freeways. There are many documented cases of these communities unsuccessfully opposing the locations of proposed warehouses in recent years. However, there is a large incentive for the cities and counties to acquire revenue from these transactions. In the absence of other offers for commercial development, and potentially due to the lack of resources of the communities opposing to the development, it has become common that logistics warehouses constitute a major portion of these neighborhoods. This places the negative impacts of air pollution, noise, and congestion from truck traffic on these areas in comparison to a zoning pattern that would space the warehouses out throughout the region. Planning and zoning requirements among cities and counties also can vary greatly, often times leading to inefficiencies in the overall arrangement of the broader logistic operations.

17.4.1 Innovative Truck Routing in Inland Southern California

Over the last several years, we have been developing a number of intelligent transportation system technologies, such as advanced fleet management tools, to mitigate the negative impacts of goods movement. These tools include, for instance, intelligent fleet scheduling software tailored for electric trucks, dynamic time-of-day fleet scheduling software, geofencing strategies, and low pollutant exposure truck routing. These tools are coupled with our research in connected and automated vehicles, with a focus on heavy-duty vehicles, leading to more efficient truck operations and smoother traffic flow.

As an example of our recent efforts, we have carried out a case study where we have employed our low pollutant exposure routing technique around the San Bernardino Airport, which is being built out as a major air cargo hub for Amazon, UPS, and FedEx here in Inland Southern California. This airport mainly supports air cargo operations and it has recently been approved to undergo a major expansion [26]. The surrounding community is largely classified as a disadvantaged community (under California SB 535, see [1]). Local residents, communities, and organizations have been expressing concerns about future employment opportunities and environmental impacts [27].

For this case study, we have applied our exposure-based routing technique, which guides heavy-duty diesel trucks (HDDTs) through a community in a way that lowers the total exposure of community members to the pollutant emissions from the truck without significantly increasing travel time [28]. The exposure-based routing technique was applied to the San Bernardino Airport area, as shown in Fig. 17.11. This area is bounded by Freeway I-215 in the west, I-10 in the south, and I-210 curving from south to north then connecting the east–west side. Corners one, two, three, and four correspond to the Northwest, Northeast, Southeast, and Southwest corners of the San Bernardino city area, respectively. The location of San Bernardino Airport west side is marked in the figure below, and we evaluated the potential HDDT trips from the four corners to and from the airport.

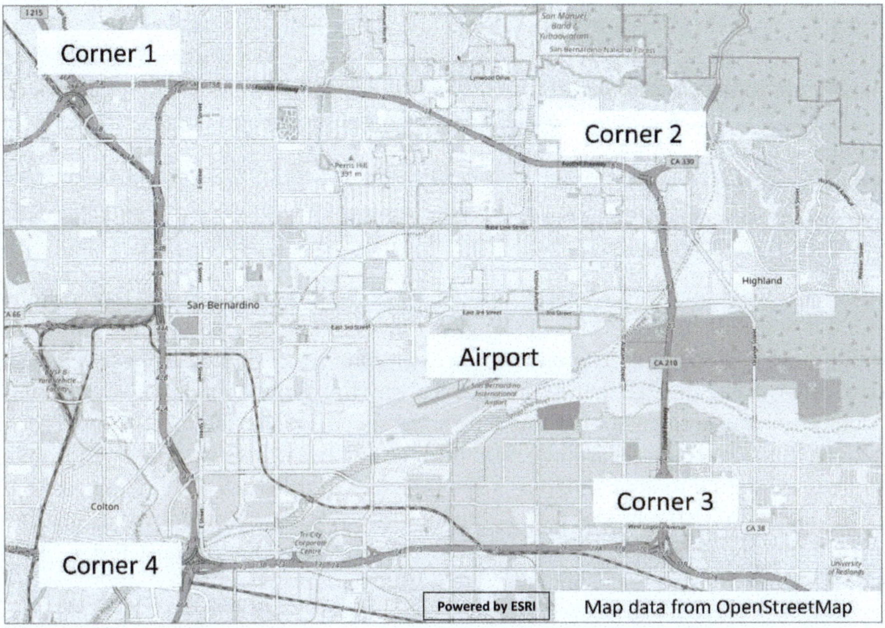

Fig. 17.11 Map of Study Area, from [11], permission for reusing this figure granted from UC Davis Institute of Trans-portation Studies; Base map from OpenStreetMap (https://www.openstreetmap.org/copyright); available under the Open Database License (http://www.opendatacommons.org)

Figure 17.12 presents the methodological framework of exposure-based routing. It involves a modeling chain that starts from vehicle emission modeling to air dispersion modeling, human exposure assessment, and finally vehicle route calculation where the output from one step is used as an input for the next step. In addition, each step also requires other inputs. The inputs and assumptions associated with each modeling step are described below.

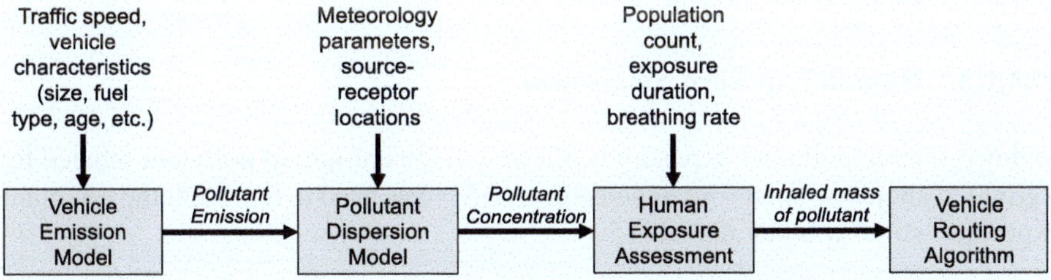

Fig. 17.12 Methodological framework of exposure-based routing (adapted from [11], permission for reusing this figure granted from UC Davis Institute of Transportation Studies)

17.4.1.1 Vehicle Emissions Modeling

The calculation of emissions was focused on fine Particulate Matter ($PM_{2.5}$), Nitrogen Oxide (NO_x), and Carbon Dioxide (CO_2) emissions from HDDTs. They were calculated using the following equation:

$$E_{i,j} = V_{i,k} \times L_i \times EF_{j,k},$$

where $E_{i,j}$ is the mass emission of pollutant j on link i; $V_{i,k}$ is the HDDT volume on link i with link speed k; L_i is the length of link i; and $EF_{j,k}$ is the emission factor of pollutant j at speed k. The calculation was done for a single heavy-duty diesel truck of model year 2012, but for all the roadway links in the modeling area. It was assumed that this truck would be traveling at the speed equal to the speed limit of each roadway link. The data regarding speed limit on roadway links was obtained from a commercial digital roadway map. Emission factors of the truck were obtained from CARB's EMFAC2017 model [29, 30], which is a regulatory model for estimating on-road mobile source emissions in California. Only running exhaust $PM_{2.5}$, NO_x and CO_2 emissions were calculated.

17.4.1.2 Air Dispersion Modeling

An atmospheric dispersion model was needed to estimate the concentration of air pollutants emitted from vehicular sources at specific receptor locations. In this study, R-LINE, a research grade dispersion model for near-roadway assessment was used [31]. Micro-meteorology data inputs for R-LINE such as temperature, wind speed, wind direction, surface friction velocity, and Monin–Obukhov length were obtained for Redlands Station from South Coast Air Quality Management District website [32]. Source height was assumed to be 2.5 m (~8.2 ft), which represents a typical height of exhaust stacks of heavy-duty diesel trucks. Receptor height was assumed to be 1 m (~3.3 ft), which represents an average height of 5-year-old children.

17.4.1.3 Human Exposure Assessment

In this research, pollutant exposure is referred to the amount of pollutant inhaled by a group of subjects. Therefore, inhaled mass (IM) was used to represent the pollutant exposure, which was calculated as

$$IM = C * Pop * t * BR,$$

where C is the pollutant concentration ($\mu g/m^3$) in a given microenvironment; Pop is the number of subjects in the microenvironment; t is the truck travel time on the road link (hour); and *BR* is the breathing rate (m^3/hour/capita) of the subjects exposed to the pollutant.

Breathing rates of population in different age groups were based on the U.S. EPA's Exposure Factors Handbook [33]. In addition, the California Office of Environmental Health Hazard Assessment's Technical Support Document of Exposure Assessment and Stochastic Analysis included detailed breathing rate scenarios [34]. It is desirable to reduce population exposure to traffic-related air pollutants because tailpipe emissions, such as $PM_{2.5}$ and NO_x, are associated with health risks in young children, older adults, patients, and even healthy adults [35–37]. Thus, in this research, both population-wide average breathing rate of 17 m^3/day and population-specific breathing rate were applied.

17.4.1.4 Vehicle Route Calculation

The Shortest Path Problem (SPP) is traditionally aimed at finding a travel route between a pair of origin–destination (OD) points that has the shortest distance or shortest travel time. However, in this research, the vehicle routing objective is to reduce inhaled mass of pollutant while limiting the increase in travel distance within a reasonable range for the trip. This is a multi-objective SPP studied by many researchers (e.g., [38]). Several methods for solving multi-objective SPP are summarized in [39]. In previous studies, we used a weighting method that transformed the multi-objective SPP into a single-objective SPP. The specific methods can be found in [40]. In this study, due to the limited number of OD pairs, we simply selected freeway routes and compared them with manually selected alternative routes that have similar travel time.

17.4.1.5 Network Characterization

Figure 17.13 shows four entry/exit points located at the four corners of the study area. The sensitive facilities or receptors considered in this study are primarily used by individuals that are most susceptible to the effects of air pollution. Daycares, schools (elementary to high schools), assisted living homes, and public parks were chosen as the sensitive facilities. The population data were projected to calendar year 2018 at census block level based on 2010 Census and 2018 American Community Survey. Population at sensitive facilities were projected based on school enrollment data and census population. Population at residential blocks are estimated based on several sources including population by age groups [41], employment data [42, 43], and school enrollment rate [44, 45].

To better understand how the R-LINE model parameters impact the output concentration values, a sensitivity analysis of road width and freeway sound barrier options in R-LINE was performed. The results showed that for the current modeling scenario, the road width and sound barrier options only have minor effects on the modeled concentration results. On the other hand, the most impactful factors are traffic

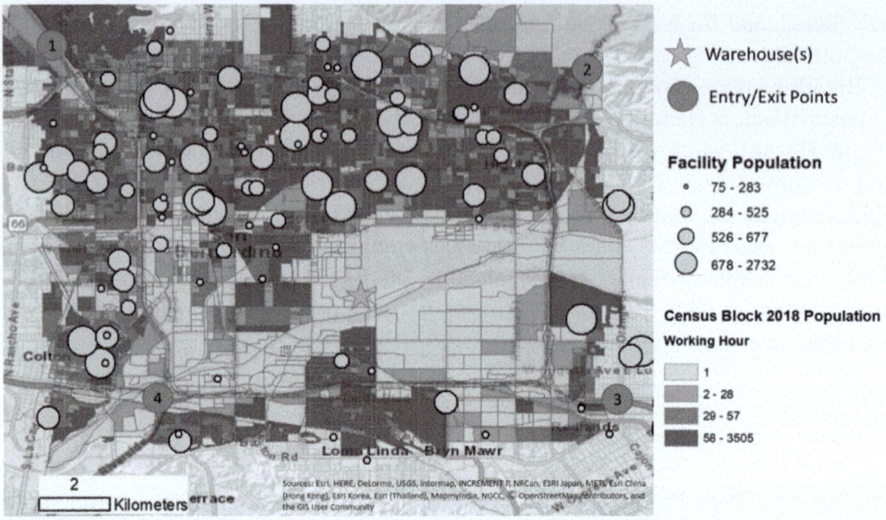

Fig. 17.13 Map of population, sensitive facilities, and truck trip attractions in San Bernardino (from [11], permission for reusing this figure granted from UC Davis Institute of Trans-portation Studies; base map accessed from ArcGIS Online, see https://www.esri.com/en-us/legal/copyright-proprietary-rights; Map image is the intellectual property of Esri and is used herein under license Copyright © 2020 Esri and its licensors. All rights reserved)

speeds, emission factors, meteorological conditions, and population distribution. The influence of varying breathing rates was also examined where three different breathing rate scenarios were applied: an averaged breathing rate of 15 m^3/day, an age-group specific breathing rates (in m^3/day), and age-group specific breathing rates normalized by average body mass (in m^3/day/kg).

We first consider modeled $PM_{2.5}$ *IM* values at sensitive facilities and census blocks based on the meteorological conditions at 10 A.M. on May 9, 2016, assuming a population-averaged breathing rate of 15 m^3/day. For instance, a $PM_{2.5}$ *IM* value of 0.23 μg/link means that there would be 0.23 μg of $PM_{2.5}$ inhaled by the nearby population after the truck traversed this roadway link in the given scenario. As air pollutants from one roadway link can reach multiple facilities/blocks within 1,500 m, the *IM* values of roadway links are generally higher for those near large sensitive facilities and densely populated census blocks. We also consider the wind direction, and it can be observed that roadway links upwind of large sensitive facilities and densely populated census blocks generally have higher *IM* values than those downwind.

Figure 17.14 shows the aggregated $PM_{2.5}$ *IM* values from both sensitive facilities and census blocks based on the meteorological conditions at 10 A.M. and 3 P.M. on May 9, 2016, assuming a population-averaged breathing rate of 15 m^3/day. The aggregated $PM_{2.5}$ *IM* values are generally higher at 10 A.M., when compared to that at 3 P.M., due to the more turbulent condition in the afternoon contributing to

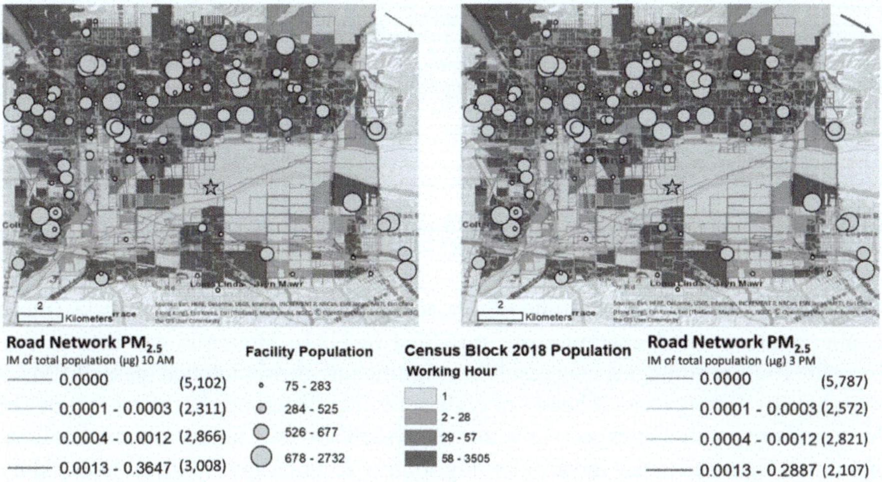

Fig. 17.14 Total inhaled mass of $PM_{2.5}$ (µg/link) at 10 A.M. (left) and 3 P.M. (right) assuming a population-averaged breathing rate of 15 m^3/day (from [11], permission for reusing this figure granted from UC Davis Institute of Transportation Studies; base map accessed from ArcGIS Online, see https://www.esri.com/en-us/legal/copyright-proprietary-rights; Map image is the intellectual property of Esri and is used herein under license Copyright © 2020 Esri and its licensors. All rights reserved)

faster dispersion of air pollutants. The comparison shows how the meteorological conditions can affect the *IM* values.

17.4.2 Low-Exposure Route Comparison

For trips that connect the four freeway corners and the air cargo hub, both a "baseline" route (typically the shortest-time route) and a low-exposure route (LER) were determined. We then weighted those truck routes based on the number of trucks coming from and going to those respective freeway corners. To estimate the number of trucks entering and exiting the four corners, we used truck flow data from the California Department of Transportation's Freeway Performance Measurement System [46].

As an example of our results, we calculated the total amount of IM of emissions for the weighted average of all four corners at 10 A.M., assuming a population-averaged breathing rate of 15 m^3/day. These results are shown in Fig. 17.15. In general, it can be seen that as compared to the baseline routes, the low-exposure routes reduce IM of PM2.5 by 46%, IM of NOx by 20%, and CO_2 emissions (and thus, vehicle fuel consumption) by 4%, on average. On the other hand, the low-exposure routes increase travel time by 29%, on average.

Has truck flow																
MY 2012 10am		Baseline Route (A)					Low Exposure Route (B)					Difference (v.s. baseline route)				
Corner #	No. of Trucks	Driving Distance (miles)	Driving duration (min)	$PM_{2.5}$ IM (ug)	NO_x IM (mg)	CO_2 (kg)	Driving Distance (miles)	Driving duration (min)	$PM_{2.5}$ IM (ug)	NO_x IM (mg)	CO_2 (kg)	Driving Distance (miles)	Driving duration (min)	$PM_{2.5}$ IM (ug)	NO_x IM (mg)	CO_2 (kg)
1	45	514.1	580.8	5390.1	485.5	836.3	418.9	685.4	2436.2	455.2	676.5	-95.2	104.6	-2953.8	-30.3	-159.8
2	11	107.6	151.0	794.6	72.8	164.6	99.2	201.0	102.5	45.0	175.9	-8.4	50.0	-692.1	-27.8	11.3
3	133	628.6	845.0	6985.4	647.0	1073.3	662.2	1155.5	4545.3	442.6	1117.2	33.6	310.5	-2440.0	-204.5	43.9
4	26	148.0	192.8	284.3	31.2	251.5	155.5	236.6	238.7	44.9	269.2	7.5	43.8	-45.6	13.7	17.7
Total	215	1398.3	1769.7	13454.4	1236.5	2325.7	1335.8	2278.6	7322.8	987.7	2238.8	-62.5	508.9	-6131.5	-248.8	-86.9
												-4%	29%	-46%	-20%	-4%

Fig. 17.15 Comparison of baseline routing to low-exposure routes at 10 A.M

These results will vary by time of day, and with other roadway networks (more details, analyses, and examples are provided in [11]). However, pro-active low-exposure routing strategies may result in a significantly lower inhaled mass of PM2.5 emissions by simply routing heavy-duty diesel trucks differently at different times of the day. From a policy perspective, it may be possible to encourage voluntary actions by truck companies to use these emerging routing technologies to divert heavy-duty truck traffic to low-impact routes, accepting a trade-off between slightly increased delivery time and reducing the exposure of PM2.5 and NOx to residents and sensitive receptors such as schools and hospitals. As an added incentive, there would be a slight reduction in their fleet average fuel consumption. Another policy approach would be to have local cities utilize their authority to designate truck routes through their communities, choosing routes that have the least air pollution impact on their residents.

As with the traffic signal management described in Sect. 17.2, these routing operations can be changed dynamically throughout the day. For example, when school children are walking to or from school, it would be best to have trucks maximally take advantage of low-exposure routing. In contrast, when there is less exposure risk (for example at night), it would be possible to switch the routing to "shortest-distance" or "shortest-time" routes so that trucks can make deliveries in the most efficient manner.

17.5 Conclusions and Future Work

In this chapter, we described a number of ITS solutions that are aimed at improving a city's mobility eco-system, with a focus on minimizing the negative impacts on disadvantaged and environmental justice communities, while enhancing the overall mobility and safety. These ITS solutions, along with many others, should be part of a Smart-City Playbook that cities across the world can use, learning about the synergies and trade-offs between different strategies.

Our research has primarily been focused on Inland Southern California, which suffers from high levels of traffic congestion, poor air quality that results from (in-part) vehicle emissions, and lack of alternative modes of transportation other than the ubiquitous personal vehicle. To better demonstrate and quantify the improvements that can be brought about by ITS solutions along a typical arterial roadway, we developed a real-world testbed in Riverside, California, called the Innovation

Corridor. This testbed runs squarely through disadvantaged neighborhoods, and ITS applications that can decrease congestion, lower emissions, and improve safety have a direct benefit on the local community. In addition, we are designing and deploying new carsharing options for these communities, so that they are not necessarily locked in to the inefficiencies of the personal car ownership model. Lastly, we are attempting to look for better solutions for the goods movement industry, where we can still enjoy the economic benefits of our goods movement system, but not necessarily suffer from all of their negative impacts.

Another key aspect of these strategies is that they can all be deployed in a way that is synergistic and dynamically adjustable to deal with changing conditions. There are often trade-offs between safety, mobility, and environmental considerations, and cities should recognize that they do not need to be locked into one solution all the time. The example applications described in this chapter can easily be adjusted by simply changing parameters within the algorithms. Any smart city should take advantage of this capability, essentially tuning their solutions to maximize benefits to their communities.

The UC Riverside team will continue to improve on the ITS solutions outlined in this chapter, deploy and test them in the real world, and develop new ideas that are squarely focused on improving the conditions of the local community.

This chapter includes figures for which specific copyright permissions were obtained. The acknowledgments below detail the sources and nature of these permissions:

(1) The City of Riverside has granted permission to use Fig. 17.1. This permission was granted via email on October 24, 2023.
(2) The base maps of Figs. 17.2, 17.8, and 17.11 were sourced from OpenStreetMap, see https://www.openstreetmap.org/copyright.
(3) Figures 17.5 and 17.6 are reprinted from the dissertation of David Oswald [reference 5] who has granted permission via email on November 16, 2023.
(4) Figures 17.7, 17.8, 17.9, 17.10, 17.11, 17.12, 17.13, and 17.14 are reprinted from a CSTACC tecnical report [reference 11]; UC Davis has granted permission for use via email on October 24, 2023.
(5) The base maps of Figs. 17.13 and 17.14 were sourced from ArcGIS Online, see https://www.esri.com/en-us/legal/copyright-proprietary-rights.

Acknowledgements Much of the material in this chapter come from a number of students who have worked on various research projects over the last 5 years. These students include Dr. David Oswald, who carried out much of the traffic signalization research along Riverside's Innovation Corridor, Yejia Liao who is working on Riverside's shared mobility system research, Dr. Jill Luo and Dr. Ayla Moretti who has developed the low-exposure truck routing algorithms.

References

1. CalEnviroScreen 4.0, OEHHA. Accessed 18 Oct 2023. [Online]. Available: https://oehha.ca.gov/calenviroscreen/report/calenviroscreen-40
2. Oswald D, Hao P (2021) Development of an innovation corridor testbed for shared electric connected and automated transportation. https://doi.org/10.7922/G21C1V6T
3. Bavarez E, Newell GF (1967) Traffic signal synchronization on a one-way street. Transp Sci 1(2):55–73. https://doi.org/10.1287/trsc.1.2.55
4. Koonce P et al (2008) Traffic signal timing manual. FHWA, Report No. FHWA-HOP-08–024
5. Oswald D, Flexible framework for co-optimizing dynamic traffic signal control: foundation for adaptive optimization strategies. Univ Calif Riverside
6. Hao P, Wu G, Boriboonsomsin K, Barth MJ (2019) Eco-Approach and Departure (EAD) application for actuated signals in real-world traffic. IEEE Trans Intell Transp Syst 20(1):30–40. https://doi.org/10.1109/TITS.2018.2794509
7. Xia H et al (2012) Field operational testing of ECO-approach technology at a fixed-time signalized intersection. In: 2012 15th international IEEE conference on intelligent transportation systems. IEEE, Anchorage, AK, USA, pp 188–193. https://doi.org/10.1109/ITSC.2012.6338888
8. Hao P, Oswald D, Wu G, Barth MJ (2023) Eco-friendly cooperative traffic optimization at signalized intersections. https://doi.org/10.7922/G26Q1VJ9
9. Shaheen S, Cohen A, Farrar E (2019) Chapter Five—Carsharing's impact and future. In: Fishman E (ed) Advances in transport policy and planning, vol 4. Academic, pp 87–120. https://doi.org/10.1016/bs.atpp.2019.09.002.
10. StratosShare. Accessed 18 Oct 2023. [Online]. Available: https://www.stratosfuel.com/
11. Blosch EC (2023) Inland Empire Regional Initiative, CSTACC. Accessed 18 Oct 2023. [Online]. Available: https://cstacc.ucdavis.edu/projects/inland-empire-regional-initiative
12. Martin EW, Shaheen SA (2011) Greenhouse gas emission impacts of carsharing in North America. IEEE Trans Intell Transp Syst 12(4):1074–1086. https://doi.org/10.1109/TITS.2011.2158539
13. Lane C (2005) PhillyCarShare: first-year social and mobility impacts of carsharing in Philadelphia, Pennsylvania. Transp Res Rec J Transp Res Board 1927(1):158–166. https://doi.org/10.1177/0361198105192700118
14. Martin E, Shaheen SA, Lidicker J (2010) Impact of carsharing on household vehicle holdings: Results from North American Shared-use vehicle survey. Transp Res Rec J Transp Res Board 2143(1):150–158. https://doi.org/10.3141/2143-19
15. Fraiberger SP, Sundararajan A (2015) Peer-to-Peer rental markets in the sharing economy. SSRN Electron J. https://doi.org/10.2139/ssrn.2574337
16. Barth MJ, Todd M (2003) UCR Intellishare. IATSS Res 27(1):48–57. https://doi.org/10.1016/S0386-1112(14)60058-3
17. Stocker A, Lazarus J, Becker S, Shaheen SA (2016) North American College/University Market Carsharing Impacts: Results From Zipcar's College Travel Study 2015. https://doi.org/10.13140/RG.2.2.31014.22085
18. Zheng J et al (2009) Carsharing in a university community: assessing potential demand and distinct market characteristics. Transp. Res. Rec. J. Transp. Res. Board 2110(1):18–26. https://doi.org/10.3141/2110-03
19. Shaheen SA (1999) Dynamics in behavioral adaptation to a transportation innovation. A case study of CarLink: A smart carsharing system. University of California, Davis
20. Shaheen SA, Martin E (2010) Demand for carsharing systems in Beijing, China: an exploratory study. Int J Sustain Transp 4(1):41–55
21. Wang M, Martin EW, Shaheen SA (2012) Carsharing in Shanghai, China: analysis of behavioral response to local survey and potential competition. Transp Res Rec 2319(1):86–95
22. Le Vine S, Lee-Gosselin M, Sivakumar A, Polak J (2013) A new concept of accessibility to personal activities: development of theory and application to an empirical study of mobility resource holdings. J Transp Geogr 31:1–10

23. Laarabi H, Needell Z, Waraich R, Poliziani C, Wenzel T (2023) BEAM: the modeling framework for behavior, energy, autonomy and mobility. ArXiv Prepr. ArXiv230802073
24. Southern California Association of Governments—SCAG, Southern California Association of Governments. Accessed 18 Oct 2023. [Online]. Available: https://scag.ca.gov/home
25. CEMDAP. Accessed 18 Oct 2023. [Online]. Available: https://www.caee.utexas.edu/prof/bhat/CEMDAP.htm
26. SBD International Airport Welcomes Amazon Air—San Bernardino International Airport (SBD). Accessed 18 Oct 2023. [Online]. Available: https://www.sbdairport.com/sbd-international-airport-welcomes-amazon-air/
27. San Bernardino Airport Communities, San Bernardino Airport Communities. Accessed 18 Oct 2023. [Online]. Available: https://sbairportcommunities.org/home
28. Boriboonsomsin K et al (2020) Geofencing as a strategy to lower emissions in disadvantaged communities. CARB Rep 2021–2101
29. Mobile Source Emission Inventory—EMFAC2017 Web Database. Accessed 18 Oct 2023. [Online]. Available: https://arb.ca.gov/emfac/2017/
30. MSEI—Modeling Tools—EMFAC Software and Technical Support Documentation|California Air Resources Board. Accessed 18 Oct 2023. [Online]. Available: https://ww2.arb.ca.gov/msei-modeling-tools-emfac-software-and-technical-support-documentation
31. Snyder M, Heist D (2013) User's Guide for R-LINE Model Version 1.2. Res. LINE Source Model-Surf Releases
32. Meteorological Data for AERMOD. Accessed 18 Oct 2023. [Online]. Available: http://www.aqmd.gov/home/air-quality/meteorological-data/data-for-aermod
33. U. US EPA (2011) Exposure factors handbook. U S Environ Prot Agency
34. Malig B (2012) Notice of adoption of technical support document for exposure assessment and stochastic analysis, OEHHA. Accessed 18 Oct 2023. [Online]. Available: https://oehha.ca.gov/air/crnr/notice-adoption-technical-support-document-exposure-assessment-and-stochastic-analysis-aug
35. Brunekreef B, Janssen NA, de Hartog J, Harssema H, Knape M, van Vliet P (1997) Air pollution from truck traffic and lung function in children living near motorways. Epidemiology 8(3):298–303
36. Gong H Jr et al (2004) Exposures of elderly volunteers with and without chronic obstructive pulmonary disease (COPD) to concentrated ambient fine particulate pollution. Inhal Toxicol 16(11–12):731–744
37. Weichenthal S et al (2012) Personal exposure to specific volatile organic compounds and acute changes in lung function and heart rate variability among urban cyclists. Environ Res 118:118–123
38. Grodzevich O, Romanko O (2006) Normalization and other topics in multi-objective optimization
39. Demir E, Bektaş T, Laporte G (2014) The bi-objective pollution-routing problem. Eur J Oper Res 232(3):464–478
40. Research|California Air Resources Board. Accessed 18 Oct 2023. [Online]. Available: https://ww2.arb.ca.gov/our-work/topics/research
41. suburbanstats.org, Current Population Demographics and Statistics for California by age, gender and race. SuburbanStats.org. Accessed 18 Oct 2023. [Online]. Available: https://suburbanstats.org/population/how-many-people-live-in-california
42. MT, Torpey E, Older workers: labor force trends and career options : Career Outlook: U.S. Bureau of Labor Statistics. Accessed 18 Oct 2023. [Online]. Available: https://www.bls.gov/careeroutlook/2017/article/older-workers.htm
43. Frederic PL, Occupational choices of the elderly : Monthly Labor Review: U.S. Bureau of Labor Statistics. Accessed 18 Oct 2023. [Online]. Available: https://www.bls.gov/opub/mlr/2017/article/occupational-choices-of-the-elderly.htm
44. Day care in the United States: Is it good or bad for kids? Accessed 18 Oct 2023. [Online]. Available: https://slate.com/human-interest/2013/08/day-care-in-the-united-states-is-it-good-or-bad-for-kids.html

45. The NCES Fast Facts Tool provides quick answers to many education questions (National Center for Education Statistics). Accessed 18 Oct 2023. [Online]. Available: https://nces.ed.gov/fastfacts/display.asp?id=4
46. Caltrans PeMS. Accessed 18 Oct 2023. [Online]. Available: https://pems.dot.ca.gov/

Chapter 18
Equitable Microtransit Services

Anuradha M. Annaswamy and Venkatesh Venkataramanan

Abstract Shared mobility services have shown that a continuum of solutions can be provided between traditional individual transport and mass transit by making use of an underlying cyber-physical substrate that provides advanced, distributed, and networked computational and communication support. Shared mobility services, also known as microtransit services, provide a way of solving the last-mile connectivity problem and can enable access to key services such as healthcare, jobs, and food for disadvantaged communities. Despite recent advances in optimizing energy efficiency in microtransit services, on-the-ground pilots have struggled to increase performance measures such as passenger miles per revenue mile or passengers per unit energy due to challenges in demand aggregation, understanding user needs, and human behavior. This chapter proposes a framework for multi-passenger transport that allows dynamic routing and behavioral dynamics of passengers and leads equitable microtransit services for members of disadvantaged communities. The implications of disadvantaged communities and how their needs may be met by the proposed framework are discussed. A numerical study based on a community in a Food Desert in the Western United States is presented.

18.1 Introduction

Mobility service in urban centers have been rapidly evolving to accommodate different travel modalities and changes in consumer patterns. Paradigm shifts such as changes in patterns due to impacts from Covid-19 and increase in remote work, increased adoption of new technology such as ride sharing, autonomous vehicles, innovative lane management systems, and e-bikes have resulted in a vastly different

---

A. M. Annaswamy (✉)
Massachusetts Institute of Technology, 77 Massachusetts Ave, Cambridge, MA 02134, USA
e-mail: aanna@mit.edu

V. Venkataramanan
National Renewable Energy Laboratory, Golden, CO, USA
e-mail: vvenkata@nrel.gov

P. Ioannou and A. A. Malikopoulos (eds.), *Transportation Mobility in Smart Cities*,
Springer Tracts on Transportation and Traffic 21,
https://doi.org/10.1007/978-3-031-64769-7_18

travel landscape as compared to even 10 years ago. Over the last decade, several new modalities of travel have emerged and gained prominence in urban transportation systems across the US. These modalities often reflect advancements in technology, changing urban lifestyles, and a growing emphasis on sustainable and efficient mobility options. A prominent selection from these modalities is listed below:

1. Ride-Sharing and Ride-Hailing Services: These are transportation services provided by on-demand, app-based ride-sharing and ride-hailing companies like Uber and Lyft. These platforms allow users to request rides from drivers using their personal vehicles, offering a convenient and flexible alternative to traditional taxis.
2. On-Demand Shuttle Services: These services aim to provide more efficient and accessible public transportation options by providing on-demand shuttle services that operate within certain neighborhoods or along predefined routes.
3. Autonomous Vehicles (AVs): AVs, including connected autonomous vehicles (CAVs) offer transportation services capable of operating independently to varying degrees. While fully autonomous vehicles (without a driver/safety personnel) have yet to be widely deployed in urban transportation, significant progress has been made in their development. Some cities such as Phoenix and San Francisco have started testing autonomous shuttles in controlled environments.
4. Mobility as a Service (MaaS): The concept of MaaS involves integrating various transportation options into a single platform, allowing users to plan, book, and pay for multi-modal journeys seamlessly.
5. Car-Sharing Services: Car-sharing services provide users with the flexibility to rent cars on an hourly or daily basis. These services promote carpooling and reduce the need for individual car ownership in urban areas.
6. E-Scooters and E-Bikes: Electric scooters and e-bikes have become increasingly popular in urban areas, providing a convenient and eco-friendly way to cover short distances.

It is important to note that the adoption and success of these new modalities can vary significantly depending on factors such as city infrastructure, regulatory environment, user preferences, and the overall urban mobility landscape. These innovations are driven by a multitude of factors such as increased computational power available for users, cloud computing, cheaper availability of technology, wider deployment of edge sensors, and increased communication leading to vehicle-to-X technologies, where "X" includes other vehicles, transportation infrastructure, power grid, and so on.

While technological advancements have created new opportunities for a wide demographic, concerns have been increasing about potential impacts on the existing inequities present in large urban centers. These inequities might arise due to a combination of historical, economic, social, and systemic factors. Historical disinvestment and discriminatory policies such as redlining contribute to social inequities significantly, resulting in inadequate infrastructure, poor housing conditions, and limited access to essential services, including public transportation. These areas often suffer from economic inequalities, often have limited financial resources, making car ownership and access to private transportation challenging. Lack of affordable transportation options can further isolate these communities from economic opportunities and essential services. Considering the lack of new development and challenges in accessing private transportation, disadvantaged communities often lack access to job centers due to their location on the outskirts of the city or away from major employment hubs. Hence, these areas are often referred to as "transportation deserts"—which are characterized by inadequate access to public transportation services. These areas might lack convenient access to bus stops, rail stations, or other transit options.

In addition to the lack of public transit and challenges with private transit, these areas often suffer from a lack of infrastructure leading to poor road conditions, lack of sidewalks, and unsafe pedestrian environments. This can make it challenging for residents in disadvantaged communities to walk or bike to transit stops, reducing their access to service, and leading to reduced willingness to use public transportation. Lack of infrastructure also extends to digital infrastructure, such as Internet-based services. Many public transit systems, convenience and subsidized services, and technology-based services rely on digital platforms for route information, real-time updates, and service updates. Lack of access to smartphones or the internet can hinder residents' ability to navigate these services effectively. This leads to a lack of adoption for these services, resulting in new services being exclusively offered in the most wealthy areas where the residents often do not have a necessity for these services but rather use them as a novelty. In some disadvantaged areas, transportation services might have limited frequency or unreliable schedules making it difficult for residents to plan their trips effectively, leading to longer wait times and less convenient travel.

Considering these challenges, a repeating theme is the difficulty in reaching public transit systems for residents in disadvantaged communities. This challenge is termed as the "First Mile/Last Mile" (FM/LM) problem, which refers to the challenge of efficiently connecting people from their origins (usually their homes) to public transportation stations or stops, and then from those stations to their final destinations. One of the solutions to this problem is microtransit services, which refers to on-demand, shared transportation services that use smaller vehicles, such as vans or minibusses, to provide flexible and personalized transportation options. While microtransit services have been proposed before to address the FM/LM problem, we propose a variant of the problem which accounts for challenges in underserved communities and propose a solution to this problem using the concept of dynamic routing.

The rest of this chapter is organized as follows: Sect. 18.2 describes typical challenges in underserved communities and introduces the proposed microtransit service, Sect. 18.3 describes the formulation of the equitable microtransit service as well as the routing algorithm in a static and dynamic context and its solution using the alternative-minimization approach. Section 18.4 demonstrates the results for a case in a disadvantaged community and discusses the utility of the proposed service, and Sect. 18.5 discusses future work and conclusions from this work.

18.2 Urban Mobility and Disadvantaged Communities

Disadvantaged communities often present a suburban design without an identifiable "center," and therefore no clear access points into and out of the neighborhoods. These neighborhoods generally lack mixed-use development, causing residents to rely heavily on access to an automobile, even more so than in the rest of the city. Barriers to reliable transportation—access to food, jobs, education, healthcare, and so much more—are concentrated in these communities and impact a greater portion of the population than elsewhere in the city/urban center. These conditions usually stem from a combination of adverse history, complicated economic trends, and market forces.

Various types of "deserts" exist in underserved communities across the country. These deserts refer to areas where residents lack access to essential goods, services, or opportunities, often due to systemic factors or economic disparities. A few common deserts prevalent in underserved communities include the following [1]:

1. Healthcare Deserts: Healthcare deserts are areas where residents have limited access to medical facilities, doctors, specialists, and essential healthcare services. This lack of access can lead to disparities in health outcomes and preventive care, particularly in low-income and rural communities.
2. Education Deserts: Education deserts are regions where residents have limited access to quality educational institutions, including schools, colleges, and training centers. This can hinder educational attainment and limit opportunities for career advancement.
3. Food Deserts: A food desert is an area, typically within urban or rural settings, where residents have limited access to affordable and nutritious food options. In a food desert, the availability of grocery stores, supermarkets, and fresh produce markets is scarce or nonexistent, making it difficult for residents to obtain healthy and well-balanced food.
4. Public Transportation Deserts: Beyond transit deserts, there are areas where residents lack access to a comprehensive public transportation network, limiting their ability to travel to work, school, and essential services.
5. Employment Deserts: Employment deserts are regions where residents have limited access to job opportunities due to economic factors or geographic isolation. This can contribute to high unemployment rates and economic disparities.
6. Affordable Housing Deserts: Affordable housing deserts are areas where residents have limited access to affordable housing options, leading to housing instability, homelessness, and overcrowding.

In this work, we plan to address the specific problem of a food desert using equitable microtransit services. Food insecurity is defined by the Department of Health as "a household-level economic and social condition of limited or uncertain access to adequate food" [1]. In 2020, 13.8 million households were food insecure at some time during the year [1]. Food insecurity does not necessarily cause hunger, but hunger is a possible outcome of food insecurity.

The United States Department of Agriculture (USDA) [2] divides food insecurity into the following two categories:

1. Low food security: "Reports of reduced quality, variety, or desirability of diet. Little or no indication of reduced food intake."
2. Very low food security: "Reports of multiple indications of disrupted eating patterns and reduced food intake."

Residents are at risk for food insecurity in neighborhoods where transportation options are limited, the travel distance to stores is greater, and there are fewer supermarkets [3]. Lack of access to public transportation or a personal vehicle limits access to food. Groups who may lack transportation to healthy food sources include those with chronic diseases or disabilities, residents of rural areas, and some racial/ethnicity groups [3, 4]. A study in Detroit found that people living in low-income, predominantly Black neighborhoods travel an average of 1.1 miles farther to the closest supermarket than people living in low-income predominantly White neighborhoods [5].

Microtransit services, also called Mobility on Demand (MoD) services, have been proposed to serve accomplish several objectives. MoD solutions of late have started to provide several alternatives other than conventional transportation options, e.g., buses, subways and taxis, for passengers to move around. These correspond to different combinations of travel time, cost, and carbon footprint per traveler there by introduce different and attractive alternatives to passengers [6, 7]. A recent article [8] illustrates clearly that the potential of shared mobility in drastically reducing the number of vehicles on the roads is very high. The article demonstrates that if ride sharing is implemented in New York City, 98% of the taxi demands could be satisfied by 3,000 four-passenger cars, which is fewer than 25% of the number of taxis currently operating in New York City with a waiting time of 2.7 min on average. In this article, we propose the use of microtransit services for addressing the needs of disadvantaged communities that may be facing food insecurities.

18.3 Equitable Microtransit Services

Dynamic routing is a main feature of ride sharing services. The work in [9] proposed dynamic routing for microtransit services with the concept of a space window. The main idea here is that the passengers are willing to walk for a certain distance, both before being picked up and after being dropped off in order to get the dynamic shuttle service. Passengers specify the maximum distances they are willing to walk and therefore the requested pickup/drop-off locations expand to pickup/drop-off regions denoted as space windows S_W, which are defined as the sets of all feasible locations that passengers are willing to walk to/from. Space windows are one of the key ingredients of the proposed dynamic shuttle service, which can allow accommodation of real street topologies and traffic congestion and the associated asymmetries between driving time and walking time, such as one-ways and pedestrian only lanes as a short walking distance for a passenger may result in significant improvements in the shuttle route.

With the above concept of a space window, the overall shared mobility service is proposed to operate in four steps: (1) *Request*: passengers request shuttle rides with specified pickup/drop-off locations, maximum distances willing to walk, and time windows of service if needed. (2) *Offer*: the shuttle server distributes offers to passengers with ride details including pickup locations, walking distances, pickup times, drop-off locations, drop-off times, and prices. (3) *Decide*: passengers decide whether to accept or decline the offers. (4) *Operate*: the shuttle server sends out operational instructions to the shuttles according to the decisions from the passengers and trips start (see Fig. 18.1).

18.3.1 Optimization: Costs and Constraints

The overall problem of dynamic routing is analogous to a Traveling Salesman Problem between regions as opposed to points. Rather than use Mixed-Integer Programming (MIP), we propose the use of an Alternating Minimization [10–12] based optimization approach in order to tackle the problem. In addition, we use this approach to carry out dynamic routing in response to real-time requests from passengers, thereby facilitates an overall dynamic framework. We provide a few preliminaries before describing the AltMin approach.

18.3.1.1 Preliminaries

Definition 18.1 (*Ride Request*) A *ride request* of the dynamic shuttle service is defined as a tuple $\{P_k, D_k, r_k^p, r_k^d, T_k^r\}$, composed of P_k the customer specified pickup location, D_k drop-off location, r_k^p and r_k^d the maximum distances willing to walk before being picked up and after being dropped off, respectively, as well as T_k^r the time of request. $P_k, D_k \in \mathbb{R}^2$ (e.g., longitude and latitude), $r_k^p, r_k^d \in \mathbb{R}_{\geq 0}$, and $T_k^r \in \mathbb{R}$. Let $S_1 = \{1, 2, \ldots, n\}$, where $n \in \mathbb{Z}_{>0}$ is the number of requests. $k \in S_1$ denotes the order of the request according to the timestamp when it is made, i.e., a smaller k value corresponds to an earlier request. Without loss of generality, we assume there is only one passenger per request, and all the requests are made at time $T_k^r = 0, \forall k \in S_1$.

Definition 18.2 (*Space Window*) A pickup/drop-off *space window* is defined as the set of all feasible pickup/drop-off locations whose distance from the requested pickup/drop-off location does not exceed the specified maximum walking distance. For each $k \in S_1$, denote $S_W(P_k, r_k^p), S_W(D_k, r_k^d) \in \mathbb{R}^2$ as the pickup and drop-off space window for the kth passenger, respectively, such that

$$S_W(P_k, r_k^p) = \{p \mid p \in \mathbb{R}^2, d(p, P_k) \leq r_k^p\}, \tag{18.1}$$

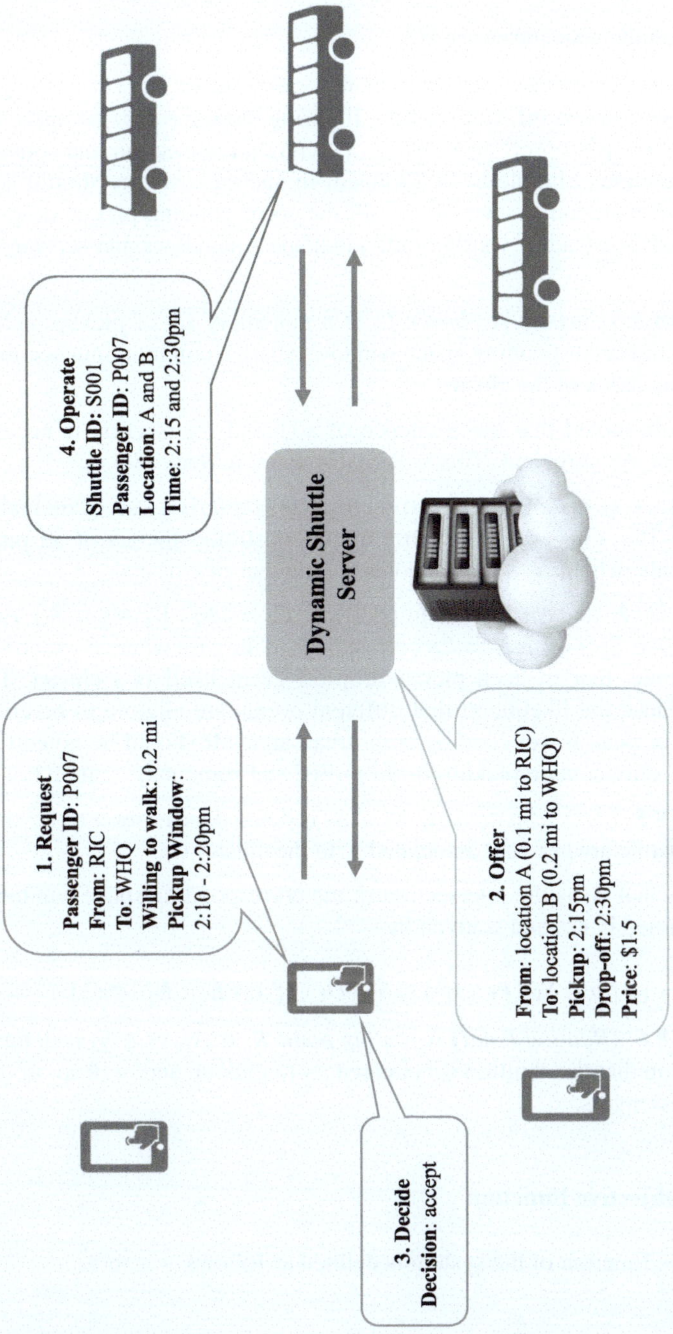

Fig. 18.1 Procedures of the dynamic shuttle service: request, offer, decide, and operate

$$S_W(D_k, r_k^d) = \{p \mid p \in \mathbb{R}^2, d(p, D_k) \leq r_k^d\}, \tag{18.2}$$

where d is the distance metric on $\mathbb{R}^2$.

Let E_k^p and E_k^d denote the events that correspond to the kth passenger being picked up and dropped off, respectively. If during the same trip, the space windows of different pickup/drop-off events share a common intersection, the corresponding events are assumed eligible to be grouped. In such a case, the shuttle is allowed to visit the intersection to take care of all the corresponding pickup/drop-off tasks simultaneously, instead of visiting individual space windows and serving the tasks one by one.

Definition 18.3 (*Cluster*) A *cluster* C_i is a nonempty set of pickup and/or drop-off events if the corresponding space windows share a common intersection, where $i \in \mathbb{Z}_{>0}$ is the index of the cluster.

It should be noted that the existence of a shared intersection is necessary for clustering, but not sufficient. This will be discussed in Sect. 18.3.2.

Definition 18.4 (*Clustering Pattern*) A *clustering pattern* is a set of mutually exclusive clusters $\{C_1, C_2, \ldots, C_N\}$, whose union constitutes the set of all pickup and drop-off events, where $N \in \mathbb{Z}_{>0}$ denotes the number of clusters.

Let $S_2 = \{1, 2, \ldots, N\}$. By definition, $\bigcup_{i \in S_2} C_i = \{E_k^p, E_k^d, \forall k \in S_1\}$. Also, $C_i \cap C_j = \emptyset, \forall i, j \in S_2, i \neq j$. $N = 2n$ corresponds to the case where no nontrivial clusters are formed, that is, each pickup/drop-off event itself is a cluster. Typically, $N < 2n$ because the likelihood that different events are eligible to be clustered is fairly high for peak travel conditions in urban areas. It should be noted that some clusters may only contain pickups or drop-offs, and some may contain both.

Definition 18.5 (*Area*) An *area* $A_i, \forall i \in S_2$ is defined as a subset of $\mathbb{R}^2$ that represents the feasible area to visit associated with the cluster C_i.

It follows that A_i is the intersection of the corresponding space windows of the events contained in C_i, and is given by

$$A_i = \{p \mid p \in S_W(P_{k_p}, r_{k_p}^p) \cap S_W(D_{k_d}, r_{k_d}^d), \forall E_{k_p}^p, E_{k_d}^d \in C_i\}. \tag{18.3}$$

Definition 18.6 (*Routing Point*) A *routing point* $R_i \in A_i, \forall i \in S_2$ is defined as the actual location that the shuttle visits in area A_i to pick up and/or drop off the corresponding passengers.

18.3.1.2 Objective Function

The objective function of the system is defined as follows

$$\mathcal{C} = \gamma_1 \sum_{i=1}^{N} t_i + \gamma_2 \sum_{k=1}^{n} (\alpha_1 WaitT_k + \alpha_2 RideT_k), \tag{18.4}$$

which represents a weighted sum of various time costs, both on the shuttle side and on the passenger side. In (18.4), $\forall i \in S_2$, t_i denotes the time taken by the shuttle to complete the ith trip segment, $\forall k \in S_1$, $WaitT_k$, $RideT_k$, denote the waiting time and riding time of passenger k, respectively, and $\gamma_1, \gamma_2, \alpha_1, \alpha_2$ are nonnegative weights.

It should be noted that the objective function defined in (18.4) is a general representation which does not depend on any specific distance metric, i.e., versatile for various distance metrics as long as represented properly. Throughout this manuscript, the underlying space topology is assumed to be a two-dimensional Euclidean space and as a result the corresponding space windows $S_w(P_k, r_k^p)/S_w(D_k, r_k^d)$ are circles centered at P_k/D_k with r_k^p/r_k^d as the radii, $\forall k \in S_1$. A similar formulation can be extended to other metrics as well, such as the Manhattan distance and metrics that correspond to real street topologies.

18.3.1.3 Constraints

The route of the dynamic shuttle service should satisfy the following constraints.

Definition 18.7 (*Legitimate Constraint*) *Legitimate constraints* specify that any passenger should be picked up before being dropped off, that being said, $\forall k \in S_1$, E_k^p should occur before E_k^d. Moreover, E_k^p and E_k^d cannot be in the same cluster as otherwise passenger k would be willing to walk directly from the requested origin to the destination instead of taking a shuttle ride.

Definition 18.8 *Capacity Constraint*) *Capacity constraints* specify that at anytime of the trip, the number of passengers on board should not exceed the capacity of the shuttle denoted as $c \in \mathbb{Z}_{>0}$.

It should be noted that the number of passengers within a trip could be larger than c, as long as they are not on board simultaneously at anytime during the trip.

Definition 18.9 (*Maximum Position Shift Constraint*) A pickup/drop-off *position shift* is defined as the absolute value of the difference between the actual occurrence of the pickup/drop-off event and that of the request event. *Maximum Position Shift (MPS) constraints* specify that the position shifts should not exceed some given upper bounds, that being said

$$|p_k - k| \le MPS^p, \quad \forall k \in S_1, \tag{18.5}$$

$$|d_k - k| \le MPS^d, \quad \forall k \in S_1, \tag{18.6}$$

where p_k and d_k denote the orders of the kth passenger in the pickup and drop-off queue, respectively. MPS^p and MPS^d are positive integer upper bounds on the pickup and drop-off position shifts, respectively, which are provided by the dynamic shuttle server as part of the service protocol.

MPS constraints are priority constraints that serve as a guarantee of the dynamic shuttle service quality, such that passengers will not experience a long waiting time or travel delay. They are especially significant in the dynamic routing scenario in order to prevent the possibility of any passenger from indefinite deferment of pickup and drop-off by the algorithm.

From the above definitions, we note that there are essentially three queues associated with the passengers, which correspond to three different sequences $\{k\}$, $\{p_k\}$, and $\{d_k\}$, where $\{k\}$ corresponds to the sequence of requests, $\{p_k\}$ and $\{d_k\}$ correspond to those of pickups and drop-offs, respectively. The sequences $\{p_k\}$ and $\{d_k\}$ are permutations of $\{k\}$ in the extreme case where each cluster consists of only one event. In general, $\{p_k\}$ and $\{d_k\}$ include both permutations of subsets of $\{k\}$ and repetitions, as the orders of different pickups/drop-offs may be tied if in the same cluster.

Definition 18.10 (*Departure Constraint*) *Departure constraints* specify that for those clusters which contain pickup events, the shuttle should not depart from the routing point before the associated passengers arrive and get on board.

In our problem formulation, legitimate, capacity and MPS constraints are named *primary constraints* as they affect the feasibility of the clustering pattern and the sequence of the clusters to be visited. While the departure constraints are called *secondary constraints* which can always be satisfied by means of forcing the shuttle to wait at the routing points and adjust the departure times, as long as the three sets of primary constraints have all been met.

In what follows, we will assume that a feasible and favorable clustering pattern is explicitly given, based on which the determination of the sequence and routing points will be explored in order to minimize the cost defined in (18.4). In order to formally state this optimization problem, we construct a N dimensional vector S to represent the sequence of the clusters to be visited, where S is a permutation of $[1, 2, \ldots, N]$, $S(i)$ indicates the order of cluster C_i in the visiting queue of the shuttle, $\forall i \in S_2$. Denote $C_0 = \emptyset$ as a dummy cluster representing the depot, whose corresponding space window is $S_W(O, 0)$ and $A_0 = O$, $R_0 = O$. Denote $R = (R_0, R_1, \ldots, R_N)$. With the objective function specified as in (18.4), and the four sets of constraints as in Definitions 2.7 through 2.10, the *master problem* for the rest of the manuscript is formally stated as

$$\min_{(S,R)\in S_f \times R_f} \mathcal{C}(S, R), \tag{18.7}$$

where S_f and R_f are the feasible domains of the decision variables S and R, respectively, that satisfy all of the constraints listed in Sect. 18.3.1.3, and $R_f = A_0 \times A_1 \times \cdots \times A_N$.

18.3.2 AltMin Algorithm

We propose an AltMin algorithm in this section to solve the master problem formulated in (18.7). Due to its simplicity and effectiveness, AltMin has long been a popular optimization method, dating back to early works in the optimization literature (e.g., [10]), and has been widely studied under various assumptions (e.g., [11, 12]). It consists of a block coordinate descent approach and operates by adjusting one block of coordinates at a time so as to attain successive reductions in the objective function. Given the presence of two sets of decision variables S and R in the master problem under consideration, and given that they are quite distinct from another, the use of such an approach is fairly natural. We therefore develop an AltMin algorithm with two phases, where in Phase 1, we optimize $\mathcal{C}$ over S, keeping R fixed, while in Phase 2, we keep S fixed and optimize $\mathcal{C}$ over R. A combination of consecutive Phase 1 plus Phase 2 constitutes one *mega iteration* in our AltMin algorithm.

More formally, let (S^*, R^*) denotes the desired decision variables that optimize $\mathcal{C}$ in (18.7). That is

$$(S^*, R^*) = \underset{(S,R)\in S_f \times R_f}{argmin} \ \mathcal{C}(S, R). \tag{18.8}$$

It should be noted that (S^*, R^*) can be derived exactly via am MIQCP formulation. The AltMin algorithm is proposed to solve (S^*, R^*) approximately via alternately iterating Phase 1 and Phase 2, where at the hth mega iteration, $\forall h \in \mathbb{Z}_{>0}$, Phases 1 and 2 correspond to updates

$$S^h = \underset{S\in S_f}{argmin} \ \mathcal{C}(S, R^{h-1}), \tag{18.9}$$

$$R^h = \underset{R\in R_f}{argmin} \ \mathcal{C}(S^h, R), \tag{18.10}$$

respectively, where R^0 is the initialization of the routing points. The mega iterations are terminated with a suitable stopping criterion.

The advantages of the AltMin algorithm are mainly that it decomposes the master problem, which is complicated and extremely computationally costly, into two subproblems, which are both fairly easy to solve. And in many circumstances, it is powerful in terms of optimality and computational efficiency [13–15]. Specifically, in our routing task, compared with the MIQCP formulation which seeks to find the optimal solution by traversing all feasible sequences, the AltMin algorithm restricts its search to a subset of the entire solution space, which is determined judiciously through successive Phase 1-Phase 2 optimizations.

One possible drawback is that the decoupling process of S and R results in the intrinsic heuristic nature of AltMin. The convergence and optimality properties of the generic AltMin algorithm can be found in various references such as [16–19], which follow under certain regularity conditions. In what follows, we show how AltMin can still be used with the mixed presence of discrete and continuous variables. We discuss details of Phases 1 and 2 of the proposed AltMin algorithm in this section and follow it with a numerical example.

To start the AltMin algorithm, the routing points need to be initialized as R^0. For simplicity and consistency, each element R_i^0 of R^0 is computed as the center of mass of A_i, $\forall i \in \{0\} \cup S_2$, in our implementation. With this initialization, we discuss details of Phase 1 in Sect. 18.3.2.1, details of Phase 2 in Sect. 18.3.2.2, and convergence analysis and stopping criterion in Sect. 18.3.2.2.

18.3.2.1 Phase 1

As mentioned above, in Phase 1, the routing points R are fixed, and the focus is on the optimization of C over S. In our problem formulation, the shuttle is assumed to travel between the areas directly following the straight lines connecting the corresponding routing points, therefore the statuses of the system of when the shuttle is at each cluster are sufficient to determine the sequence of visit. The status of the system when the shuttle is at each cluster is uniquely determined by: 1) the cluster it is currently at, and 2) the other clusters that have been visited. With this assumption, we define the state vectors of the system and derive the underlying recurrence relation that corresponds to Phase 1 using these states. This is described below.

State Vector

Let an $N + 1$ dimensional vector $(L, k_1, k_2, \ldots, k_N)$ denote the state of the system where

- $L \in \{0\} \cup S_2$ is the index of the cluster that the shuttle is currently at. $L = 0$ corresponds to the shuttle being at the depot, and $1 \leq L \leq N$ corresponds to the shuttle being at cluster C_L.
- $k_i \in \{0, 1\}$, $\forall i \in S_2$ is a flag that indicates whether the shuttle has visited cluster C_i or not, with $k_i = 1$ if the shuttle has visited or is currently at cluster C_i, and $k_i = 0$ otherwise.

The definition of the state vector implies that the *initial state* of the shuttle could only be $(0, 0, 0, \ldots, 0)$. And due to the open-ended setting of our formulation, the shuttle is eligible to finish the trip with any feasible cluster except for C_0. Therefore, a *terminal state* is expressed as $(L, 1, 1, \ldots, 1)$, $\forall L \in S_2$.

Feasibility of State Vectors

A state vector is feasible if and only if: (1) it is consistent by definition, (2) it satisfies all primary constraints specified in Sect. 18.3.1.3. Each of these cases is detailed below.

Consistency

A state vector should be consistent by definition, essentially: if $L = 0$, it follows that $k_i = 0$, $\forall i \in S_2$; otherwise if $1 \leq L \leq N$, it follows that $k_L = 1$.

Capacity Constraints

Recalling for each $i \in S_2$, ℓ_i is the net load, which corresponds to the difference between the number of passengers been picked up and that of passengers been dropped off, at cluster C_i. It follows that capacity constraints are satisfied if

$$\sum_{i \mid k_i=1,\, i\in S_2} \ell_i \le c. \tag{18.11}$$

The left-hand side of (18.11) computes the sum of net loads of the shuttle from departure until the current state, therefore equates to the number of passengers who are currently on board.

MPS Constraints

$\sum_{i \mid k_i=1,\, i\in S_2\setminus\{L\}} \ell_i^p$ and $\sum_{i \mid k_i=1,\, i\in S_2\setminus\{L\}} \ell_i^d$ correspond to the number of total finished pickups and drop-offs prior to visiting the current cluster C_L, respectively. Therefore, the MPS constraints can be verified as follows

$$\left|\left(\sum_{i \mid k_i=1,\, i\in S_2\setminus\{L\}} \ell_i^p + 1\right) - k\right| \le MPS^p, \quad \forall k \mid E_k^p \in C_L, \tag{18.12}$$

$$\left|\left(\sum_{i \mid k_i=1,\, i\in S_2\setminus\{L\}} \ell_i^d + 1\right) - k\right| \le MPS^d, \quad \forall k \mid E_k^d \in C_L. \tag{18.13}$$

Legitimate Constraints

The legitimate constraints are accommodated via the fact that a state is feasible only if there exist at least one feasible *next state*, as long as it is not a terminal state. A next state of a given state is defined as one that is eligible to occur immediately after the current state. A state vector $(L', k_1', k_2', \ldots, k_N')$, $L' \neq 0$, where L' is the index of the *next cluster* of the current state, can be a next state if and only if

1. The cluster $C_{L'}$ has not been visited yet, i.e., $k_{L'} = 0$. This condition will automatically guarantee that all passengers to be picked up/dropped off have not been picked up/dropped off yet, i.e., the events contained in $C_{L'}$ have not yet occurred.
2. The passengers to be dropped off have already been picked up, i.e., $\forall E_{k_d}^d \in C_L$, $k_{c\ell_{k_d}^p} = 1$.
3. The status of the clusters will remain the same except for that of cluster $C_{L'}$, i.e., $k_{L'} = 1$ while $k_i' = k_i, \forall i \in S_2$ and $i \neq L'$.

This screening of feasible next states guarantees that the legitimate constraints of all passengers are satisfied. The secondary constraints, which are defined in Definition 2.10, are not considered in Phase 1 as they do not affect the feasibility of the sequence of clusters to be visited.

Recurrence Relation

Once the feasibility verifications of all possible states have been conducted, the next step is optimality considerations, which is accomplished via a backward recursion. Let $V(L, k_1, k_2, \ldots, k_N)$ be the optimal system cost measured in terms of $\mathcal{C}$ defined in (18.4) of all subsequent decisions from the current state $(L, k_1, k_2, \ldots, k_N)$ onwards till the end of the trip, i.e., any feasible terminal state is reached. It should be noted that V is only defined for the states which are feasible. It is clear that $\mathcal{C}^* = V(0, 0, 0, \ldots, 0)$ and $V = 0$ for any terminal state. The recurrence relation of $V(L, k_1, k_2, \ldots, k_N)$ is therefore given by

$$V(L, k_1, k_2, \ldots, k_N) = \begin{cases} \min\limits_{\forall L'}\left[t(L, L') \cdot M + V(L', k_1', k_2', \ldots, k_N')\right] & \text{if not a terminal state} \\ 0 & \text{otherwise} \end{cases}. \tag{18.14}$$

The term $t(L, L') \cdot M$ denotes the subcomponent of $\mathcal{C}$ that corresponds to the incremental time cost incurred due to the travel of the shuttle from cluster C_L to $C_{L'}$. It is distributed over the first, second, and third term in (18.4), since it includes time costs incurred by the shuttle, the passengers waiting to be picked up, and those that are riding on the shuttle. Therefore M is given by

$$M = \gamma_1 + \gamma_2\left(\alpha_1 \sum_{i \mid k_i=0,\, i\in S_2} \ell_i^p + \alpha_2 \sum_{i \mid k_i=1,\, i\in S_2} \ell_i\right). \tag{18.15}$$

It should be noted that $t(L, L') = \frac{||R_L - R_{L'}||_2}{v_s} + t_a$ is the time taken for the shuttle to travel from C_L to $C_{L'}$. It is possible to calculate $t(L, L')$ offline as the routing points are fixed prior to this phase.

The main analytical property of the solution derived via the modified BHK algorithm to Phase 1 is summarized in Lemma 18.1.

Lemma 18.1 (Uniqueness and Optimality of Phase 1) *$\forall h \in \mathbb{Z}_{\geq 0}$, given R^h, S^{h+1} is uniquely determined in (18.9) via Phase 1 and it is optimal.*

Proof $\forall h \in \mathbb{Z}_{\geq 0}$, given R^h, S^{h+1} is derived via the modified BHK algorithm detailed in Sect. 18.3.2.1. The algorithm essentially examines all feasible states and seeks the sequence with minimal cost in a backward recursive manner, therefore S^{h+1} is optimal and uniquely determined. □

18.3.2.2 Phase 2

In Phase 2, the sequence of the clusters to be visited, S, is fixed, and the focus is on the optimization of $\mathcal{C}$ over the routing points, R, whose determination process can be formulated as a Quadratically Constrained Quadratic Programming (QCQP). The QCQP in Phase 2 is convex and can be solved to optimality via standard solvers such as Gurobi in our implementation. It should be noted that the primary constraints need not to be examined in this phase as the sequence is given while the secondary constraints are to be accommodated. The QCQP formulation is detailed as follows.

For each cluster $C_i, \forall i \in \{0\} \cup S_2$, its routing point should be within the corresponding area, i.e., $R_i \in A_i$. And $t_i, w_k^p, w_k^d, \forall i \in S_2, k \in S_1$ are defined earlier. Similarly, rotational and shifted explaining variables are introduced to transform the quadratic equality representations of the time cost terms to SOC constraints as follows,

$$R_i^r = \frac{R_{S(i-1)} - R_{S(i)}}{v_s}, \quad \forall i \in S_2, \tag{18.16}$$

$$R_k^{s,p} = \frac{R_{c\ell_k^p} - P_k}{v_p}, \quad R_k^{s,d} = \frac{R_{c\ell_k^d} - D_k}{v_p}, \quad \forall k \in S_1. \tag{18.17}$$

Therefore,

$$||R_i^r||_2^2 \leq t_i^2, \quad \forall i \in S_2, \tag{18.18}$$

$$||R_k^{s,p}||_2^2 \leq w_k^{p2}, \quad ||R_k^{s,d}||_2^2 \leq w_k^{d2}, \quad \forall k \in S_1, \tag{18.19}$$

where $t_i \geq 0$ and $w_k^p, w_k^d \geq 0$. In order to satisfy the departure constraints, the walking time of any passenger to be picked up should not exceed the departure time of the shuttle from the cluster that contains the corresponding pickup event, which is rewritten as

$$w_k^p \leq \sum_{i=1}^{c\ell_k^p} t_i + S(c\ell_k^p)t_a, \quad \forall k \in S_1. \tag{18.20}$$

Using the above notations, we rewrite (18.4) as

$$C = \sum_{i=1}^{N} \left[\gamma_1 + \gamma_2\left(\alpha_1 g_i^1 + \alpha_2 g_i^2\right)\right](t_i + t_a) + \gamma_2 \sum_{k=1}^{n} \left(\alpha_3^p w_k^p + \alpha_3^d w_k^d\right), \tag{18.21}$$

where

$$g_i^1 = n - \sum_{j=0}^{i-1} \ell_{S(j)}^p, \quad \forall i \in S_2 \tag{18.22}$$

is the number of passengers who have not been picked up yet therefore are waiting when the shuttle is traveling on the ith trip segment, and

$$g_i^2 = \sum_{j=0}^{i-1} \ell_{S(j)}, \quad \forall i \in S_2 \tag{18.23}$$

is the number of passengers who are on board when the shuttle is traveling on the ith trip segment.

With the above definitions, the underlying problem in Phase 2 is formulated as a QCQP which minimizes $\mathcal{C}$ in (18.21), subject to constraints defined in (18.16)–(18.19). The uniqueness and optimality of the solution to the QCQP formulation is given by Lemma 18.2.

Lemma 18.2 (Uniqueness and Optimality of Phase 2) $\forall h \in \mathbb{Z}_{>0}$, *given* S^h, R^h *is uniquely determined in (18.10) via Phase 2 and it is optimal.*

Proof $\forall h \in \mathbb{Z}_{>0}$, given S^h, R^h is derived by solving the QCQP formulation detailed in Sect. 18.3.2.2. $\forall i \in \{0\} \cup S_2$, A_i is strictly convex, therefore R_f is strictly convex as well. For each time cost term in (18.21), it is either an L_2 norm between two routing points (t_i) or an L_2 norm between one routing point and one requested location which is constant (w_k^p or w_k^d), therefore strictly convex as well. Thus a nonnegative weighted sum of the time cost terms, i.e., $\mathcal{C}$, is strictly convex, as a result the QCQP formulation is strictly convex, thus R^h is optimal and uniquely determined. □

18.3.3 Extension to Dynamic Routing

In this section, the algorithm for dynamic routing is briefly discussed using its static counterpart as a subcomponent. Dynamic routing is more flexible in comparison to static routing in terms of how to accommodate new requests and achieve better system level performance. In the dynamic routing scheme, the shuttle server searches for new requests periodically and makes suitable updates on its current route to take care of the new requests immediately as long as all constraints can be satisfied and a reduction on the cost of the system can be achieved.

The main difference between dynamic and static routing is that any update on the route can affect those passengers who have either been assigned pickups but have not been picked up yet or already been on the shuttle, such that their actual pickups/drop-offs might be delayed. We denote these passengers as *existing passengers*. However, as long as the delays are below certain upper bounds specified in the initial ride offers provide by the shuttle server, the delays are considered acceptable.

Overall, one can view dynamic routing as a generalized version of its static counterpart, with a periodical update on the static route in response to real-time requests from new passengers. Below, we summarize how we modify the AltMin algorithm described in Sect. 18.3.2 to accommodate this update. It should be noted that the same assumption is made that the updated clustering pattern is explicitly given beforehand.

1. Modified dummy initial cluster C_0.
   In static routing, the dummy initial cluster is set as $C_0 = \emptyset$. In dynamic routing, C_0 needs to be modified in order to accommodate the passengers who are already on board when updating the route. In particular, C_0 consists of the pickup events of the passengers who are on board (if any), with the corresponding A_0 as the shuttle's exact location when the update occurs.

2. Frozen routing points of existing but unexecuted pickup events.
   The clustering pattern, sequence of the clusters to be visited and routing points are all subject to change when any update occurs. An additional constraint in dynamic routing should be included that the routing points corresponding to the pickup events of the existing passengers who are yet to be picked up have to be frozen, and are no longer decision variables. Otherwise, additional walking times may be imposed on these passengers. Though the overall walking distance might still be within the specified maximum walking distance r_k^p, the *double walk* may cause huge negative externalities to the riding experiences of these passengers, which is not desirable. This can be easily accomplished by setting $S_W(P_k, r_k^p) = S_W(R_{c\ell_k^p}, 0), \forall k$ that corresponds to an existing passenger who has not yet been picked up. It should be noted that the negotiated drop-off locations of the existing passengers are not necessarily frozen as the walks to the final destinations have not yet occurred, thus no such double walk will happen because of the updates on the current route (Fig. 18.2).
3. Consistent references of the time frame and queues.
   Static routing only takes care of one batch of requests, while dynamic routing conducts static routing periodically, where consecutive batches of requests are accommodated. The references of time frame and the three queues associated with the passengers, i.e., $\{k\}$, $\{q_k\}$ and $\{d_k\}$ should be consistent throughout different batches. The consistency can be guaranteed by simply using the absolute references for both the time frame and the three queues. More specifically, for any passenger k, $\forall k \in \mathbb{Z}_{>0}$, k, p_k and d_k are his/her absolute sequences in the entire request, pickup and drop-off queue, respectively, instead of the relative sequences in the current batch. And then the MPS constraints are able to guarantee that no passenger is likely to experience any indefinite deferment due to the updates on the routes. Similarly, the waiting and riding time of any passenger should be derived via collapsing between the absolute timestamps when he/she gets picked up and requests the shuttle ride, and between those when he/she gets dropped off and picked up, respectively.

An illustrative example of static vs dynamic routing for various phases is presented in Fig. 18.3.

Dynamic routing has been demonstrated to be advantageous over its static counterpart when the requests arrive frequently, more specifically, when the time intervals between different batches of requests are shorter than those it takes to serve all existing passengers [20]. However, static routing is easier to operate and understand, passengers are not subject to any delays on their pickups or drop-offs, and the drivers will not suffer from fatigue due to frequent updates on the shuttle routes (certainly this issue could be alleviated significantly if autonomous driving vehicles could come into play). Therefore, if the requests are not frequent compared to the duration of the existing trips, static routing suffices. However, in any scenario with a high travel demand, dynamic routing may be unavoidable and necessary.

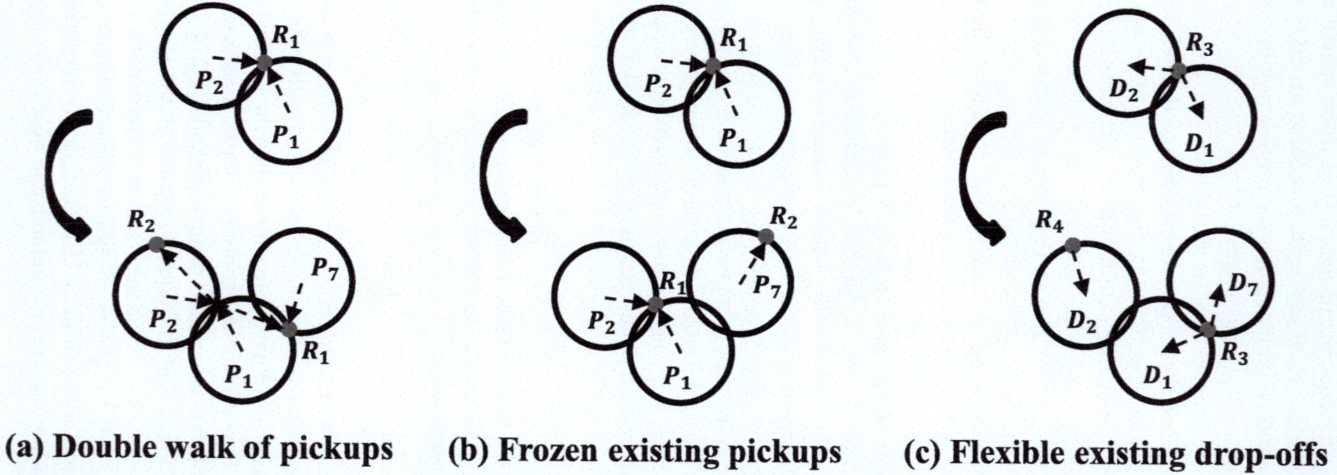

Fig. 18.2 Demonstrations of double walks. **a** Double walks caused by updating R_1 which serves as the negotiated pickup locations for passengers 1 and 2, who have not been picked up yet. **b** Double walks are prevented by freezing R_1. **c** R_3 is flexible as it serves as the negotiated drop-off location for passenger 1 and 2, whose updates will not result in any double walk

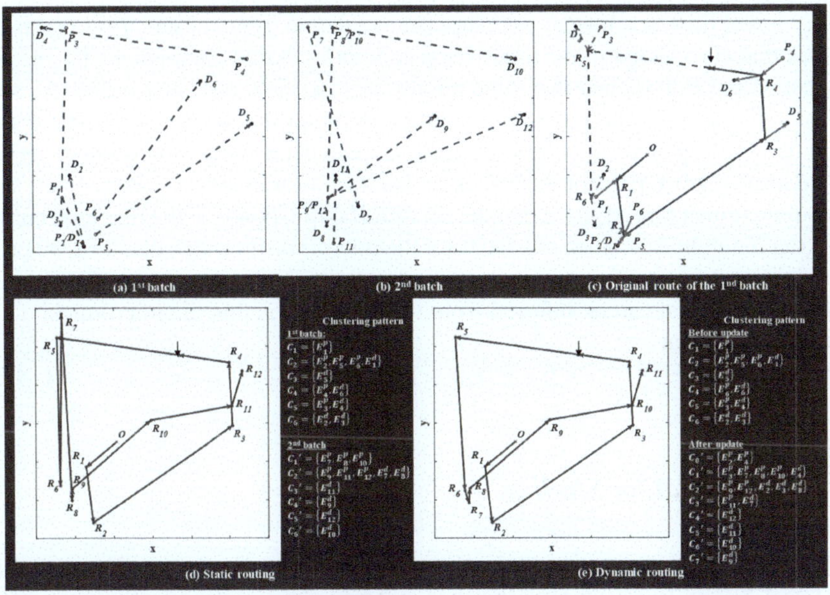

Fig. 18.3 Comparison between static and dynamic routes at various phases

18.4 Utility of Microtransit Services Using Dynamic Routing for a DAC

In the previous section, we addressed the problem of designing a shuttle service that accommodates various (O, D) pairs that correspond to multiple passengers so that an overall time cost is minimized. In this section, we shift the focus from the shuttle to its passengers. In particular, the question that we address here is whether such a dynamic routing service designed as in Sect. 18.3 will be accepted by the passenger. Of particular interest is when the passengers are from a DAC. For this purpose, we will use utility theory [21].

Suppose there are two travel services A and B, where A corresponds to the microtransit service and B is an alternative, with corresponding utilities U_A and U_B with $\Delta U = U_A - U_B$. The probability p_A of accepting service A is then given by

$$p_A = \frac{1}{1 + exp\,(-\Delta U)}. \tag{18.24}$$

The next question is, what is a suitable choice for a utility function for a DAC passenger. The specific context we will evaluate this question is for travel to and from a grocery store for the DAC. We broadly define this function as $f_1(\tau) + f_2(\rho)$, where f_i are increasing functions of their arguments, τ is time, ρ is a monetary-outlay. It is clear that the utility of the service increases as τ and ρ decreases, as it implies a shorter travel time at a lower cost. Suppose we define f_{1A}, f_{1B} and f_{2A}, f_{2B} as the utilities that correspond to the time-based and outlay-based components for the two services A and B. It is clear that if the microtransit services provide a utility f_{1A} such that $f_{1A} > f_{1B}$, then the passenger is likely to adopt the service the outlay-based utility is also higher with the microtransit service. That is, the microtransit service should impose any tariffs for their rides so that it is lower than other alternatives, such as a public transit fare. If, for example, service B corresponds to the financial burden due to an individual car, then any tariff for the microtransit service that is of the order of mass transit fare will clearly make the microtransit offer more attractive. Overall, a general point that can be inferred from utility theory is that the microtransit tariff should be such that

$$f_{2A} \geq f_{2B} - [f_{1A} - f_{1B}]. \tag{18.25}$$

18.4.1 Simulation Studies

In this simulation study, we will focus on a historically disadvantaged community (DAC) of North Tulsa, an area that has suffered socio-economically since being isolated from the rest of the city when Interstate 244 was built in 1975. The area consists of primarily low- and moderate-income households, is acknowledged as a "food desert," and has the lowest life expectancy across Tulsa County. Tulsa's Creating Access to Nutrition (C.A.N.) Plan [22], a study commissioned by the Indian Nations Council of Governments (INCOG) Area Agency on Aging program in 2019, provides recent insight into North Tulsa's food access problem and documents the primary methods used by the community's older adults and people with disabilities to get to the grocery store. A portion of the target area identified by the C.A.N. plan is shown in Fig. 18.4.

The C.A.N. Plan found that North Tulsa residents with disabilities and/or over 60 years of age primarily access fresh food via: (i) personal automobile, (ii) rides with friends or family, (iii) Tulsa Lift paratransit service, (iv) fixed-route public transit, and (v) private residence transportation services. Of these options, survey respondents indicated a strong preference toward the use of personal automobiles, whether on their own or that of friends or family, as other options do not provide easy access to the grocery store. Overall, 55% of survey respondents indicated that they had difficulty buying food always or most of the time. Respondents also indicated that if public transportation options were made more convenient and accessible, they'd be more inclined to use those services. One of the near-term recommendations from the C.A.N. Plan was to offer a pilot transit program offering "flexible transit routes capable of serving multiple riders according to the most efficient route," which is essentially a microtransit service.

However, microtransit services have two major barriers to adoption—providing value across both time considerations and cost considerations as compared to other modalities. There are certainly other challenges in microtransit adoption such as technological challenges, integration with current offerings, user reluctance, and so

Table 18.1 Comparison of shuttle operation cost and ride times per passenger for static and dynamic routing in DAC

	Static routing	Dynamic routing	
		Without space window	With space window
Shuttle Operation Cost [$]	8.67	5.27	5.11
Waiting Time ($WaitT_k$) [minutes]	19.42	8.09	6.98
Riding Time ($RideT_k$) [minutes]	10.58	15.16	15.47

on, however, these issues can be resolved with time provided that the service itself has sufficient utility to the consumers. Hence, we focus on designing a microtransit service that offers better utility in terms of travel times than public transit, and better utility in terms of cost than private transit. In this simulation study, we will focus on the improving the utility of travel times, as publicly available datasets for comparing costs between transit offerings is hard to come by, and newer transit offerings in DACs tend to be subsidized heavily to incentivize adoption at least in the pilot stages. This ensures that cost considerations can be a lower priority at least during the pilot stages of such a microtransit service.

The objective function for the proposed microtransit service follows (18.4) where the first term represents the cost for the shuttle i, and the second term is the cost function for passenger k. While walking time to the routing point is also part of the time cost, for the numerical study reported here, we are only including waiting time and ride time.

Figure 18.4 shows the comparison between using a static route and a dynamic route with a space window approach for providing a microtransit service. The static route leads to higher travel times and inefficiencies in terms of energy spent for mobility because of the lack of adaptability of the rides. As illustrated in the figure, the static route often leads to criss-crossing of routes and leads to higher travel times as the shuttle has to traverse the same route multiple times to ferry passengers.

On the other hand, the dynamic routing with space window approach presents an optimized method to reduce the combined cost of wait and ride times, which can be tweaked by the parameter γ_2. By suitably varying the parameters, it is possible to increase the utility of the proposed microtransit service to be greater than public transit systems and incentivize ridership.

Table 18.1 illustrates that dynamic routing offers significant advantages in terms of reducing travel times, which ultimately increases the utility of the proposed microtransit service. For the final case, the weights for the dynamic routing with space window have been tweaked to minimize the waiting time for the passengers, $WaitT_k$. The weights used for generating these results are presented in Table 18.2:

The utility of the proposed microtransit service can be understood using the probability defined by Eq. 18.25. Plotting the probability of acceptance gives us an idea

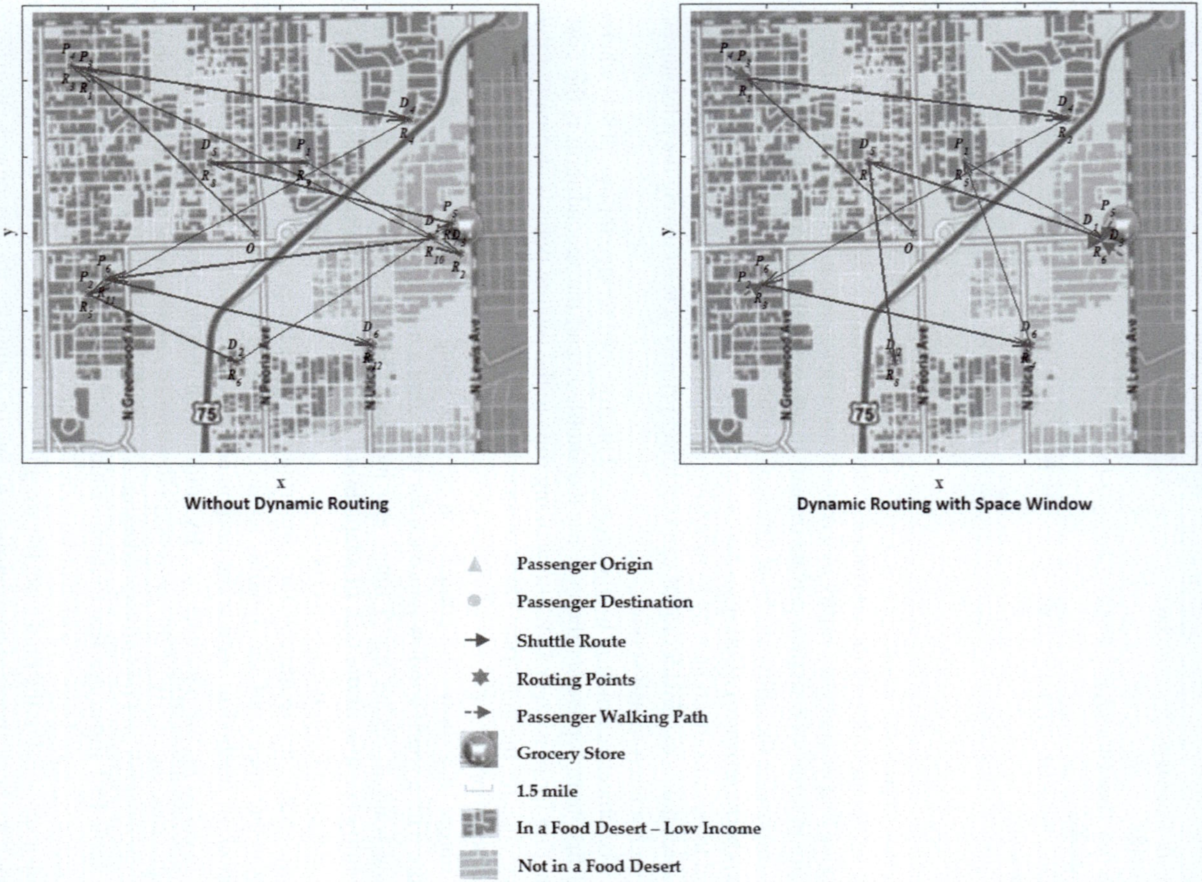

Fig. 18.4 Dynamic route-based microtransit service for DAC, adapted from [22] with permission

Table 18.2 Weights used for simulations

	Dynamic routing without space window	Dynamic routing with space window
Weights	$\gamma_1 = 1, \gamma_2 = 1, \alpha_1 = 1, \alpha_2 = 10, \alpha_3 = 0.05$	$\gamma_1 = 1, \gamma_2 = 1, \alpha_1 = 10, \alpha_2 = 10, \alpha_3 = 0.05$

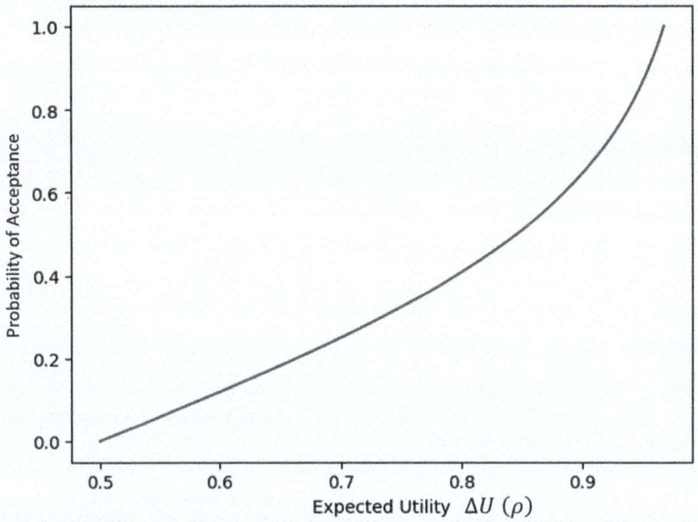

Fig. 18.5 Representative probability of acceptance for proposed microtransit service as compared to public transit travel times

for understanding the utility offered by the proposed microtransit service. To obtain a representative graph, average travel transit times for the proposed area have been obtained from Google Maps data, and ΔU has been calculated. The resulting graph is shown in Fig. 18.5. The figure shows that the probability of acceptance steeply increases with increasing utility, as expected from the results.

18.5 Summary

A shared mobility service is proposed in this chapter as an option for travel for passengers in a disadvantaged community. A dynamic routing algorithm is proposed that minimizes a time cost, which includes waiting time, walking time, and travel time of all passengers who use the service. Using utility theory, the willingness of the community members to adopt this service is also discussed. A numerical example is presented to illustrate the proposed microtransit service using data from a DAC in North Tulsa that lives in a food desert.

The work reported here is only the first step of many more that remains to be done. A more careful articulation of alternatives to the proposed microtransit service and their utility functions is needed. The suitability of prospect theory, a variation of utility theory, that takes into account the risk attitudes of the DAC passengers should be studied. The suitability of monetary tariffs for DAC passengers and other non-monetary alternatives for using the microtransit service needs to be explored. While this article articulates the need and feasibility of a microtransit service for DAC and shows the conditions under which such a service may be highly attractive, significantly more research needs to be done before deploying this service in the field.

Acknowledgements We gratefully acknowledge several useful discussions with Han Zheng and Ethan Kita. We would like to acknowledge Han Zheng's help in carrying out the numerical simulations reported in the chapter.

References

1. Food insecurity. https://health.gov/healthypeople/priority-areas/social-determinants-health/literature-summaries/food-insecurity
2. Definitions of food security. Accessed 10 March 2022. https://www.ers.usda.gov/topics/food-nutrition-assistance/food-security-in-the-u-s/definitions-of-food-security/
3. Ver Ploeg M, Breneman V, Farrigan T, Hamrick K, Hopkins D, Kaufman P, Lin BH, Nord M, Smith TA, Williams R et al (2009) Access to affordable and nutritious food: measuring and understanding food deserts and their consequences: report to congress, Tech. Rep.
4. Zenk SN, Schulz AJ, Israel BA, James SA, Bao S, Wilson ML (2005) Neighborhood racial composition, neighborhood poverty, and the spatial accessibility of supermarkets in metropolitan detroit. Am J Public Health 95(4):660–667
5. Seligman HK, Laraia BA, Kushel MB (2010) Food insecurity is associated with chronic disease among low-income nhanes participants. J Nutrit 140(2):304–310
6. Martin EW, Shaheen SA (2011) Greenhouse gas emission impacts of carsharing in north America. IEEE Trans Intell Transp Syst 12(4):1074–1086
7. Yu B, Ma Y, Xue M, Tang B, Wang B, Yan J, Wei Y-M (2017) Environmental benefits from ridesharing: a case of Beijing. Appl Energy 191:141–152
8. Alonso-Mora J, Samaranayake S, Wallar A, Frazzoli E, Rus D (2017) On-demand high-capacity ride-sharing via dynamic trip-vehicle assignment. In: Proceedings of the national academy of sciences, p 201611675
9. Guan Y, Annaswamy AM, Tseng HE (2019) A dynamic routing framework for shared mobility services. ACM Trans Cyber-Phys Syst 4(1):1–28
10. Ortega JM, Rheinboldt WC (2000) Iterative solution of nonlinear equations in several variables. SIAM
11. Auslender A (1976) Optimisation: méthodes numériques
12. Luo Z-Q, Tseng P (1993) Error bounds and convergence analysis of feasible descent methods: a general approach. Ann Oper Res 46(1):157–178
13. Csisz I, Tusnády G et al (1984) Information geometry and alternating minimization procedures. Stat Decis
14. Niesen U, Shah D, Wornell GW (2009) Adaptive alternating minimization algorithms. IEEE Trans Inf Theory 55(3):1423–1429

15. Wang Y, Yang J, Yin W, Zhang Y (2008) A new alternating minimization algorithm for total variation image reconstruction. SIAM J Imag Sci 1(3):248–272
16. Jain P, Netrapalli P, Sanghavi S (2013) Low-rank matrix completion using alternating minimization. In: Proceedings of the forty-fifth annual ACM symposium on Theory of computing. ACM, pp 665–674
17. Grippo L, Sciandrone M (2000) On the convergence of the block nonlinear gauss-seidel method under convex constraints. Oper Res Lett 26(3):127–136
18. Tseng P (2001) Convergence of a block coordinate descent method for nondifferentiable minimization. J Optim Theory Appl 109(3):475–494
19. Lin C-J (2007) Projected gradient methods for nonnegative matrix factorization. Neural Comput 19(10):2756–2779
20. Annaswamy AM, Guan Y, Tseng EH, Hao Z, Thao P, Yanakiev D (2017) Transactive control in smart cities. In: Proceedings of the IEEE, Special issue on smart cities (accepted)
21. Ben-Akiva M, Bierlaire M (1999) Discrete choice methods and their applications to short term travel decisions. In: Handbook of transportation science. Springer, pp 5–33
22. Miller C, McManus R, Baeza P, Dinoa P, Simmons T et al (2019) Tulsa C.A.N. Plan. Tech. Rep, Indian Nations Council of Governments

Chapter 19
Smart Cities and Transportation: Governance and Institutional Frameworks

Andre Comandon, Marlon G. Boarnet, John Cadiz, and Andrea Holmes

Abstract This chapter examines the institutional environment in which new transportation technologies are being introduced within the United States. We focus on four broad technology categories: Shared mobility, integrated mobility, priced mobility, and autonomous mobility to illustrate the potential transformations to the transportation system and the institutional obstacles to adoption at scale. We identify two overarching themes to conceptualize how transportation innovations interact with the existing technology. First, the omnipresence of transportation in the public sphere and in people's daily lives sets transportation technologies apart from other areas of rapid innovation that are more clearly in the private or public sector. Second, transportation technologies have the potential to significantly reshape land use, which greatly expands the scope of change to how cities function and develop. We use these themes to synthesize the literature and formulate a set of recommendations to steer innovations in the direction of equitably distributed benefits.

Cadiz and Holmes performed this work while graduate students at the USC Sol Price School of Public Policy.

A. Comandon (✉) · M. G. Boarnet · A. Holmes
METRANS Transportation Consortium and Sol Price School of Public Policy, 650 Childs Way, Los Angeles, CA 90089, USA
e-mail: comandon@usc.edu

M. G. Boarnet
e-mail: boarnet@usc.edu

A. Holmes
e-mail: holmesa@usc.edu

J. Cadiz
UC Davis Institute of Transportation Studies, 1605 Tilia Street, Davis, CA 95616, USA
e-mail: jpcadiz@ucdavis.edu

P. Ioannou and A. A. Malikopoulos (eds.), *Transportation Mobility in Smart Cities*, Springer Tracts on Transportation and Traffic 21,
https://doi.org/10.1007/978-3-031-64769-7_19

19.1 Introduction

Smart mobility operates differently than other technologies underpinning smart cities. While all smart city technologies face obstacles that arise from the technologies being spearheaded in the private sector but relying on public infrastructure for deployment, transportation also operates in the public sphere. This means that firms can introduce new technologies independently, with little coordination with governments and citizens (e.g., transportation network companies or TNCs, micromobility). This is not the case for utilities (where infrastructure ownership is centralized and outside the public realm) or public services that rely on communication alone and which, while they can be public, are not necessarily so (e.g., tools for accountability, fees and taxation, information dissemination). Transportation, on the other hand, occurs in public space, almost always in view of others, and passing through streets, walkways, and transit stations that are locations of commerce, interaction, and at times speech and assembly (see, e.g., [1, 2]). To travel is to see and be seen in a place of potential interaction among persons, in ways that are not inherent in using other infrastructure systems (e.g., water or electric power).

The other distinctive feature of smart mobility is that it interacts and can significantly alter land use patterns. Parking alone illustrates the impact of transportation and how innovations could disrupt the status quo. About 14% of Los Angeles County's land is devoted to surface parking and still 8% in the space-constrained Bay Area [3]. Innovations in curbside management, increases in shared mobility, and the introduction of autonomous vehicles could greatly reduce how much parking is required and how it is used. Changes in parking management and requirements, in turn, can affect housing cost and density, the demand for commercial space, and how much on-street parking is necessary to accommodate idle cars and trucks. Smart technologies that modify the demand for parking—either long term (e.g., by creating more trips) or short term (e.g., by requiring quick curbside pickups or drop-offs) can have impacts well beyond the market for the transportation technology itself.

The above examples illustrate the myriad issues new transportation technologies create for the local governments primarily responsible for planning transportation infrastructure and service. The uncertainty that the rapid pace of technological development produces often clashes with the much slower pace of changes in land use and vehicle fleets [4]. The mobile nature of transportation technologies adds to the interaction with land use and planning challenges by extending the impact of new technologies across jurisdictional boundaries. One city may allow e-scooters while the next does not. This can hamper the development of technologies if it relies on a standardized infrastructure on which local governments are unwilling to coordinate. In short, new transportation technologies cannot be deployed in isolation and with disparate sets of rules to govern how they function at scale.

This chapter focuses on the institutional obstacles that arise from smart mobility's operation in public spaces and its interactions with land use. This chapter focuses not on the technologies, but on the governance and institutional frameworks that will influence the overall adoption, impact, and potentially unintended consequences of new smart mobility. First, we discuss the scope of what we will study in exploring the governance of smart transportation technologies Three technologies are driving innovation in transportation:

1. Location aware technology (Global Positioning System—GPS).
2. High-bandwidth communication networks accessed through smartphones and vehicles.
3. High-capacity computing to process data generated by the combination of GPS and high-bandwidth communication technologies.

We focus on four broad transportation domains that these technologies have and will continue to disrupt:

1. Shared Mobility: The combination of GPS and smartphones has allowed sharing that ranges from docked systems to on-demand rideshare.
2. Integrated mobility: Transportation is shifting from purchasing travel on single modes to purchasing trips, at times across multiple modes, using services that can span the public and private sectors.
3. Priced mobility: Pricing transportation is not new. Transit has been priced since its inception. Roads, commonly thought of as free, have long been tolled. Yet the nature of pricing is changing due to technology. Fare collection is electronic and can be based on time of day, location, and even route, and in-vehicle systems can track and hence price travel based on a more comprehensive level than is possible with on-road transponders. This has the potential to open new directions in pricing.
4. Autonomous mobility: Self-driving technology promises substantial disruptions to transportation, changing the cost structure of driving and with it the land use patterns, use of vehicles, and safety of the vehicles.

For each domain, we outline how technology is affecting usage and discuss how institutions can frame ongoing deployment to prioritize the public interest. We then synthesize cross-cutting issues that transportation governance must deal with before formulating a set of policy recommendations to serve as possible first steps toward developing governance and institutional approaches for smart mobility.

19.2 Smart Cities and Transportation: Background and Institutional Obstacles

19.2.1 Shared Mobility

19.2.1.1 Background

Shared mobility refers to shared transportation resources (like shared cars) and the technologies that streamline access to these resources for users (e.g., smartphone interfaces). The transportation services that fall under shared mobility include Transportation Network Companies (TNC) or ride-hailing firms (e.g., Lyft and Uber), micromobility (e.g., e-scooters), and car sharing.

Shared mobility is a distinct kind of transportation technology because it relies largely on existing technology and requires little new physical infrastructure. The innovation is in how the services are distributed and accessed. Users use an app on their smartphone to schedule a ride, or rent a car, bike, or scooter. The technology's relatively easy integration with existing transportation systems and regulations means that governments have primarily been enablers of the technology [5]. Governments' role is to ensure access does not conflict with other road and sidewalk uses [6]. The enabling role of institutions has led to ride hailing, as the most widespread form of shared mobility, doing little to reduce congestion [7, 8] or increase carpooling [9], while creating concerns for safety [9], parking usage [10], and equitable access [11].

Other forms of shared mobility are more sustainable in current implementations. Bike sharing and car sharing decrease the amount users drive overall and reduce the number of cars on the road [12]. An early study of car-sharing programs summarized evidence that found reductions in private vehicle ownership and, in some cases, reductions in driving [13]. For these reasons, vehicle sharing could significantly decrease congestion and pollution without any other technological changes [14, 15].

19.2.1.2 Institutional Obstacles

The main institutional obstacles to the expansion of shared mobility are rules governing land use and data collection. While shared mobility does not require new immediate public investments, the technology's expansion can quickly be limited by existing land use around parking and curb space management. The availability of parking directly affects people's decisions to use shared mobility. People use ride hailing more in areas where parking is expensive and scarce [16]. At the same time, one study showed with data from Seattle that increased reliance on ride hailing leads to induced demand for parking. Parking occupancy is estimated to not decline until shared mobility usage is over twice as high as current levels [10].

The increasing demands on curb space from shared mobility services, deliveries, and parking are in conflict with most existing rules that offer often-free parking and fixed prices [17]. Cities have begun to experiment with new rules to increase adaptability in how curb spaces are used. Some of these rules, like demand-based parking pricing, will have indirect effects on shared mobility by increasing the cost of parking. Other programs that reform the prioritization of curb space toward active shared mobility (e.g., shared bikes) in Washington D.C. [18] or that modify the hierarchy of curb uses to prioritize shared mobility as in San Francisco [19], will have a more direct impact by using parking policy to incentivize shared mobility.[1]

[1] The Santa Monica Zero-Emission Delivery Zone pilot is another example which, while not shared mobility, intends to test how the shift in curb space use can benefit zero-emission vehicles and,

Parking regulation and linking parking usage to real-time communication is especially important for car sharing. Car sharing can rely on two-way trips (returning the car to its original location), one-way trips (driving between two designated locations), or free-floating (parking anywhere within a designated area). For all forms of car sharing, parking availability is crucial for how cars are distributed, reducing uncertainty about availability, and incentivizing use through reduced parking cost [20]. Research on free-floating car sharing is limited due to the small number of existing programs. Simulations show that parking policies like bundling the cost of the car rental with guaranteed parking or increasing the cost of parking for non-shared cars can lead to higher rates of car sharing [21, 22]. One study, based on data from a program in Austin, shows that higher cost and scarcity of parking increases the probability that a vehicle will be rented four-fold [23].

The wider adoption of shared mobility, especially modes that are not car reliant, depends on broader changes to transportation infrastructure and land use. The lack of infrastructure to increase safety while using e-scooters and shared bikes is an important deterrent for potential users, especially those in marginalized areas with lower quality infrastructure [24]. Therefore, while shared mobility can be implemented with little or no new street infrastructure, shared mobility fits within a much broader effort to change transportation infrastructure through road diets [18], the creation of linked transportation hubs [25], the integration of shared mobility in long-range planning [26], and, generally, increasing the safety of roads.

Some cities have linked their efforts to regulate shared mobility with an expansion of data collection. Some of these efforts focus on investments in complementary technology that can assist in tracking curb space usage and enforcing rules [27] or increasing the efficiency and equitability of shared mobility hub siting [18, 28]. Not uncommonly, city regulations are limited or superseded by higher-level governments. San Francisco, for example, identified the regulations of TNC at the state level as an impediment on the city's ability to steer the industry toward more equitable goals and highlighted the need for supplementary local rules or reforms at the state level [19, 29]. One approach is to make the licensing of TNCs and other shared mobility providers conditional on sharing some of their data [19, 26]. Data sharing and governance have also been identified as a crucial factor in addressing issues of accessibility and discrimination that private providers can skirt without oversight [11, 30].

---

indirectly, shared mobility. For more information: https://www.santamonica.gov/zero-emission-delivery-zone accessed August 28, 2022.

19.2.2 Integrated Mobility

19.2.2.1 Background

Integrated mobility is a set of technologies that enable users to pay for multiple transportation services as a package, or in the same purchase. The earliest example of integrated mobility was farecards that allowed the user to travel on multiple transit systems. The ability of the farecard to communicate with the transit terminal means the terminal can adjust fares based on the agency, the distance traveling, or time of day, giving the user greater simplicity and transit agencies greater flexibility [31]. An important feature of integrated farecards (and significant drawback if not included) is the ability to automatically apply discounts for riders.

This limitation has led to the development of a next generation of payment technology. Some agencies, like LADOT in Los Angeles, are partnering with other mobility services to develop a mobility wallet that will greatly enhance the ability to provide subsidies and discounts to mobility disadvantaged populations.[2] A mobility wallet takes advantage of the connectivity of smartphones (or alternative devices) to allow the users to pay fares for all participating transit agencies and private mobility services without having to worry about different discounts offered. The wallet can also be designed to allow governments to easily credit users' accounts or update their discounts without using banks as intermediaries.

Mobility as a Service (MaaS) expands on the idea of a mobility wallet by integrating information on schedules and connections, booking, and payment across all modes of transportation within a single platform [32, 33]. Several MaaS programs exist, most of them in Europe, that offer a means to pay for a trip that includes public transit, ride hailing, and micromobility within a single app. Move PGH (a public-private partnership launched in 2021 in Pittsburgh) is the first MaaS program in the United States and gives users access to shared mobility options (e.g., scooters and cars) and public transit through a single app [34]. Some programs, like Whim in Helsinki, Finland, have started offering subscriptions that give subscribers unlimited use of all mobility services for a flat monthly fee [25, 32].

Mobility as a Service builds on shared mobility but is a distinct service with broader needs. Shared mobility is a set of services provided by separate firms. Each service can function on its own and usually is available only where demand is most concentrated [11]. MaaS, in contrast, relies on a comprehensive coverage of services that, unlike shared mobility services, are both dense and well connected [25, 35]. In addition, while shared mobility is largely the purview of private firms, governments tend to be more involved in MaaS, mobility wallets, and integrated farecards because they rely heavily on the integration with public transit systems [36].

[2] The Universal Basic Mobility Pilot was launched in April 2022. The program will provide 2,000 residents of South Los Angeles with a smartphone that links to a mobility account that can be used on public transit and private mobility services like Uber. Accessed on August 27, 2022: https://ladot.lacity.org/sites/default/files/press-releases/press-release-ladot-launches-universal-basic-mobility-pilot.pdf.

19.2.2.2 Institutional Obstacles

Integrated farecard programs, as a precursor to MaaS, are a rich source of insights on obstacles to the integration of different systems. The California Integrated Travel Project (Cal-ITP) has identified access to reliable transit information, reducing friction in payment methods, and standardizing eligibility for discounts as priority areas [37]. In addition, interoperability across networks, fare standardization and data infrastructure are major obstacles that require institutional support [35]. Even if information is available and accurate, coordination between services that run on a fixed schedule, like buses, is essential for people traveling over longer distances [38].

The integrated nature of MaaS heightens the need for seamless data sharing and, therefore, concerns about data privacy and security [39]. Standardizing data formats and regulating access so that transit agencies and private mobility providers can easily share their data are essential to the service [25]. Many models currently exist for building an integrated data system, including platforms like the General Transit Feed Specification that provides real-time transit information. MaaS, however, requires payment and service coordination across the public and private sector, a task that may be better handled by a third-party entity unaffiliated with governments or participating businesses [40]. The inclusion of discounts and subsidies requires further integration with administrative data, which are subject to strict privacy laws.

Alternatively, some have proposed a framework that enables data to work across systems [41]. Instead of establishing a common data format and centralizing data, firms provide the language to read their data and allow the framework to generate the information relevant to the end user without having to collect data. This solution, however, requires high-level coordination and authority to develop and implement the framework (which is proposed to work across European countries). The use of such a framework effectively solves the fragmentation issue by allowing a single farecard to "talk" to all fare collectors.

The most significant limitation to MaaS, however, and the reason the technology is more developed in Europe, is that MaaS relies on a dense, interconnected transportation system [42]. Europe developed a dense network of rail, public transit, and active transportation that does not exist in many parts of the US. Reaching a critical mass that would enable a substantial uptake of MaaS requires coordinated investments in infrastructure and transportation services. In other words, governments are not only enablers of the technology, but also participants in its development [6].

19.2.3 Road and Congestion Pricing

19.2.3.1 Background

In 2018, the overall cost of congestion was estimated at $87 billion [44]. Congestion pricing charges drivers for the delay externality that they impose on other drivers. Road pricing, in contrast to congestion pricing, is simply the idea that people should pay for the usage of roads without a link to externalities.

Road pricing is put in place primarily to raise revenue (usually for the purpose of road construction and maintenance). Traditionally, road pricing was implemented with the use of tolls. Congestion pricing, in contrast, aims to improve the flow of traffic (or reduce emissions) by reducing the volume of cars on the road at any given time in a specified area. The differing goals of congestion and road pricing mean that local governments often have different motivations than higher-level government. Cities in California, for example, are more focused on congestion pricing to reduce central city traffic and pollution while the state has piloted road charges (based on vehicle miles traveled) as a fee that could replace the fuel tax [45, 46].

There are five main tools for implementing congestion or externality pricing [47–49]:

- **Cordon Pricing:** This pricing program charges road users when they pass through (inbound, outbound, or both) a designated cordon around an area, such as a central business district. Examples include Stockholm and Singapore.
- **Area Pricing:** Like cordon pricing, this method charges users that enter or exit an area, but also anyone who drives entirely within the area. The main example is London.
- **Highway Pricing:** This pricing program charges road users for using specific lanes on a segment of highway or all lanes on a segment of highway. Highway pricing is commonly used throughout California and the United States.
- **Corridor Pricing:** This pricing program charges road users for driving on certain roadways in a specific corridor. Vehicles are charged each time they pass a tolling point and tolling points are designed so the charge is closely related to the total number of miles driven on a priced roadway. This differs from highway pricing in that the tolls are located on local arterials, in addition to highways.
- **Distance-Based Pricing:** This method charges the user based on the distance traveled by a vehicle. The primary existing use of this method is applied to trucks in Austria, Germany, and Switzerland.

In addition, each of the pricing methods can be refined by varying the fee based on the time of the day and vehicle type [48].

The technology to implement congestion pricing is well-developed. Since the use of toll booths, several alternatives have become more common.

- **Automatic license plate recognition** uses cameras to read the plate of passing cars and process them automatically to generate either a bill or deduct the fee from an account. The programs in London and Stockholm rely on this technology.
- **Dedicated short range communications** use a transponder equipped inside vehicles to communicate with a receiver on the road and charge the user automatically. Singapore and some highway tolls in the US (like FasTrak in California) rely on this technology.

- **GPS and Cellular technologies** rely on the information collected by a GPS unit installed in the vehicle or a cell phone connected to the vehicle to calculate the fee. This technology has only been applied to trucks in Europe.

19.2.3.2 Institutional Obstacles

As existing programs in London, Stockholm, and Singapore show, technology is not the main obstacle to successfully implementing congestion pricing. The main obstacles are drivers' perceived price fairness and redistribution of revenues, the uncertainty about the effectiveness of the system, and, in the case of multi-jurisdictional systems, jurisdictions' willingness to participate in the program. A number of studies have examined the public's reluctance to support road charges and strategies for overcoming resistance (e.g., [50–54]).

Support for road pricing depends on two sets of institutional arrangements. The first institutional hurdle is technical in nature but critical to generating buy in from users. The technology used to set up a cordon and how people pay for the tolls, for example, tend to be context specific and are primarily a matter of communicating with stakeholders [55]. The second institutional hurdle for congestion pricing is in regulating revenue sharing. This set of rules can be more contentious when the public (through a referendum) or other governments have veto power (e.g., New York City, [54]). The congestion pricing programs in London, Singapore, and Stockholm all generated substantial revenues within two years of implementation [48, 49]. Even in cases where congestion pricing covers a single jurisdiction, other jurisdictions, and constituents have a stake in the implementation since drivers from outside the congestion area are paying into the system [52]. Cities are most likely to advocate for tolling if they stand to gain directly in the form of revenues. From an equity perspective, cities through which freeways pass (or are adjacent to freeways) should be the direct beneficiaries of revenues because they bear a disproportionate share of the pollution externalities [49, 52]. People can quickly change their opinion favorably when the framing of the road charge aligns with their goals (Stockholm emphasized environmental impacts) or are guaranteed benefits to mitigate cost (e.g., better public transit service).

For systems that rely on location-based technologies, the institutional needs become more complex. In contrast to localized congestion pricing which can rely on non-invasive technologies (license plate recognition or short range communication), road pricing works best when using a device that tracks how many miles a driver travels. Comprehensive coverage is not easy based on fixed infrastructure, making the use of devices installed in the vehicle preferable.

Road user charge programs that use in-vehicle tracking are limited in scope. In an Oregon pilot program, about 700 people volunteered to pay a VMT-based fee instead of the state's fuel tax [56]. California initiated a pilot project that included 5,000 participants [45]. The Oregon program, like the California pilot project, allows for manual entry of miles traveled to offer the highest level of privacy. The alternative is to equip cars with GPS enabled devices that count miles traveled.

A statewide, comprehensive program, however, would not be able to function if most users opted for manual entry because each entry requires human verification [48]. Technological innovation in this sector, therefore, focuses on protecting the privacy of drivers [55, 57]. A critical concern is that a device that records miles traveled would have a record of locations and potentially much more, that would be in the hands of a government agency and any third-party entity involved in the processing of the data (as is the case in Oregon). One solution, which the European Union has legislated, is to mandate “thick clients.” Thick clients are devices that process all the required information internally (i.e., onboard the car) and only share the information pertinent to the fee (total mileage or the fee amount) so that no location data ever leaves the users’ cars [48].

Thick-client onboard units are already in use and provide an appealing solution to road pricing generally. Onboard units give the government the most flexibility in enacting congestion or user charges and, from the users’ perspective, the units have the potential to address important equity issues by automating discounts. Institutional hurdles include the creation of security standards, privacy regulations regarding the use and access of data, program structure (including the use of revenues), and substantial investments in equipping cars and building the infrastructure to collect information.

The high barriers to implementing onboard recording have led some researchers to look for alternatives. The Virtual Trip Lines (VTL) technology is a form of dedicated short range communication that relies on the communication between a collection point and the vehicle (through smartphones, for example). The technology is similar to that already in use by the California FasTrak program [58]. While this solution overcomes many institutional obstacles like privacy by strictly controlling what data is needed, VTL are not as comprehensive as an onboard device because the technology relies on fixed collection points (fixed infrastructure that communicates with the phone), which may not have universal coverage, especially outside freeway networks.

19.2.4 Connected and Autonomous Vehicles (CAV)

19.2.4.1 Background

Vehicle autonomy operates on a spectrum ranging from minimal assistance to the driver to full automation. The U.S. Department of Transportation (USDOT) adopted the SAE J3016 taxonomy to define the different levels of automation. Levels 0 through 2 assist the driver who must always be supervising the car and able to take control. These features have become standard in many cars. As of 2022, vehicles at

Level 3 and above existed, but were not allowed on roads outside specific areas and were not for sale to individuals.[3]

Level 3 vehicles offer conditional autonomy, freeing the driver from actively driving under certain conditions, but requiring a person to take control when the conditions are not met (e.g., off freeways). As of 2022, test vehicles at Level 4 have been deployed in some locations (e.g., Waymo One in Phoenix and pilot programs in San Francisco). These vehicles operate as taxis without a driver on public roads. While the technology to move from Level 3 vehicles to Level 4 is expected to be well established in the 2020s, the transition to Level 5 and a fully automated system is not expected to happen for decades (optimistic estimates are for the 2040s) [59]. Level 5 autonomous vehicles can operate under any conditions, fully replacing cars with a human driver.

Vehicles can reach full autonomy without being connected. The Waymo vehicles, for example, carry all the hardware and software on the vehicles and do not communicate with a centralized system. However, transitioning to a fully autonomous system is expected to require the additional deployment of a cooperative infrastructure. That is, vehicles must be able to communicate with each other, with the infrastructure around them, and with users, adding the "connected" aspect to autonomy [59]. A connected system removes some of the sensing and computing load borne by the car by making information available remotely. The ability of the car to connect to a centralized system that communicates real-time information enhances coordination (e.g., parking use) and makes information about the local context available (e.g., accidents) [60, 61].

19.2.4.2 Institutional Obstacles

The transition to autonomous vehicles will be gradual. Features will be added to commercially available vehicles that will increase the level of assistance the vehicle provides over time. While this gives more time for institutions to adapt, there is also a risk that decisions made in the early days of the technology will significantly constrain future innovations or required changes to the technology [62]. Even among the most proactive cities, there is a lack of discussion of these specific parameters and how to establish an institutional framework to govern them [14]. Institutional discrepancies are already worsened by the devolution of rulemaking at the state level, which creates contradictions between states and federal rules [63, 64]. The discrepancies between state regulations have allowed Waymo to deploy driverless taxi technology in Arizona first, but the lack of regulation elsewhere may limit how (and if) Waymo will operate as the company scales up operations. Local government variation in what technology is allowed could further exacerbate discrepancies.

[3] Based on the information retrieved on August 25, 2022 on: https://www.nhtsa.gov/technology-innovation/automated-vehicles-safety. Note that Tesla vehicles with full self-driving are classified as SAE Level 2. See Tesla's FSD Recall Impacts AV Industry, SEA International, at https://www.sae.org/news/2023/03/tesla-fsd-recall, accessed July 15, 2023.

The gradual introduction of autonomous technology means that a full range of vehicles, from non-autonomous to fully autonomous, will share the road. In a mixed environment where human-driven and autonomous vehicles share the road, the challenges are compounded and the benefits of AVs are reduced [65]. Some have proposed to segregate autonomous and non-autonomous vehicles to maximize the benefits of driverless operation [66]. Research shows that CAV will have fewer benefits and significant costs unless the vehicles are integrated with a shared mobility infrastructure [67, 68]. The International Transport Forum, based on simulations, found that the lack of sharing would lead to greater vehicle miles traveled, urban sprawl, and decreased mobility for lower income people [67]. Some large European cities are planning for CAV integration while prioritizing transit and non-motorized mobility to reduce the number of cars [14]. While automated buses have similar limitations to human-driven buses, researchers have focused on automation as a solution to the last-mile problem. Flexible shuttles that offer on-demand service can act as an effective feeder into the established transit system but the demands of sharing vehicles on routing penalizes the technology compared to personal car travel [69].

The human element compounds these issues. The public is able to make decisions when it comes to safety of the vehicle, which are unlikely to match the programmed procedure vehicles follow in case of emergency [70]. These perceptions extend to automated public transit vehicles, which many people would not necessarily see as preferable alternatives [71]. This poses a technological problem and an ethical one. Should humans have ultimate control even if human oversight introduces greater safety risks? In shaping institutions, government must contend with variations in people's perception and openness to the new technology [72].

Automated vehicle developers emphasize the safety advantages of removing the human factor in driving. Machines are faster to react and more consistent. This makes AV risk averse. The programming prioritizes safety and testing supports this claim. Waymo on-road testing and simulations show that nearly all accidents were caused by road violations or errors by humans [73]. While this is a great benefit for safety, the programming of the cars may inadvertently reduce people's adoption of the technology if people perceive the cars are overly cautious and impeding their travel. The algorithms, for example, may paralyze the car when faced with unruly pedestrians, a feature that is beneficial to pedestrians and pedestrian-oriented planning, but detrimental to the flow of traffic [74]. When overlaid with rules governing human-driven cars, contradictions in how humans and artificial intelligence make decisions may lead to conflicts and increase governments' reluctance to allow autonomous cars on shared roads.

In Table 19.1, we summarize the key characteristics of the four smart transportation domains, and the primary institutional obstacles and governance issues in each domain. Table 19.1 serves to highlight the factors that we discussed in the previous subsections. In the next section, we discuss governance issues which, in varying ways, cut across the smart transportation domains.

Table 19.1 Examples, characteristics, and institutional obstacles for smart city transportation

Category	Examples	Characteristics	Main institutional obstacles
Shared mobility	• TNC/ride hailing • Micromobility • Bikeshare • Carshare	• Single mode • Docked or dockless • Per ride or per time period (i.e., per hour/day) • Service provided by: • Platform apps (e.g., for TNC) • Shared provider (carshare, bikeshare) • Public, private, non-profit	• Parking mgmt. (for docked vehicles or for shared ride pickup/drop-off) • Curb property rights • Infrastructure safety (e.g., travel speed differences) • Data sharing • Lack of information on impacts (e.g., congestion, GHG) • Rule enforcement
Integrated mobility	• Integrated payments (e.g., for public transit) • MAAS	• Payments across multiple systems (e.g., public transit) • Payments across multiple modes and public/private providers (MAAS) • Cashless/electronic payments elements of sharing (MAAS)	• Interoperability of data systems • Trust across agencies and providers • Common data format and communication protocols • Dense networks (MAAS)
Priced mobility	• Road pricing • Congestion pricing • VMT fee • Parking pricing • TNC congestion pricing* • GHG emission fee*	• Priced by cordon/area, highway/corridor, distance • Priced by time of day/congestion level • Demand management • Externality management • Electronic payment (toll collection)	• Revenue sharing/use of revenues • Jurisdictional coordination • Privacy of location data • Political: equity, opposition to tolls, unpopularity of high toll prices
Autonomous Mobility	• Autonomous vehicles • Connected vehicles • Driver-assist technology	• Hardware and software to support driving • Human-in-the-loop (Levels I-IV) or fully driverless (Level V) • Potential safety improvements	• Divergence of national and state regulations • Questions of safe piloting on streets • Mixed autonomous/human-driven traffic • Path dependencies—early actions

* Not yet implemented in U.S

19.3 Governance, Institutions, and Challenges in Smart Transportation

19.3.1 Standards and Regulations

Much of transportation is governed by standards and regulations. Standards are the set of rules that dictate how the transportation system functions, from the design of vehicles and roads to how roads can be used. While standards and regulations can often be used interchangeably in that a standard can be regulatory (i.e., the standard mandates a specific behavior, like adding seat belts to cars), standards include a wider set of rules, including those set voluntarily by industries and that require no government intervention, like computer programing standards [75].

Regulations are rules that the government enforces based on the law. In the transportation sector, many of these rules ensure the safety of cars and the roads people drive, cycle and walk on. New transportation technologies raise four main issues for the creation of standards and regulations that support innovation.

19.3.1.1 Hierarchy of Governments

First, the hierarchy of government creates many opportunities for contradictions and conflicting rules. The ability of vehicles to cross many jurisdictions, local and state, puts much of the regulatory mandate on the federal government to ensure safety standards are consistent and infrastructure operates as expected nationally. State agencies complement federal agencies. States have autonomy in establishing some standards, speed limits for example, but most have converged on a narrow set of standards for the rules that govern driving and infrastructure building.

New technologies, however, have created some wider discrepancies in regulation. The lack of a federal framework for regulating many of the new transportation technologies and their associated innovations (e.g., location data collected by TNCs) has left states to create their own set of regulations. California has been comparatively proactive in regulating new technologies and has generally been stricter. This resulted in California companies, like Waymo, to test their technologies in different, more permissive, states.

Local governments are bound by federal and state regulations but have considerable authority over transportation through their zoning and planning powers. The ability to regulate street design and parking requirements, for example, has far reaching consequences for much of the transportation system and the ability of new transportation technologies to function universally. Municipalities have the authority to support car sharing by providing dedicated parking (e.g., New York City, [76]) or completely or partially banning micromobility devices (e.g., New Orleans, LA, and Austin, TX. See also, [77, 78]). Transit agencies have narrower regulatory powers but have also become important players in the transportation system as multi-modal mobility has become a priority for state and federal governments.

19.3.1.2 Industrial Interests and Industry Capture

Second, the transportation sector must contend with industrial interests. This is a longstanding issue. The US Federal Motor Vehicle Safety Standards were created in 1966 to dictate a set of standards for vehicles which the National Highway Traffic Safety Administration (NHTSA) enforces. The NHTSA is illustrative of the difficulty of institutionalizing standards. The agency was developed based on the principle that automobile safety was a technical issue that could be resolved through a set of rules on how to design and operate cars. While the agency was successful in reducing traffic fatalities [79], the enforcement power of the agency was systematically challenged by automakers who made enforcement political rather than technical [80]. The shortfalls were emblematic of the difficulties of translating an agency's mandate to actions. Rather than fundamental flaws in the agency's structure, the interference coming from powerful carmaker lobbies, judicial review, and changes in political administrations all played a role in undermining new rules [81].

The influence of car makers and other actors is a defining feature of the governance structure of the transportation sector. Standards do not flow from the federal level to lower levels and industry. The increasing complexity of vehicles and the entry of new firms with little to no background in the transportation sector has created a field where many rules originate locally and within the private sector, including some standards that are wholly private, enforced without state interventions for the purpose of coordination (e.g., computer programming standards). Other standards are enforced by the state but written by specialized groups, like AES, a global association of engineers that focuses on mobility, which developed many of the standards for automated vehicles [75, 82].

19.3.1.3 Rapid Pace of Innovation

Third, new technologies have considerably increased the pace at which regulations would have to be created to keep up with new developments. The rapid pace of innovation in the private sector, and their ability to deploy rapidly, means that governments rarely have the opportunity to plan how to regulate new technologies and coordinate across levels before a service has become embedded in a city [83]. This creates issues as a technology often emerges locally. Uber started in San Francisco for example, only to be regulated at the state level, pre-empting the ability of the local government to devise regulations adapted to the context in which the technology operates. Autonomous vehicle development is fast paced and extremely fragmented with dozens of private entities engaged in pilots across many states.

The government hierarchy creates another layer of issues for the pace of regulations. California's proactive approach to regulating AVs led to an early but fragmented regulatory approach. The Department of Motor Vehicles oversees pilots and testing starting in 2014, the California Public Utilities Commission (which regulates TNC) regulates commercial deployment from 2018, while the National Highway Traffic Safety Administration created its own guidelines for testing and deployment. After several years of revisions (from 2015 to 2018), the California and federal organizations achieved greater consistency with federal guidelines and continue to revise regulations regularly [82, 84].[4] The federal government tried to introduce legislation in 2017 that would have pre-empted state-level discrepancies, but neither the Self Drive Act nor the AV START Act of 2017 had moved through the legislature as of 2023.

19.3.1.4 Technologies that Span Sectors

Fourth, new technologies are no longer sector specific. New technologies rely on a mix of physical infrastructure and digital infrastructure. Micromobility quickly becomes complicated when considering the intersection of mobility rules (where e-scooters can travel), infrastructure (provision of parking, sidewalks, and bicycle lanes), and communication (privacy of locational data). Some cities have been relatively quick to adapt. Los Angeles, for example, implemented a data governance framework that mandates data sharing protocols for all micromobility providers that help with enforcing rules, assisting the city in planning where to add infrastructure and reaching a broader population. Other technologies, like autonomous vehicles, are not constrained to a single jurisdiction and require greater coordination that usually impedes rapid responses. While the federal government has taken a leading role in establishing standards for security [85], making sure that the private sector meets safety standards, there is a critical gap where communication networks are concerned and which transportation agencies alone will not be able to fill because communication is outside their mandate.

19.3.2 Data Governance

Smart cities require data governance, which is an evolving field. Data governance pertains to the rules governing the collection, use, management, and sharing of data (the technical aspects) and how those rules are applied and changed [86]. Most activity has focused on the regulation of technical aspects, especially security and privacy, for good reasons. Large surveys indicate that the public is generally concerned about a security vulnerability in the system (e.g., hacking of autonomous vehicles) and the large amounts of personal data collected and shared [72, 87]. However, data governance aims to create a broader framework that includes security and privacy, but that extends to integrating data across services and ensuring equitable access. We highlight three areas of data governance that will touch transportation.

[4] The California DMV initially required that a person be behind the wheel at all times in autonomous vehicles, but the federal level issued directives that were more permissive creating confusion for firms operating in California. The regulation was updated in 2018 to align with the federal level and with other states competing for investments.

19.3.2.1 Cybersecurity and Privacy

The US federal government passed the Internet of Things (IoT) Cybersecurity Improvement Act of 2020, which sets standards for securing networks against cyber-attacks. The bill creates mandates for the National Institute of Standards and Technology (NIST) and the Office of Management and Budget (OMB). The NIST must develop standards for IoT devices, and the OMB is charged with reviewing agency standards to ensure compliance. While the bill only applies to devices owned by or connected to federal agencies, regulations would cover much of the transportation sector that is often linked to federal agencies.

The State of California anticipated the federal regulation and passed SB-327 in 2020. The bill requires manufacturers of connected devices to take reasonable steps to secure the data collected. The bill also includes text on the data collection and the deemed responsibilities in safeguarding such data. In The United States, California is at the forefront of data governance. As of 2023, it was the only state also to have passed data privacy laws (SB-1121 and the California Privacy Rights Act). Importantly, these laws preempt and supersede any local laws, ensuring a statewide standard.

In addition to privacy laws, some agencies like the California Public Utilities Commission (CPUC) have taken a leading role in regulating mobility technologies and data [5, 29]. One of the goals is to establish clear standards for data formats so that private mobility firms can be held accountable more easily and their service integrated with the broader transportation system. These initiatives provide a starting point but tend to create gaps with local governments and agencies. For example, standardizing data formats may put an increased burden on smaller agencies that lack the capacity to meet these standards or deter startups that operate in different states with different standards.

19.3.2.2 Integration

Security and privacy focus on the data management of one firm. Much less has been done to facilitate the integration of data across multiple mobility services, which would increase both the volume and complexity of data. It is unclear to what extent existing data governance frameworks are adapted to the demands of massive data flows and coordination with local jurisdictions [88]. Research based on a focus group of state DOT officials, federal agencies, and experts identified changing the organizational structure of governance and the lack of executive support as primary obstacles to coordination which intersects with the technical upgrading (of both technology and staffing) of state and local agencies that will be needed to operate in a new data-rich environment [89] (see also [30, 90]).

19.3.2.3 Algorithms and Artificial Intelligence

Another set of tools that do not fit within a security-focused or sector-specific framework is the regulation of algorithms and artificial intelligence. The algorithms that private firms develop to automate management of rides, placement of shared bikes, and pricing are sets of rules that constrain what the users (both service providers and consumers) can see and do. These algorithms have gone largely ungoverned and risk reproducing the biases embedded in existing institutions and practices [91]. Algorithms are proprietary technologies beyond the reach of many regulations but must conform to enforced rules (e.g., non-discrimination) giving institutions a prominent role in shaping how algorithms are developed. Artificial intelligence highlights the issue of regulations keeping pace with innovations and how data cut across multiple sectors. While artificial intelligence is central to the development of CAV, regulation is unlikely to focus on this application alone with implications for how AV technology evolves and behaves on roads [92, 93].

19.4 Policy Recommendations Part 1: Overview

19.4.1 Government Capacity

Many governments lack capacity to manage complex data systems. In part the question is which level of government, in a hierarchical system, can best coordinate smart city data needs in the transportation sector. One model would be a cross-agency coordinating entity, like the Transportation Operations Coordinating Committee (TRANSCOM) that spans agencies across the New York metropolitan area. While TRANSCOM centralizes data management operations to coordinate responses to traffic disruptions (e.g., incidents and construction), the organization has no centralized authority. Policies about how to best achieve the organization's functions are decided based on consensus among members.

In part due to the absence of clear coordination, cities in the U.S. have been taking the lead. Los Angeles has an agreement with shared mobility providers that details which data are shared with the city. San Francisco has argued for a similar arrangement with TNCs. In these cases, the city established an application programming interface (API) that allows city analysts to access the service's data under certain conditions. The technology has much wider applications [40]. For example, cars equipped with a device to record locations can share data on road and parking usage, among other things. A strict protocol needs to be in place to determine who can access the information, how much data can be accessed, and under what conditions. Given the number of potential applications, research is needed on the state of the technology and what the pros and cons are to help agencies make decision about creating standards.

19.4.2 *Property Rights and Sharing (of the Commons and of Property)*

Property rights remain poorly articulated, in part because the importance of property rights has been overlooked. Property rights in two domains will be fundamental for smart cities and transportation: curb space and data.

The curb is often treated as a commons, used by whichever car or dockless vehicle drives up to drop off a passenger or delivery, pick up persons or goods, or park. When the curb is uncongested, that fact that property rights to the curb are not assigned is not a problem. For decades vehicles have dropped off or picked up persons or goods, and with or without parking management (e.g., via parking meters) this has often worked well. Yet congestion and new ride-hailing and dockless modes create use conflicts for the curb. Imagine the difficulty of finding a ride-hailing vehicle after a major event ends as crowds seek to navigate across limited pickup/drop-off space.

Several cities are implementing new rules for curb space management. As examples, San Francisco and Washington DC are among a small number of cities piloting flexible curb space management in the US and globally [18, 19, 94].[5] These programs are shifting how curb space is regulated (to accommodate the increasing demands beyond parking). The expansion of Waymo and free-floating car-sharing services is adding another dimension to curb space management. Real-time management through infrastructure-to-user or infrastructure-to-vehicle communication could support the expansion of car sharing and automated vehicles through curb management that prioritizes some uses, allows reservations, and tracks violations of standards.

Car sharing is the more developed sector that illustrates the potential of this innovation and already uses technology that lends itself to a pilot. Existing providers like GIG carshare in Sacramento and Oakland, California rely on agreements with the city to secure access to parking spaces where the users can leave a car for free [95]. This works when the fleet of vehicles is small (the largest program in Sacramento has 300 vehicles available) but could quickly hit limitations as the program expands.

A pilot to empirically evaluate the ability of parking pricing to shift demand from personal cars to shared cars could inform new rules for curb space management. Such a pilot would require a short-term investment in connected parking meters. These would be parking meters that can send and receive information to a centralized system and adjust pricing automatically based on a set of rules, in some ways expanding on current real-time parking pricing programs which are not adapted for car sharing, such as SFpark in San Francisco [96]. Free floating car sharing relies on a home zone, an area where cars can be dropped off freely. Connected meters within a home zone could take into account the number of shared cars in use to prioritize parking availability for the users by increasing the price of parking for non-shared cars or

[5] Among other programs, Transportation for America funded projects in Boston, Minneapolis, and Bellevue to develop and pilot curb management programs. See https://t4america.org/2019/12/10/t4america-selects-3-cities-to-launch-curbside-management-pilots/, accessed 25/07/2023.

reserving spaces, the location of which would be communicated to the users through the car-sharing app.

Data sharing is the second area where property rights need clarification. New mobility providers—TNCs and dockless providers—create a large amount of data on the location, use, and travel route of their vehicles. Private owner-operated cars increasingly generate the same data, either through onboard GPS technology (collected by car manufacturers) or, more commonly, through GPS enabled and network-connected smart phone apps (collected by the app owners) [97]. Those data could allow answers to important public questions. Does ride hailing increase or decrease congestion or greenhouse gas emissions? Do dockless vehicles block the right of way on sidewalks, creating hazards for disabled persons? Can traffic be routed away from congested locations, in real time, in ways that are both efficient and equitable? In theory, the detailed location data inherent in shared services (TNCs and dockless mobility) allow a way to answer those questions. Beyond that, new mobility providers have data that allow a ready assessment of whether those providers are, or are not, complying with regulations on, for example, appropriate docking of vehicles and use of street and curb space.

Industry has often resisted allowing data to be used to either aid in regulatory enforcement or to answer controversial questions about externalities. The uncertainty around the appropriate split of private data and public interest, and the fact that cities are often regulating data in a fragmented environment, to date have limited data access and the use of new data streams by the public sector.

19.4.3 Integration

The public sector is in a unique position to coordinate fare integration. A potential early opportunity is a mobility wallet. The distinct advantage of mobility wallets over other forms of payment (e.g., using a touchless bank card to pay for transit) is the ability to bundle travel services and automate the application of discounts. Mobility wallets do not have to be integrated with MaaS (which adds the ability to plan and book journeys across multiple modes of transportation). The wallet can integrate sets of rules that automatically apply discounts if, for example, a traveler makes a transfer that qualifies, and incentivizes choices by discounting choosing bike share or public transit over driving (through TNC or carshare). The ability to automate discounts, provide subsidies (as with the Universal Basic Mobility pilot program in Los Angeles), and reach marginalized population (e.g., unbanked people) makes a mobility wallet a powerful tool for achieving transportation equity goals [37].

Some existing programs, notably Clipper in the Bay Area and TAP in Los Angeles County, offer payment options that integrate many features of a mobility wallet. The development of a versatile mobility wallet would build on these existing programs and expand their reach and streamline access. For example, linking the wallet to administrative data can automate the qualification for discounts or subsidies and remove undue burden on users. The function of the wallet can also be expanded by anticipating the use of other technologies like road pricing or curb space management digital infrastructure. Should road pricing integrate on-vehicle data recording, for example, the infrastructure for a mobility wallet could provide a private and secured means of paying fees. The Virtual Trip Lines technology, which is a faster means of deploying road use pricing, would benefit greatly from the integration of a mobility wallet.

19.4.4 Pricing of TNCs

There is some evidence that associates TNCs with increased traffic congestion, due in large part to the deadhead miles that TNCs drive without a passenger (San Francisco County Transportation Authority, 2018). That potential for congestion, and the unique data capabilities of TNCs compared to general traffic, create the potential for TNC congestion pricing.

The GPS tracking for TNCs allows each ride to identify a start and end point, and the time of the ride is known. There is already a platform for transparent communication between the service and the customer. While TNC providers already implement a form of congestion pricing based on peak demand times (often called surge pricing), a congestion pricing pilot would add a fee based on location and would charge a congestion toll based on the impact on other road users, not on a profit-maximizing algorithm of a private firm. Like other real-time congestion pricing schemes, a TNC congestion fee would be based on the delay imposed on other users under congested road conditions.

Research based on simulations finds that such a fee outperforms applying congestion pricing with a cordon that charges all TNC vehicles with no or one rider and a policy that prevents cruising within the designated area [98]. While the goal is to mitigate the contribution of TNCs to congestion, such a pilot would have the benefit of generating data on using smartphones to implement congestion pricing and people's behavior in response to the fee, both of which would be valuable inputs for designing a broader congestion pricing scheme.

19.4.5 Privacy

Data privacy has long been part of discussions of technology applications in transportation, dating to electronic toll collection. Fries et al. [99] note that users of early congestion toll facilities were often concerned about data privacy, although the concern appeared to be declining at the time of their article. Fries et al. [99] describe several approaches to masking individual data and protocols for data confidentiality that, in operating toll systems such as London and Singapore, were accepted by the public despite initial public concerns.

New technologies raise the potential to broaden the scope of data privacy issues. As mentioned earlier, the public (agencies, governments) have an interest in the large amount of data being collected by new mobility providers such as TNCs or dockless or shared vehicles. Some authors have suggested that traditional protocols of data security and user anonymity can be used to protect privacy and confidentiality (see, e.g., [100]). Yet new approaches, such as integrated payments or MaaS, will work best when combined with user information such as account billing, personal data that might allow applications of discounts or incentives, and billing information. Traditional encryption may handle many concerns, but users will likely still be concerned about whether their travel data could be hacked by malicious actors or accessed by law enforcement agencies. Of course, these concerns are not unique to transportation, but they still pose challenges. Traditional approaches of clear and strict user privacy and confidentiality protocols, communication efforts to reassure the public, and data security to protect against breaches are the best approach. We note that sensitive data have long been handled by government agencies, including those involved in tax collection, health care, and census data. Transportation agencies, typically isolated from the security and privacy concerns of data collection in those more sensitive domains, will have to learn and likely adapt approaches from those other domains and entities.

19.4.6 Land Use

Much effort has focused on permitting automated vehicles and establishing standards to ensure the safety of users and other people using roads. One area that has been largely overlooked is land use [101]. Automated vehicles bring the potential to greatly lower the user-experienced cost of driving. If users can work, rest, or watch entertainment while being driven in an automated vehicle, the time cost of long trips will drop substantially. This raises the possibility that automated vehicles will increase sprawl and lead to long commute trips. If those are in single-occupant vehicles, the impact on roads and the environment could be negative [103].

Alternatively, a future of shared autonomous vehicles might bring a countervailing force for more compact urban areas. Shared mobility and its connection to transit, like shared automated vehicles, rely largely on compact urban development [25, 102]. Hence the nature of land use impacts, sprawling or more compact, will depend importantly on whether autonomous vehicles are single occupant or shared [103].

The net effect will be the result of countervailing factors, i.e., of the balance of longer car trips due to increased driver convenience or shorter trips across multiple modes due to interactions between autonomous technology and less need for parking and street space. Most modeling studies suggest that the net effect, given status quo approaches to regulation, land use, and infrastructure, will be to increase driving and hence, likely, to decentralized cities and urban areas [104].

Note that no single result will likely dominate. Urban development and transportation today exists on a broad continuum, from cities with high mode shares for transit (e.g., Hong Kong moves over 80 percent of trips via transit), bicycle (Amsterdam's mode share is above 30 percent) to cities where driving mode shares are above 60 or 70 percent, as in much of the US, Canada, and Australia (For an overview, see [105]). Generally, policies and infrastructure that favors shared rides, Maas, non-car modes such as transit, bicycling, micromobility, and walking, and prices and development patterns that incentivize that multi-modal and shared future are modeled and conjectured to lead to AV adoption consistent with reduced driving and more compact urban form (e.g., [103]). Cities that pursue those policies—a broad range of incentives and infrastructure that make non-car and shared modes more convenient while fully pricing the externalities of single occupant automobiles—will likely have a more sustainable transition to an autonomous vehicle future.

19.4.7 Summary

The recommendations above involve (not in the order they were discussed): building capacity, integration of systems and payments, TNC pricing, property rights (to urban space and to data), privacy, and land use. This is a broad range, and finding an overarching theme that unifies these policy recommendations would be difficult if not impossible. Let us return to the institutional factors highlighted in the previous section: hierarchies of governments and inconsistent regulations, industry interests and industry capture, the pace of change, and the fact that smart mobility spans sectors that are currently often siloed. These factors might suggest a need for strong central government regulation, particularly to lessen inconsistencies related to hierarchical government regulation and difficulties integrating across sectors and governments. Yet such a view is at odds with the nature of the recommendations we cited above. First, transportation has always been highly diverse, from walking to transit to bicycling to car cities with much in between. That reflects differences in local context, history, and development paths. Arguing for a "one size fits all approach" would not only be counter to that reality but would miss the importance of local context. Second, of the recommendations cited above, issues of privacy, rights to data, land use, and pricing are intensely political, with inherent tensions (e.g., between public interest in rights to data and the need for user privacy). Those political tensions have existed in the current transportation realm for decades.

In the next section, we analyze the shared mobility domains in relation to the governance needs. One conclusion of that analysis is that resolving the political tensions inherent in smart mobility will require experimentation, learning, and progress on a case-by-case, city-by-city basis. We turn our attention now to that analysis.

19.5 Policy Recommendations, Part 2: A Way Forward

The range of technologies, changes, and issues associated with smart cities and transportation is broad. To organize the themes, we assess each of the four mobility changes (shared, integrated, priced, and autonomous mobility) in relation to three broad groups of policies. We further breakdown priced mobility into two subcategories (tolls and on-vehicle) to emphasize the very different implications of technological approaches to road charging.

We first group policies as regulatory approaches and standards. These approaches are largely non-spatial and focus on transportation as an industry and a consumer good or service. The regulatory approaches and standards are often limited in their view of impacts on urban space and urban land use.

We next group policies as those dealing with externalities, the use of the commons (e.g., congested road or sidewalk space), and property rights. These policies often focus on longstanding transportation concerns. Congestion pricing and congestion management have been discussed in practice and the literature for decades. Other externalities, such as air quality or noise, also have spawned large literatures.

Lastly, we group policies related to the uniquely urban and spatial elements of transportation. Two characteristics are important. First, urban transportation occurs in the public realm, on streets, sidewalks, and nearby rights of way where persons see and interact with others. Second, transportation shapes land use (see, e.g., [106]).

For each change in mobility from smart technology, we rank the relationship to each governance issue as "low," "medium," or "high" in Tables 19.2, 19.3, and 19.4. Table 19.2 ranks mobility changes in relation to regulations and standards. Table 19.3 ranks mobility changes in relation to externalities and property rights. Table 19.4 ranks mobility changes in relation to urban, spatial, and land use issues. A ranking of "low" occurs when the mobility changes create concerns that, while salient, arc not a new issue (meaning the concern is well understood and has attention in the policy environment) and is generally limited to one type of operator. A ranking of "high"

Table 19.2 Mobility changes with respect to regulatory and standards issues

Regulatory issue → mobility change ↓	Hierarchy of governments	Industry interests/industry capture	Rapid pace	Sector spanning technologies
Shared mobility	Medium	High	High	Medium
Integrated mobility	Medium	Medium	Medium	High
Priced mobility: Tolls	Low	Low	Low	Low
Priced mobility: On-vehicle	High	High	Medium	Medium
Autonomous mobility	Medium	High	Medium	Low

Table 19.3 Mobility changes with respect to externality and property rights issues

Issues of the commons → mobility change ↓	Externalities	Property rights
Shared mobility	High	High
Integrated mobility	High	High
Priced mobility: Tolls	High	Low
Priced mobility: On-vehicle	High	High
Autonomous mobility	High	High

Table 19.4 Mobility changes with respect to urban space and land use issues

Urban space and land use → mobility change ↓	Publicness	Land use
Shared mobility	High	High
Integrated mobility	High	High
Priced mobility: Tolls	Medium	Medium
Priced mobility: On-vehicle	High	Medium
Autonomous mobility	High	High

involves concerns in new or evolving policy areas, often with multiple operators or multiple sectors (possibly spanning public and private providers), in contexts where the concern is not often discussed or overlooked, and in areas with high potential for impact or harm. "Medium" is intermediate between "low" and "high." The rankings are our judgement and are not intended to reflect anything more than our assessment after our review of the literature.

One conclusion from Tables 19.2, 19.3, and 19.4 is that the mobility changes that will be created by smart cities create particularly salient and pressing concerns in the realm of externalities and property rights and for urban space and land use. This is not to downplay the importance of standard and regulations, but to elevate the often overlooked ways that smart cities and transportation will require new and innovative governance responses related to externalities and particularly property rights and also to the public and land use elements of urban space. We suggest three approaches that can help chart a way forward.

19.5.1 Build Capacity

We highlighted capacity building efforts in the previous section. Building capacity is not unique to the issues of property rights and urban space and land use, but without capacity agencies and governments will have difficulty making progress. We suggest capacity building along two dimensions.

Build capacity for data and smart transportation: This is the most obvious dimension. This includes the efforts discussed above to coordinate data across agencies, either voluntarily (e.g., New York's TRANSCOM) or with standards and protocols at the state (e.g., the California ITP) or the local level (e.g., the Mobility Data Specification).

Build capacity to regulate property rights and data privacy: This will require new thinking. The concepts for regulating curb space and access to infrastructure include questions of basic economics, property rights, prioritization, and pricing. These concepts are under-discussed and under-developed but not completely foreign in transportation [107]. Property rights also will require analysis and political agreements on the relative roles of public interest in travel data and the right to and need for privacy. While technical approaches to privacy have been discussed, transportation agencies are only beginning to gain experience with the concepts of rights to data and the privacy needs created by modern data streams.

The latter of the two capacity building efforts—learning to handle the inherently political questions of property rights to space, infrastructure, and data and the associated issues of access, speech, and privacy—will be the most challenging. Because the resolution of those questions, in a democratic society, should be political, agencies will need to develop capacity in the associated regulatory concepts and the ability to communicate those concepts and interests to the public.

19.5.2 Pursue Pilot Programs

There are already nascent pilot programs related to smart transportation: GIG carshare which prioritizes parking for shared mobility in Sacramento, the zero-emission delivery zone experiments in Santa Monica and Los Angeles which prioritize cub space for zero-emission delivery vehicles, the Mobility Data Specification which conditions micromobility operating licenses on data sharing with government agencies, and discussions (but no implementation to date) of TNC congestion pricing. All of these pilots combine issues of property rights, urban space, and privacy. In that respect, they are all learning opportunities. Each of those pilots are efforts to use technology to better manage innovations associated with smart mobility. In each case, the pilot seeks to reduce externalities (typically congestion, emissions, and at times urban land use) while building a regulatory approach that balances access to public space and data with regulatory approaches to privately provided transportation, all seeking a broader public (rather than strictly private) interest. These pilot programs are likely the best path forward both to build capacity within agencies and governments and to build a broader understanding of and learning around new smart transportation issues.

19.5.3 Use Those Pilot Programs to Help Articulate Improved Understanding of Urban, Spatial, and Land Use Governance Issues

We suggest that the best approach forward is not an attempt to find a single centralized approach to regulate smart transportation. In a realm where local context and political preferences vary, there is no single approach. Instead, the nascent pilot programs are the best tool to elevate a discussion of issues of property rights, urban space, and urban form. These pilots will not be the only approach, but they are a start. There will also be a need for research to focus on how transportation shapes urban growth patterns and urban space, and the implications for sustainability and equity. Research should also focus on methods for protecting user privacy while sharing data across the private and public sectors. These topics—urban development and privacy/data sharing—are not new. What is new, and what needs to be fostered, is building learning pathways from the research to practice. Pilot programs are one of the best ways to do that. Higher-level governments should build incentives for pilot programs related to new mobility, with a focus on exploring regulatory and governance approaches to questions of property rights, urban space, and land use.

19.6 Conclusion

Smart transportation is a broad realm, with innovations that will arise rapidly, spanning multiple sectors, and often with timing, locations, and pacing that is unexpected. Some of the issues raised by smart mobility are longstanding—particularly questions of traditional externality pricing, technical capacity, and integration across agencies and sectors. Yet, unique to transportation, smart city technologies raise questions of property rights, privacy, and urban space and land use. The appropriate governance responses to those issues will be political agreements that balance competing interests. Hence we suggest that the best role is one of experimentation, discussion, and learning, being alert to the issues that we described in this chapter.

References

1. Jamme H-T (2023) Productive frictions: a theory of mobility and street commerce grounded in Vietnam's motorbike-centric urbanism. J Am Plan Assoc :1–15. https://doi.org/10.1080/01944363.2022.2155687
2. Loukaitou-Sideris A, Ehrenfeucht R (2009) Sidewalk as space of dissent. In: Loukaitou-Sideris A, Ehrenfeucht R (eds) Sidewalks: conflict and negotiation over public space. The MIT Press, Cambridge, Massachusetts
3. Chester M, Helmrich A, Li R (2022) Inventorying San Francisco bay area parking spaces: technical report describing objectives, methods, and results. Mineta Transportation Institute, San Jose, CA. https://doi.org/10.31979/mti.2022.2123

4. Litman T (2021) New mobilities: smart planning for emerging transportation technologies. Island Press, Washington, D.C
5. Dudley G, Banister D, Schwanen T (2017) The rise of Uber and regulating the disruptive innovator. Polit Q 88(3):492–499. https://doi.org/10.1111/1467-923X.12373
6. Smith G, Hensher DA (2020) Towards a framework for Mobility-as-a-Service policies. Transp Policy 89:54–65. https://doi.org/10.1016/j.tranpol.2020.02.004
7. Henao A, Marshall WE (2019) The impact of ride-hailing on vehicle miles traveled. Transportation 46(6):2173–2194. https://doi.org/10.1007/s11116-018-9923-2
8. Machado C, de Salles Hue N, Berssaneti F et al (2018) An overview of shared mobility. Sustainability 10(12). https://doi.org/10.3390/su10124342
9. Morris EA, Pratt A, Zhou Y et al (2019) Assessing the experience of providers and users of transportation network company ridesharing services. Center for Connected Multimodal Mobility, Clemson, SC
10. Clark BY, Brown A (2021) What does ride-hailing mean for parking? Associations between on-street parking occupancy and ride-hail trips in Seattle. Case Stud Transp Policy 9(2):775–783. https://doi.org/10.1016/j.cstp.2021.03.014
11. Barajas JM, Brown A (2021) Not minding the gap: does ride-hailing serve transit deserts? J Transp Geogr 90. https://doi.org/10.1016/j.jtrangeo.2020.102918
12. Shaheen S, Totte H, Stocker A (2018) Future of mobility white paper. Institute of Transportation Studies at UC Berkeley, Berkeley, CA. https://doi.org/10.7922/G2WH2N5D. https://escholarship.org/uc/item/68g2h1qv#author. (Accessed 14 Sept 2022)
13. Millard-Ball A, Murray G (2005) Car-sharing as a parking management strategy. Transportation Research Board, Washington, D.C
14. Grindsted TS, Christensen TH, Freudendal-Pedersen M et al (2022) The urban governance of autonomous vehicles—In love with AVs or critical sustainability risks to future mobility transitions. Cities 120. https://doi.org/10.1016/j.cities.2021.103504
15. Merlin LA (2019) Transportation sustainability follows from more people in fewer vehicles, not necessarily automation. J Am Plan Assoc 85(4):501–510. https://doi.org/10.1080/01944363.2019.1637770
16. Henao A, Marshall WE (2019) The impact of ride hailing on parking (and vice versa). J Transp Land Use 12(1):127–147. https://doi.org/10.5198/jtlu.2019.1392
17. Dowling R (2018) Smart mobility: disrupting transport governance? In: Marsden G, Reardon L (eds) Governance of the smart mobility transition. Emerald Publishing Limited, Bingley, UK. https://doi.org/10.1108/978-1-78754-317-120181004
18. DDOT (2021) Move DC: The District of Columbia's multimodal long-range transportation plan. District Department of Transportation, Washington, D.C
19. SFMTA (2020) Curb management strategy. San Francisco Municipal Transportation Agency, San Francisco, CA. https://www.sfmta.com/sites/default/files/reports-and-documents/2020/02/curb_management_strategy_report.pdf. (Accessed 12 Sept 2022)
20. Ferrero F, Perboli G, Rosano M et al (2018) Car-sharing services: an annotated review. Sustain Cities Soc 37:501–518. https://doi.org/10.1016/j.scs.2017.09.020
21. Balac M, Ciari F, Axhausen KW (2017) Modeling the impact of parking price policy on free-floating carsharing: case study for Zurich, Switzerland. Transp Res Part C Emerg Technol 77:207–225. https://doi.org/10.1016/j.trc.2017.01.022
22. Jian S, Liu W, Wang X et al (2020) On integrating carsharing and parking sharing services. Transp Res Part B Methodol 142:19–44. https://doi.org/10.1016/j.trb.2020.09.013
23. Thakuriah P, Tilahun N, Zellner M (2017) Seeing cities through big data: research, methods and applications in urban informatics. In: Springer geography. https://doi.org/10.1007/978-3-319-40902-3
24. Pan A, Shaheen S (2021) Strategies to overcome transportation barriers for rent burdened Oakland residents. California Growth Strategic Council. https://escholarship.org/uc/item/327773q9. (Accessed 16 Sept 2022)
25. SCAG (2022) Mobility as a service. Feasibility white paper. Southern California Association of Governments, Los Angeles, CA

26. McCoy K, Andrew J, Glynn R et al (2018) Integrating shared mobility into multimodal transportation planning: improving regional performance to meet public goals. Federal Highway Administration, Washington, D.C
27. NACTO (2017) Curb appeal: curbside management strategies for improving transit reliability. National Association of City Transportation Officials. https://nacto.org/wp-content/uploads/2017/11/NACTO-Curb-Appeal-Curbside-Management.pdf. (Accessed 12 Sept 2022)
28. Sandoval R, Van Geffen C, Wilbur M et al (2021) Data driven methods for effective micromobility parking. Transp Res Interdiscip Perspect 10. https://doi.org/10.1016/j.trip.2021.100368
29. Matute J, Cohen D'Agostino M, Brown A (2020) Sharing mobility data for planning and policy research. Policy Brief. UC Office of the President. https://escholarship.org/uc/item/88p873g4. (Accessed 12 Sept 2022)
30. Creutzig F (2021) From smart city to digital urban commons: institutional considerations for governing shared mobility data. Environ Res Infrastruct Sustain 1(2). https://doi.org/10.1088/2634-4505/ac0a4e
31. Giuliano G, Moore JE, Golob J (2000) Integrated smart-card fare system: results from field operational test. Transp Res Rec J Transp Res Board 1735(1):138–146. https://doi.org/10.3141/1735-17
32. Jittrapirom P, Caiati V, Feneri A-M et al (2017) Mobility as a service: a critical review of definitions, assessments of schemes, and key challenges. Urban Plan 2(2):13–25. https://doi.org/10.17645/up.v2i2.931
33. Sochor J, Arby H, Karlsson ICM et al (2018) A topological approach to mobility as a service: a proposed tool for understanding requirements and effects, and for aiding the integration of societal goals. Res Transp Bus & Manag 27:3–14. https://doi.org/10.1016/j.rtbm.2018.12.003
34. Move PGH and City of Pittsburgh (2022) Move PGH Mid-Pilot Report, Pittsburgh, PA. https://apps.pittsburghpa.gov/redtail/images/19169_Move_PGH_Mid_Pilot_Report_[FINAL]_v2.pdf (accessed 17 July 2023)
35. Arias-Molinares D, Carlos García-Palomares J (2020) Shared mobility development as key for prompting mobility as a service (MaaS) in urban areas: the case of Madrid. Case Stud Transp Policy 8(3):846–859. https://doi.org/10.1016/j.cstp.2020.05.017
36. Audouin M, Finger M (2019) Empower or Thwart? Insights from Vienna and Helsinki regarding the role of public authorities in the development of MaaS schemes. Transp Res Procedia 41:6–16. https://doi.org/10.1016/j.trpro.2019.09.003
37. Baud J (2020) Evaluation of Cal-ITP proposed initiatives. Cal-ITP, Sacramento, CA
38. Du H, Gao Z, Ren H (2016) Competition and regulation in a new integrated transit system across jurisdictional borders. J Adv Transp 50(8):1831–1852. https://doi.org/10.1002/atr.1432
39. Belanche-Gracia D, Casaló-Ariño LV, Pérez-Rueda A (2015) Determinants of multi-service smartcard success for smart cities development: a study based on citizens' privacy and security perceptions. Gov Inf Q 32(2):154–163. https://doi.org/10.1016/j.giq.2014.12.004
40. D'Agostino M, Pellaton P, Brown A (2019) Mobility data sharing: challenges and policy recommendations. Issue Paper, White Paper. UC Davis Policy Institute for Energy, Environment, and the Economy, Davis, CA
41. Pasquale GD (2017) Interoperability framework and governance enabling Mobility as a Service. In: ITS conference, Strasbourg
42. Falconer R, Zhou T, Felder M (2018) Mobility-as-a-service: the value proposition for the public and our urban systems. Arup. https://www.arup.com/perspectives/publications/research/section/maas-the-value-proposition. (Accessed 20 Jul 2022)
43. Arnott R, Small K (1994) The economics of traffic congestion. Am Sci 82(5):446–455
44. INRIX I (2019) INRIX: congestion costs each American 97 hours, $1,348 A Year. INRIX. https://inrix.com/press-releases/scorecard-2018-us/. (Accessed 29 Apr 2022)
45. CalSTA (2017) California road charge pilot program. California State Transportation Agency, Sacramento, CA
46. Wachs M, King H, Agrawal AW (2018) The future of California transportation revenue. Mineta Transportation Institute, San Jose, CA

47. Cottingham DN, Beresford AR, Harle RK (2007) Survey of technologies for the implementation of national-scale road user charging. Transp Rev 27(4):499–523. https://doi.org/10.1080/01441640701214304
48. de Palma A, Lindsey R (2011) Traffic congestion pricing methodologies and technologies. Transp Res Part C Emerg Technol 19(6):1377–1399. https://doi.org/10.1016/j.trc.2011.02.010
49. SCAG (2022) Mobility innovations and pricing: an initiative to elevate equity in planning. Southern California Association of Governments, Los Angeles, CA. https://scag.ca.gov/sites/main/files/file-attachments/scag-mobility-innovations-and-pricing-report-final.pdf?1648504727. (Accessed 26 Aug 2022)
50. Bonsall P, Shires J, Maule J et al (2007) Responses to complex pricing signals: theory, evidence and implications for road pricing. Transp Res Part Policy Pract 41(7):672–683. https://doi.org/10.1016/j.tra.2006.06.001
51. Börjesson M, Eliasson J, Hugosson MB et al (2012) The Stockholm congestion charges-5 years on. Effects, acceptability and lessons learnt. Transp Policy 20:1–12. https://doi.org/10.1016/j.tranpol.2011.11.001
52. King D, Manville M, Shoup D (2007) The political calculus of congestion pricing. Transp Policy 14(2):111–123. https://doi.org/10.1016/j.tranpol.2006.11.002
53. Olsson AR, Davis DE (2017) Expanding the scope of sustainability planning: lessons from Stockholm's congestion charging policy. Urban Plan 2(4):81–92. https://doi.org/10.17645/up.v2i4.1028
54. Schaller B (2010) New York City's congestion pricing experience and implications for road pricing acceptance in the United States. Transp Policy 17(4):266–273. https://doi.org/10.1016/j.tranpol.2010.01.013
55. Gu Z, Liu Z, Cheng Q et al (2018) Congestion pricing practices and public acceptance: a review of evidence. Case Stud Transp Policy 6(1):94–101. https://doi.org/10.1016/j.cstp.2018.01.004
56. McMullen BS, Ke Y, Wang H (2016) Importance of recognizing locational differences in assessing the impact of a road user charge: Oregon case study. Transp Res Rec J Transp Res Board 2597(1):75–81. https://doi.org/10.3141/2597-10
57. Goodin G, Baker RT, Taylor L (2009) Mileage-based user fees: defining a path toward implementation; phase 2: an assessment of institutional issues. Texas Transportation Institute. https://rosap.ntl.bts.gov/view/dot/17914
58. Bayen A, Forscher T, Shaheen S (2018) Preserving privacy in road user charge architectures. Policy Brief. UC Berkeley, Berkeley, CA. https://escholarship.org/uc/item/9ws6j5fk. (Accessed 22 Sept 2022)
59. Martínez-Díaz M, Soriguera F (2018) Autonomous vehicles: theoretical and practical challenges. Transp Res Procedia 33:275–282. https://doi.org/10.1016/j.trpro.2018.10.103
60. Bagloee SA, Tavana M, Asadi M et al (2016) Autonomous vehicles: challenges, opportunities, and future implications for transportation policies. J Mod Transp 24(4):284–303. https://doi.org/10.1007/s40534-016-0117-3
61. Kondor D, Santi P, Le D-T et al (2020) Addressing the "minimum parking" problem for on-demand mobility. Sci Rep 10(1). https://doi.org/10.1038/s41598-020-71867-1
62. Guerra E (2016) Planning for cars that drive themselves: metropolitan planning organizations, regional transportation plans, and autonomous vehicles. J Plan Educ Res 36(2):210–224. https://doi.org/10.1177/0739456X15613591
63. Davis DE (2018) Governmental capacity and the smart mobility transition. In: Marsden G, Reardon L (eds) Governance of the smart mobility transition. Emerald Publishing Limited, Bingley, UK. https://doi.org/10.1108/978-1-78754-317-120181007
64. Fagnant DJ, Kockelman K (2015) Preparing a nation for autonomous vehicles: opportunities, barriers and policy recommendations. Transp Res Part Policy Pract 77:167–181. https://doi.org/10.1016/j.tra.2015.04.003
65. Cugurullo F, Acheampong RA, Gueriau M et al (2021) The transition to autonomous cars, the redesign of cities and the future of urban sustainability. Urban Geogr 42(6):833–859. https://doi.org/10.1080/02723638.2020.1746096

66. Wang J, Lu L, Peeta S et al (2021) Optimal toll design problems under mixed traffic flow of human-driven vehicles and connected and autonomous vehicles. Transp Res Part C Emerg Technol 125:102952. https://doi.org/10.1016/j.trc.2020.102952
67. ITF (2015) Urban mobility system upgrade: how shared self-driving cars could change city traffic. International Transportation Forum. http://www.internationaltransportforum.org/Pub/pdf/15CPB_Self-drivingcars.pdf. (Accessed 28 Apr 2022)
68. Reardon L (2020) Smart mobility as a catalyst for policy change towards low carbon mobility? In: Paulsson A, Hedegaard Sørensen C (eds) Shaping smart mobility futures: governance and policy instruments in times of sustainability transitions. Emerald Publishing, Bingley, UK
69. Levin MW, Odell M, Samarasena S et al (2019) A linear program for optimal integration of shared autonomous vehicles with public transit. Transp Res Part C Emerg Technol 109:267–288. https://doi.org/10.1016/j.trc.2019.10.007
70. Kalra N, Paddock SM (2016) Driving to safety: how many miles of driving would it take to demonstrate autonomous vehicle reliability? Transp Res Part Policy Pract 94:182–193. https://doi.org/10.1016/j.tra.2016.09.010
71. Kassens-Noor E, Kotval-Karamchandani Z, Cai M (2020) Willingness to ride and perceptions of autonomous public transit. Transp Res Part Policy Pract 138:92–104. https://doi.org/10.1016/j.tra.2020.05.010
72. Manfreda A, Ljubi K, Groznik A (2021) Autonomous vehicles in the smart city era: an empirical study of adoption factors important for millennials. Int J Inf Manag 58. https://doi.org/10.1016/j.ijinfomgt.2019.102050
73. Schwall M, Daniel T, Victor T et al (2020) Waymo public road safety performance data. https://doi.org/10.48550/ARXIV.2011.00038
74. Millard-Ball A (2018) Pedestrians, autonomous vehicles, and cities. J Plan Educ Res 38(1):6–12. https://doi.org/10.1177/0739456X16675674
75. Schepel H (2005) The constitution of private governance: product standards in the regulation of integrating markets. In: International studies in the theory of private law, vol 4. Hart Pub., Oxford, UK
76. Martin E, Stocker A, Nichols A et al (2021) Roundtrip carsharing in New York City: an evaluation of a pilot program and system impacts. Transportation Sustainability Research Center, Berkeley, CA, UC Berkeley
77. Fearnley N (2020) Micromobility—regulatory challenges and opportunities. In: Paulsson A, Sørensen CH (eds) Shaping smart mobility futures: governance and policy instruments in times of sustainability transitions. Emerald Publishing Limited, Bingley, UK. https://doi.org/10.1108/978-1-83982-650-420201010
78. Latinopoulos C, Patrier A, Sivakumar A (2021) Planning for e-scooter use in metropolitan cities: a case study for Paris. Transp Res Part D Transp Environ 100. https://doi.org/10.1016/j.trd.2021.103037
79. Waller P (2002) Challenges in motor vehicle safety. Annu Rev Public Health 23:93–113
80. Harfst DL, Mashaw JL (1990) The struggle for auto safety. Harvard University Press, Cambridge, MA
81. Nader R (1991) The struggle for auto safety. Polit Sci Q 106(3). https://doi.org/10.2307/2151752
82. Canis B (2021) Issues in autonomous vehicle testing and deployment. Congressional Research Service Report. R45985
83. Marchant GE, Allenby BR, Herkert JR (eds) (2011) The growing gap between emerging technologies and legal-ethical oversight. Springer, Netherlands, Dordrecht, The International Library of Ethics, Law and Technology. https://doi.org/10.1007/978-94-007-1356-7
84. McCauley R (2016) Three concerns automakers have with California's autonomous vehicle regulations. Government Technology. https://www.govtech.com/fs/3-concerns-automakers-have-with-californias-autonomous-vehicle-regulations.html. (Accessed 18 Jul 2023)
85. USDOT (2021) Automated vehicles—comprehensive plan. United States Department of Transportation, Washington, D.C

86. Choenni S, Bargh MS, Busker T et al (2022) Data governance in smart cities: challenges and solution directions. J Smart Cities Soc 1(1):31–51. https://doi.org/10.3233/SCS-210119
87. Kyriakidis M, Happee R, de Winter JCF (2015) Public opinion on automated driving: results of an international questionnaire among 5000 respondents. Transp Res Part F Traffic Psychol Behav 32:127–140. https://doi.org/10.1016/j.trf.2015.04.014
88. Stickel JR, Vandervalk A (2014) Data business plans and governance programs: aligning transportation data to agency strategic objectives. Transp Res Rec J Transp Res Board 2460(1):154–163. https://doi.org/10.3141/2460-17
89. Hall J (2017) Data governance at state departments of transportation. In: MWAIS 2017 proceedings, vol 24. https://aisel.aisnet.org/mwais2017/24. (Accessed 23 June 2023)
90. Harrison F, Michael G, Gary A (2016) Leadership guide for strategic information management for state departments of transportation. Transportation Research Board, Washington, D.C. https://doi.org/10.17226/23480
91. Löfgren K, Webster CWR (2020) The value of Big Data in government: the case of 'smart cities'. Big Data & Soc 7(1). https://doi.org/10.1177/2053951720912775
92. Bartolini D (2015) Municipal fragmentation and economic performance of OECD TL2 Regions. In: OECD regional development working papers 2015/02. OECD Publishing, Paris. https://doi.org/10.1787/5jrxqs60st5h-en
93. Santoni De Sio F (2021) The European Commission report on ethics of connected and automated vehicles and the future of ethics of transportation. Ethics Inf Technol 23(4):713–726. https://doi.org/10.1007/s10676-021-09609-8
94. Marsden G, Docherty I, Dowling R (2020) Parking futures: curbside management in the era of 'new mobility' services in British and Australian cities. Land Use Policy 91:104012. https://doi.org/10.1016/j.landusepol.2019.05.031
95. Martin E, Pan A, Shaheen S (2020) An evaluation Of free-floating carsharing In Oakland, California. Recent Work. UC Berkeley Transportation Sustainability Research Center, Berkeley, CA. https://escholarship.org/uc/item/3j722968. (Accessed 15 Aug 2023)
96. Pierce G, Shoup D (2013) SFpark: pricing by demand. https://www.accessmagazine.org/fall-2013/sfpark-pricing-parking-demand/. (Accessed 15 Aug 2023)
97. Raphael R (2022) Who is collecting data from your car? The Markup, 27 July. https://themarkup.org/the-breakdown/2022/07/27/who-is-collecting-data-from-your-car. (Accessed 20 Jul 2023)
98. Zhang K, Nie YM (2022) Mitigating traffic congestion induced by transportation network companies: a policy analysis. Transp Res Part Policy Pract 159:96–118. https://doi.org/10.1016/j.tra.2022.01.017
99. Fries RN, Gahrooei MR, Chowdhury M, Conway AJ (2012) Meeting privacy challenges while advancing intelligent transportation systems. Transp Res Part C 25:34–45
100. Hahn D, Munir A, Behzadan V (2021) Security and privacy issues in intelligent transportation systems: classification and challenges. IEEE Intell Transp Syst Mag 13(1):181–196. https://ieeexplore.ieee.org/stamp/stamp.jsp?tp=&arnumber=8686209. (Accessed 20 Jul 2023)
101. Rode P, Heeckt C (2019) National transport policy and cities: key policy interventions to drive compact and connected urban growth. Coalition for Urban Transition, London
102. González-González E, Nogués S, Stead D (2019) Automated vehicles and the city of tomorrow: A backcasting approach. Cities 94:153–160. https://doi.org/10.1016/j.cities.2019.05.034
103. Guerra E (2015) When autonomous cars take to the road. Planning Magazine. May 2015. https://www.planning.org/planning/2015/may/autonomouscars.htm. (Accessed 16 Jul 2023)
104. Soteropoulos A, Berger M, Ciari F (2019) Impacts of automated vehicles on travel behaviour and land use: an international review of modelling studies. Transp Rev 39(1):29–49
105. Fountas G, Sun Y, Akizu-Gardoki O, Pomponi F (2020) How do people move around? National data on transport modal shares for 131 Countries. World 1:34–43. https://doi.org/10.3390/world1010003
106. Giuliano G, Agarwal A (2017) Land use impacts of transportation investments. In: Giuliano, Hanson (eds) The geography of urban transportation, 4th edn. Guilford Press, New York
107. Klein D, Moore A, Reja B (1997) Curb rights: a foundation for free enterprise in urban transit. Brookings Institution, Washington, D.C

Index

P. Ioannou and A. A. Malikopoulos (eds.), *Transportation Mobility in Smart Cities*, Springer Tracts on Transportation and Traffic 21,
https://doi.org/10.1007/978-3-031-64769-7

T

U

V

W

The manufacturer's authorised representative in the EU is Springer Nature Customer Service Centre GmbH, Europaplatz 3, 69115 Heidelberg, Germany. If you have any concerns regarding our products, please contact ProductSafety@springernature.com

Printed and bound by CPI Group (UK) Ltd, Croydon, CR0 4YY
01/10/2026
02225963-0007